MW00709546

HEALTH RISK ASSESSMENT
Dermal and Inhalation Exposure and Absorption of Toxicants

EDITED BY

Rhoda G. M. Wang, Ph.D.
Staff Toxicologist
Worker Health and Safety Branch
Department of Pesticide Regulation
California Environmental Protection Agency
Sacramento, California

James B. Knaak, Ph.D.
Product Stewardship Scientist
Occidental Chemical Corporation
Niagara Falls, New York

Howard I. Maibach, M.D.
Professor of Dermatology
Department of Dermatology
University of California
San Francisco, California

CRC Press
Boca Raton   Ann Arbor   London   Tokyo

Library of Congress Cataloging-in-Publication Data

Health risk assessment: dermal and inhalation exposure and
  absorption of toxicants / edited by Rhoda G. Wang, James B. Knaak,
  Howard I. Maibach.
     p. cm. -- (Dermatology)
   Includes bibliographical references and index.
   ISBN 0-8493-7357-3
   1. Dermatotoxicology. 2. Gases, Asphyxiating and poisonous-
-Toxicology. I. Wang, Rhoda G. M., 1947- . II. Knaak, James B.
III. Maibach, Howard I. IV. Series: Dermatology (Boca Raton, Fla.)
   [DNLM: 1. Air Pollutants, Environmental--toxicity. 2. Dose
-Response Relationship, Drug. 3. Environmental Pollutants-
-toxicity. 4. Risk Factors. 5. Skin Absorption. 6. Toxicology-
-methods. QV 602 H434]
   RL803.H43   1992
616.5--dc20
DNLM/DLC
for Library of Congress   92-20111
                                                     CIP

This book represents information obtained from authentic and highly regarded sources. Reprinted material is quoted with permission, and sources are indicated. A wide variety of references are listed. Every reasonable effort has been made to give reliable data and information, but the author and the publisher cannot assume responsibility for the validity of all materials or for the consequences of their use.

Neither this book nor any part may be reproduced or transmitted in any form or by any means, electronic or mechanical, including photocopying, microfilming, and recording, or by any information storage and retrieval system, without permission in writing from the publisher.

All rights reserved. Authorization to photocopy items for internal or personal use, or the personal or internal use of specific clients, is granted by CRC Press, Inc., provided that $.50 per page photocopied is paid directly to Copyright Clearance Center, 27 Congress Street, Salem, MA, 01970 USA. The fee code for users of the Transactional Reporting Service is ISBN 0-8493-7357-3/93 $0.00 + $.50. The fee is subject to change without notice. For organizations that have been granted a photocopy license by the CCC, a separate system of payment has been arranged.

The copyright owner's consent does not extend to copying for general distribution, for promotion, for creating new works, or for resale. Specific permission must be obtained from CRC Press for such copying.

Direct all inquiries to CRC Press, Inc. 2000 Corporate Blvd., N.W., Boca Raton, Florida, 33431.

© 1993 by CRC Press, Inc.

International Standard Book Number 0-8493-7357-3

Library of Congress Card Number 92-20111

Printed in the United States of America 1 2 3 4 5 6 7 8 9 0

Printed on acid-free paper

CRC Series in
DERMATOLOGY: CLINICAL AND BASIC SCIENCE
Edited by Dr. Howard I. Maibach

The CRC Dermatology Series combines scholarship, basic science, and clinical relevance. This comprehensive reference focuses on dermal absorption, cholinesterase inhibition, adverse reproductive effects, and carcinogenicity.

Forthcoming Titles:

Pigmentation and Pigmentary Disorders
Norman Levine

Human Papillomavirus Infections in Dermatovenereology
Gerd Gross and Geo von Krogh

Bioengineering of the Skin: Water and the Stratum Corneum
Peter Elsner, Enzo Berardesca, and Howard I. Maibach

Bioengineering of the Skin: Cutaneous Blood Flow and Erythema
Enzo Berardesca, Peter Elsner, and Howard I. Maibach

Protective Gloves for Occupational Use
Gunh Mellstrom, J.E. Walhberg, and Howard I. Maibach

Hand Eczema Book
Torkil Menne and Howard I. Maibach

Handbook of Contact Dermatitis
Christopher J. Dannaker, Daniel J. Hogan, and Howard I. Maibach

The Irritant Contact Dermatitis Syndrome
Pieter Van der Valk, Pieter Coenrads, and Howard I. Maibach

Skin Cancer: Mechanisms and Relevance
Hasan Mukhtar

The Contact Urticaria Syndrome
Arto Lahti and Howard I. Maibach

PREFACE

Risk assessment involves the extrapolation of known toxicological endpoints from animal test results to humans. One of the major difficulties is how to relate a toxicological response observed in animals through oral administration of a test material to that of humans exposed through dermal contact. Unlike well-designed animal studies, human exposure, especially occupational exposure to toxicants, varies during each exposure interval, such that steady state levels in blood and tissues are rarely achieved. In decades past, regulatory agencies regard 100% absorption as a feasible number to be applied equally to all routes of exposure. However, experimental results indicate that even via oral routes, 100% absorption is hardly achievable. Disregarding the various routes of exposure may lead to erroneous extrapolation of toxicological response from animals to humans, resulting in inappropriate risk assessment.

Since human exposure to toxicants in the environment is predominantly non-oral, bringing this issue to focus is of imminent concern. This book emphasizes aspects of dermal exposure to toxicants with respect to skin metabolism, absorption, and pharmacokinetic modeling, as well as the linking of dermal exposure with the various toxicological endpoints it produces. However, it also addresses and compares other routes of exposure, expecially inhalation exposure. The issue of external dose, such as that found in air, on the skin or administered per os, and the internal dose which reflects the actual dose reaching the tissues and target organs by means of pharmacokinetic modeling, has likewise been addressed. Physiologically based pharmacokinetics and pharmacodynamic models are needed to describe the comparative metabolic pathways, such as the metabolic rate of individual toxicants, V_{max}, and tissue concentrations, K_m, thus, allowing tissue dose predictions in humans under various routes of exposure conditions. Internal dose vs. external dose is a critical issue in toxicology and risk assessment. Methods are being developed and refined for better estimation of internal dose during various routes of exposure in different species, including man. Developing comparable external dose is an absolute must for successful risk assessment.

Another prominent feature of this book is that it brings to focus epidemiologic studies of humans exposed to toxicants under various routes. Since the best representative species for humans is the human, this area should be given the attention it deserves. There is hardly any well-documented literature relating route of exposure and definitive dose to epidemiologic study results. The difficulties appear to be that the major route of exposure has not been determined, and that no recorded dose, either external or internal, exists. Efforts have been made in Section 5 of this book to highlight and address these issues.

The intended audience of this book is expected to be very broad, encompassing academic and laboratory researchers, scientists at government agencies and in the private sector, as well as students at all levels of higher education.

We are confident that when the dose issue is resolved, perhaps in the coming decade, the accuracy of human risk assessment will take a giant step forward.

Rhoda G. M. Wang
Jim B. Knaak
Howard I. Maibach

EDITORS

Rhoda G. M. Wang, Ph.D., is founder and first slate president of the Carcinogenesis Specialty Section Organizational Committee of the Society of Toxicology. Dr. Wang was the recipient of a cancer research grant from the prestigious Cancer Research Fund in Europe. She has made substantial contributions in illucidating the mechanism of combined drug treatment and toxicological responses.

Since 1981, Dr. Wang has been focusing on the route of exposure and toxicological response. She is the editor of four books pertaining to exposure assessment, risk assessment and toxicology, and she has authored more than 60 publications in the fields of toxicology, pharmacology, and biochemistry.

James B. Knaak, Ph.D., is a Product Stewardship Scientist in Corporate Environmental Affairs, Occidental Chemical Corporation, Niagara Falls, New York. Dr. Knaak graduated from the University of Wisconsin with a Ph.D. in Biochemistry and Dairy Husbandry, and he has had a lengthy working history in pesticide product development.

He is an associate editor of the *Bulletin of Environmental Contamination and Toxicology,* and an assistant adjunct professor in the Department of Biochemistry, Nutrition, Pharmacology and Toxicology, School of Veterinary Medicine, University of California, Davis. Dr. Knaak has presented numerous lectures at scientific meetings and has authored over 60 publications.

Howard I. Maibach, M.D., is Professor of Dermatology at the School of Medicine, University of California, San Francisco.

Dr. Maibach graduated from Tulane University, New Orleans, Louisiana (A.B. and M.D.) and received his research and clinical training at the University of Pennsylvania, Philadelphia. He received an honorary doctorate from the University of Paris Sud in 1988.

Dr. Maibach is a member of the International Contact Dermatitis Research Group, the North American Contact Dermatitis Group, and the European Environmental Contact Dermatitis Group. He has published more than 1000 papers and 40 textbooks.

CONTRIBUTORS

Mohamed S. Abdel-Rahman, Ph.D., F.C.P.
Professor of Pharmacology and Director
 of Toxicology
Department of Pharmacology and
 Toxicology
University of Medicine and Dentistry of
 New Jersey-New Jersey Medical School
Newark, New Jersey

Rajesh Agarwal, Ph.D.
Assistant Professor
Department of Dermatology
Case Western Reserve University
Cleveland, Ohio

Mohammed Ali Al-Bayati, Ph.D.
Research Toxicologist
Institute of Toxicology and
 Environmental Health
University of California, Davis
Davis, California

Yolanda Banks Anderson, Ph.D.
Special Assistant, Office of the Director
Health Effects Research Laboratory
U.S. Environmental Protection Agency
Research Triangle Park, North Carolina

Curtis N. Barton, Ph.D.
Mathematical Statistician
Division of Mathematics
Food and Drug Administration
Washington, D.C.

Linda S. Birnbaum, Ph.D.
Director, Environmental Toxicology
 Division
Health Effects Research Laboratory
U.S. Environmental Protection Agency
Research Triangle Park, North Carolina

Kenneth B. Bischoff, Ph.D.
Professor
Department of Chemical Engineering
University of Delaware
Newark, Delaware

Jerry N. Blancato, Ph.D.
Research Biologist
Exposure Assessment Research Division
U.S. Environmental Protection Agency
Las Vegas, Nevada

Gregory G. Bond, Ph.D.
Manager
Department of Environment, Health and
 Regulatory Affairs
Dow Chemical Co.
Midland, Michigan

Christopher J. Borgert, Ph.D.
Post Doctoral Associate
Center for Environmental and Human
 Toxicology
University of Florida
Alachua, Florida

Robert L. Bronaugh, Ph.D.
Supervisory Pharmacologist
Division of Toxicological Studies
Food and Drug Administration
Laurel, Maryland

Daniel A. W. Bucks, Ph.D.
Research Scientist II
Department of Drug Transport
Penederm Inc.
Foster City, California

Chao W. Chen, Ph.D.
Senior Statistician
Office of Health and Environmental
 Assessment
U. S. Environmental Protection Agency
Washington, D.C.

Steven W. Collier, M.S.
Research Chemist
Division of Toxicological Studies
Food and Drug Administration
Laurel, Maryland

Ralph R. Cook, M.D.
Corporate Director of Epidemiology
Health and Environmental Sciences
Dow Chemical Co.
Midland, Michigan

Linval R. DePass, Ph.D., DABT
Department Head
Department of Toxicology
Syntex Research
Palo Alto, California

James J. Freeman, Ph.D.
Staff Toxicologist
Division of Toxicology
Exxon Biomedical Sciences, Inc.
East Millstone, New Jersey

Raymond D. Harbison, Ph.D.
Professor and Director
Center for Environmental and Human
  Toxicology
University of Florida
Alachua, Florida

Kim-Chi Hoang, M.S.
Environmental Engineer
Office of Health and Environmental
  Assessment
U. S. Environmental Protection Agency
Washiington, D.C.

David W. Hobson, Ph.D.
Research Leader
Medical Research and Evaluation Facility
Battelle Memorial Institute
Columbus, Ohio

Robert C. James, Ph.D.
Associate Scientist
Center for Environmental and Human
  Toxicology
University of Florida
Alachua, Florida

James B. Knaak, Ph.D.
Product Stewardship Scientist
Corporate Environmental Affairs
Occidental Chemical Corp.
Niagara Falls, New York

Gerald G. Krueger, M.D.
Professor of Medicine
Division of Dermatology
University of Utah School of Medicine
Salt Lake City, Utah

Hon-Wing Leung, Ph.D.
Associate Director of Applied Toxicology
Health, Safety, and Environment
Union Carbide Corporation
Danbury, Connecticut

Howard I. Maibach, M.D.
Professor of Dermatology
Department of Dermatology
University of California
San Francisco, California

Francis N. Marzulli, Ph.D.
Consultant in Toxicology
Bethesda, Maryland

Richard H. McKee, Ph.D.
Toxicology Associate
Division of Toxicology
Exxon Biomedical Sciences, Inc.
East Millstone, New Jersey

Hasan Mukhtar, Ph.D.
Professor
Department of Dermatology
Case Western Reserve University
Cleveland, Ohio

Roy C. Myers, B.S., DABT
Senior Group Leader
Department of Acute Toxicology/Bushy
  Run Research Center
Union Carbide Chemicals and Plastics
  Co. Inc.
Export, Pennsylvania

Dennis J. Paustenbach, Ph.D., DABT
Vice President and Chief Technical
  Officer
McLaren/Hart Environmental Engineering
National Director, ChemRisk Division
Alameda, California

Lynn K. Pershing, Ph.D.
Research Associate Professor
Division of Dermatology
University of Utah School of Medicine
Salt Lake City, Utah

Anthony P. Polednak, Ph.D.
Senior Research Scientist
Department of Epidemiology and Public
 Health
Yale School of Medicine
New Haven, Connecticut

Otto G. Raabe, Ph.D.
Professor
Institute of Toxicology and
 Environmental Health, and Department
 of Biochemistry, Nutrition,
 Pharmacology and Toxicology, School
 of Veterinary Medicine
University of California, Davis
Davis, California

Jim E. Riviere, D.V.M., Ph.D.
Director and Burroughs Wellcome
 Distinguished Professor of
 Pharmacology
Cutaneous Pharmacology and Toxicology
 Center
College of Veterinary Medicine
North Carolina State University
Raleigh, North Carolina

Stephen M. Roberts, Ph.D.
Assistant Professor of Toxicology
Center for Environmental and Human
 Toxicology
University of Florida
Alachua, Florida

James L. Schardein, M.S.
Associate Director of Research
Vice President
Division of Reproductive and
 Developmental Toxicology
International Research and Developmental
 Corp.
Mattawan, Michigan

Gloria A. Skowronski, Ph.D.
Research Associate
Department of Pharmacology and
 Toxicology
University of Medicine and Dentistry of
 New Jersey-New Jersey Medical School
Newark, New Jersey

Curtis Travis, Ph.D.
Director
Center for Risk Analysis
Oak Ridge National Laboratory
Oak Ridge, Tennessee

Rita M. Turkall, Ph.D.
Chairman of the Clinical Laboratory
 Sciences Department School of Health
 Related Professions and Associate
 Professor
Department of Pharmacology and
 Toxicology
University of Medicine and Dentistry of
 New Jersey-New Jersey Medical School
Newark, New Jersey

Rochelle W. Tyl, Ph.D.
Assistant Research Director
Program Director, Reproductive and
 Developmental Toxicology
Center for Life Sciences and Toxicology
Research Triangle Institute
Research Triangle Park, North Carolina

Rhoda G. Wang, Ph.D.
Worker Health and Safety Branch
Department of Pesticide Regulation
California Environmental Protection
 Agency
Sacramento, California

Elizabeth K. Weisburger, Ph.D., D.Sc.
Consultant
Bethesda, Maryland

Ronald C. Wester, Ph.D.
Associate Research Dermatologist
Department of Dermatology
University of California
San Francisco, California

Raymond G. York, Ph.D.
Associate Director
Division of Reproductive and
 Developmental Toxicology
International Research and Development
 Corp.
Mattawan, Michigan

TABLE OF CONTENTS

SECTION 1: PHARMACOKINETCS, METABOLISM, AND PERCUTANEOUS PENETRATION OF THE SKIN

SECTION 3: DOSE, TOXICITY, AND RISK ASSESSMENT

SECTION 4: NEW MODELS IN SKIN RESEARCH

SECTION 5: EPIDEMIOLOGIC STUDIES OF ROUTE OF EXPOSURE AND HEALTH EFFECTS

Pharmacokinetics, Metabolism, and Percutaneous Penetration of the Skin

Chapter 1

PHYSIOLOGICALLY BASED PHARMACOKINETIC MODELING TO PREDICT TISSUE DOSE AND CHOLINESTERASE INHIBITION IN WORKERS EXPOSED TO ORGANOPHOSPHORUS AND CARBAMATE PESTICIDES

James B. Knaak, Mohammed A. Al-Bayati, and Otto G. Raabe

TABLE OF CONTENTS

0-8493-7357-3/93/$0.00 + $.50
© 1993 by CRC Press, Inc.

I. INTRODUCTION

Numerous studies have been conducted dealing with the percutaneous absorption of ^{14}C-labeled pesticides.[1-12] Of these studies only a few describe the development of percutaneous physiologically based pharmacokinetic (PB-PK) models.[11,12] The toxicity of pesticides coupled with their complex metabolic pathways presents the toxicologist with toxicological, biochemical, and analytical problems in the development of PB-PK models. This chapter addresses these problems as they relate to the development of percutaneous PB-PK models for pesticides and in particular toxic organophosphorus and carbamate insecticides in the work place. For purposes of this discussion, a review of the components of a PB-PK model is presented as they pertain to chlorinated solvents. This review will be used to set the stage for the development of information needed to construct, run, and validate a dermal model for predicting tissue dose and AChE inhibition in workers.

II. REVIEW OF CURRENT PB-PK MODELS INVOLVING VOLATILE SOLVENTS

The inhalation[13-17] and percutaneous[18-20] uptake and elimination of gases and voltaile solvents used in medicine and industry have been studied and modeled. The chlorinated solvents are suitable candidate compounds for PB-PK model development because of their low toxicity and the availability of extremely sensitive gas chromatographic methods. Gas-uptake and retention studies are easily performed with small animals in standard glass inhalation chambers. The eqilibration of solvent between breathing air, blood, and tissues may be readily attained during the work day, making it possible to measure the effects of various solvent concentrations, K_m, on V_{max} *in vivo* under steady-state conditions. Partition coefficients (blood/air, tissue/air, and tissue/blood) for the solvents of interest are easily obtained *in vitro* by directly measuring their concentrations in air in closed vials and in the blood or tissue homogenates when equilibrium is reached.[21,22]

A. PB-PK MODEL

A typical inhalation PB-PK dosimetry model is given in Figure 1, consisting of tissue compartments and lung space. Flows, Q, are alveolar (alv), total cardiac (t), and those to the fat, liver, muscle and richly perfused tissues (f, l, m and r, respectively). Concentrations, C, are those in arterial blood (art), venous blood (ven), and in the venous blood leaving the tissues (vf, vl, vr, and vm). The venous blood concentrations are dependent on the chemical concentrations in the tissues and tissue/blood partition coefficients (P_t). Metabolic removal of chemical from the liver of this model is described kinetically by a maximum rate, V_{max}, and the Michaelis constant, K_m.

B. MASS BALANCE EQUATIONS

On the basis of the model of interest, inhalation, oral, or percutaneous absorption, a set of mass balance differential equations is written describing the rate of change of the concentration of the chemical in each tissue through time. The volumes of the tissues, cardiac output, blood flow rates through tissues, tissue/blood partition coefficients, and metabolism are considered.[19,23]

Dermal absorption:

$$V_{sk}\, dC_{sk}/dt = K_p A\, (C_{air} - C_{sk}/P_{sk/air}) + Q_{sk}\, (C_a - C_{sk}/P_{sk/b}) \qquad (1)$$

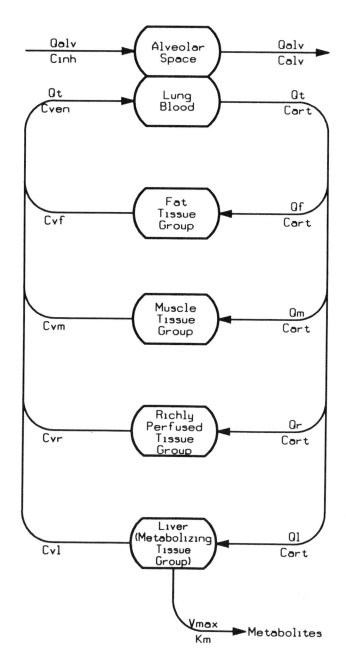

FIGURE 1. Diagram of a physiologically based pharmacokinetic model.

Gas exchange compartment:

$$Q_{alv}C_{inh}dt + Q_bC_{ven}dt = Q_{alv}C_{alv}dt + Q_bC_{art}dt \tag{2}$$

or:

$$Q_{alv}(C_{inh} - C_{alv}) = Q_b(C_{art} - C_{ven}) \tag{3}$$

$$C_{art} = P_bC_{alv} \tag{4}$$

Gavage-absorption from the gut:

$$dA_1/dt = Q_1(C_{art} - C_{v1}) - dA_m/dt - KD_o e^{-kt} \tag{5}$$

Aterial blood:

$$C_{art} = (Q_{aiv}C_{inh} + Q_bC_{ven})/(Q_b + Q_{aiv}/P_b) \tag{6}$$

Mixed venous blood:

$$C_{ven} = 1/Q_b(\Sigma Q_iC_{vi}) \tag{7}$$

Tissue/blood exchanges:

$$Q_iC_{art}dt = dA_i + Q_iC_{vi}dt \tag{8}$$

$$dA_i/dt = Q_i(C_{art} - C_{vi}) \tag{9}$$

$$C_{vi} = C_i/P_i \tag{10}$$

$$C_i = A_i/V_i \tag{11}$$

Metabolism:

$$dA_m/dt = V_{max} C_{vl}/(K_m + C_{vl}) + K_fV_lC_{vl} \tag{12}$$

$$dA_l/dt = Q_l(C_{art} - C_{vl}) - dA_m/dt \tag{13}$$

The differential equations are converted into FORTRAN format for use with computer simulation programs such as Advanced Continuous Simulation Language (ACSL), or with SimuSolv. The source code consists of: (1) an INITIAL section containing physiological, chemical-specific, metabolic, and calculated parameters as well as timing commands; (2) a DYNAMIC section containing the algorithm; and (3) a DERIVATIVE section containing the differential equations. The nature of the parameters is discussed in the following sections.

C. PHYSIOLOGICAL PARAMETERS

The physiological parameters were compiled by Arms and Travis and published by the EPA.[24] Table 1 gives the values of the parameters used by Andersen et al.[17] for the B6C3F1 mouse, F344 rat, and human.

A new compilation initiated by the Physiological Parameters Work Group, International Life Science Institute-Risk Science Institute (ILS RSI) and supported by the ILS RSI, the American Industrial Health Council (AIHC), U.S. EPA, and other organizations is being worked on by Dr. Stan Lindstedt, Northern Arizona University, to include physiological parameters (blood flows, volumes, tissue mass, cardiac output, ventilation rate, lymph/bile flow), key species (B6C3F1 mouse, Sprague-Dawley rat, Beagle dog, and man), and tissues (lung, liver, kidney, fat, muscle, bone including marrow, brain, gonads, endocrine organs, heart, and placenta), and a set of allometric relationships governing body weight, tissue size, and blood flow rates. Values from other species will be included in the compilation as available and appropriate.

Structural measurements will include "equilibrium blood volume" (estimated as capillary volume); pulmonary diffusing capacity (a measure of the lung's total capacity to transport airborne substances into the blood). Physiological measurements will include pul-

TABLE 1
Physiological Parameters

	B6C3F1	F344 rats	Human
Weights (kg)			
Body	0.0345	0.233	70.0
Lung ($\times 10^{-3}$)	0.410	2.72	772.0
Percentage of Body Weight			
Liver	4.0	4.0	3.14
Rapidly perfused	5.0	5.0	3.71
Slowly perfused	78.0	75.0	62.1
Fat	4.0	7.0	23.1
Flows (l/hr)			
Alveolar ventilation	2.32	5.10	348.0
Cardiac output	2.32	5.10	348.0
Percentage of Cardiac Output			
Liver	0.24	0.24	0.24
Rapidly perfused[a]	0.52	0.52	0.52
Slowly perfused[b]	0.19	0.19	0.19
Fat	0.05	0.05	0.05

[a] Organs.
[b] Muscles.

From Andersen, M. E., Clewell, H. J., III, Gargas, M. L., Smith, F. A., and Reitz, R. H., *Toxicol. Appl. Pharmacol.*, 87, 185, 1987.

monary ventilation, cardiac output, organ blood flows to each of the organs, and, if possible, maximum and minimum values. Mammalian scaling equations ($Y = aM^b$) will include default values, structural scaling for skeleton and skin, control functions and fat, volumes or capacities, functional scaling for frequencies (rates) and times, and volume rates.

The work is expected to be reviewed and completed some time in 1992 and should reduce inconsistencies in the parameters now being used. The dog was added to the list of animals because of the need for an intermediate species between rodent and man. The lipid content of organs was considered.

D. CHEMICAL-SPECIFIC CONSTANTS

The chemical-specific constants include tissue solubilities, biotransformation rate constants, and specific binding constants to macromolecules such as proteins, hemoglobin, and DNA. Blood and tissue solubilities are represented by partition coefficients. The simplest distribution system representing blood and tissue in common use today involves *n*-octanol and water in a closed system with very little head space for volatiles to collect. The two phases are separated, and the concentration of chemical in each phase is determined by sensitive procedures such as gas chromatography or radioassay.

TABLE 2
Partition Coefficients for Methylene Chloride
and Tetrachloroethylene

	B6C3F1 Mouse	F344 Rats	Human
Methylene chloride[a]			
Blood/air	8.29	19.4	9.7
Liver/blood	1.71	0.732	1.46
Lung/blood	1.71	0.732	1.46
Rapidly perfused/blood	1.71	0.732	1.46
Slowly perfused/blood	0.960	0.408	0.82
Fat/blood	14.3	6.19	12.4
Tetrachloroethylene[b]			
Blood/air	16.9	18.9	10.3
Liver/blood	4.16	3.72	6.82
Rapidly perfused/blood[c]	4.16	3.72	6.82
Slowly perfused/blood[d]	1.18	1.06	7.77
Fat/blood	121.9	121.7	159.0

[a]   Reitz, R. H., Nolan, R. J., and Schumann, A. M., *Pharmacokinetics in Risk Assessment: Drinking Water and Health, 8,* National Academy Press, Washington, D.C., 1987, 391.

[b]   Travis, C. C., White, R. K., and Arms, A. D., *The Risk Assessment of Environmental Hazards: A Textbook of Case Studies,* John Wiley & Sons, New York, 1986, Chap. 23. Tissue/blood values obtained by dividing tissue/air values by blood/air values.

[c]   Organs.

[d]   Muscles.

Because these two phases do not adequately represent the distribution of a chemical between blood and tissue in the body, the vial equilibration technique of Sato and Nakajima[21] was modified by Gargas et al.[22] to accommodate tissues and tissue homogenates. The vial equilibration technique relies on measurements of volatile chemical in the head space of vials containing the test material, as well as in paired reference vials. In this procedure, a measurement of chemical in the test material is not required. The partition coefficients obtained for methylene chloride and tetrachloroethylene using this method are given in Table 2 for the rat, mouse, and man.[13,23]

E. METABOLIC PATHWAYS

Methylene chloride ($MeCl_2$) is extensively metabolized by liver microsomal P-450 enzymes (MFO) and cytosolic glutathione-S-transferases.[13] Liver P-450 contents are assayed according to the method of Omura and Sato[25] and expressed as nanomole per gram microsomal protein on the basis of a millimolar extinction coefficient of $91m/M$, while glutathione-S-transferase (GST) activity is assayed in the cytosol using 1-chloro-2,4-dinitrobenzene (CDNB).[26]

Both pathways release 2 mol of halide ion per mole of $MeCl_2$ consumed. The oxidative pathway is saturated at inhaled concentrations of a few hundred parts per million, while the GSH pathway is unsaturated up to 10,000 ppm. The metabolic pathway for methylene chloride is given in Figure 2. Formyl chloride is formed in the cytochrome P-450 pathway and chloromethyl glutathione in the cytosolic (GSH) pathway. Carbon dioxide is produced by both pathways, but only the MFO pathway produces carbon monoxide and carboxyhemoglobin. The 48-h fate of an orally administered dose of 1 and 50 mg/kg ^{14}C-methylene chloride in rats is given in Table 3.[14]

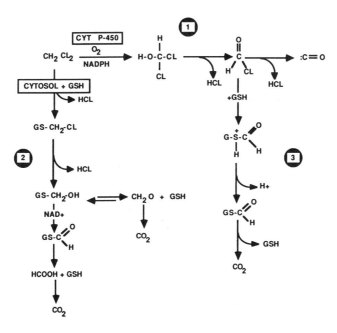

FIGURE 2. Metabolic pathway for methylene chloride. (From Reitz, R. H., Nolan, R. J., and Schumann, A. M., *Pharmacokinetics in Risk Assessment: Drinking Water and Health, 8,* National Academy Press, Washington, D.C., 1987, 391. With permission.)

TABLE 3
Fate of [^{14}C] Methylene Chloride in Rats 48 h after a Single Oral Dose

Dose (mg/kg)	Percentage of dose recovered in								Total % recovered
	Expired air as		CO	Urine	Feces	Carcass	Skin	Cage wash	
	CH_2Cl_2	CO_2							
1	12.3	35.0	30.9	4.5	0.93	5.84	1.56	0.53	91.6
50	72.1	6.33	11.9	1.96	0.25	2.40	1.15	0.08	96.1

From McKenna, M. J. and Zempel, J. A., *Fed. Cosmet. Toxicol.,* 19, 73, 1980.

F. METABOLIC RATE CONSTANTS

Metabolic rate constants have traditionally been measured *in vivo* by relating intake to the elimination of simple end products such as carbon dioxide that may be easily traced to the administered chemical. In the case of *in vivo* measurements, the metabolite must be produced in the first metabolic step or shortly after and be stable. This condition may exist for terminal carbons labeled with ^{14}C that are not incorporated into other metabolites or natural products.

Because of the complexity of many metabolic pathways, *in vitro* studies involving tissue slices, homogenates, P-450 microsomal, and cytosolic enzymes are often used to estimate the *in vivo* kinetic rate constants. Direct use of *in vitro* metabolic constants in pharmacokinetic models to predict *in vivo* kinetic behaviors that cover a range of concentrations has been evaluated for only a few chemicals (Dedrick et al.,[27] Dedrick and Forrester,[28] Lin et al.,[29] and Gearhart et al.[30]). The V_{max} and K_m values used in the methylene chloride and tetrachloroethylene PB-PK models are given in Table 4. Gargas[31] recently published the data in Table 5, comparing V_{max} values for 14 solvents obtained using *in vivo* and *in vitro* techniques.

TABLE 4

V_{max}, K_m Values for Methylene Chloride and Tetrachloroethylene

in vivo Metabolic constants	B6C3F1	F344 rats	Humans
Methylene chloride[a]			
V_{max} (mg/h)	1.054	1.50	118.9
K_m (mg/l)	0.396	0.771	0.580
K_f (h^{-1})	4.017	2.21	0.53
Tetrachloroethylene[b]			
V_{max} (mg/h)	0.11	0.068	3.5
K_m (mg/l)	0.4	0.3	0.3
K_f (h^{-1})	1.84	2.73	0.0

[a] Andersen, M. E., Clewell, H. J., III, Gargas, M. L., Smith, F. A., and Reitz, R. H., *Toxicol. Appl. Pharmacol.*, 87, 185, 1987.

[b] Travis, C. C., White, R. K., and Arms, A. D., *The Risk Assessment of Environmental Hazards: A Textbook of Case Studies,* John Wiley & Sons, New York, 1986, chap. 23.

TABLE 5

V_{max} Comparisons

Compound	V_{max}, μmol/h/335 g Rat		
	in vivo[a]	*in vitro* (fed)[b]	*in vitro* (fasted)[b]
Benzene	19.62	8.0	16.56
Toluene	37.85	10.57	18.34
m-Xylene	35.06	12.26	18.66
Styrene	37.48	16.64	20.98
Chloroform	26.46	11.5	25.07
Carbon tetrachloride	1.21	1.11	2.68
1,1-Dichloroethane	35.25	11.15	25.48
1,1,1-Trichloroethane	1.46	0.29	0.53
1,1,2-Trichloroethane	26.83	12.26	25.48
1,1,1,2-Tetrachloroethane	18.00	4.73	14.92
1,1,2,2-Tetrachloroethane	33.25	7.77	18.20
1,1-Dichloroethylene	35.99	18.16	30.62
Trichloroethylene	38.92	11.04	25.94
Tetrachloroethylene	0.50	0.292	0.87

[a] Gargas, M. L., Andersen, M. E., and Clewell, H. J., III, *Toxicol. Appl. Pharmacol.*, 86, 341, 1986.

[b] Nakajima, T. and Sato, A., *Toxicol. Appl. Pharmacol.*, 50, 549, 1979.

In most cases, the V_{max} values from the *in vivo* studies exceeded the values from *in vitro* studies (fed or fasted rats). Both methods require sufficient solvent to saturate the enzymes involved.

III. THE STATUS OF WORKER EXPOSURE, CHOLINESTERASE INHIBITION, AND PERCUTANEOUS ABSORPTION STUDIES INVOLVING ORGANOPHOSPHORUS AND CARBAMATE INSECTICIDES

Considering the large amount of information published on percutaneous penetration (partition coefficients, percutaneous absorption, skin binding, skin metabolism, influence of age on percutaneous absorption, etc.), the number of articles dealing with the development of percutaneous PB-PK models and their application to worker problems is small. The reason for this is largely due to the lack of a strong interdiscipinary team of scientists to carry out the necessary studies.

A 1980 research conference and workshop on minimizing occupational exposure to pesticides pointed out the need for new methods for measuring exposure, absorption, and adverse health effects in workers.[32] The progress of this work was reviewed in 1989 by Knaak et al.[33] and is partially reviewed in this section in relationship to the development of PB-PK models for predicting tissue dose and AChE inhibition in exposed workers.

A. DERMAL EXPOSURE

A large amount of uncertainty exists in measuring pesticide residues on the skin of workers. Exposure is measured in micrograms per square centimeter of skin using the patch technique. The exposed area in many cases is limited to the hands, lower arms, face, and neck regions. The patch technique developed by Durham and Wolfe[34] has never been completely validated under field conditions. Evaporative losses from the patch; the size, number, and placement of patches; and the ability of the patch to measure the amount and distribution of the residue reaching clothing and skin appear to be the main problems associated with the use of the patch as a dosimeter. The fluorescent tracer technique developed by Fenske[35] decreases the uncertainties involved associated with the distribution of the residue, but does not address the problem of evaporative losses. According the Fenske,[35] good correlation was obtained between malathion (O,O-dimethyl S[1,2-*di*(ethoxycarbonyl) ethyl] phosphorodithioate) exposure measurements using the fluroescent tracer technqiue and the excretion of alkyl phosphates in urine. A plot of the fluorescent tracer exposure and malathion metabolites (alkyl phosphates) for 20 workers is presented in Figure 3. Peak excretion occurred within the first 24 h. Fluorescent tracer measurements and 24-h urine values correlated highly with an R^2 value of 0.84.

The factors affecting exposure of applicator and mixer-loader groups[36] are (1) the type of application equipment used, (2) the number of gallons applied per unit of time, (3) the concentration of the pesticide, (4) the loading equipment, and (5) the protective clothing worn by the worker. In the case of field workers, exposure is related to: (1) the production rate of the worker, (2) the amount of contact with foliage, and (3) the amount of dislodgable residue on foliage measured in micrograms per square centimeter.

In California, Popendorf[37] used the patch method to develop a mathematical relationship between foliar residues on peaches and citrus and the amount of residue transferred to the skin and clothing of workers. A similar relationship for citrus developed by Nigg et al.[38] in Florida compared favorably with the Popendorf[37] values. O'Malley[39] found this relationship underestimated residues transferred to workers when applied to foliar residues on grapes in California.

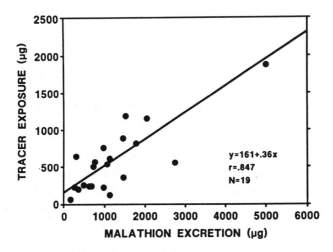

FIGURE 3. Regression analysis of fluorescent tracer exposure and 24-h malathion metabolite excretion. (From Fenske, R. A., *Biological Monitoring for Pesticide Exposure: Measurement, Estimation, and Risk Reduction*, American Chemical Society, 1989, chap. 6. With permission.)

A large number of investigators have tried to use urinary metabolites as a means of estimating exposure and the absorbed dose. In most cases urinary values correlated poorly if at all with exposure measurements.[40] This problem has never been resolved, but it may be related to the uncertainties associated with measuring exposure and the appropriate metabolites in urine. The best results were obtained with 2,4-D, a phenoxy herbicide, eliminated intact in urine.[41]

B. DERMAL DOSE-AChE RESPONSE

Field studies with agricultural chemicals and workers indicate that OP and carbamate residues in micrograms per square centimeter of skin are responsible for some of the illnesses encountered in California agriculture.[33] The quantitative relationship between these residues, their toxic foliar conversion products (e.g., oxons), and red cell cholinesterase inhibition was investigated in rats by Knaak et al.[33]

Pharmacokinetic rat dermal absorption studies using ^{14}C-ring labeled parathion (O,O-diethyl O-(4-nitrophenyl) phosphorothioate) indicated that a substantial amount (>35%) of parathion is lost to air after application.[6] The work suggested that evaporative losses should be measured in animals and estimated for man by PB-PK models. The importance of this loss in the case of workers is largely unknown.

Knaak et al.[33] used dislodgeable residues and field worker ChE response data to estimate a safe level for parathion on citrus. The rat dermal toxicity of parathion (ED_{50}, in micrograms per square centimeter) in relationship to the ED_{50} of a second organophosphorus pesticide (OP) was used to estimate a safe level for the second pesticide. This procedure assumed that parathion and the second OP were transferred via foliar dust to skin, absorbed in a similar manner, and the toxicity of the second OP was equivalent in rat and man. The safe levels established by this procedure have largely held up under field conditions.

C. ANIMAL STUDIES

A number of percutaneous absorption studies have been conducted using ^{14}C-labeled pesticides.[5-7] The studies provide material balance and metabolic fate data. A topical dose of 40 μg/cm^2 and a surface area of approximately 5 to 10% of the body surface (12 to 25 cm^2) was used by Knaak et al.[6,7,12] to study the absorption of parathion, carbaryl (1-naphthyl methylcarbamate), thiodicarb (dimethyl N,N′[thio-*bis*(methylimino)] carbamoyloxy thioacetimidate), and isofenphos (1-methylethyl-2-[ethoxy-II (1-methylethyl)-amino] phosphinothioyl]oxy]benzoate) through the back skin of the rat. The concentration used approximates the upper bounds on the concentrations found on clothing of field workers in California. The topically applied dose was left on the skin for a period of 7 d. During this time period, the simultaneous absorption and elimination of the topically applied dose was followed in major organs. Oral and i.v. studies with these pesticides indicated that they are rapidly absorbed, metabolized, and eliminated primarily in urine. Peak blood concentrations were reached approximately 12 h. after a topical dose was applied. The elimination rate of a topically applied pesticide closely follows its rate of absorption. For these small quantities (approximately 40 $\mu g/cm^2$, 12.5 cm^2, or 600 μg per animal), a steady-state concentration of pesticide/metabolites in blood and other tissues was not reached over the course of the study. Figures 4, 5, 6, and 7 give the absorption-elimination curves for parathion, carbaryl, and isofenphos in tissues. The metabolism of these compounds have been extensively studied. Parathion is partially metabolized to paraoxon (0,0-diethyl 0-(4-nitrophenyl) phosphate), and the two OPs are subsequently hydrolyzed to give *p*-nitrophenol, diethylthiophosphate, and diethylphosphate in urine.[42] Carbaryl is metabolized to 1-naphthyl glucuronide, 1-naphthyl sulfate, 5,6-dihydro- 5,6-dihydroxyl carbaryl glucuronide, and 4-hydroxycarbaryl glucuronide.[43] Isofenphos is initially metabolized to des N-isopropyl isofenphos, isofenphos oxon, and des N-isopropyl isofenphos oxon with further metabolism to the glycine conjugate of salicylic acid.[12]

D. HUMAN STUDIES

The EPA has never required the development of dermal absorption data using human volunteers. Studies of this nature, however, have been conducted by industry and universities alike. The most notable studies have been those conducted by Feldmann and Maibach[2] using low concentrations of ^{14}C-labeled pesticides (4 $\mu g/cm^2$) applied to small areas (2.8 to 20 cm^2) on the ventral forearm of male volunteers. Urine was collected over a period of 5 d after topical application of the pesticide. Recovery in urine was corrected based on urinary values from a separate i.v. study in male volunteers. Urinary excretion from a topical application varied from a low of 8.2% to a high of 73.9% for malathion and carbaryl, respectively.

A recent dermal absorption study by Wester et al.[44] using human volunteers and ^{14}C-ring-labeled isofenphos showed that losses to air are greater on human skin than on the rat skin as determined by Knaak et al.[12] for isofenphos.

IV. DATA REQUIRED FOR DEVELOPING PERCUTANEOUS PB-PK MODELS FOR ORGANOPHOSPHORUS AND CARBAMATE INSECTICIDES

The development of a percutaneous PB-PK model requires: (1) pharmacokinetic data on evaporation from skin (K_{vap}), skin permeability (K_p); (2) a description of the metabolic pathway which includes enzymes, their location, and metabolites; (3) partition coefficients (P_{sk}, P_l, etc.) on the pesticide and metabolites; (4) linear (K_a) or Michaelis Menten (V_{max}, K_m) metabolic kinetic data; and (5) elimination data (K_e). Published animal and human studies may provide information on skin absorption, the metabolic pathway, and some information on the enzymes and cofactors involved in the pathway. The studies, however, may not provide skin permeability data, tissue partition coefficients, or kinetic data on evaporation, metabolism, and elimination.

New methods are needed for: (1) developing partition coefficient data on nonvolatile pesticides and metabolites, and (2) analyzing for them in blood and tissues during and after

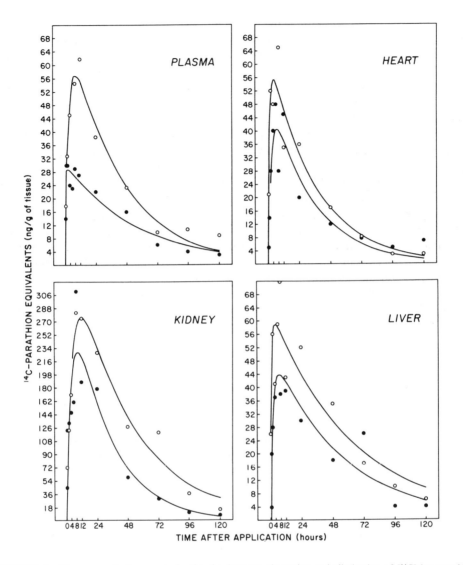

FIGURE 4. Time-concentration curves for the simultaneous absorption and elimination of [^{14}C]ring-parathion equivalents in plasma, heart, kidney, and liver. ●, adult males; ○, adult females. The mean coefficient of variation for the tissue values at each time interval was 37%. (From Knaak, J. B., Yee, K., Ackerman, C. R., Zweig, G., Fry, D. M., and Wilson, B. W., *Toxicol. Appl. Pharmacol.*, 76, 252, 1984. With permission.)

a study. The methods discussed in the following sections may provide a path to the development of suitable protocols.

A. PERMEABILITY CONSTANT

The permeability constant (K_p) is related to the affinity of the chemical for skin (K_m, partition coefficient), diffusivity (D), and the skin thickness (δ) by:

$$K_p = DK_m/\delta = \text{centimeters per hour} \qquad (14)$$

K_p is determined according to the following equation:

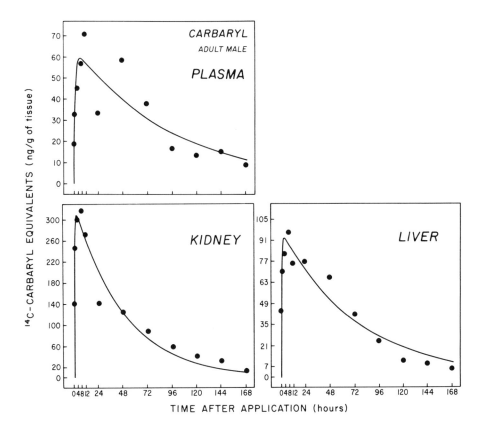

FIGURE 5. Time-concentration curves for the simultaneous absorption and elimination of [^{14}C]ring-carbaryl equivalents in plasma, kidney, and liver of adult male rats. The mean coefficient of variation for the tissue values at each time interval was 27%. (From Knaak, J. B., Yee, K., Ackerman, C. R., Zweig, G., Fry, D. M., and Wilson, B. W., *Toxicol. Appl. Pharamcol.*, 76, 252, 1984. With permission.)

$$K_P \ = \ ABS/A_s*C_s*t \tag{15}$$

where ABS is the total absorbed dose in milligrams, A_s is the surface area exposed in square centimeters, t is the time in hours, and C_s is the concentration gradient across the skin in milligrams per cubic centimeter.

This equation assumes an infinite dose has been applied to the skin and the value for C_s remains relatively constant during the study. In the pesticide studies performed by Feldmann and Maibach,[2] Knaak et al.,[6,7] and others, a finite dose was applied to the surface of the skin. Under these conditions C_s decreased exponentially as the topical dose was absorbed or was lost by evaporation. In estimating K_p, Knaak et al.[6] compensated for these changes by using $^1/_2$ ABS and $t^1/_2$ (h) for elimination from plasma. This procedure prevented the occurence of large changes in C_s due to absorption and evaporative losses by reducing the length of the experimental period, thereby producing a better estimate of K_p.

In percutaneous absorption studies, skin permeability (K_p, centimeters per hour) is directly related to $\log_{10}P$ and inversely related to water solubility (percent of weight per unit volume). According to Wester and Maibach,[45] the partition coefficient, K_m, by itself gives only a limited indication as to whether a compound will be absorbed through human skin. The reason for this may be related to experimental conditions used to generate the data. Metabolic transformations *in vivo* may also enhance percutaneous absorption. In

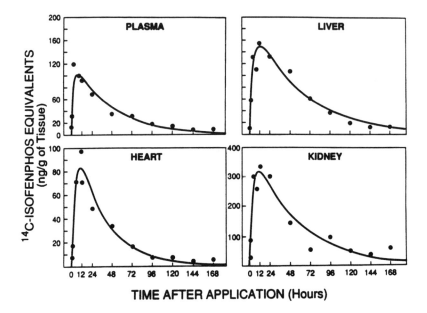

FIGURE 6. Time-concentration curves for the simultaneous absorption and elimination of [^{14}C]ring-isofenphos equivalents in plasma, heart, kidney, and liver. Data were fitted to the Bateman function ($Y = AB/(B - C)*[exp(-CX) - exp(-By)]$). (From Knaak, J. B., Al-Bayati, M., Raabe, O. G., and Blancato, J. N., *Prediction of Percutaneous Penetration: Methods, Measurements, and Modelling*, IBC Technical Services Ltd., London, 1990, 1. With permission.)

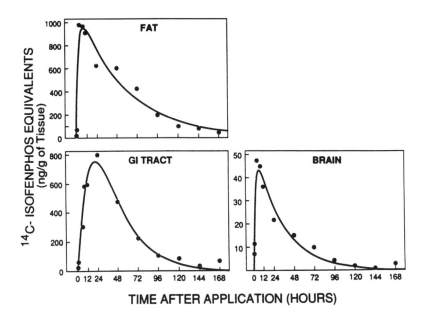

FIGURE 7. Time-concentration curves for the simultaneous absorption and elimination of [^{14}C]ring-isofenphos equivalents in brain, fat, and gastrointestinal tract. Data fitted to the Bateman function ($Y = AB/(B - C)*[exp(-Cx) - exp(-By)]$). (From Knaak, J. B., Al-Bayati, M., Raabe, O. G., and Blancato, J. N., *Prediction of Percutaneous Penetration: Methods, Measurements, and Modelling*, IBC Technical Services Ltd., London, 1990, 1. With permission.)

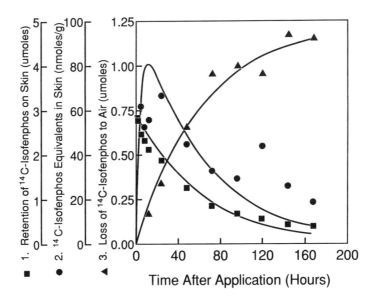

FIGURE 8. Physiologically based pharmacokinetic model: retention of [^{14}C]ring-isofenphos on skin, loss to air, transfer to skin. (From Knaak, J. B., Al-Bayati, M., Raabe, O. G., and Blancato, J. N., *Prediction of Percutaneous Penetration: Methods, Measurements, and Modelling,* IBC Technical Services Ltd., London, 1990, 1. With permission.)

percutaneous PB-PK models the permeability constant K_p (centimeters per hour) is used to describe the absorption of a topically applied dose. The following equations were used by Knaak et al.[12] for a PB-PK model on the percutaneous absorption of isofenphos by the rat:

$$dA_{surf}/dt = K_p*A*(C_{sk}/P_{a/sk} - C_{surf}) - K_a A_{surf} \tag{16}$$

$$A_{surf} = \text{topical dose in pmol} \tag{17}$$

$$C_{sk} = A_{sk}/V_{sk}, A_{sk} = \text{amount in skin} \tag{18}$$

$$C_{surf} = A_{surf}/V_{sk}, A_{surf} = \text{amount on skin} \tag{19}$$

$$dA_{air}/dt = K_a*A_{surf}, K_a = \text{rate of loss to air} \tag{20}$$

$$dA_{sk}/dt = K_p*A*(C_{surf} - C_{sk}/P_{a/sk}) + Q_{sk}*(CA - C_{sk}/P_{sk/b}) \tag{21}$$

Figure 8 gives the curves from the model, showing the simultaneous loss of isofenphos to air, absorption by skin, and retention of isofenphos on the surface of the skin.

B. PARTITION COEFFICIENTS

The partition coefficients used in PB-PK models are simple concentration ratios, expressed as volume per volume or weight per volume in blood/gas, tissue/blood, and tissue/air at equilibrium. The vial equilibration procedures of Sato and Nakajima[21] and Gargas et al.[22] work well for obtaining partition coefficients on volatile solvents. Jepson et al.[46] developed a vial procedure for nonvolatile and intermediate volatile chemicals in biological tissues. Parathion, lindane, paraoxon, diisopropylfluorophosphate, and perchloroethylene were evaluated. Vapor pressures for these chemicals ranged from 9×10^{-6} to 14 mmHg. Small pieces (<1.0 g) of fat, muscle, liver, and skin were incubated for 24 h with shaking in 20 ml of saline solution containing the chemical of interest (e.g., 4 and 8 ppm parathion).

TABLE 6
Effect of Heat-Inactivated Horse Serum
on the Distribution Quotient[a]

Horse serum in Waymouth's medium (%)	f	q[b]
0	0.494 ± 0.027	332
2.5	0.628 ± 0.019	193
5	0.711 ± 0.025	132
10	0.711 ± 0.018	96
17.5	0.847 ± 0.021	59
50	0.907 ± 0.011	34

Note: $f = C_m/C_o$ (ambient conc of parathion per parathion conc in total volume).

$q = C_c/C_m$ (conc of parathion in hepatocytes per ambient conc of parathion).

[a] Hepatocyte and [^{35}S] parathion concentrations were 5×10^5 cells per milliliter and 1×10^{-5} M, respectively. Average of two duplicate experiments.

[b] Based on the reported average hepatocyte volume of 6186 μm^3.

From Nakatsugawa, T., Bradford, W. L., and Usui, K., *Pestic. Biochem. Physiol.*, 14, 13, 1980.

A 2.0-ml aliquot of saline was analyzed for the chemical after centrifugation and ultrafiltration. The partition coefficients for blood, fat, muscle, liver, and skin were 54.2, 5365, 136, 270, and 160, respectively. The fat, muscle, liver, and skin values were divided by the blood values for use in PBPK models.

In a recent study, Nakatsugawa et al.[47] developed distribution quotients, q values, for the distribution of parathion between Waymouth's medium and intact hepatocytes. Parathion is rapidly taken up by isolated rat liver hepatocytes from Waymouth's medium and released back to the medium as parathion or metabolized products. The q values vary according to the concentration of the hepatocytes and the concentration of the parathion in the medium. The addition of heat-inactivated horse serum in Table 6 resulted in a decrease in the q values from a high value of 332 for Waymouth's medium alone to a value of 34 for a 50% mixture. Blood plasma is equivalent to Waymouth's medium and 5% horse serum. The combination of Waymouth's medium and 5% horse serum gives a q value of 132 for parathion. According to Nakatsugawa et al.[47] the uptake by hepatocytes occurs over a few seconds. The q value, 332, developed for parathion using Waymouth's medium is equivalent to the partition coefficient (270) determine in the vial method of Jepson et al.[46]

In situ parathion liver perfusion studies, by Nakatsugawa et al.[47] showed that the intact liver lobule and its hepatocytes function like a reverse-phase chromatographic column in which parathion enters the hepatocyte from a mobile liquid phase, is metabolized, or partitions back out to the liquid phase with the more lipid-soluble parathion eluting after the water soluble metabolites. The relationship between translobular migration velocity ($-$Log V) and hydrophobicity (Log P) was determined by Murakami et al.[48] for parathion and four other chemicals. Log P values were estimated from standard Log P vs. HPLC retention time plots using chromatographic retention time data.[49]

Knaak et al.[12] showed that relatively small structural changes in isofenphos resulted in large changes in the partition coefficient and the chemical properties of the pesticide. Desulfuration of isofenphos to its oxon (P = S to P = O conversion) decreases its octanol-water partition coefficient from 25,000:1 to 600:1 (Log P 4.39 to 2.78), enhances its toxicity, and increases its rate of hydrolysis. The toxicity of parathion (paraoxon), malathion (malaoxon, O,O-dimethyl S-[1,2-di(ethoxycarbonyl) ethyl] phosphorothioate) and several other OPs are also changed by desulfuration. The octanol/water partition coefficients for these compounds have not been reported.

C. METABOLIC ENZYMES

A knowledge of the enzymes involved in the metabolism of OPs and carbamates, the kinetics of the reactions, and the distribution of the metabolic products are required for PB-PK modeling.

In pharmacokinetics, the liver is treated as a well-stirred system to simplify the events taking place and the mathematics involved. Studies by Nakatsugawa et al.,[50] Xu et al.,[51] and others using infusion techniques have demonstrated the heterogeneity of the liver. Different enzymes are located in the periportal and centrilobular hepatocytes. The enzymes in the periportal hepatocytes are the first to absorb parathion according to Nakatsugawa et al.[50] using fluorescein diacetate (FDA) as a surrogate material. The efficiency of the metabolism of low doses of parathion is a function of the ability of hepatocytes and their enzyme systems to extract parathion from blood plasma.

Hepatocyte microsomal P-450 enzymes play an important role in the metabolism of OP pesticides. Parathion, isofenphos, malathion, and other phosphorothioates are desulfurated (P = S to P = O) to yield toxic oxons.[52] Hepatic enzymes also hydrolyze P-O-aryl, P-O-alkyl, and P-HN-alkyl bonds yielding in some cases less toxic, more water-soluble products. The Michaelis-Menten parameters (V_{max}, K_m) for the metabolism of isofenphos to isofenphos oxon, des N-isopropyl isofenphos, and des N-isopropyl isofenphos oxon were determined by Knaak et al.[53] using gas chromatography. The oxons are substrates/inhibitors of the B-esterases and substrates of the A-esterases. The A-esterases found in the cytosol hydrolyze malathion (S-(1,2-dicarbethoxyethyl) O,O-dimethyl phosphorodithioate) at the P-S linkage to yield O,O-dimethyl phosphorothioate. These esterases are insensitive to the action of DFP (diisopropyl phosphorofluoridate) and other organophosphorus insecticides at high concentrations. Malathion is also hydrolyzed by a B-esterase (carboxylesterases) to give the malathion monoacid. This esterase is sensitive to low concentrations of DFP and paraoxon. The B-esterases are predominantly found in the microsomes, according to Bhagwat and Ramachandran.[54] Enzymes capable of hydrolyzing DFP and related acetylcholinesterase inhibitors have recently been renamed as organophosphate acid (opa) anhydrases.[55]

D. ENZYME INHIBITION

Organophosphorus compounds in the presence of substrate react with esterases to produce relatively stable phosphorylated enzymes.[56,57] The reactions according to Main[57] may be described by the following two equations:

$$EH + AX \underset{k_1}{\overset{K_a}{\rightleftharpoons}} EHAX \overset{k_2}{\underset{HX}{\searrow}} EA \qquad (22)$$

$$EH + CB \underset{k_{-1}}{\overset{k_1}{\rightleftharpoons}} EHCB \overset{k_2'}{\underset{HB}{\searrow}} EC + HOH \overset{k_3'}{\longrightarrow} EH + AOH \qquad (23)$$

where EH is free enzyme, AX is an ester of an organophosphorus acid, EHAX is the enzyme OP complex, EA is the phosphorylated enzyme and HX is the first leaving product, CB is the substrate, EHCB is the enzyme substrate complex, EC is the acetyl-enzyme, HB the leaving group, and EH the regenerated enzyme. K_a is the dissociation constant for the enzyme-inhibitor complex. Main[57] worked out a procedure for evaluating K_a, k_2, and k_i. In conditions where enzyme $<<$ inhibitor and enzyme $<<$ substrate, the enzyme catalyzing the substrate is progressively removed by the action of the organophosphorus inhibitor present in the substrate, causing the reaction curve to deviate from linearity. The curve is analyzed by drawing tangents (slopes $=$ v) at time t. A plot of log v against t gives the first order inhibition rate in min^{-1}, where the slope of the line is

$$p = \frac{2.3 \log (V_o/V)}{t} = \frac{k_2}{1 + (K_{a/i})} \tag{24}$$

To determine K_a and k_2, a set of four or five log v vs. t plots are obtained using an appropriate range of inhibitor. The slopes, p, and inhibitor concentrations are used to construct plots of $1/p$ vs. $1/[i]$. A plot of $1/p$ vs. $1/[i]$ gives the values for $1/k_i$ (slope), $1/k_2$ (y intercept), and $-1/K_a$ (x intercept) according to the following equation:

$$1/p = \frac{K_a}{k_2} \times \frac{1}{i} + \frac{1}{k_2} \tag{25}$$

The bimolecular rate constant, k_i, is related to K_a and k_2 by:

$$k_i = k_2/K_a \tag{26}$$

The OP oxons inhibit (phosphorylate) tissue butyrylesterases (BChE), and carboxylesterases (CaE),[58] and acetylcholinesterase (AChE). The CaE are more sensitive to inhibition than nervous system AChE. The presence of these enzymes in liver suggests they have a protective role in preventing AChE inhibition. Plasma CaE functions in a similar manner by reacting more rapidly with the oxons. The recovery of these enzymes is dependent upon the synthesis of a new enzyme in the liver and not on the dissociation of a phosphorylated enzyme.[58] In the reaction of cholinesterase with organophosphorus compounds, a change takes place called "aging" of the phosphorylated enzyme. This phenomenon was demonstrated when phosphorylated enzymes could not be reactivated by nucleophilic reagents after storage. The change is due to the loss of an alkyl group from an alkoxy group attached to the phosphorus. The following equation describes this change:

$$\tag{27}$$

N-methylcarbamates such as carbaryl, aldicarb (2-methyl-2(methylthio) propionaldehyde O-(methyl carbamoyl) oxime), and methomyl (S-methyl-N-[(methylcarbamoyl)oxy] thioacetamidate) cause reversible inhibition of cholinesterase. The enzyme inhibitor complex leads to acylation of the enzyme, followed by deacylation. The affinity of the carbamate for the enzyme, rate of conversion of enzyme inhibitor complex to acylated enzyme, and the stability of the acylated enzyme must be considered in PB-PK models for carbamates.

E. PHARMACODYNAMICS

The most important task in the percutaneous PB-PK modeling of OPs and carbamates involves using the physiological parameters, permeability constants, tissue partition coef-

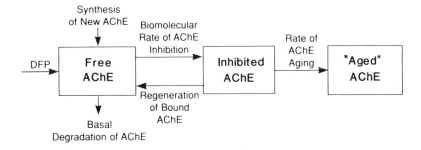

FIGURE 9. Model for AChE inhibition, aging, regeneration, synthesis, and degradation. (From Gearhart, J. M., Jepson, G. W., Clewell, H. J., III, Andersen, M. E., and Conolly, R. B., *Toxicol. Appl. Pharmacol.*, 106, 295, 1990. With permission).

ficients, and metabolic and enzymic inhibition kinetic data to develop a model to predict the effects of the absorbed tissue dose on the activity of acetylcholinesterase in blood and nervous tissue of exposed workers.

Gearhart et al.[30] recently developed a physiologically based pharmacokinetic and pharmacodynamic model for the inhibition of AChE, BChE, and CaE by an i.v. dose of 1.0 mg/kg of DFP. DFP is a toxic, nonpesticidal organophosphorous ester capable of inhibiting acetylcholinesterases by direct phosphorylation. It is hydrolyzed by liver A-esterases (opa) at the P-F bond to diisopropylphosphate.

Figure 9 describes the inhibition of AChE. V_{max} and K_m values for the hydroysis of DFP in brain, liver, kidney, richly perfused tissue, venous, and arterial blood were obtained by measuring the release of fluoride ion. The bimolecular inhibition rate constants for the reaction of DFP with AChE, BChE, and CaE were obtained by measuring the decrease in enzyme activity after the addition of DFP.

The bimolecular inhibition rate constant was evaluated as:

$$-k_i = (\ln(V/V_o)/t)(K_m + [S])/K_m \times [DFP] \tag{28}$$

where k_i is the rate constant in (molar per minute)$^{-1}$, [S] is the substrate concentration (micromolar), V is the reaction velocity (micromoles per minute per gram of tissue), V_o is the AChE activity at time zero (micromoles per minute per gram of tissue), t is the duration (minute) of incubation, [DFP] is the concentration of DFP (micromolar), and K_m is the Michaelis constant (micromolar). Slopes of the tangents to the curve provided the velocity of the substrate hydrolysis reaction. In this procedure a value for K_m is needed. The K_m value for the hydrolysis of acetylcholine bromide by AChE is 320 μM, while V_{max} is 22.7 μmol/min/ml.[59] According to Main,[57] the K_a values for inhibitors are of the same magnitude as the K_m values for substrates.

The tissue partition coefficients were determined by the vial equilibration technique of Sato and Nakajiima[21] and Gargas et al.[22] Tissue/blood partition coefficients were calculated by dividing the tissue/air value by that for blood/air.

The mass balance differential equation for DFP in the brain was represented by:

$$V_{Br} * dC_{Br}/dt = Q_{Br} * (C_{ABr} - C_{VBr}) \tag{29}$$

$$- (V_{maxBr} * C_{VBr})/(K_{mBr} + C_{VBr}), \text{ hydrolysis by DFPase} \tag{30}$$

$$- K_{AChE} * C_{AEBr} * C_{Br}, \text{ inhibition of AChE} \tag{31}$$

$$- \; K_{CaE} * C_{CEBr} * C_{Br}, \; \text{inhibition of carboxylesterase} \qquad (32)$$

$$- \; K_{BChE} * C_{BEBr} * C_{Br}, \; \text{inhibition of butyrylesterase} \qquad (33)$$

where:

V_{Br}	=	Volume of brain (liter)
C_{Br}	=	Concentration of DFP in the brain compartment (milligrams per liter)
Q_{Br}	=	Blood flow to the brain (liters per hour)
C_{ABr}	=	DFP in arterial arterial blood brain (milligrams per liter)
C_{VBr}	=	DFP in venous blood leaving brain (milligrams per liter)
V_{maxBr}	=	Maximum rate of DFPase hydrolysis (milligrams per hour)
K_{mBr}	=	Michaelis constant for DFPase in brain (milligrams per liter)
K_{AChE}	=	Bimolecular rate constant for DFP reaction with AChE (micromolar per hour)$^{-1}$
C_{AEBr}	=	AChE concentration in brain (micromolar)
K_{CaE}	=	Bimolecular rate constant for DFP reaction with CaE (micromolar per hour)$^{-1}$
C_{CEBr}	=	CaE concentration in brain (micromolar)
K_{BChE}	=	Bimolecular rate constant for DFP reaction with BChE (micromolar per hour)$^{-1}$
C_{BEBr}	=	BChE concentration in brain (micromolar)

The differential equation used to calculate inhibited acetylcholinesterase activity in tissue was

$$V_{Br} * d_{AEBr}/dt \; = \; (K_{AChE} * C_{AEBr} * C_{BrM}), \; \text{inhibition of AChE} \qquad (34)$$

$$- \; (K_{RABr} * A_{EBr}), \; \text{regeneration} \qquad (35)$$

$$- \; (K_{AABr} * A_{EBr}), \; \text{aging} \qquad (36)$$

where:

V_{Br}	=	Volume of brain (liters)
K_{AChE}	=	Bimolecular inhibition rate constant (micromolar per hour)$^{-1}$
C_{AEBr}	=	Free AChE in brain (micromolar)
C_{BrM}	=	DFP in brain (micromolar)
K_{RABr}	=	Rate of regeneration of inhibited AChE (per hour)
A_{EBr}	=	Inhibited AChE (micromolar)
K_{AABr}	=	Rate of aging of inhibited AChE (per hour)

The model nicely simulated the concentration of free DFP in the plasma and brain of mice (Figure 10). In Figure 11, the injected DFP rapidly inhibited AChE to about 20% of control values. Recovery was rapid and nearly complete by 45 h. The model was used to simulate the concentration of DFP in human brain after a 5-min, 50-ppm exposure.

In the case of the phosphorothioate insecticides such as parathion and isofenphos, V_{max}, K_m values for the desulfuration of the insecticide to its oxon are required. Additional kinetic values may also be needed to describe competitive reactions involving the oxons, metabolic enzymes (e.g., N-dealkylation, deamination, and hydrolysis), AChE, and other tissue esterases (CaE and BChE). An example of this is given in Figure 12, where ^{14}C-ring-labeled isofenphos in the rat is dealkylated to des N-isopropyl isofenphos (DNI), desulfurated to isofenphos oxon (IO), and further N-dealkylated or desulfurated to des N-isopropyl isofen-

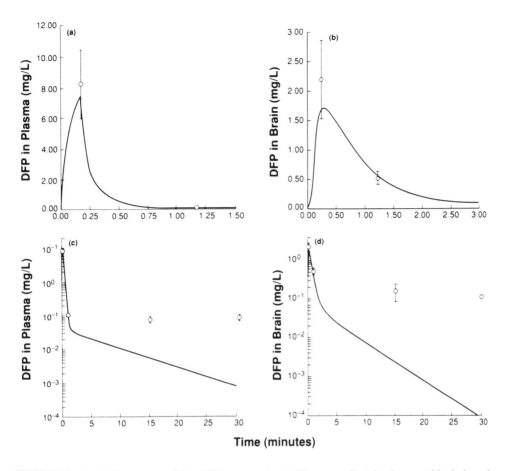

FIGURE 10. (a–d) Time course of free DFP concentration (milligram per liter) in plasma and brain in male mice (Dublin ICR) after tail vein injection of 1 mg DFP/kg. Each datum represents the mean (± one standard deviation of the mean) of five animals.[62] Solid line depicts computer simulation generated with the PB-PK model. (From Gearhart, J. M., Jepson, G. W., Clewell, H. J., III, Anderson, M. E., and Conolly, R. B., *Toxicol. Appl. Pharmacol.*, 106, 295, 1990. With permission).

phos oxon (DNIO), the active inhibitor of AChE. Competitive reactions to AChE phosphorylation by DNIO involve its deamination, hydrolysis by A-esterases, and phosphorylation of tissue CaE and BChE. A PB-PK model for the dermal and i.v. administration of isofenphos was recently prepared by Raabe et al.[60] using Figure 12 to estimate the concentration of DNIO in liver, kidney, and other slowly and rapidly perfused tissues. The model is currently being revised to include a brain compartment for the metabolites and the kinetics for the bimolecular rate reaction of DNIO and AChE in blood and brain, CaE in liver, and BChE in plasma and liver. The rat model output will be validated using *in vivo* rat blood, liver, and brain AChE, BChE, and CaE values.

Recent studies in the dog[60] with isofenphos indicate the pathway to DNIO is similar to that in the rat, while the detoxification pathway in the dog favors the formation of the glycine conjugates of the intact OPs rather than hydrolysis to alkyl phosphates. Models may be directly extrapolated to man or other animals, such as the dog, provided the metabolic pathways are the same. Flows and organ volumes relate across species as a power of body weight. The allometric exponent is generally equal to one for volumes and is a fraction of body weight for flows.

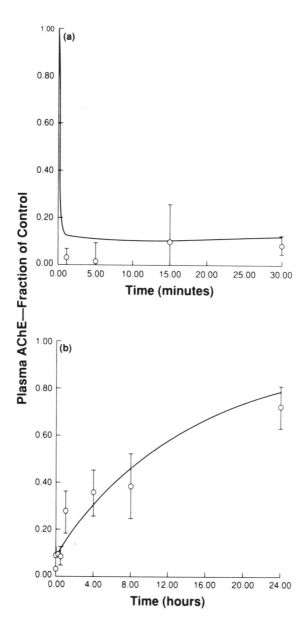

FIGURE 11. (a,b) Time course of plasma AChE activity in male mice (Dublin ICR) after tail vein injection of 1 mg DFP/kg. Data are expressed as a fraction of control activity. Each datum represents the mean ($\pm$ one standard deviation of the mean) of five animals. Solid line depicts computer simulation. (From Martin, B. R., *Toxicol. Appl. Pharmacol.*, 77, 275, 1985. With permission).

In some cases the model, or a portion of the model, may be validated using human subjects. Human studies involving EPA-registered pesticides may be conducted when approved by human subject committees.[2,44]

V. DISCUSSION

Since the synthesis of the first carbamate and OP insecticides, dermal models were desired for predicting the effects of these insecticides on the cholinesterase activity of exposed

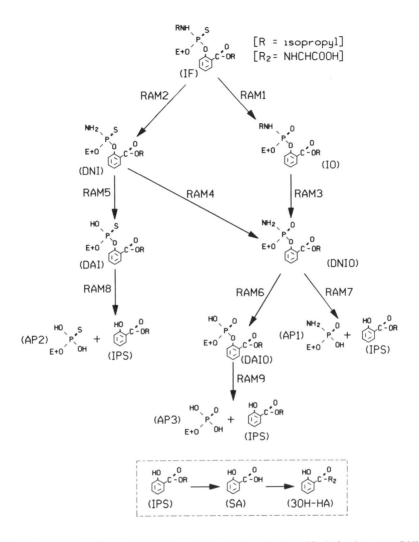

FIGURE 12. Metabolic pathway for isofenphos in the rat: IF, isofenphos; IO, isofenphos oxon; DNI, des N-isopropyl isofenphos; DNIO, des N-isopropyl isofenphos oxon; DAIO, deaminated DNIO; DAI, deaminated DNI; AP1, aminoethylphosphate; AP2, O-ethylthiophosphate; AP3, monoethylphosphate; IPS, isopropyl salicyclate; SA, salicyclic acid; and 3 OH-HA, 3 hydroxy hippuric acid. DAIO, DAI, and IPS were not found in *in vivo* or in *in vitro* studies. V_{max} and K_m values were determined for RAM 1, 2, 3, and 4. (From Raabe, O. G., Knaak, J. B., and Al-Bayati, M. A., Improved Models of Risk for Exposure to Organic Toxicants and Carcinogens, Continuation Application, EPA Proj. No. CR-816332-01-0, University of California, Davis, June 30, 1991. With permission).

workers. PB-PK models as described in this chapter are capable of combining all the necessary information into one model for simulating the pharmacokinetic events and predicting AChE inhibition in blood and nervous tissue.

Validation of a model requires appropriately designed *in vivo* studies to measure the metabolites formed and the inhibition/recovery of AChE and other esterases. The size of the dose that may be used in validation studies are limited by the toxicity of the OP and carbamate pesticides. Large dosages of OPs and carbamates, sufficient in size to saturate metabolic enzymes, may result in sick animals and poor estimates of the kinetics involved.

The future of percutaneous PB-PK modeling appears to be good for predicting tissue dose and AChE inhibition in workers exposed to OP and carbamate pesticides despite the small number of scientists involved in modeling and the substantial amount of information required for model development and validation. A team of scientists (e.g., toxicologists, biochemists, bioanalytical chemists, modelers, physicians, etc.) is needed to develop exposure, dermal absorption and metabolic pathway data, V_{max} and K_m values, partition coefficients, enzyme inhibition kinetics, and other data necessary for developing suitable models and validating the models with human studies. The described field worker studies[33] provide the necessary information on exposure, while the human studies of Feldmann and Maibach,[2] Wester et al.,[44] and others provide information for model validation. Suitable computer software, ACSL and SimuSolv, has been developed for PB-PK modeling in the laboratory. This software is currently available to run on IBM and Digital mainframes and on IBM 386SX personal computers (PC) as well as PC clones.

Training courses conducted by Chemical Industry Institute of Toxicology members and sponsored by the Society of Toxicology are available several times a year. To be successful in developing models, laboratory data development and PB-PK modeling must be carried out simultaneously. Industry and the Office of Research and Development, U.S. EPA, are currently sponsoring studies designed to provide information and improved protocols for model development. Neither of the EPA's regulatory programs, OPP or OTS, require the development of PB-PK models for registering pesticides or for data development under current test rules.

ACKNOWLEDGMENTS

We thank Dale Uyeminami, Fiorella Gielow, and Chengxian Gao for their support in conducting the isofenphos studies and in reviewing the chapter. The isofenphos work was supported by EPA Cooperative Agreement CR-816332-01-0. The views and opinions expressed in this chapter are solely those of the authors and do not necessarily reflect those of the Agency.

REFERENCES

1. **Maibach, H. I., Feldmann, R. J., Milby, T. H., and Serat, W. F.,** Regional variation in percutaneous penetration in man, *Arch. Environ. Health,* 23, 208, 1971.
2. **Feldmann, R. J. and Maibach, H. I.,** Percutaneous penetration of some pesticides and herbicides in man, *Toxicol. Appl. Pharmacol.,* 88, 126, 1974.
3. **Spencer, T. S., Hill, J. A., Feldmann, R. J., and Maibach, H. I.,** Evaporation of diethyltoluamide from human skin *in vivo* and *in vitro, J. Invest. Dermatol.,* 72(6), 317, 1979.
4. **Reifenrath, W. G., Hill, J. A., Robinson, P. B., McVey, D. L., et al.,** Percutaneous absorption of Carbon 14-labeled insect repellents in hairless dogs, *J. Environ. Pathol. Toxicol.,* 4(1), 249, 1980.
5. **Shah, P. V., Monroe, R. J., and Guthrie, F. E.,** Comparative rates of dermal penetration of insecticides in mice, *Toxicol. Appl. Pharmacol.,* 59, 414, 1981.
6. **Knaak, J. B., Yee, K., Ackerman, C. R., Zweig, G., Fry, D. M., and Wilson, B. W.,** Percutaneous absorption and dermal dose-cholinesterase response studies with parathion and carbaryl in the rat, *Toxicol. Appl. Pharmacol.,* 76, 252, 1984.
7. **Knaak, J. B., Ackerman, C. R., and Wilson, B. W.,** Percutaneous Absorption and Dermal-Dose Cholinesterase Response Studies with Thiodicarb in the Adult Female Rat, California Department of Food and Agriculture, Sacramento, CA, 1984.
8. **Van Lier, R. B.,** The use of monkey percutaneous absorption studies, in *Dermal Exposure Related to Pesticide Use: Discussion of Risk Assessment,* Honeycutt, R. C., Zweig, G., and Ragsdale, N. N., Eds., ACS Symposium Series 273, Washington, D.C., 1985, chap. 6.
9. **Marco, G. J., Simoneaux, B. J., Williams, S. C., Cassidy, J. E., Bissig, R., and Muecke, W.,** Radiotracer approaches to rodent dermal studies, in *Dermal Exposure Related to Pesticide Use: Discussion of Risk Assessment,* Honeycutt, R. C., Zweig, G., and Ragsdale, N. N., Eds., ACS Symposium Series 273, Washington, D.C., 1985, chap. 4.

10. **Longacre, S. L., DiDonato, L. J., Wester, R. C., Maibach, H. I., Hurt, S. S., and Costlow, R. D.,** Dinocap dermal absorption in female rabbits and rhesus monkey: implications for humans, in *Biological Monitoring for Pesticide Exposure: Measurement, Estimation, and Risk Reduction,* Wang, G. M., Franklin, C. A., Honeycutt, R. C., and Reinert, J. C., Eds., ACS Symposium Series 382, Washington, D.C., 1989, chap. 11.

11. **Shah, P. V., Fisher, H. L., Sumler, M. R., and Hall, L. L.,** Dermal absorption and pharmacokinetics of pesticides in rats, in *Biological Monitoring for Pesticide Exposure: Measurement, Estimation, and Risk Reduction,* Wang, G. M., Franklin, C. A., Honeycutt, R. C., and Reinert, J. C., Eds., ACS Symposium Series 382, Washington, D.C., 1989, chap. 14.

12. **Knaak, J. B., Al-Bayati, M., Raabe, O. G., and Blancato, J. N.,** *In vivo* percutaneous absorption studies in the rat: pharmacokinetics and modelling of isofenphos absorption, in *Prediction of Percutaneous Penetration: Methods, Measurements and Modelling,* Scott, R. C., Guy, R. H., and Hadgraft, J., Eds., IBC Technical Service Ltd., London, 1990, 1.

13. **Reitz, R. H., Nolan, R. J., and Schumann, A. M.,** Development of multispecies, multiroute pharma-cokinetic models for methylene chloride and 1,1,1-trichloroethane (methyl chloroform), in *Pharmacokinetics in Risk Assessment: Drinking Water and Health, 8,* National Academy Press, Washington, D.C., 1987, 391.

14. **McKenna, M. J. and Zempel, J. A.,** The dose-dependent metabolism of [^{14}C] methylene chloride following oral administration to rats, *Fed. Cosmet. Toxicol.,* 19, 73, 1980.

15. **McKenna, M. J., Zempel, J. A., and Braun, W. H.,** The pharmacokinetics of inhaled methylene chloride in rats, *Toxicol. Appl. Pharmacol.,* 65, 1, 1982.

16. **Angelo, M. J. and Pritchard, A. B.** Simulations of methylene chloride pharmacokinetics using a phys-iologically based model, *Regul. Toxicol. Pharmacol.,* 4, 329, 1984.

17. **Andersen, M. E., Clewell, H. J., III, Gargas, M. L., Smith, F. A., and Reitz, R. H.,** Physiologically based pharmacokinetics and the risk assessment process for methylene chloride, *Toxicol. Appl. Pharmacol.,* 87, 185, 1987.

18. **Jakobson, I., Wahlberg, J. E., Holmberg, B., and Johansson, G.,** Uptake via blood and elimination of 10 organic solvents following epicutaneous exposures of anesthetized guinea pigs, *Toxicol. Appl. Phar-amcol.,* 63, 181, 1982.

19. **McDougal, J. N., Jepson, G. W., Clewell, H. J., III, MacNaughton, M. G., and Andersen, M. E.,** A physiological pharmacokinetic model for dermal absorption of vapors in the rat, *Toxicol. Appl. Phar-macol.,* 85, 286, 1986.

20. **McDougal, J. N., Jepson, G. W., Clewell, H. J., III, Gargas, M. L., and Andersen, M. E.,** Dermal absorption of organic chemical vapors in rats and humans, *Fundam. Appl. Toxicol.,* 14, 299, 1990.

21. **Sato, A. and Nakajimak, T.,** A vial-equilibration method to evaluate the drug metabolizing enzyme activity for volatile hydrocarbons, *Toxicol. Appl. Pharmacol.,* 47, 41, 1979.

22. **Gargas, M. L., Burgess, R. J., Voisard, D. J., Carson, G. H., and Andersen, M. E.,** Partition coefficients of low molecular weight volatile chemicals in various liquids and tissues, *Toxicol. Appl. Pharmacol.,* 98, 87, 1989.

23. **Travis, C. C., White, R. K., and Arms, A. D.,** A physiologically based pharmacokinetic approach for assessing the cancer risk of tetrachloroethylene, in *The Risk Assessment of Environmental Hazards: A Textbook of Case Studies,* Paustenbach, D., Ed., John Wiley & Sons, New York, 1986, chap. 23.

24. **Arms, A. D. and Travis, C. C.,** Reference Physiological Parameters in Pharmacokinetic Modeling, Rep. No. EPA/600/6–88/004, National Technical Information Service, Springfield, VA.

25. **Omura, T. and Sato, R.,** The carbon monoxide binding pigment of liver microsomes. Evidence for its hemoprotein nature, *J. Biol. Chem.,* 239, 2370, 1964.

26. **Habig, W. H., Pabst, M. J., and Jakoby, W. B.,** Glutathione-S-transferase. The first step in mercapturic acid formation, *J. Biol. Chem.,* 249, 7130, 1974.

27. **Dedrick, R. L., Forrester, D. D., and Ho, D. H. W.,** *In vitro-in vivo* correlation of drug metabolism, Deamination of 1-*b*-D-arabinofuranosylcystosine, *Biochem. Pharmacol.,* 21, 1, 1972.

28. **Dedrick, R. L. and Forrester, D. D.,** Blow flow limitation in interpreting Michaelis constants for ethanol oxidation *in vivo, Biochem. Pharmacol.,* 22, 1133, 1973.

29. **Lin, J. H., Hayashi, M., Awazu, S., and Hanano, M.,** Correlation between *in vitro* and *in vivo* drug metabolism rate: oxidation of ethoxybenzamide in rat, *J. Pharmacokinet. Biopharm.,* 6(4), 327, 1978.

30. **Gearhart, J. M., Jepson, G. W., Clewell, H. J., III, Andersen, M. E., and Conolly, R. B.,** Physio-logically based pharmacokinetic and pharmacodynamic model for the inhibition of acetylcholinesterase by diisopropylfluorophosphate, *Toxicol. Appl. Pharmacol.* 106, 295, 1990.

31. **Gargas, M. L.,** Chemical-specific constants for physiologically based pharmacokinetic models, *CIIT Act.,* 11(3), 1991.

32. Residues of pesticides and other contaminants in the total environment, *Residue Rev.,* p. 75, 1980.

33. **Knaak, J. B., Iwata, Y., and Maddy, K. T.,** The worker hazard posed by re-entry into pesticide-treated foliage: development of safe reentry times, with emphasis on chlorthiophos and carbosulfan, in *The Risk Assessment of Environmental Hazards: A Textbook of Case Studies,* Paustenbach, D. J., Ed., John Wiley & Sons, New York, 1989, chap. 24.

34. **Durham, W. F. and Wolfe, H. R.,** *Bull. WHO,* 26, 75, 1962.

35. **Fenske, R. A.,** Validation of environmental monitoring by biological monitoring: fluorescent tracer technique and patch technique, in *Biological Monitoring for Pesticide Exposure: Measurement, Estimation, and Risk Reduction,* Wang, G. M., Franklin, C. A., Honeycutt, R. C., and Reinert, J. C., Eds., ACS Symposium Series 382, Washington, D.C., 1989, chap. 6.

36. **Nigg, H. N. and Stamper, J. H.,** Biological monitoring for pesticide dose determination: historical perspecitves, current practices, and new approaches, in *Biological Monitoring for Pesticide Exposure: Measurement, Estimation, and Risk Reduction,* Wang, G. M., Franklin, C. A., Honeycutt, R. C., and Reinert, J. C., Eds., ACS Symposium Series 382, Washington, C.C., 1989, chap. 1.

37. **Popendorf, W. J.,** Advances in the unified field model for reentry hazards, in *Dermal Exposure Related to Pesticide Use: Discussion of Risk Assessment,* Honeycutt, R. C., Zweig, G., and Ragsdale, N. N., Eds., ACS Symposium Series 273, Washington, D.C., 1985, chap. 23.

38. **Nigg, H. N., Stamper, J. H., and Queen, R. M.,** The development and use of a universal model to predict tree crop harvester pesticide exposure, *Am. Ind. Hyg. Assoc. J.,* 45, 182, 1982.

39. **O'Malley, M.,** Priority Investigations Involving Phosalone in Fresno and Madera Counties, 1987, California Department of Food and Agriculture, Rept. HS-1487, Worker Health and Safety Branch, Sacramento, CA, 1987.

40. **Bradway, D. E., Lores, E. M., and Edgerton, T. R.,** Minimizing occupational exposure to pesticides: recent development in methodology for monitoring pesticide metabolites in human urine, *Residue Rev.,* 75, 1980, 51.

41. **Lavy, T. L. and Mattice, J. D.,** Monitoring field applicator exposure to pesticides, in *Dermal Exposure Related to Pesticide Use: Discussion of Risk Assessment,* Honeycutt, R. C., Zweig, G., Ragsdale, N. N., Eds., ACS Symposium Series 273, Washington, D.C., 1985, chap. 11.

42. **Dauterman, W. C.,** Biological and nonbiological modifications of organophosphorus compounds, *Bull. WHO,* 44, 133, 1971.

43. **Sullivan, L. J., Eldridge, J. M., Knaak, J. B., and Tallant, M. J.,** 5,6-Dihydro-5,6-dihydroxycarbaryl glucuronide as a significant metabolite of carbaryl in the rat, *J. Agric. Food Chem.,* 20, 980, 1972.

44. **Wester, R. C., Maibach, H. I., Melendres, J., Sedik, L., Knaak, J. B., and Wang, R.,** *In vivo* and *in vitro* percutaneous absorption and skin evaporation of isofenphos in man, *Fundam. Appl. Pharmacol.,* in press.

45. **Wester, R. C. and Maibach, H. I.,** Structure-activity correlations in percutaneous absorption, in *Percutaneous Absorption, Mechanisms-Methodology-Drug Delivery,* Bronaugh, R. L. and Maibach, H. I., Eds., Marcel Dekker, Inc., New York, 1985, chap. 8.

46. **Jepson, G. W., Hoover, D. K., Black, R. K., McCafferty, J. D., Mahle, D. A., and Gearhart, J. M.,** Partition coefficient determination for nonvolatile and intermediate volatility chemicals in biological tissues, Abstract No. 996, 1992 Annual meeting of Society of Toxicology, Seattle, WA, February 23–27, 1992.

47. **Nakatsugawa, T., Bradford, W. L., and Usui, K.,** Hepatic disposition of parathion: uptake by isolated hepatocytes and chromatographic translobular migration, *Pestic. Biochem. Physiol.,* 14, 13, 1980.

48. **Murakami, N., Uchida, M., Sugimoto, T., and Nakatsugawa, T.,** Relationship between translobular migration in perfused rat liver and hydrophobicity of dialkyl dithiolanylidenemalonates and related compounds, *Xenobiotica,* 17(2), 241, 1987.

49. **Krikorian, S. E., Chorn, T. A., King, J. W., and Elizy, M. W.,** Determination of the partition coefficients of organophosphorus compounds using high-performance liquid chromatography, Report CRDEC-CR-88015 School of Pharmacy, University of Maryland.

50. **Nakatsugawa, T. and Timoszyk, J.,** Fluorescence labeling and sorting of hepatocyte subpopulations to determine the intralobular heterogeneity of paraoxon-metabolizing enzymes in DDE-treated and control rats, *Pestic. Biochem. Physiol.,* 30, 113, 1988.

51. **Xu, X., Tang, B. K., and Pang, K. S.,** Sequential metabolism of salicylamide exclusively to gentisamide 5-glucuronide and not gentisamide sulfate conjugates in single pass *in situ* perfused rat liver, *J. Pharmacol. Exp. Ther.,* 253, 965, 1990.

52. **Fukuto, T. R.,** Mechanism of action of organophosphorus and carbamate insecticides, *Environ. Health Perspects.,* 87, 245, 1990.

53. **Knaak, J. B., Al-Bayati, M. A., Raabe, O. G., and Blancato, J. N.,** *In vitro* Metabolism of Isofenphos by Human and Animal Liver, in ACS National Meet., April 1990.

54. **Bhagwat, V. M. and Ramachandran, B. V.,** Malathion A and B esterases of mouse liver. I. Separation and properties, *Biochem. Pharamcol.,* 24(18), 1713, 1975.

55. **Chester, N. A. and Landis, W. G.**, Activity of organophosphate Acid Anhydrase in *Rangia Cuneata*, Rep. CRDEC-TR 88045, Chemical Research Development and Engineering Center, Aberdeen Proving Ground, MD, 1988.
56. **Aldridge, W. N.**, The nature of the reaction of organophosphorus compounds and carbamates with esterases, *Bull. WHO*, 44, 25, 1971.
57. **Main, A. R.**, Kinetics of active-site directed irreversible inhibition, in *Essays in Toxicology*, Hayes, W. L., Ed., Academic Press, New York, 1973, 59.
58. **Chambers, H., Brown, B., and Chambers, J. E.**, Noncatalytic detoxification of six organophosphorus compounds by rat liver homogenates, *Pestic. Biochem. Physiol.*, 36, 000, 1990.
59. **Krupka, R. M., Hastings, F. L., Main, A. R., and Iverson, F.**, Nitrophenyl acetate hydrolysis by acetylcholinesterase: a Correction, *J. Agr. Food Chem.*, 22, 1150, 1974.
60. **Raabe, O. G., Knaak, J. B., and Al-Bayati, M. A.**, Improved Models of Risk for Exposure to Organic Toxicants and Carcinogens, Continuation Application, EPA Proj. No. CR-816332-01-0, University of California, Davis, June 30, 1991.

Chapter 2

THE APPLICATION OF PHARMACOKINETIC MODELS TO PREDICT TARGET DOSE*

Jerry N. Blancato and Kenneth B. Bishoff

TABLE OF CONTENTS

*   Although the research described in this chapter has been supported by the U.S. Environmental Protection Agency, it has not been subjected to Agency review and therefore does not necessarily reflect the views of the Agency and no official endorsement should be inferred.

I. INTRODUCTION

The number of new and old chemical compounds to which the human race is exposed has quickly become astronomical. Many of these compounds are of tremendous benefit. Many of these same beneficial compounds have later been shown to, under some or all exposure conditions, cause deleterious effects. In addition, many therapeutic drugs which are used to combat a wide variety of diseases may, in the wrong dose, be toxic and perhaps lethal.

Methods of risk assessment have been slow to develop. In the case of nontherapeutic chemicals, the need for such assessment was not obvious for many years. In the case of therapeutic drugs, the only true means was to simply perform numerous trial and error studies with the patient being one of the experimental instruments. As time progressed, animal studies became the foremost means of examining the effects of chemicals on living tissue. The primary criticism levied against these types of studies is that it is not clear how to extrapolate the findings in laboratory animals, often quite small, to humans that are much larger and often are exposed under vastly different conditions than are the animals.

Such animal studies are also time-consuming and often very expensive. Industry, government, and academia have had to scrupulously examine the benefits-to-cost ratio of these studies. Again, the limiting factors appear to be the questions of extrapolation and the compatatibility of different metabolic processes in the different species.

Tissue culture studies provide a faster more efficient testing means. However, fundamental limitations exist with these as well. First, an *in vitro* system may present different metabolic pathways than those same cells would have *in vivo*. Second, as has become recognized, it is very difficult with *in vitro* test systems to examine all the transport effects on disposition, etc. *In vitro* studies are not, for example, necessarily reliable models of transmembrane passage. Third, it is not always obvious how to translate the meaning and values of parameters derived from *in vitro* studies to *in vivo* conditions. It seems readily apparent that more practical risk assessment methods must be developed. Techniques are being formulated which will provide adequate information as to what safe levels of exposure can occur if any. There is a need for studies that can predict quickly, cheaply, and reliably whether substances have a safe level of exposure and at what level of exposure they become toxic. It is to this end that pharmacokinetic modeling is being used as a tool in the risk assessment process.

A. PHARMACOKINETICS

Pharmacokinetics is the study of the time course distribution of foreign chemicals and their metabolic products in the body. Pharmacokinetic models are designed to describe the pharmacokinetics and to predict drug or chemical disposition at other dose, times, exposure regimens, and species. These models are usually of two related varieties. One type, often called the classical model, uses rate constants that, although can be shown to have physiologic meaning, are for practical purposes fitted to the available data. As such, these models do not lend themselves for easy and reliable prediction of disposition between greatly different doses and species. They have been widely used in clinical medicine to help establish optimum therapeutic doses for drugs. The second type, called physiologic models, uses actual physiologic constants such as blood flows, organ volumes, and metabolic rates. Although curve-fitting routines are sometimes used to determine values for constants that are chemical- and experiment-specific, the values for many of the constants such as blood flow are determined from known biological data or independent experiments. As such, the models can be adjusted to predict for different species and for widely different exposure regimens. Frequently the models are formulated in conjunction with experimental data for the chemical of interest.

Based on these, other experiments, and other biological data, the model can be reformulated to apply to humans exposed under various scenarios of interest. It is because of this flexibility that physiologic models are being increasingly used in risk assessments. It is well known that many physiologic processes are nonlinear and that the characteristics of these nonlinear processes may differ between dose levels and between species. Basically the physiologic models enable the risk assessor to account, in a quantitative fashion, for differences in pharmacokinetics that occur among species and among dose levels and regimens.

The risk assessor is often required to extrapolate from high-dose animal experiments to low-dose human exposure conditions. Understanding and quantifying the pharmacokinetic nonlinearities can greatly reduce some of the uncertainties associated with the extrapolation process. It is important to note that differences in incidence of pathology between species and even individuals is probably due to more than just pharmacokinetic differences. Different organisms have different repair mechanisms, immunosurveillance, cell numbers, and sensitivities. However, quantifying pharmacokinetic differences reduces at least one kind of the uncertainty that exists with most assessments.

II. MODEL DEVELOPMENT

Physiologically based pharmacokinetic models describe the disposition of xenobiotics and their metabolic byproducts throughout the body. These models are essentially a system of differential equations describing the rate of change of concentration of the chemicals of interest within the various tissues of the body. The results of either the numerical or analytical solution of these differential equations are the instantaneous concentrations of the chemical in the tissues.

An organ is considered to have three subcompartments: a vascular, an interstitial or extracellular, and an intracellular region. The vascular region is actually the blood contained in the organ's capillaries. The chemical of interest is assumed to be in equilibrium with the venous blood. It should also be remembered that for many organs, not all of the blood entering that organ via the arterial flow passes through the capillaries. Some is shunted, depending upon the state of the arterioles, directly into the organ's venous outflow. In addition the capillaries are very narrow and thin-walled and thus have a great surface area while occupying a comparatively small volume.

The interstitial regions are typically composed of fluid, fibers, and amorphous fluid material. The exact proportions of each of these depends upon the exact tissue and its function. Blood, for example, has a large liquid matrix compared to bone which, while still maintaining some extracellular fluid, is composed of a large number of fibers encased in a calcified matrix.

Intracellular regions are frequently where toxic substances exert their deleterious effects and thus may be the actual region of interest for an accurate risk assessment. Some measure of the toxin, such as its concentration or amount present inside the cell or intereacting with one of its constituents, is the parameter upon which the risk assessor will ultimately base his/her risk characterization and prediction of response. Thus, the resolution of the pharmacokinetic model should be great enough to predict the intracellular concentration or other measure of the ultimate toxin. Often, the actual toxic species is not the parent compound which entered the body, but rather some product that resulted when the entering compound was biotransformed by endogenous metabolic processes. In fact, it is becoming increasingly apparent that sometimes the basis for determining response is not simply the amount or concentration of a toxin, but the amount which interacts with endogenous substances. Examples of this latter point include the amount of adduct formed or, in other cases, the amount of toxin which reversibly binds with endogenous cell surface and intracellular receptors. As more information becomes available about the exact mechanism of action of toxic compounds, the resolution of the pharmacokinetic models will increase.

A. GENERAL MATHEMATICAL DESCRIPTION

Physiologically based pharmacokinetic models can be described in a variety of ways, depending upon the complexity that needs to be incorporated. Various transport and metabolic processes can be mathematically described. Simple and facilitated diffusion and active transport can all be considered when developing a model. Likewise chemical and physiological processes such as ionization in gastric fluids, binding to extracellular and cellular proteins, are all important and in many cases need to be explicitly described for accurate modeling. A simple, but general, case is described here which includes only the diffusion component of transport. In a later section, results are presented of a model which includes metabolic processes and binding to intracellular components.

For any organ the three subcompartments may be described according to:

$$V_V \frac{dC_V}{dt} = Q(C_{Bu} - C_V) - N_{V \rightarrow E} \tag{1}$$

$$V_E \frac{dC_E}{dt} = N_{V \rightarrow E} - N_{E \rightarrow I} \tag{2}$$

$$V_I \frac{dC_I}{dt} = N_{E \rightarrow I} \tag{3}$$

V: Volume of the various compartments of organ.
O: Blood flow to organ.
C_{Bl}, C_E, C_I: Concentration of toxin in arterial blood, extracellular region, and intracellular region.
t: Time.
N: Flux.
subscript V: Vascular compartment.

Integration of the differential equations by numerical methods results in approximate values for the concentration, C, at specific times, t. The total flux, N, is the sum of the flux due to diffusion, J, and that due to convection. Typically in these cases, the convection component is assumed to be neglible, thus:

$$N = J \tag{4}$$

To complete the system for the body, the following equation is written for the systemic venous blood:

$$V_{V,Bl} \frac{dC_{V,Bl}}{dt} = \Sigma\, QC_V - Q_{Bl}C_{V,Bl} \tag{5}$$

Q_{bl}:   Systemic blood flow or cardiac output.
$C_{V,Bl}$:   Concentration of toxin in systemic venous blood.
$V_{V,Bl}$:   Volume of systemic venous blood.
t: Time.

The first term on the right side of Equation 5 represents the sum of the flows times concentration from the vascular regions of all the body's individual organs.

B. FLOW-LIMITED ASSUMPTION

The most commonly used physiologically based pharmacokinetic models assume each organ to be homegenous and thus the discussed subcompartments are all combined into one. This implies that the transfer of a chemcial across the membrane, i.e., capillary wall and cell membrane, is very rapid compared to the tissue perfusion rate. Under this condition, the permeability across the membrane is assumed to be very large. Therefore, the slowest or rate-limiting step in the process of drug distribution must be its delivery by the circulatory flow.

The flow-limited situation can be thought of as the three subcompartments being in a quasiequilibrium with one another and as such Equation 1, 2, and 3 are added resulting in:

$$V_V \frac{dC_V}{dt} + V_E \frac{dC_E}{dt} + V_I \frac{dC_I}{dt} = Q(C_{BI} - C_V) \tag{6}$$

C, the concentration in the whole organ can be described as:

$$\frac{C_V \cdot V_V + C_E \cdot V_E + C_I \cdot V_I}{V} = C \tag{7}$$

where the total organ volume (V) is described as:

$$V_V + V_E + V_I = V \tag{8}$$

At steady-state levels, the ratio of the concentration in the organ to the concentration in the vascular compartment is constant and as a result:

$$R = \frac{C}{C_V} \tag{9}$$

Then by substituting Equations 7, 8, and 9, Equation 6 becomes:

$$V \frac{dC}{dt} = Q\left(C_{BI} - \frac{C}{R}\right) \tag{10}$$

Equation 10 is the equation often observed in most of the currently used physiologically based pharmacokinetic models in toxicology. It should be remembered then that the model described as in Equation 10 assumes flow limitation and implies that the three subcompartments of an organ are in equilibrium. This type of model would not be considered to have high cellular resolution. Such a model is macroscopic, i.e., describing concentration only to an organ level and only for flow-limited cases.

C. MEMBRANE-LIMITED ASSUMPTION

Any higher resolution model would most often assume some noninstantaneous transfer of the toxin across a membrane, such as the capillary membrane or the cell membrane. A model describing transmembrane transport as the rate-limiting step is referred to as a model with membrane-limited assumptions. This implies that the permeability of the membrane is low when compared to the perfusion of the tissue. If the rate-limiting step is the transfer of chemical across the capillary membrane, a two subcompartment model is assumed. One subcompartment consists of the vascular space or capillary space of the organ, and the other is a combination of the extracellualr and intracellular spaces. For any organ, the formulas used to describe such a case are

$$V_{Bl} \frac{dC_{Bl}}{dt} = Q(C_{Bl} - C_T) - N_{Bl \to T} \tag{11}$$

$$V_T \frac{dC_T}{dt} = N_{Bl \to T} \tag{12}$$

$$C = \frac{V_{Bl}C_{Bl} + C_T V_T}{V_{Bl} + V_T} \tag{13}$$

C, C_{Bl}, C_T: Concentration of toxin in blood of capillaries or organ, in whole organ, and in tissue of organ.

V_{Bl}, V_T: Volume of capillaries and tissues of organ.

$N_{Bl \to T}$:   Flux of material from capillaries into tissues of organ.

A slight variation on the previously discussed case is when transport across the cell membrane is the rate-limiting step. In this case, one subcompartment is a combination of the capillary blood and the extracellular spaces (this implies that the capillary blood and extracellular spaces are in equilibrium with respect to the toxin), and the second subcompartment is the intracellular space. This example is described by:

$$V_E \frac{dC_E}{dt} = Q(C_{Bl} - C_E) - N_{E \to I} \tag{14}$$

$$V_I \frac{dC_I}{dt} = N_{E \to I} \tag{15}$$

$$C = \frac{V_E C_E + V_I C_I}{V_E + V_I} \tag{16}$$

C, C_E, C_{Bl}, C_I: Concentration of toxin in whole organ, extracellular space, arterial blood, and intracellular space.

V, V_E, V_I: Volume of whole organ, extracellular space of organ, and intracellular space.

$N_{E \to I}$:   Flux from extracellular to intracellular region of organ.

D. MICROPHARMACOKINETIC PHYSIOLOGIC MODEL

The concepts discussed in the previous section can be mathematically expanded to result in a description of disposition into cellular regions which depends upon parameters which have more physiologic meaning than those enumerated in the previous equations. For this purpose, the organ of interest will be initially subcompartmentalized into three distinct regions: the capillary pool, the extravascular space, and the intracellular space. The blood in the capillary bed represents only a fraction of the total arterial flow to the organ; thus equilibrium with the arterial blood may not be rapidly reached. Over time, different capillaries in an organ are open. Usually not all are open at one time. As a result, not all the capillaries would be receiving the chemical from the arterial flow at any one time. Thus it is to be expected that at the first pass of arterial blood containing a toxin, only some capillaries would receive blood rich in toxin. Therefore if all of the capillaries were collectively sampled at that time, the concentration of toxin in the blood of the collective capillary sample would be less than the concentration in the arterial sample. For modeling purposes, the capillaries are assumed to be such a collective sample rather than just a sample of one individual capillary reaching instantaneous equilibrium with the artery that just supplied its blood. Toxin is then able to diffuse across the collective capillary membrane into the extracellular space and from there across the cell membrane into the intracellular space.

From the collective capillary pool:

$$V_V \frac{dC_V}{dt} = Q(C_A - C_V) - PA_{Cp}\left(C_V - \frac{C_E}{R_E}\right) \qquad (17)$$

C_A, C_V, C_E: Concentration of toxin in arterial blood, collective capillary pool, and extracellular region.
Q: Arterial blood flow to organ.
V_V: Volume of collective capillary pool of organ.
PA_{Cp}: Permeability area product of capillary membrane of collective capillary pool of organ.

where:

$$R_E \approx \left(\frac{C_E}{C_V}\right)_{ss} \qquad (18)$$

The extracellular region is described according to:

$$V_E \frac{dC_E}{dt} = PA_{Cp}\left(C_V - \frac{C_E}{R_E}\right) - PA_M\left(C_E - \frac{C_I}{R_I}\right) \qquad (19)$$

V_E: Volume of extracellular space of organ.
PA_M: Permeability area product of cell membranes of organ.
C_I: Concentration of toxin in intracellular region of organ.

where:

$$C_E \approx \frac{C_I}{R_I} \qquad (20)$$

The last term on the right side of Equation 19 is developed analogously to that in Equation 17.

With the capillary membrane being much more permeable than the cell membrane (in this case), Equations 17 and 19 can be added together with the following result:

$$V_V \frac{dC_V}{dt} + V_{VE} \frac{dC_E}{dt} = Q(C_A - C_V) - PA_M\left(C_E - \frac{C_I}{R_I}\right) \qquad (21)$$

Substituting Equation 18 into Equation 21 results in:

$$V_V \frac{dC_V}{dt} + V_E R_E \frac{dC_V}{dt} = Q(C_A - C_V) - PA_M\left(C_E - \frac{C_I}{R_I}\right) \qquad (22)$$

which, with rearrangement becomes:

$$(V_V + V_E R_E) \frac{dC_V}{dt} = Q(C_A - C_V) - PA_M\left(C_E - \frac{C_I}{R_I}\right) \qquad (23)$$

From Equation 18 and assuming quasi steady-state:

$$C_V = \frac{C_E}{R_E} \tag{24}$$

Equation 24 is then substituted into the left side of Equation 23, resulting in:

$$(V_V + V_E R_E)\left(\frac{1}{R_E}\right)\frac{dC_E}{dt} = Q(C_A - C_V) - PA_M\left(C_E - \frac{C_I}{R_I}\right) \tag{25}$$

which, with further rearrangement, becomes:

$$\left(\frac{V_V}{R_E} + V_E\right)\frac{dC_E}{dt} = Q(C_A - C_V) - PA_M\left(C_E - \frac{C_I}{R_I}\right) \tag{26}$$

The volume of the collective capillary pool (V) is quite small compared to the extracellular and intracellular volumes. Thus:

$$\frac{V_V}{R_E} \Rightarrow 0 \tag{27}$$

and again using the relationship from Equation 24, Equation 26 becomes:

$$V_E\frac{dC_E}{dt} = Q\left(C_A - \frac{C_E}{R_E}\right) - PA_M\left(C_E - \frac{C_I}{R_I}\right) \tag{28}$$

It then follows that for the intracellular space:

$$V_I\frac{dC_I}{dt} = PA_M\left(C_E - \frac{C_I}{R_I}\right) \tag{29}$$

V_I: Volume of intracellular space of organ.
        All other symbols previously defined.

Then, for the whole organ, the concentration would be described by:

$$C = \frac{V_E C_E + V_I C_I}{V_E + V_I} \tag{30}$$

Equations 28 and 29 clearly describe the disposition of toxin in two pertinent and important regions of an organ. As such, this model can now be described as a micropharmacokinetic model. The concentration as described in Equation 29 may then be used as the input for another series of equations that might, for example, describe binding to crucial intracellular receptors regulating cellular physiologic processes or at some cellular organelles such as the mitochondria. In this manner, the endpoint of the micropharmacokinetic model would serve as the input function for a pharmacodynamic model describing alterations in endogenous cellular physiologic processes that might occur as a result of exposure to a toxin. In other cases the concentration described in Equation 29 is the starting point for a series of equations describing some other endpoint such as protein adduct concentrations which may then serve as a biomarker of exposure.

Equations 28 and 29 contain parameters that are specific for each region of the organ being described. The values for some of these, such as the volumes, can be estimated from literature values or be somewhat easily determined in the laboratory. Others such as the membrane permeation and area product and the partition coefficient (R_E) may have to be determined by either very extensive and sophisticated laboratory experiments or by a variety of mathematical techniques using the model equations and experimentally determined time course values for the intra- and extracellular concentrations.

Often experimental pharmacokinetic and toxicological studies reveal that one organ of the body stands unique when compared to the rest. Many neurotoxins, for example, are differentially retained for long periods of time in nervous tissue while other organs exhibit no such unique pharmacokinetic or toxicologic behavior. In cases as described in Blancato,[1] a pharmacokinetic model composed of high-resolution micropharmacokinetic equations for the nervous system is combined with lower resolution macropharmacokinetic equations for the rest of the body. In this manner, neither the mathematical nor laboratory techniques are unnecessarily taxed for organs in which the material is quickly eliminated or of little toxicological interest. The hybrid macroscopic equations are in the model to complete the mass balance and to provide the necessary input for the organ of interest. Depending upon the available data on the kinetic behavior in some of these nontoxicologically affected organs, many may be collapsed together, further simplifying the mathematics and even resulting in the design of more efficient, cost-effective experiments and monitoring studies.

It may be, for example, that the nonaffected organs can be lumped together in such a way that only a few macroscopic equations describing the blood, the liver, and perhaps the kidney remain. The liver is usually an important organ for biotransformation and thus is usually included in the macroscopic portion of the model. Likewise the kidney, an important organ of elimination, is often included.

Bernareggi and Rowland[2] describe a method for aiding in the selection of compartments for lumping. For each tissue, a distribution rate constant is calculated according to:

$$k_\tau = \frac{Q/V}{R} \tag{31}$$

k_τ: Distribution rate constant per minute.
Q: Blood flow to organ.
V: Volume of organ.
R: Equilibrium distribution ratio between tissue and blood. Same as the "partition coefficient term" as used in previous equations in this chapter.

Organs with similar values in the distribution rate constant are lumped together as the chemicals in these tissues are assumed to exhibit a similar kinetic profile. In addition, anatomic and physiologic factors are also considered. As discussed earlier, the liver is kept as a separate compartment irrespective of its distribution rate constant.

The blood is, of course, the input medium for the organ of toxicological interest and may be described in more classical pharmacokinetic terms with the key parameters fit from available data. The micropharmacokinetic equations of the model than take the output from the blood equation and describe in extensive detail the disposition in the organ of toxicological interest. In this manner, after the model is developed and validated, exposed persons may be monitored for the time course disposition in the blood. From these monitored data and the model, concentrations at the actual target may be predicted. Take for example, a compound which can be monitored in the blood and is not eliminated by the lungs during exhalation. In this case, the blood concentration might be described by:

$$C_{Bl,t} = C_{Bl,ss}(1 - e^{-kt}) \tag{32}$$

$C_{Bl,t}$:   Concentration of toxin in blood at time t.
$C_{Bl,ss}$:   Steady-state concentration of toxin in blood.
k: Rate constant for elimination from blood.

The steady-state blood concentration is monitored as is the elimination profile after exposure ceases. From these data the elimination rate constant would be estimated. Then a model is formulated using Equation 32 for the whole body and Equations 28 and 29 written for the specific organs of toxicological interest. Such a model describes the concentration in the cells of the toxicologically affected organ.

Of course if steady-state conditions do not apply, a different equation is written. Also, if the compound is eliminated by the lungs, the earlier equation is then altered and combined with others so that arterial blood can be described. This would be necessary because while the arterial blood concentration is the needed input for the organs of interest, venous blood is, especially outside a medical laboratory setting, usually easier to monitor than is arterial blood. In fact, under actual exposure conditions, monitoring the exhaled breath may be even more preferable. In such a case, equations describing the pulmonary process would be needed in order to use monitored data for deriving necessary parameters. Those parameters are then used in the equation that describes arterial blood concentrations which are, in turn, inputted into Equation 28. Similarly, urinary elimination of parent compound or one of its metabolites could also be monitored and used with the appropriate euqations.

In other instances, however, data regarding the blood, urine, or exhaled air may not be available or desirable to attain. The exposure period may have not yet begun or have long ceased. For instance, the compound may be too toxic to expose individuals for testing purposes. In these cases, internal dose or potential dose may have to be determined from either measurements or estimates of concentrations in environmental media. In these instances, the absorption across the pertinent portals of entry into the body must be described. The toxin passes across the body's barrier membranes and eventually into the blood. Equations describing this process then have as their output the arterial blood concentration which is input into the micropharmacokinetic equations for the organ or organs of interest. The next section describes the implementation of a micropharmacokinetic model for the neurotoxin 2,5-hexanedione.

III. MICROPHARMACOKINETIC MODEL IMPLEMENTATION

A wide variety of compounds are known to produce nervous system toxicity. The neuropathies caused by the inorganic substances lead[3] and cadmium chloride[4] are well-documented and described.

Over the last 20 years there have been numerous reports of neuropathies induced by exposure to aliphatic hydrocarbons.[5] Of particular interest to this study is the neuropathy induced by exposure to hexacarbons. Reviews by Spencer et al.[6,7] provide an extensive overview of the investigations concerning the hexacarbons. Generally, it is accepted that the compound *n*-hexane can be the parent compound absorbed by the body. It is metabolized to methy-N-butyl ketone and then to 2,5-hexanedione (2,5-HD), a diketone. Most investigators believe that neurologic damage occurs at sites of 2,5-hexanedione presence.

The neuropathy resulting from chronic exposure to these compounds is generally referred to as central-distal neuropathy as described by Spencer and Schaumburg.[7] The morphological changes include giant axonal swelling, focal accumulation of neurofilaments, and paranodal myelin retraction. Humans and animals exposed develop distal hindlimb paralysis. In animal studies these neurologic deficits accompany the morphologic changes.

A variety of different theories exist regarding the biochemical pathway of toxicity. Reports by Sabri and Ochs[8] and DeCaprio et al.[9] are just a few which list possible sites of hexacarbon toxicity. The one common denominator in all of them is that the 2,5-HD species is taken as the key compound in the metabolic scheme. With this in mind, pharmacokinetic schemes previosuly formulated by Angelo[10] and formulated for this study were undertaken. It was believed that these schemes could be useful in predicting the pharmacokinetics of these toxic compounds, both at the macroscopic and microscopic levels. It is known from the work of Angelo,[10] Couri and Nachtman,[13] that 2,5-HD is metabolized in the body. However, at short times it is reasonable to assume that a large amount of parent compound (2,5-HD) still remains circulating in the body. This fact is confirmed by the work of Angelo.[10] Due to the nature of the 2,5-HD molecule, it can be expected to be part of the intracellular radio-label pool (when exposure occurred to radio-labeled 2,5-HD), at least at short times. Angelo's studies showed that the bulk of 2,5-HD disappeared from the plasma in less than 8 h and the percentage of 2,5-HD bound to albumin was less than 10%.

For this study male Sprague-Dawley rats were injected intraperitoneally with radio-labeled 2,5-HD at a dose level of 8.0 mg/kg body weight. Animals were then sacrificed by decapitation at various times postinjection. The brains were quickly removed and subjected to homogenation and differential centrifugation. Details of the experimental protocol can be found elsewhere.[1]

The metabolic scheme formulated by Angelo was applied in the interpretation of the results of this study. That is, the C^{14} counted can be part of one of three species. First, it can be in the parent compound; second, in a soluble metabolite formed outside the brain, probably the liver; and third, it is C^{14} incorporated into the biomolecules of the cell. In order to model these three, it had to first be determined if the 2,5-HD bound with the subcellular fractions. If it did not, then it could be safely assumed that no parent compound remained associated with the subcellular fractions after 8 h.

The results of binding studies[1] performed as part of these experiments showed that essentially no 2,5-HD (<1%) bound to the subcellular fractions which were isolated. This then means that the radio label which remained within these fractions for long times resulted from metabolite and metabolite incorporation.

Results of these studies show a very quick appearance of radio label within the subcellular compartments. C^{14} was observed within 2 min of injection. The radio label showed a differential distribution within the cell. At short times (0 to 2 h) the highest concentration of radio label is found in the cytosolic (S-4) fraction. Due to the rapid appearance of intracellular label, one would expect that the radioactive species would include the parent compound. Angelo's data[11] showed the presence of 2,5-HD within the brain and its almost total elimination by 8 h. The decrease of label from this fraction indicates movement of C^{14} to other fractions and/or its elimination from the cell.

Inspection of Figure 1 reveals a multiphase elimination profile. The persistence of the label beyond the time expected from simple linear elimination kinetics and from the findings regrading the parent 2,5-HD, as determined by Angelo,[10] indicates presence of species other than the parent compound.

Profiles in the nonsoluble cellular fractions show the concentration of label in these fractions gradually increases for the first 2 h and then very gradually decreases over the next 17 weeks. The half-life of radio label within this fraction was found to be 36 d. Some radio label is still present even after 17 weeks. As discussed previously, the radio label in these fractions must be some species other than the parent compound, 2,5-HD.

The data derived from this study[1] and that of Angelo lead to the hypothesis that some metabolic derivative of 2,5-HD shows a prolonged residual time within subcellular organelles, particularly the insoluble fractions. Given the centrifugation separation scheme used in this study, those fractions contain mitochondrial and microsomal elements. The elimination

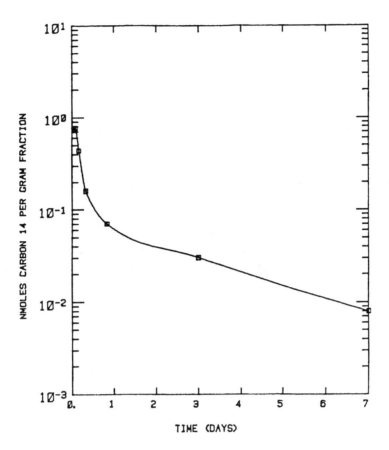

FIGURE 1. Experimental data — C^{14} in S4 or soluble fraction.

of the label from these insoluble fractions is slower than the average protein turnover rates of 4 to 10 d reported for the brain by Chee and Dahl.[11] This could be due to several factors. First, some radio-labeled molecule becomes incorporated within these fractions. Elimination only could occur when it becomes "unincorporated". The label may be incorporated with some component whose turnover is much slower than the average protein turnover rate. Second, the "unincorporated" species may be a molecule that, because of its physical properties, may not easily pass out of the cell. Thus, it may acutally "reincorporate" within the subcellular fractions and thus account for the long residence time of label within these fractions.

In summary, it appears that based on the comparison of the *in vivo* half-life of 2,5-HD with the half-lives calculated in this study, more than one species exists within the soluble fraction at short times. Also, based on these findings, some long-term association between label and intracellular precipitable fractions occurs. This association is significant when considering the potential sites of toxicity.

Based on the interpretations of these findings, a micropharmacokinetic model was formulated. Figure 2 shows the intracellular fractions that are part of the micropharmacokinetic model. Figure 3 shows the chemical species of the model. A is the parent 2,5-HD in the plasma, the extracellular fluid and the intracellular fluid. B is the metabolite formed in the liver and located in the plasma and the extracellular and intracellular fluids. B'P is that species bound to the protein fractions within the cell.

The model equations were solved using Gear's method for stiff differential equations on a Digital Equipment Corporation (DEC) computer at the University of Delaware. Figure

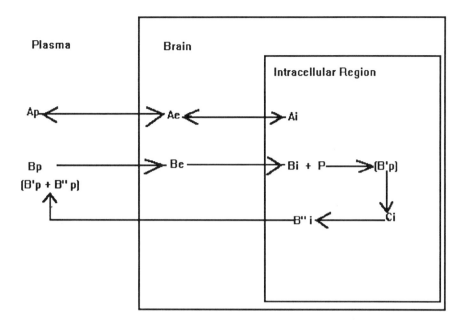

FIGURE 2. Chemical species and their location in the micropharmacokinetic model.

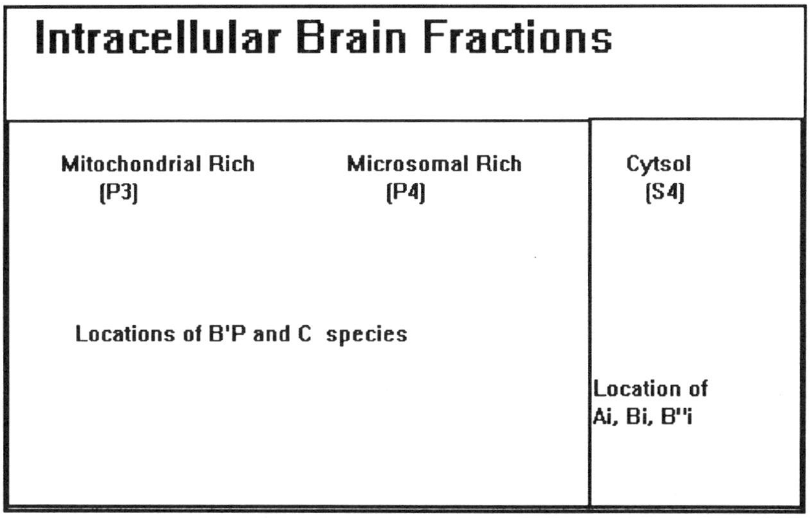

Intracellular Brain Fractions

Mitochondrial Rich [P3]	Microsomal Rich [P4]	Cytsol [S4]
Locations of B'P and C species		Location of Ai, Bi, B''i

FIGURE 3. Subcellular compartments of the brain.

4 shows the model output compared to actual data for the insoluble fraction for the first 7 d postinjection. Although not presented here, model output for the other fractions up to 17 weeks postinjection are also available.

The data (+) represent total label as determined experimentally. Thus the data are actually for both the bound (B'P) and the incorporated (C) label. The model output (TPP) corresponding to these data mirrors them well. The model also shows the expected profile for B'P and C taken separately. Protein turnover rates in the brain average about 10 d. Therefore, it is reasonable to expect incorporation to occur during this time period. The

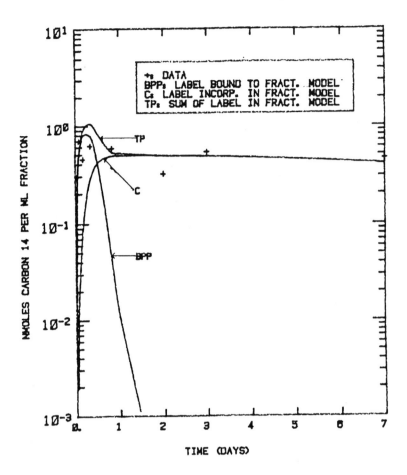

FIGURE 4.   Model output and actual data in insoluble fraction.

model shows the rapid accumulation of C, presumably at the expense of $B'P$. Its peak occurs at about 6 h, after the peaking of unbound B. According to the model, the total incorporated label peaks at between 6 to 10 h. There are no data to compare with at these times. However, this peak time is reasonable when considering that the model is in good agreement throughout the rest of the time period. The concentration of $B'P$ continues to drop to negligible amounts after 2 d. This is reasonable when it is realized that its parent, B', is almost negligible by the second day.

In summary, good agreement was observed between model predictions and the data obtained from the associated experiments. The model describes the observed rapid entry into and the slow, but continued, elimination of radio label from subcellular compartments. The model also described well the data on which the macroscopic model developed by Angelo.[12]

This work shows that micropharmacokinetic models can be developed to describe the disposition of foreign substances within subcellular compartments. The internal cell structures can be considered as compartments within the cell, rather than treating the whole internal cell or the whole organ as one single compartment.

Such models are useful to biochemists and toxicologists because they describe concentrations of foreign substances and their metabolites at possible crucial intracellular sites, at the experimental exposure levels, and at other exposure levels and scenarios by simulation. As more data become available which elucidate more understanding about biologic mechanisms, these models can be restructured to take these into account.

IV. CLOSING

Pharmacokinetic models such as the ones presented here and termed micropharmaco-kinetic models have been applied[1,12] to describe and predict the concentration of toxins at subcellular levels. As more becomes known about the mechanisms by which chemicals cause toxicity, the cellular sites of interest will be identified in more and more cases. Models such as those described here can be formulated to calculate the concentration at these sites. The utility of this approach is that more rational scaling or extrapolation between doses, dosage regimens, and species can be easily and accurately performed. In general it would be expected that a typical physiologically based pharmacokinetic model, termed the macroscopic model, would be formulated for a compound of interest. Such a macroscopic model and the associated experimental studies would identify the suitable targets for micropharmacokinetic modeling. The macromodel provides estimates of the concentration in the extracellular spaces of the toxicologically vulnerable organs in addition to the disposition in the other organs and the whole body. The micromodel describes and predicts concentration at subcellular sites of toxicity.

For modeling of this type to be truly successful and useful, rational scaling across species, ultimately up to man, must be possible. Thus, high priority should be given to the development of methods for formulating macro- and micropharmacokinetic models using data derived from a combination of studies conducted in experimental test species and tissue culture studies. Several important processes and biological parameters have to be considered when scaling across species lines. The numbers of cells, total cross-sectional areas of the cell membranes, chemical and physical cellular characteristics, and pathways and rates of metabolism are those which readily come to mind. The proper use of *in vitro* culture systems to determine the values for these and other key model parameters needs to be further developed and implemented.

Finally work needs to be done to use the best mathematical techniques and theories to augment and improve techniques for model formulation, simplification, and solution. Such improvements will not only result in more rational and accurate solutions, but will also help in developing ways to use models to form cost-effective and efficient monitoring strategies. A combination of models composed of high-resolution descriptions of the organs which may be affected by the toxin with rationally clustered descriptions of the other organs which may not be targets of toxicity, albeit important kinetically, are a necessary addition to the risk assessment process. In sum, progress needs to be made to better identify and describe the mechanisms of toxicity and the cellular targets of that toxicity. Also, progress needs to be achieved in developing better methods to formulate and solve models which put to use this gained mechanistic knowledge. Research must therefore progress in the arenas of experimental biology as well as those of mathematics and engineering.

REFERENCES

1. **Blancato, J. N.,** *Micropharmacokinetics of the Intracellular Disposition of the Neuotoxin, 2,5-Hexane-Dione,* University Microfilms International, Ann Arbor, MI, 1985.
2. **Bernareggi, A. and Rowland, M.,** Physiologic modeling of cyclosporin kinetics in rat and man, *J. Pharmacokinet. Biopharm.,* 19(1), 21, 1991.
3. **Dyck, P. J.,** Lead neuropathy: random distribution of segmental demyelination among "old internodes" of myelinated fibers, *J. Neuropathol. Exp. Neurol.,* 6(3)4, 570, 1977.
4. **Sato, P.,** An altrastructural study of chronic and cadminum chloride induced neuropathy, *Acta Neuropathol.,* 41(3), 185, 1978.

5. **Gaultier, M.,** Polyneuritis and aliphatic hydrocarbons, *Eur. J. Toxicol.,* 64, 294, 1973.
6. **Spencer, P. S., Qischoff, M. C., and Schaumburg, H. H.,** On specific molecular confirmation of neurotoxic aliphatic hydrocarbon compounds causing central peripheral distal axonopathy, *Toxicol. Appl. Pharmacol.,* 44, 17, 1978.
7. **Spencer, P. S., Schaumburg, H. H., Sabri, M. I., and Veronesi, B.,** The enlarging view of hexacarbon neurotoxidity, *CRC Crit. Rev. Toxicol.,* 7, 278, 1980.
8. **Sabri, M. I. and Ochs, S.,** Relation of ATP and creative phosphate to fast axoplasmic transport in mammalian nerve, *J. Neurochem.,* 18, 1509, 1972.
9. **DeCaprio, A. P., Olajos, E. J., and Weber, P.,** Covalent binding of a neurotoxic *n*-hexane metabolite: conversion of primary amines to substituted pyrrole adducts by 2,5-hexanedione, *Toxicol. Appl. Pharmacol.,* 654, 440, 1982.
10. **Angelo, M. J.,** The Pharmacokinetics of the Neurotoxin 2,5-Hexanedione: Distribution, Elimination and Model Simulations, Ph.D. thesis, University of Delaware, 1981.
11. **Chee, Y. Y. and Dahl, J. L.,** Measurement of protein turnover in rat brain, *J. Neurochem.,* 30, 1485, 1978.
12. **Blancato, J. N. and Bischoff, K. B.,** Subcellular pharmacokinetics of 2,5-hexane dione, paper presented at the North American Symp. on Risk Assessment and the Biological Fate of Xenobiotics, November 1985.
13. **Couri and Nachtman,** unpublished results, 1979.

Chapter 3

CYTOCHROME P450-DEPENDENT METABOLISM OF DRUGS AND CARCINOGENS IN SKIN

Hasan Mukhtar and Rajesh Agarwal

TABLE OF CONTENTS

0-8493-7357-3/93/$0.00 + $.50
© 1993 by CRC Press, Inc.

I. INTRODUCTION

The enzymes which metabolize foreign compounds including drugs and other xenobiotics are known as drug metabolizing enzymes. They were first discovered in liver, but thereafter virtually all extrahepatic tissues including skin[1-8] have been shown to possess varying levels of such activity. The drug and xenobiotic metabolism is a general phenomenon which plays a major role in the removal of foreign compounds from the body. The major role of such enzymes is to convert lipid-soluble foreign chemicals into water-soluble metabolites which can be eliminated easily from the body. This enzyme system therefore diminishes the biologic activity of foreign compounds. Ironically, this enzymatic process also generates chemically reactive metabolites which can covalently bind to cellular macromolecules such as DNA and thereby result in cell toxicity, mutagenicity, and often cancer induction.[9]

Skin is the largest body organ which functions as a major environmental interface and therefore is directly and continuously exposed to a vast number of foreign agents, some of which are capable of altering the metabolism of cells within the skin and those which traffic through cutaneous tissue.[10] In recent years it became clear that skin is also an active site of diverse types of metabolic activities.[1,11]

II. HISTORY, BACKGROUND, AND GENERAL CONCEPTS

Most of the foreign compounds, also known as xenobiotics, are lipophilic in nature, a property which enables them to penetrate into the lipid membranes, and, therefore, they are transported by lipoproteins in the blood. These lipophilic compounds are substrates for drug metabolizing enzymes. Drug or xenobiotic metabolism can be subdivided into two distinct phases. In Phase I, a polar reactive group is introduced into the molecule to convert it into a suitable substrate for Phase II enzymes. Phase I reactions include cytochrome P450 (P450)-dependent monooxygenases. The term P450 originated when a reduced pigment that has an absorption band with a λ max at 450 nm after binding to carbon monoxide was identified in 1958 by Klingenberg[12] and Garfinkle.[13] This pigment was further characterized as a P450 hemoprotein by Omura and Sato.[14] More extensive characterization of solubilized enzyme was the subject of classical papers published by Omura and Sato.[15,16] The role of P450 in the microsomal mixed function monooxygenase system was also established around the same time.[17] In the past several years it became evident from the protein purification studies that P450 was not a single entity, but in fact a multiplicity of forms which exist in mammals and other species. Studies using purified P450s reconstituted with phospholipid and NADPH-cytochrome P450-oxidoreductase[18] have shown that individual forms of P450 exhibit highly specific yet overlapping substrate specificities.[19]

In recent years considerable interest has developed on studying the P450-dependent reactions in hepatic and extrahepatic tissues of mammals. At present, 154 P450 genes and 7 putative pseudogenes have been described, and a nomenclature system for the P450 super family is recommended.[20] Our present knowledge of these enzymes indicates that, with two exceptions, each P450 gene produces a single protein. In general, P450-dependent reactions result in detoxification of pharmacologically active drugs and other foreign compounds into their inactive form(s). With certain compounds, however, this defense mechanism fails and a highly reactive intermediate of oxygenation such as epoxide results. These reactive epoxide derivatives usually attack hydrophilic sites on cellular biomolecules DNA, RNA, and protein, resulting in toxicity, cell death, mutation, and cell transformation often leading to cancer induction. For example, it has been established that the cancer-causing effects of skin carcinogens, the polycyclic aromatic hydrocarbons (PAHs), relate to their metabolism by the P450 and other enzymes.[9] The isozyme of P450 which plays a major role in this process is known as P450IA1 (or P450c; in the earlier literature, because of its different absorption maxima, this was known as P-448). This particular isozyme and its dependent catalytic activities are highly induced in rodent and human skin *in vivo*[1] and in epidermal cells *in vitro*[1] by the exposure to PAHs such as benzo(a)pyrene (BP), 3-methylcholanthrene (3-MC), benz(a)anthracene (BA), and β-naphthoflavone (β-NF). This enzyme-induction response results in increased rates of substrate turnover and since reactive metabolites of these compounds are carcinogenic, the likelihood of neoplasia increases in tissues with inducible P450IA1.[1-9] Metabolic products formed by Phase I reactions undergo conjugation with glucuronide, sulfate, glutathione, etc., a process known as Phase II reactions.[21] NAD(P)H:quinone reductase is another example of a Phase II enzyme where a reactive quinone derivative is reduced to its inactive form.[22] The products of Phase II reactions, with rare exception, are hydrophilic and therefore more readily excreted from the body. Skin possesses both Phase I and Phase II enzymes and may therefore alter chemicals that enter it.[1,22]

Depending on the particular reaction and the nature of various unstable intermediates, different types of Phase I reactions can occur in skin.[1-8] However, in a typical Phase I reaction, hydroxylation of the substrate occurs at a carbon atom. In the first step, the substrate binds to the protein moiety of the oxidized P450, followed by one electron reduction of the heme iron (Fe^{3+}) provided by NADPH-P450 reductase. The reduced P450-substrate complex, thus formed, binds to molecular oxygen and utilizes another electron from NADPH-P450 reductase. During this process, the substrate is attacked by the oxygen molecule and thereby one oxygen atom is inserted into the substrate which is released from the enzyme as the hydroxylated product, while the second oxygen atom yields water. Based on this process, this enzyme reaction has been termed as a ''mixed-function oxidase''- or a ''mono-oxygenase''- reaction.

In addition, P450 also oxidizes heteroatoms such as nitrogen and sulfur. Aliphatic double bonds or aromatic hydrocarbons are oxidized by P450, resulting in the formation of epoxides either as labile intermediates or as stable products. These epoxides can either be hydrolyzed by water or catalyzed by epoxide hydrolase, leading to dihydroxy metabolites commonly known as dihydrodiols or ''diols.''[9] The broad range of compounds metabolized by P450 include exogenous compounds like the majority of drugs such as nifedipine, mephenytonin, codeine, midazolam, cyclosporine, etc. and carcinogens including PAHs, nitrosamines, hydrazines, arylamines, etc. as well as endogenous substances such as fatty acids, prostaglandin, leukotrienes, steroid hormones, ketones, etc.[1]

III. SPECTRAL EVIDENCE FOR P450 IN SKIN

First spectral evidence for the occurrence of P450 in skin came in 1976 when Pohl et al.[23] reported that microsomes prepared from the skin of Swiss-Webster mice possess P450 in this tissue. In this study topical application of 3-MC produced no induction, whereas 2,3,7,8-tetrachlorodibenzo-*p*-dioxin (TCDD) resulted in significant induction of P450. Subsequent study from our laboratory showed that epidermis is the major compartment for P450 in skin.[24] Using skin from 4-d-old rats treated with a single topical application of Aroclor 1254, which is known to induce various P450 isozymes in liver,[25] only epidermal microsomes prepared from control and Aroclor 1254-treated rats exhibited a reproducible CO-difference spectrum with absorbance peaks at 453 and 451 nm, respectively.[24] The P450 content of epidermal microsomes prepared from Aroclor 1254-pretreated animals showed an induction of 87% when compared to that from control animals. Application of 3-MC,[26] β-NF,[26] therapeutic crude coal tar,[27] and BA[28] to murine skin also resulted in significant induction of cutaneous P450. Using HPLC separation and absorption of heme-protein at 405 nm,

Mukhtar et al.[29] demonstrated that solubilized epidermal microsomes from control animals also possess a single peak and that it coeluted with P450IIB1 (P450b). However, solubilized microsomes prepared from animals pretreated topically with 3-MC showed the presence of at least one new peak, the retention time of which was found to be similar to that of P450IA1 (P450c). These studies, while suggesting that mammalian epidermis possesses multiple isozymes of P450, also strongly suggested that skin is capable of metabolizing a wider range of both endogenous and exogenous substrates.

IV. PURIFICATION AND CHARACTERIZATION OF CUTANEOUS P450

Although several studies have been done in the past to purify and characterize basal and induced P450s from extracutaneous tissues of animals treated with a variety of xenobiotics,[30] purification and characterization of any P450 form from cutaneous tissue was not accomplished until recently due to its extremely low content (2 to 5% of liver) and labile nature of the hemoprotein in this tissue. Ichikawa et al.[28] have partially purified P450 from skin microsomes of BA-treated male Balb/c mice. This purified preparation in the reconstituted system efficiently catalyzed BP hydroxylation, suggesting that it may be P450IA1. In another study, Raza et al.[26] purified P450 from epidermal and liver microsomes of neonatal rats pretreated with skin application of β-NF. In this study one major form of P450 was purified from epidermal microsomes by hydrophobic affinity column chromatography. The purified epidermal P450 showed a major band at 54 kDa on SDS-PAGE comigrating with hepatic P450IA1 purified under identical conditions. The specific content of epidermal P450 thus purified was 1.59 nmol/mg protein, corresponding to about 44-fold purification. The purified preparation efficiently catalyzed BP hydroxylation when reconstituted with purified NADPH-cytochrome P450 reductase and phospholipid; this activity was inhibited by α-NF as well as by antibodies to P450IA1. Peptide fingerprint analysis of the purified epidermal and liver P450IA1 showed identical 1-7-1 reacting epitopes. Furthermore, partial *N*-terminal amino acid sequence analysis of the purified epidermal P450 showed complete homology with the known sequence of hepatic P450IA1. These findings suggested that P450, induced in epidermis following skin application of β-NF to rats, is enzymatically and immunochemically identical to rat liver P450IA1 and that it may be a product of the same gene family.

V. USE OF P450 ISOZYME-SPECIFIC ANTIBODIES IN PHENOTYPING P450 SPECIES IN SKIN

Polyclonal[31] and monoclonal[32] antibodies have been developed to various P450s in order to identify and quantify various P450s in animals exposed to xenobiotic agents. Similarly, these antibodies have also been employed to quantify the expression of P450 isozymes in skin and their contribution to the metabolism of substrates by skin microsomes.[1] Highly specific monoclonal antibodies (MAbs) 2-66-3 and 1-7-1 directed against purified rat liver P450 isozyme IIB1 and IA1, induced by phenobarbital and 3-MC, respectively, were employed by Khan et al.[33] to assess the monooxygenase enzyme activities in neonatal rat epidermis after topical application of 3-MC. Radioimmunoassay of epidermal microsomes from control animals with ^{35}S-labeled MAb 2-66-3 showed significant binding, whereas with MAb 1-7-1, negligible binding was observed. On the contrary, with epidermal microsomes prepared from carcinogen-treated animals, significant binding occurred with ^{35}S-labeled MAb 1-7-1. Similarly, histochemical staining of epidermis from control animals showed no immunoreactivity either with MAb 1-7-1 or MAb 2-66-3, whereas epidermis obtained from 3-MC-treated rats showed significant immunoreactivity with MAb 1-7-1.

Additionally Western blot analysis of epidermal microsomes prepared from control animals showed no immunoreactivity towards MAb 1-7-1 or MAb 2-66-3, whereas epidermal microsomes from 3-MC-treated animals showed distinct immunoreactivity with MAb 1-7-1. MAbs 2-66-3 and 1-7-1 also significantly inhibited monooxygenase activities in microsomes prepared from control and 3-MC-treated animals, respectively. The results of this study suggested that multiple isozymes of P450 exist in neonatal rat epidermis and that isozyme IIB1-specific epitopes are present in control tissue, whereas isozyme IA1 is expressed predominantly after treatment with 3-MC. Similarly, in another study, Khan et al.[27] assessed on the induction response of crude coal tar in rat skin using immunoblot analysis of epidermal and hepatic microsomes with MAb 1-7-1, which revealed strong immunoreactivity in both tissues. On the contrary, with MAb 2-66-3, only hepatic microsomes showed significant immunoreactivity. In addition, crude coal tar treatment resulted in suppression of immunoreactivity in hepatic microsomes towards MAb 1-98-1, a highly specific antibody directed against rat liver P450 isozyme IIE1, a form induced by ethanol. These studies suggested that crude coal tar on topical application onto the skin of neonatal rats results in the induction of P450 isozyme IA1 in epidermis and isozyme IIB1 and IA1 in liver, and that this induction is associated with the suppression of P450 isozyme IIE1 in liver. Using antibodies raised against various purified isozymes of P450, Whitter et al.[34] have shown that antibodies to phenobarbital-inducible hepatic P450 specifically bind to the epidermal cells, to the outer root sheath of hair follicles, and to the sebaceous glands of untreated rats. However, antibodies directed against β-NF- and 3-MC-inducible hepatic P450 showed very weak, if any, staining in these structures in untreated animals. On the contrary, topical application of 3-MC resulted in dramatic increase in staining in hair follicles and sebaceous glands towards MAb 1-7-1.

Antibodies have also been reported to be also instrumental in the isolation and characterization of many P450 cDNAs and their genes.[32] The use of vaccinia virus-mediated cDNA expression offers a new and powerful approach in assessing the catalytic activities of human P450s,[35] a procedure which also circumvents the problems inherent with antibody preparations. By expressing cDNAs encoding for specific P450s in hepatoma cells using recombinant vaccinia viruses, which are virtually devoid of endogenous P450s, it is possible to accurately define which P450 form(s) metabolizes specific drugs or activate carcinogens.[36] This novel approach has not yet been utilized to phenotype P450 expression in the metabolism of drugs and/or carcinogens in cutaneous tissue.

VI. P450-DEPENDENT DRUG METABOLIZING ENZYMES IN SKIN

In order to determine drug metabolism in skin, dermis is separated from the epidermis either by scraping, short-term heat treatment, trypsinization, or by incubation with dithiothreitol.[24] Drug metabolizing enzyme activities for xenobiotics detected in the skin are generally much lower than those in other tissues.[1] Moreover, homogenization of the skin requires large shearing forces which further diminishes the xenobiotic metabolism. In addition, skin homogenization is often performed after freezing the tissue at $-170°C$ in liquid nitrogen.[37] Several studies in the last 20 years have shown the presence of P450-dependent drug metabolizing enzymes in skin. Since in various laboratories drug metabolizing enzymes have been studied in microsomes prepared from whole skin, epidermis, dermis, hair follicles, cell culture systems, etc., interlaboratory reproducibility has been another challenging problem.[1-5] In addition, in most cases, the enzyme activities are expressed based on wet weight or dry weight of the tissue, or on protein content of total homogenate, postmitochondrial supernatant, cytosol, or microsomes. Cutaneous drug metabolizing enzyme activities are measurable in considerable amounts only after induction by xenobiotics, and therefore exact data on constitutive levels of such activities in intact skin have been difficult to obtain. Norden[38] for the first time showed that following topical application of BP, "metabolite fluorescence" was demonstrable in epidermis, hair follicles, and sebaceous glands of guinea pig skin with a change in the absorption maximum of the fluorescence signal in treated skin and suggested that this change in fluorescence signal is due to hydroxylation of BP in the skin. In addition, a wide range of PAHs are also metabolized by the P450-dependent mono-oxygenase pathway, and therefore the enzyme involved in the process was termed aryl hydrocarbon hydroxylase (AHH).

A. MAJOR SITE FOR DRUG METABOLISM IN SKIN

Although most of the evidences suggest epidermis as the major compartment for drug metabolizing enzyme activity,[24] some controversy still exists in literature.[39] Studies using dithiothreitol-separated epidermis[40] from neonatal rats[24] and mice[41] have unequivocally demonstrated the highest specific activity of monooxygenases and other drug metabolizing enzymes in the epidermal compartment. In addition, marker enzyme analysis has also supported the fact that epidermal microsomes in particular are more important for carcinogen metabolism.[42,43] These results have also been confirmed in studies employing isolated epidermal cells.[44-47] Pohl et al.[45] and Coomes et al.[46,47] reported higher levels of AHH and other drug metabolizing enzyme activities in sebaceous cells isolated from skin by elutriation techniques. These findings further confirmed an earlier study in which a histochemical fluorimetric technique was employed to demonstrate and measure BP hydroxylation in intact skin.[48]

B. INDUCTION OF DRUG METABOLIZING ENZYMES IN SKIN

The pretreatment of animals with a wide range of xenobiotics, either applied topically or administered systemically, have been shown to result in significant induction of cutaneous AHH activity and other drug metabolizing enzyme activities including 7-ethoxycoumarin *O*-deethylase (ECD) and 7-ethoxyresorufin *O*-deethylase (ERD).[1-8] However, the inducibility of AHH, ECD, and ERD activities in skin was significantly more after topical application of xenobiotics when compared to any other mode of exposure.[33,49]

A 10-fold increase in cutaneous AHH activity in rats pretreated topically with 3-MC was reported by Schlede and Conney.[50] Similarly, Levin et al.[51] and Alvares et al.[52] demonstrated the presence of inducible AHH activity in human foreskin treated in organ culture with BA. Bickers et al.[53,54] showed that topical application of Aroclor 1254 and microscope immersion oil to murine skin resulted in several-fold induction of cutaneous AHH activity. In addition, Thompson and Slaga[55] also showed that AHH activity in mouse skin is induced by topical application of PAHs. In other studies Wiebel et al.[27] showed that topical application of BA to the skin of several strains of mice resulted in AHH induction in skin.[56] Briggs and Briggs[57] demonstrated that the topical application of selected adrenocorticosteroids, used in dermatologic therapy, also induces cutaneous AHH activity in mice. This study suggested that inducible enzyme activity in the skin may have the capacity to metabolize topically applied therapeutic agents and that therapeutic efficacy of such agents may be related to the drug metabolizing enzyme activity in cutaneous tissue. Mukhtar and Bickers[58] studied the mixed-function oxidase enzyme activities in the skin of neonatal rats and compared their levels in epidermis and dermis with that in other body tissues. In this study, topical application of BP or Aroclor 1254 resulted in AHH induction in each tissue studied, however the rate of increase in enzyme activity was greater in skin as compared to that in other tissues. Skin microsomes exhibited 7, 16, and 28% of the corresponding specific activity of hepatic AHH in control, BP-, or Aroclor 1254-pretreated rats, respectively.[58] Similarly, several other studies[59-61] from our laboratory also showed that a therapeutic preparation of crude coal tar and some of its constituents induce epidermal AHH, ECD, and ERD activities. In another study, Merk et al.[62] studied the effect of topical application of antifungal agent clotrimazole to the skin of neonatal rats on induction response of monooxygenase activities in epidermal and hepatic microsomes. A single topical application of clotrimazole (10 mg/100 g body weight) resulted in a 53% increase in hepatic P450 content and 58% induction of hepatic p-nitrophenol hydroxylase, an enzyme catalyzed principally by the ethanol-inducible P450 isozyme. On the other hand, in case of epidermal microsomes, clotrimazole treatment resulted in significant induction of only epidermal ECD activity, whereas p-nitrophenol hydroxylase activity was undetectable.

C. EXOGENOUS SUBSTRATE METABOLISM IN SKIN

P450-dependent metabolism of xenobiotics in the skin is known to catalyze the biotransformation of a wide range of reactions such as deamination; O-, N-, and S-dealkylation; hydroxylation of alkyl and aryl hydrocarbons; epoxidation; and N-oxidation.[63] The exogenous substrates metabolized by P450-dependent drug metabolizing enzymes in skin are provided in Table 1. Dealkylation reactions in particular have been shown to result in the formation of products in various skin preparations as identified by spectrofluorimetric techniques.[45-47,64,65] The presence of AHH activity has been demonstrated in almost all skin preparations from various species examined.[1] As discussed later in this chapter, the enzyme has also been shown to be involved in PAH-induced carcinogenesis in murine skin.[9] Bickers et al.[66] studied the BP metabolism in skin microsomes from Sprague-Dawley rats and AHH-responsive C57BL/6N and AHH-nonresponsive DBA/2N mouse strains, treated topically with 3-MC or Aroclor 1254. Similarly, Das et al.[41] reported the metabolism of BP and BP-7,8-diol in the skin of neonatal Balb/C mice. Their results showed that epidermis is the major site of PAH metabolism, enzyme-mediated mutagenesis, and enzyme-mediated binding of carcinogens to DNA. Asokan et al.[67,68] compared the effects of topical application of nitrated and nonnitrated arenes on cutaneous monooxygenase activities and showed that after topical application to rodents, nitroarenes induce cutaneous AHH, ECD, and ERD activities. In another study, Bickers et al.[69] showed that microsomes from epidermis of neonatal Sprague-Dawley rats and Balb/C mice, and keratinocytes cultured from Balb/C mouse skin can convert 2-aminoanthracene to mutagenic metabolites.

D. ENDOGENOUS SUBSTRATE METABOLISM IN SKIN

Mukhtar et al.[70] showed that incubation of leukotriene B_4 (LTB_4) with epidermal microsomal suspension from neonatal rat, adult guinea pig, or human keratinocytes resulted in the formation of 20-OH-LTB_4 and 20-$COOH$-LTB_4. Similarly in another study, Mukhtar et al.[71] reported that epidermal P450 can also hydroxylate the sex steroid hormone testosterone at the 6β-, 7α-, and 16α-positions. However, these metabolism in skin were not affected by topical application of known inducers of cutaneous AHH activity. Vanden Bossche et al.[72] showed that epidermal microsomes from neonatal rats can hydroxylate all *trans*-retinoic acid by a P450-dependent pathway. Henke et al.[73] showed that arachidonic acid is metabolized by murine keratinocytes to the prostaglandin H synthase products prostaglandins E_2 and D_2. This findings, in addition to the study of Pentland and Needleman[74] which showed that prostaglandin F_2 enhances keratinocyte proliferation, indicates that epidermal cell-derived eicosanoids may play a significant role in the control of cell proliferation.

E. Ah RECEPTOR AND ITS ROLE IN DRUG METABOLISM IN SKIN

AHH inducibility has been thought to require binding of the inducer to a receptor, present in cytosolic fraction, that transports the receptor-inducer complex to the nucleus. Poland

TABLE 1
Exogenous Substrate Specificity of
Cutaneous Cytochrome P450

Substrate	Specific activity (% of liver)[a]
Aliphatic oxidation	
Coumarin	<1
Methoxycoumarin	≈1
Propoxycoumarin	≈1
Butoxycoumarin	≈1
7-Ethoxycoumarin	2–5
7-Ethoxyresorufin	7–12
Diphenyloxazole	2–3
p-Nitroanisole	2–3
Aminopyrine	≈1
Aldrin	1–2
Aromatic oxidation	
Benzo(a)pyrene	3–9
Benzo(a)pyrene 7,8-diol	9–11
7,12-Dimethylbenz(a)anthracene	3–7
3-Methylcholanthrene	2–4
Benz(a)anthracene	2–4
Dibenzanthracenes	1–4
7-Methylbenz(a)anthracene	1–3
Chrysene	2–5
5-Methylchrysene	3–6

[a] Specific activity range within indicated percentage of corresponding liver values in various species and strains studied. Control animals were compared with control animals, whereas data obtained from inducer pretreated animals were compared with the corresponding treated animals.

and Glover[75] studied the effects of Aroclor-1254 in various inbred mouse strains and showed that in those species of mice in which AHH is unresponsive to PAHs, Aroclor-1254 is capable of inducing the AHH enzyme activity. This finding suggested that inbred mouse strains with nonresponsive AHH possess the genes necessary for the expression of AHH activity. However, a defective cytosolic receptor with diminished affinity for most inducers could explain their relative refractoriness to enzyme induction.[76] In a later study, Okey et al.[77] provided clear evidence for such an AHH receptor in the liver that was shown to share some of the characteristics of steroid hormone receptors. In recent years it is believed that induction of P450IA1 and thereby monooxygenases requires the presence of this cytosolic receptor protein known as Ah receptor. It binds and translocates the inducer receptor complex to the nucleus and therefore results in binding of the ligand receptor complex to specific regions flanking the 5'-end of the P450IA1 gene, hence enhancing the transcription rates for its mRNA.[78] Recently Poland et al.[79] have purified Ah receptor from C57BL/6J mouse hepatic cytosol, determined its N-terminal amino acid sequence, and raised polyclonal antibodies. The existence of the Ah receptor in the human squamous cell carcinoma cell line A-431 has also been demonstrated by Harper et al.[80] The identification of the receptor in skin cells has further strengthened the concept of receptor-mediated DNA modification in cutaneous tissue.

F. EFFECT OF ULTRAVIOLET RADIATION ON CUTANEOUS DRUG METABOLISM

Only limited studies have assessed the effects of ultraviolet (UV) radiation on drug and carcinogen metabolism in skin. As reported by Pohl and Fouts,[81] exposure of female hairless mice to short-wavelength UV-radiation (254 nm) and to a sunlamp (280 to 750 nm) resulted in the increase in cutaneous ECD activity. Similarly, in another study, Gorez et al.[82] exposed hairless mice to UV light for 16 h a day for a period of 24 weeks (mean daily dose in this experiment was UVA, 106 J/cm^2; UVB, 0.06 J/cm^2) and reported an increase in AHH, aminopyrine N-demethylase and ECD activities, and the content of P450 in liver, whereas no effect was seen on the cutaneous drug metabolizing enzymes. Mukhtar et al.[83] examined the effect of cutaneous exposure to UVB on the inducibility of microsomal AHH, ECD, and ERD activities; on the patterns of BP metabolism; and on the enzyme-mediated binding of ^{3}H-BP to epidermal DNA. In this study, the exposure of animals to UVB (400 to 1600 mJ/cm^2) resulted in a dose-dependent increase in each of these cutaneous enzyme activities. In addition, the combination of crude coal tar and UVB exposure resulted in additive effects on cutaneous AHH, ECD, and ERD activities, and on BP metabolism. Similarly, Das et al.[84] studied the cutaneous drug metabolizing enzyme activities in SKH-1 hairless mice chronically exposed to UVB radiation to induce squamous cell carcinomas. Their results showed that repetitive exposure of mammalian skin to UVB radiation can significantly alter the activity and the inducibility of drug and carcinogen metabolizing enzymes.

VII. P450-DEPENDENT DRUG METABOLIZING ENZYMES DURING EPIDERMAL DIFFERENTIATION

In epidermis, keratinocytes account for more than 90% of the cells. Keratinocyte differentiation is a tightly controlled and a spatially regulated process. It is known that various pharmacological agents, hormones, and matrix components influence this phenomenon.[85] The relationship of drug metabolism and epidermal differentiation has gained considerable interest in recent past. DiGiovanni et al.[86] assessed the role of extracellular Ca^{2+} concentration on the metabolism of BP and 7,12-dimethylbenz(a)anthracene during the process of keratinocyte differentiation. This study indicated that extracellular Ca^{2+} concentration and possibly the differentiation state dramatically affect the metabolism of PAHs in adult murine epidermal keratinocytes. In subsequent studies these authors[87] established a correlation between DNA adduct formation with the state of mouse epidermal differentiation. In another study, Cameron et al.[88] showed that the metabolism of arachidonic acid is altered in the state of differentiation in density gradient-separated mouse epidermal cells. In addition, Guo et al.[89] also showed that AHH activity is higher in differentiated than in germinative cutaneous neonatal rat keratinocytes. In another study Reiners et al.[90] estimated ECD and ERD activities per cell in density-separated murine epidermal keratinocytes and in a system in which the keratinocyte differentiation was modulated as a function of extracellular Ca^{2+} concentration. Using these two systems these authors provided evidence that constitutive activities of the monooxygenases in murine keratinocytes vary with the stage of differentiation. Berghard et al.[91] demonstrated the requirement of serum and extracellular Ca^{2+} for the induction of monooxygenase activity in human keratinocytes. Their study showed that under *in vitro* conditions, which were selected for an undifferentiated basal-like phenotype, enzyme induction does not occur unless a terminal differentiation signal is applied by raising extracellular Ca^{2+} concentration and/or by treatment with serum. Though these studies have provided some information about the alterations in drug metabolizing capacity of the skin in terms of epidermal differentiation, additional studies are still needed in order to better define the relationship between stage of epidermal differentiation and xenobiotic metabolism.

VIII. METABOLISM OF PAHs IN SKIN AND ITS RELEVANCE TO CUTANEOUS CARCINOGENESIS

PAHs which are carcinogenic in nature are actively metabolized by drug metabolizing enzymes in the skin. In addition, some of the PAHs are potent inducers of cutaneous AHH activity and thereby increase their own metabolism as well as the metabolism of other skin carcinogens which are metabolized under identical conditions.[9]

BP, the well-studied PAH, is metabolized in skin *in vivo* and *in vitro* into a number of metabolites including BP-phenols, BP-quinones, BP-diols, and BP-tetrols. The first step in the metabolism of BP in skin is the P450-dependent stereospecific epoxidation of 7,8 double bond which results in the formation of two isomeric BP 7,8-epoxides, which are then converted by epoxide hydrolase, a non-P450-dependent microsomal enzyme, to stable *trans*-BP 7,8-diols. A substantial amount of these BP 7,8-diols is further epoxidized stereospecifically at the 9,10 position by P450-dependent monooxygenases. Although this process results in four isomers of BP-7,8-diol-9,10-epoxide, the major metabolite thus formed in the skin is the anti-*trans*-BP 7,8-diol 9,10-epoxide (BPDE-2), which is a highly reactive BP metabolite and preferentially binds *in vivo* to the amino group of desoxyguanosine of DNA in the skin, and also forms adducts with desoxyadenosine.[9] Based on this specific DNA binding, which correlates well with skin tumorigenicity, BPDE-2 is regarded as the ultimate carcinogenic metabolite of BP.

In cutaneous chemical carcinogenicity tests, BPDE-2, although more active than the other distereoisomers, has been shown to be less carcinogenic than the parent compound BP.[9] On the contrary, (-)-7β,8α-diol of BP, the precursor of BPDE-2, has higher or at least comparable tumorigenic potency in skin, suggesting that this compound is the proximate skin carcinogen.[9] In numerous other studies, wide range of carcinogenic PAHs have been shown to be metabolized by P450-dependent drug metabolizing enzymes in the skin.[92]

IX. P450-DEPENDENT DRUG METABOLIZING ENZYMES IN HUMAN SKIN

Levin et al.[51] for the first time demonstrated that P450-dependent AHH activity is present in human foreskin and that it can be induced by xenobiotics. In a subsequent study, Alvares et al.[52] provided further evidence that PAHs can induce AHH activity by two- to five-fold in human skin maintained in organ culture. Later Bickers and Kappas[93] showed that following crude coal tar application, the AHH activity in the skin of human volunteers is also elevated by two- to five-fold. In another study, Bickers et al.[94] assayed AHH activity in freshly keratomed human epidermis from normal volunteers and from visibly uninvolved skin of patients with psoriasis and showed that both the epidermis samples possess comparable AHH activity as well as BP metabolism. In another study, Don et al.[95] compared the BP metabolism and its subsequent binding to DNA in fibroblasts cultured from normal skin and from tumor-bearing skin of a patient with multiple basal cell carcinoma located exclusively in an area of the left upper trunk that was chronically exposed to PAHs. This study showed that differences in BP metabolism in normal and tumor-bearing skin. Merk et al.[96] showed that BP metabolism can be induced in cultured human keratinocytes by prior incubation of cells with BA. AHH activity induced by BA was found to be substantially inhibited by MAb 1-7-1, whereas MAb 2-66-3 was devoid of inhibitory effect, suggesting the presence of P450IA1 isozyme in human skin following induction by BA.

Recently Holtzman et al.[97] studied the enzymatic mechanisms involved in arachidonic acid oxygenation in homogeneous cell suspensions obtained from trypsinization of epidermis from healthy subjects. Incubation of these cells with arachidonic acid resulted in the predominant generation of 12-hydroxyeicosatetraenoic acid (12-HETE) as a mixture of 12S- and 12R-stereoisomers. Subcellular fractionation showed that >99% of the 12-HETE was generated by enzyme activity associated equally with mitochondrial and microsomal pellets. This activity was found to be increased by the addition of NADPH and inhibited by carbon monoxide. Based on the results of the study, the authors concluded that human epidermis possesses an active membrane-bound monooxygenase system for eicosanoid metabolism, and that the system is more operative in mitochondrial and microsomal fractions of the epidermis. In addition to studies demonstrating the presence of P450-dependent drug metabolizing enzymes in human skin and keratinocytes in culture, several studies have shown that human hair roots also possess measurable AHH[98,99] and ERD activities.[100] Furthermore, these enzyme activities are induced in human hair roots following the use of a therapeutic shampoo containing crude coal tar.[101,102] Merk et al.[101] showed that human hair follicles, in addition to metabolism of BP to BP-diols, BP-quinones, and BP-phenols, can also metabolize BP 7,8-diol. Since 7-ethoxyresorufin is more selectively metabolized by P450IA1, it has been suggested that the measurement of ERD in human hair roots may be a useful marker for identifying potentially induced species of P450 isozymes in human populations.[100] Recently Alexandrov et al.[103] reported a new sensitive fluorometric assay to assess the metabolism of (-)-BP-7,8-diol by freshly isolated human hair follicles. This rapid and noninvasive procedure, which supports the presence of P450-dependent monooxygenase system in human hair follicles, also provides a new means for evaluating human subjects for their capability to metabolize BP-7,8-diol to its carcinogenic metabolite anti-*trans*-BPDE.

ACKNOWLEDGMENTS

The studies in authors' laboratories were supported by USPHS grants ES-1900 and P-30-AR-39750 and by research funds from the Department of Veterans Affairs. We thank Ms. Sandra Evans for preparing the manuscript.

REFERENCES

1. **Mukhtar, H. and Khan, W. A.,** Cutaneous cytochrome P450, *Drug Metab. Rev.,* 20, 657, 1989.
2. **Kappus, H.,** Drug metabolism in the skin, in *Pharmacology of the Skin,* Vol. 2, Greaves, M. W. and Shuster, S., Eds., Springer-Verlag, New York, 1989, 123.
3. **Bickers, D. R.,** Drugs, carcinogen and steroid hormone metabolism in the skin, in *Biochemistry and Physiology of the Skin,* Goldsmith, L. A., Ed., Oxford University Press, New York, 1983, 1169.
4. **Kao, J. and Carver, M. P.,** Cutaneous metabolism of xenobiotics, *Drug Metab. Rev.,* 22, 363, 1990.
5. **Goerz, G. and Merk, H.,** Animal models for cutaneous drug-metabolizing enzymes, in *Models in Dermatology,* Vol. 3, Maibach, H. I. and Lowe, N. J., Eds., S. Karger, Basel, 1987, 93.
6. **Bickers, D. R., Das, M., and Mukhtar, H.,** Pharmacological modification of epidermal detoxification systems, *Br. J. Dermatol.,* 115, (Suppl. 31), 9, 1986.
7. **Noonan, P. K. and Wester, R. C.,** Cutaneous biotransformations and some pharmacological and toxicological implications, in *Dermatotoxicology,* 2nd ed., Marzulli, F. N. and Maibach, H. I., Eds., Hemisphere, Washington, D.C., 1983, 71.
8. **Pannatier, A., Jenner, P., Testa, B., and Etter, J. C.,** The skin as a drug-metabolizing organ, *Drug Metab. Rev.,* 8, 319, 1978.
9. **Conney, A. H.,** Induction of microsomal enzymes by foreign chemicals and carcinogenesis by polycyclic aromatic hydrocarbons: GHA Clowes memorial lecture, *Cancer Res.,* 42, 4875, 1982.
10. **Bronaugh, R. L., Stewart, R. F., and Storm, J. E.,** Extent of cutaneous metabolism during percutaneous absorption of xenobiotics, *Toxicol. Appl. Pharmacol.,* 99, 534, 1989.
11. **Collier, S. W., Sheikh, N. M., Sakr, A., Lichtin, J. L., Stewart, R. F., and Bronaugh, R. L.,** Maintenance of skin viability during *in vitro* percutaneous absorption/metabolism studies, *Toxicol. Appl. Pharmacol.,* 99, 522, 1989.

12. **Klingenberg, M.,** Pigments of rat liver microsomes, *Arch. Biochem. Biophys.,* 75, 376, 1958.
13. **Garfinkle, D.,** Studies on pig liver microsomes. I. Enzyme and pigment composition of different microsomal fractions, *Arch. Biochem. Biophys.,* 77, 493, 1958.
14. **Omura, T. and Sato, R.,** A new cytochrome in liver microsomes, *J. Biol. Chem.,* 237, 1375, 1961.
15. **Omura, T. and Sato, R.,** The carbon monoxide-binding pigment of liver microsomes. I. Evidence for its hemoprotein nature, *J. Biol. Chem.,* 239, 2370, 1964.
16. **Omura, T. and Sato, R.,** The carbon monoxide-binding pigment of liver microsomes. II. Solubilization, purification and properties, *J. Biol. Chem.,* 239, 2379, 1964.
17. **Cooper, D. Y., Levin, S., Narasimhulu, S., Rosenthal, O., and Estabrook, R. W.,** Photochemical action spectrum of the terminal oxidase of mixed function oxidase systems, *Science,* 147, 400, 1965.
18. **Ryan, D. E. and Levin, W.,** Purification and characterization of hepatic microsomal cytochrome P450, *Pharmacol. Ther.,* 45, 153, 1989.
19. **Levin, W.,** The 1988 Bernard B. Brodie Award lecture. Functional diversity of hepatic cytochrome P450, *Drug Metab. Dispos.,* 18, 824, 1990.
20. **Nebert, D. W., Nelson, D. R., Coon, M. J., Estabrook, R. W., Feyereisen, R., Fujii-Kuriyama, Y., Gonzalez, F. J., Guengerich, F. P., Gunsalus, I. C., Johnson, E. F., Loper, J. C., Sato, R., Waterman, M. R., and Waxman, D. J.,** The P450 superfamily: update on new sequences, gene mapping, and recommended nomenclature, *DNA and Cell Biol.,* 10, 1, 1991.
21. **Jakoby, W. B. and Ziegler, D. M.,** The enzymes of detoxification, *J. Biol. Chem.,* 265, 20715, 1990.
22. **Khan, W. A., Das, M., Stick, S., Javed, S., Bickers, D. R., and Mukhtar, H.,** Induction of epidermal NAD(P)H:quinone reductase by chemical carcinogens: a possible mechanism for the detoxification, *Biochem. Biophys. Res. Commun.,* 146, 126, 1987.
23. **Pohl, R. J., Philpot, R. M., and Fouts, J. R.,** Cytochrome P450 content and mixed-function oxidase activity in microsomes isolated from mouse skin, *Drug Metab. Dispos.,* 4, 442, 1976.
24. **Bickers, D. R., Dutta-Choudhury, T., and Mukhtar, H.,** Epidermis: a site of drug metabolism in neonatal rat skin. Studies on cytochrome P450 content and mixed-function oxidase and epoxide hydrolase activity, *Mol. Pharmacol.,* 21, 239, 1982.
25. **Alvares, A. P., Bickers, D. R., and Kappas, A.,** Polychlorinated biphenyls: a new type of inducer of cytochrome P-448 in the liver, *Proc. Natl. Acad. Sci. U.S.A.,* 70, 1321, 1973.
26. **Raza, H., Bickers, D. R., and Mukhtar, H.,** Purification and molecular characterization of β-naphtho-flavone-inducible from rat epidermis cytochrome P450, *J. Invest. Dermatol.,* 98, 233, 1992.
27. **Khan, W. A., Park, S. S., Gelboin, H. V., Bickers, D. R., and Mukhtar, H.,** Monoclonal antibodies directed characterization of epidermal and hepatic cytochrome P450 isozymes induced by skin application of therapeutic crude coal tar, *J. Invest. Dermatol.,* 93, 40, 1989.
28. **Ichikawa, T., Hayashi, S., Nosiro, M., Takada, K., and Okuda, K.,** Purification and characterization of cytochrome P450 induced by benz(a)anthracene in mouse skin microsomes, *Cancer Res.,* 49, 806, 1989.
29. **Mukhtar, H., Das, M., Steele, J. D., and Bickers, D. R.,** HPLC separation of multiple isozymes of epidermal cytochrome P450 in neonatal Sprague-Dawley rats, *Clin. Res. (Abstr.),* 34, 991A, 1986.
30. **Okey, A. B.,** Enzyme induction in the cytochrome P450 system, *Pharmacol. Ther.,* 45, 241, 1990.
31. **Thomas, P. E., Bandiera, S., Reik, L. M., Maines, S. L., Ryan, D. E., and Levin, W.,** Polyclonal and monoclonal antibodies as probes of rat hepatic cytochrome P450 isozymes, *Fed. Proc.,* 46, 2563, 1987.
32. **Gelboin, H. V. and Friedman, F. K.,** Monoclonal antibodies for studies on xenobiotic and endobiotic metabolism, *Biochem. Pharmacol.,* 34, 2225, 1985.
33. **Khan, W. A., Park, S. S., Gelboin, H. V., Bickers, D. R., and Mukhtar, H.,** Epidermal cytochrome P450: immunochemical characterization of isoform induced by topical application of 3-methylcholanthrene to neonatal rats, *J. Pharmacol. Exp. Ther.,* 249, 921, 1989.
34. **Whitter, T. B., Guengerich, F. P., and Baron, J.,** Effects of topical application of 3-methylcholanthrene on xenobiotic metabolizing enzymes in rat skin, *Fed. Proc. (Abstr.),* 44, 1115, 1985.
35. **Batula, N., Sagara, J., and Gelboin, H. V.,** Expression of P_1-450 and P_3-450 DNA coding sequences as enzymatically active cytochromes P450 in mammalian cells, *Proc. Natl. Acad. Sci. U.S.A.,* 84, 4073, 1987.
36. **Aoyama, T., Yamano, S., Guzelian, P. S., Gelboin, H. V., and Gonzalez, F. J.,** Five of 12 forms of vaccinia-expressed human hepatic cytochrome P450 metabolically activate aflatoxin B_1, *Proc. Natl. Acad. Sci. U.S.A.,* 87, 4790, 1990.
37. **Alvares, A. P., Bickers, D. R., and Kappas, A.,** Induction of drug-metabolizing enzymes and aryl hydrocarbon hydroxylase by microscope immersion oil, *Life Sci.,* 14, 853, 1974.
38. **Norden, G.,** The role of appearance, metabolism and disappearance of 3,5-benzpyrene in the epithelium of mouse skin, *Acta Pathol. Microbiol. Scand. Suppl.,* 96, 1, 1953.
39. **Finnen, M. J., Herdman, M. L., and Shuster, S.,** Distribution and sub-cellular localization of drug metabolizing enzymes in the skin, *Br. J. Dermatol.,* 113, 713, 1985.

40. **Epstein, E. H., Jr., Munderloh, N. H., and Fukuyama, K.**, Dithiothreitol separation of newborn rodent dermis and epidermis, *J. Invest. Dermatol.*, 78, 207, 1979.

41. **Das, M., Bickers, D. R., and Mukhtar, H.**, Epidermis: the major site of cutaneous benzo(a)pyrene and benzo(a)pyrene 7,8-diol metabolism in neonatal Balb/C mice, *Drug Metab. Dispos.*, 14, 637, 1986.

42. **Moloney, S. J., Fromson, J. M., and Bridges, J. W.**, The metabolism of 7-ethoxycoumarin and 7-hydroxycoumarin by rat and hairless mouse skin strips, *Biochem. Pharmacol.*, 31, 4005, 1982.

43. **Moloney, S. J., Fromson, J. M., and Bridges, J. W.**, Cytochrome P450 dependent deethylase activity in rat and hairless mouse skin microsomes, *Biochem. Pharmacol.*, 31, 4011, 1982.

44. **Bickers, D. R., Marcelo, L., Dutta-Choudhury, T., and Mukhtar, H.**, Studies on microsomal cytochrome P450 monooxygenases and epoxide hydrolase in cultured keratinocytes and intact epidermis from BALB/c mice, *J. Pharmacol. Exp. Ther.*, 223, 163, 1982.

45. **Pohl, R. J., Coomes, M. W., Sparks, R. W., and Fouts, J. R.**, 7-Ethoxycoumarin O-deethylation activity in viable basal and differentiated keratinocytes isolated from the skin of the hairless mouse, *Drug Metab. Dispos.*, 12, 25, 1984.

46. **Coomes, M. W., Norling, A. H., Pohl, R. J., Muller, D., and Fouts, J. R.**, Foreign compound metabolism by isolated skin cells from the hairless mouse, *J. Pharmacol. Exp. Ther.*, 225, 770, 1983.

47. **Coomes, M. W., Sparks, R. W., and Fouts, J. R.**, Oxidation of 7-ethoxycoumarin and conjugation of umbelliferone by intact, viable epidermal cells from the hairless mouse, *J. Invest. Dermatol.*, 82, 598, 1984.

48. **Wattenberg, L. W. and Leong, J. L.**, Tissue distribution studies of polycyclic hydrocarbon hydroxylase activity, in *Concepts in Biochemical Pharmacology*, Vol. 28/2, Brodie, B. B. and Gillette, J. R., Eds., Springer-Verlag, New York, 1971, 422.

49. **Vizethum, W., Ruzicka, T., and Goerz, G.**, Inducibility of drug-metabolizing enzymes in the rat skin, *Chem. Biol. Interact.*, 31, 215, 1980.

50. **Schlede, E. and Conney, A. H.**, Induction of benzo(a)pyrene hydroxylase activity in rat skin, *Life Sci.*, 9, 1295, 1970.

51. **Levin, W., Conney, A. H., Alvares, A. P., Merkatz, I., and Kappas, A.**, Induction of benzo(a)pyrene hydroxylase in human skin, *Science*, 176, 419, 1972.

52. **Alvares, A. P., Kappas, A., Levin, W., and Conney, A. H.**, Inducibility of benzo(a)pyrene hydroxylase in human skin by polycyclic hydrocarbons, *Clin. Pharmacol. Ther.*, 14, 30, 1973.

53. **Bickers, D. R., Kappas, A., and Alvares, A. P.**, Differences in inducibility of cutaneous and hepatic drug metabolizing enzymes and cytochrome P450 by polychlorinated biphenyls and 1,1,1-trichloro-2,2-bis(p-chlorophenyl)ethane (DDT), *J. Pharmacol. Exp. Ther.*, 188, 300, 1974.

54. **Bickers, D. R., Eiseman, J., Kappas, A., and Alvares, A. P.**, Microscope immersion oils: effects of skin application on cutaneous and hepatic drug metabolizing enzymes, *Biochem. Pharmacol.*, 24, 779, 1975.

55. **Thompson, S. and Slaga, T. J.**, Mouse epidermal aryl hydrocarbon hydroxylase, *J. Invest. Dermatol.*, 66, 108, 1976.

56. **Wiebel, F. J., Leutz, J. C., Diamond, L., and Gelboin, H. V.**, Aryl hydrocarbon (benzo[a]pyrene) hydroxylase in microsomes from rat tissue: differential inhibition and stimulation by benzoflavones and organic solvents, *Arch. Biochem. Biophys.*, 144, 78, 1976.

57. **Briggs, M. M. and Briggs, M.**, Induction by topical corticosteroids of skin enzymes metabolizing carcinogenic hydrocarbons, *Br. J. Dermatol.*, 88, 75, 1973.

58. **Mukhtar, H. and Bickers, D. R.**, Comparative activity of the mixed-function oxidases, epoxide hydratase, and glutathione-S-transferase in liver and skin of the neonatal rat, *Drug Metab. Dispos.*, 9, 311, 1981.

59. **Bickers, D. R., Wroblewski, D., Dutta-Choudhury, T., and Mukhtar, H.**, Induction of neonatal rat skin and liver aryl hydrocarbon hydroxylase by coal tar and its constituents, *J. Invest. Dermatol.*, 78, 227, 1982.

60. **Mukhtar, H. and Bickers, D. R.**, Evidence that coal tar is a mixed inducer of microsomal drug-metabolizing enzymes, *Toxicol. Lett.*, 11, 221, 1982.

61. **Mukhtar, H., Link, C. M., Cherniack, E., Kushiner, D. M., and Bickers, D. R.**, Effect of topical application of defined hydroxylase and 7-ethoxycoumarin deethylase activities, *Toxicol. Appl. Pharmacol.*, 64, 541, 1982.

62. **Merk, H., Khan, W. A., Kuhn, C., Bickers, D. R., and Mukhtar, H.**, Effect of topical application of clotrimazole to rats on epidermal and hepatic monooxygenase activities and cytochrome P450, *Arch. Dermatol. Res.*, 281, 198, 1989.

63. **Gonzalez, F. J.**, The molecular biology of cytochrome P450s, *Pharmacol. Rev.*, 40, 243, 1989.

64. **Damen, F. J. M. and Mier, P. D.**, Cytochrome P450-dependent-O-dealkylase activity in mammalian skin, *Br. J. Pharmacol.*, 75, 123, 1982.

65. **Pannatier, A., Testa, B., and Etter, J. -C.**, Aryl ether O-dealkylase activity in the skin of untreated mice *in vitro*, *Xenobiotica*, 11, 345, 1981.

66. **Bickers, D. R., Mukhtar, H., and Yang, S. K.,** Cutaneous metabolism of benzo(a)pyrene: comparative studies in C57BL/6N and DBA/2N mice and neonatal Sprague-Dawley rats, *Chem. Biol. Interact.,* 43, 263, 1983.

67. **Asokan, P., Das, M., Rosenkranz, H. S., Bickers, D. R., and Mukhtar, H.,** Topically applied nitropyrenes are potent inducers of cutaneous and hepatic monooxygenases, *Biochem. Biophys. Res. Commun.,* 129, 134, 1985.

68. **Asokan, P., Das, M., Bik, D. P., Howard, P. C., McCoy, G. D., Rosenkranz, H. S., Bickers, D. R., and Mukhtar, H.,** Comparative effects of topically applied nitrated arenes and their nonnitrated parent arenes on cutaneous and hepatic drug and carcinogen metabolism in neonatal rats, *Toxicol. Appl. Pharmacol.,* 86, 33, 1986.

69. **Bickers, D. R., Mukhtar, H., Meyer, C. W., and Speck, W. T.,** Epidermal enzyme mediated mutagenesis of the skin carcinogen 2-aminoanthracene, *Mutat. Res.,* 147, 37, 1985.

70. **Mukhtar, H., Bik, D. P., Ruzicka, T., Merk, H. F., and Bickers, D. R.,** Cytochrome P450-dependent omega-oxidation of leukotriene B$_4$ in rodent and human epidermis, *J. Invest. Dermatol.,* 93, 231, 1989.

71. **Mukhtar, H., Athar, M., and Bickers, D. R.,** Cytochrome P-450-dependent metabolism of testosterone in rat skin, *Biochem. Biophys. Res. Commun.,* 145, 749, 1987.

72. **Vanden Bossche, H., Willemsens, G., and Janssen, P. A. J.,** Cytochrome-P450-dependent metabolism of retinoic acid in rat skin microsomes: inhibition by ketoconazole, *Skin Pharmacol.,* 1, 176, 1988.

73. **Henke, D., Danilowicz, R., and Eling, T.,** Arachidonic acid metabolism by isolated epidermal basal and differentiated keratinocytes from the hairless mouse, *Biochim. Biophys. Acta,* 876, 271, 1986.

74. **Pentland, A. P. and Needleman, P.,** Modulation of keratinocyte proliferation *in vitro* by endogenous prostaglandin synthesis, *J. Clin. Invest.,* 77, 246, 1986.

75. **Poland, A. and Glover, E.,** Genetic expression of aryl hydrocarbon hydroxylase by 2,3,7,8-tetrachlorodibenzo-p-dioxin: evidence of a receptor mutation in genetically nonresponsive mice, *Mol. Pharmacol.,* 11, 389, 1975.

76. **Poland, A., Glover, E., and Kende, A. S.,** Stereospecific high-affinity binding of 2,3,7,8-tetrachlorodibenzo-p-dioxin by hepatic cytosol, *J. Biol. Chem.,* 251, 4936, 1976.

77. **Okey, A. B., Bondy, G. P., Mason, M. E., Kahl, G. F., Eisen, H. J., Guenther, T. M., and Nebert, D. W.,** Regulatory gene product of the Ah locus. Characterization of the cytosolic inducer-receptor complex and evidence for its nuclear translocation, *J. Biol. Chem.,* 254, 11636, 1979.

78. **Neuhold, L. A., Gonzelez, F. J., Jaiswal, A. K., and Nebert, D. W.,** Dioxin inducible enhancer region upstream from the P$_1$-450 gene and interaction with a heterologous SV 40 promoter, *DNA,* 5, 403, 1986.

79. **Poland, A., Glover, E., and Bradfield, C. A.,** Characterization of polyclonal antibodies to the Ah receptor prepared by immunization with a synthetic peptide hapten, *Mol. Pharmacol.,* 39, 20, 1991.

80. **Harper, P. A., Golas, C. L., and Okey, A. B.,** Characterization of the Ah receptor and aryl hydrocarbon hydroxylase induction by 2,3,7,8-tetrachlorodibenzo(p)dioxin and benz(a)anthracene in the human A-431 squamous cell carcinoma line, *Cancer Res.,* 48, 2388, 1988.

81. **Pohl, R. J. and Fouts, J. R.,** Xenobiotic metabolism in skin of hairless mice exposed to UV-radiation (UV), Aroclor 1254, or chlordane, *Pharmacologist,* 19, 200, 1977.

82. **Goerz, G., Merk, H., Bolsen, K., Tsambaos, D., and Berger, H.,** Influence of chronic UV-light exposure on hepatic and cutaneous monooxygenases, *Experimentia,* 39, 385, 1983.

83. **Mukhtar, H., DelTito, B. J., Matgouranis, P. M., Das, M., Asokan, P., and Bickers, D. R.,** Additive effects of ultraviolet B and crude coal tar on cutaneous carcinogen metabolism: possible relevance to the tumorigenicity of the Goeckerman Regimen, *J. Invest. Dermatol.* 87, 348, 1986.

84. **Das, M., Bickers, D. R., Santella, R. M., and Mukhtar, H.,** Altered patterns of cutaneous xenobiotic metabolism in ultraviolet- B-induced squamous cell carcinoma in SKH-1 hairless mice, *J. Invest. Dermatol.,* 84, 532, 1985.

85. **Eckert, R. L.,** The structure and function of skin, in *Pharmacology of the Skin,* Mukhtar, H., Ed., CRC Press, Boca Raton, 1991, in press.

86. **DiGiovanni, J., Gill, R. D., Nettikumara, A. N., Colby, A. B., and Reiners, J. J., Jr.,** Effect of extracellular calcium concentration on the metabolism of polycyclic aromatic hydrocarbons by cultured mouse keratinocytes, *Cancer Res.,* 49, 5567, 1989.

87. **Baer-Dubowska, W., Morris, R. J., Gill, R. D., and DiGiovanni, J.,** Distribution of covalent DNA adducts in mouse epidermal subpopulations after topical application of benzo(a)pyrene and 7,12-dimethylbenz(a)anthracene, *Cancer Res.,* 50, 3048, 1990.

88. **Cameron, G. S., Baldwin, J. K., Jasheway, D. W., Patrick, K. E., and Fischer, S. M.,** Arachidonic acid metabolism varies with the state of differentiation in density gradient-separated mouse epidermal cells, *J. Invest. Dermatol.,* 94, 292, 1990.

89. **Guo, J. -F., Brown, R., Rothwell, C. E., and Bernstein, I. A.,** Levels of cytochrome P450-mediated aryl hydrocarbon hydroxylase (AHH) are higher in differentiated than in germinative cutaneous keratinocytes, *J. Invest. Dermatol.,* 94, 86, 1990.

90. **Reiners, J. J., Jr., Cantu, A. R., and Pavone, A.,** Modulation of constitutive cytochrome P450 expression *in vivo* and *in vitro* in murine keratinocytes as a function of differentiation and extracellular Ca^{2+} concentration, *Proc. Natl. Acad. Sci. U.S.A.,* 87, 1825, 1990.

91. **Berghard, A., Gradin, K., and Toftgard, R.,** Serum and extracellular calcium modulate induction of cytochrome P450IA1 in human keratinocytes, *J. Biol. Chem.,* 265, 21086, 1990.

92. **MacNicoll, A. D., Grover, P. L., and Sims, P.,** The metabolism of a series of polycyclic hydrocarbons by mouse skin maintained in short-term organ culture, *Chem. Biol. Interact.,* 29, 169, 1980.

93. **Bickers, D. R. and Kappas, A.,** Human skin aryl hydrocarbon hydroxylase-induction by coal tar, *J. Clin. Invest.,* 62, 1061, 1978.

94. **Bickers, D. R., Mukhtar, H., Dutta-Choudhury, T., Marcello, C. L., and Voorhees, J. J.,** Aryl hydrocarbon hydroxylase, epoxide hydrolase and benzo(a)pyrene metabolism in human epidermis. Comparative studies in normal subjects and patients with psoriasis, *J. Invest. Dermatol.,* 83, 51, 1984.

95. **Don, P. S. C., Mukhtar, H., Das, M., Berger, N. A., and Bickers, D. R.,** Benzo(a)pyrene metabolism, DNA-binding and UV-induced repair of DNA damage in cultured skin fibroblasts from a patient with unilateral basal cell carcinoma, *Br. J. Dermatol.,* 120, 161, 1989.

96. **Merk, H. F., Jugert, F., and Khan, W. A.,** Human skin cytochrome P450: catalytic, electrophoretic and immunochemical analysis, *Skin Pharmacol.,* 1, 61, 1988.

97. **Holtzman, M. J., Turk, J., and Pentland, A.,** A regiospecific monooxygenase with novel stereopreference is the major pathway for arachidonic acid oxygenation in isolated epidermal cells, *J. Clin. Invest.,* 84, 1446, 1989.

98. **Merk, H., Rumpf, M., Bolsen, K., Wirth, G., and Goerz, G.,** Inducibility of aryl hydrocarbon-hydroxylase activity in human hair follicles by topical application of liquor carbonis detergents (coal tar), *Br. J. Dermatol.,* 111, 279, 1984.

99. **Vermorken, A. J. M., Goos, C. M. A. A., Roelofs, H. M. J., Henderson, P. T. H., and Bloemendal, H.,** Metabolism of benzo(a)pyrene in isolated human scalp hair follicles, *Toxicology,* 14, 109, 1979.

100. **Merk, H. F., Mukhtar, H., Schutte, B., Kaufman, I., Das, M., and Bickers, D. R.,** 7-Ethoxyresorufin-O-deethylase activity in human hair roots: a potential marker for toxifying species of cytochrome P450 isozymes, *Biochem. Biophys. Res. Commun.,* 148, 755, 1987.

101. **Merk, H. F., Mukhtar, H., Kaufman, I., Das, M., and Bickers, D. R. ,** Human hair follicle benzo(a)pyrene and benzo(a)pyrene 7,8-diol metabolism: effect of exposure to a coal tar-containing shampoo, *J. Invest. Dermatol.,* 88, 71, 1987.

102. **Hukkelhoven, M. W. A. C., Vromans, L. W. M., van Pelt, F. N. A. M., Keulers, R. A. C., and Vermorken, A. J. M.,** *In vivo* induction of aryl hydrocarbon hydroxylase in human scalp hair follicles by topical application of a commercial coal tar preparation, *Cancer Lett.,* 23, 135, 1984.

103. **Alexandrov, K., Rojas, M., Goldberg, M., Camus, A. -M., and Bartsch, H.,** A new sensitive fluorometric assay for the metabolism of (-)-7,8-dihydroxy-7,8-dihydrobenzo(a)pyrene by human hair follicles, *Carcinogenesis,* 11, 2157, 1990.

Chapter 4

PERCUTANEOUS ABSORPTION

Ronald C. Wester and Howard I. Maibach

TABLE OF CONTENTS

0-8493-7357-3/93/$0.00 + $.50
© 1993 by CRC Press, Inc.

I. INTRODUCTION

The skin is a primary area of body contact with the environment and the route by which many chemicals enter the body. Introduction of chemicals into the body via the skin occurs both through passive contact with the environment and through direct application of chemicals on the body. These direct applications include medical therapy (skin disease, transdermal drug delivery) and cosmetics. In most instances the toxicity of chemicals is slight, perhaps because the bioavailability (rate and amount of absorption) of the chemical is too low to cause an immediate response; however, some chemicals applied to the skin undoubtedly have produced toxicity. Moreover, there is a continuing discovery of potentially toxic chemicals that come in contact with the skin and a continual growing awareness of how chemicals enter the body through the skin.

This chapter summarizes the methodology used to study percutaneous absorption, and this chapter will also summarize results from absorption studies. The impression which will be left in the reader's mind is that there are many variables in percutaneous absorption. The variables included in the study design influence the final results. The interpretation of such studies should be restricted to the limits of the study design. The methodology and supportive information discussed here should help formulate good study design.

II. POWDERED HUMAN STRATUM CORNEUM

This is an *in vitro* model that utilizes the partition coefficient of the chemical contaminant in water or other vehicle with that of powdered human stratum corneum. Adult foot calluses are ground with dry ice and freeze-dried to form a powder. That portion of the powder that passed through a 40-mesh, but not an 80-mesh, sieve is used. The chemical (radio-labeled) as a solution in 1.5 ml of water or other vehicle is mixed with 1.5 mg of powdered human stratum corneum and the mixture is allowed to set for 30 min. The mixture is centrifuged and the proportions of chemical bound to human stratum corneum and that remaining in water are determined by scintillation counting or some other analytical method.[1]

The capacity of skin and soil for cadmium was studied with binding studies. Cadmium chloride in water (116 ppb) was partitioned against 1 mg of soil, and against 1 mg of powdered human stratum corneum. Table 1 shows the percent dose in water and in matter (soil or powdered human stratum corneum), and Table 2 gives the partition coefficient of cadmium chloride. Soil has a relatively higher affinity for cadmium than does stratum corneum. This correlates with data in studies where the skin absorption is greater from water than from soil (soil-binding capacity relative to skin).

Another example of the use of powdered human stratum corneum is shown in Table 3. Here, the ability of soap and water to decontaminate skin is shown for alachlor.[2]

III. *IN VITRO* PERCUTANEOUS ABSORPTION METHOD

The most commonly used *in vitro* technique involves placing a piece of excised skin in a diffusion chamber, applying radioactive compound to one side of the skin, and then assaying for radioactivity in the collection vessel on the other side.[2] Excised human or animal skin may be used, and the skin can be wholly intact or separated into epidermis or dermis. Artificial membranes can be used in place of skin to measure diffusion kinetics. The advantages of the standard *in vitro* technique are that the method is easy to use and the results are obtained quickly. The disadvantage is that the fluid in the collection bath which bathes the skin is saline, which may be appropriate for studying hydrophilic compounds, but is not suitable for hydrophobic compounds. Table 4 shows that absorption of triclocarban in a

TABLE 1
The Partitioning of Cadmium Chloride between Water and Powdered Human Stratum Corneum and between Water and Soil

	Percent dose	
Water	Stratum corneum	Total
68.6 ± 5.6	33.2 ± 3.8	101.8 ± 3.3
Water	Soil	Total
9.3 ± 1.4	82.5 ± 1.0	91.8 ± 1.8

Note: 116 ppb Cd/1.0 ml water/1.0 mg stratum corneum or 1.0 mg soil mixed for 30 min followed by centrifugation: n = 3.

TABLE 2
Partition Coefficient of Cadmium Chloride in Human Powdered Stratum Corneum/Water and Soil/Water

Test material	Partition coefficient[a]
Stratum corneum	3.61×10^1
Soil	1.03×10^5

[a] Partition

$$\text{coefficient} = \frac{\text{Conc of } CdCl_2 \text{ in 1000 mg of HPSC (soil)}}{\text{Conc of } CdCl_2 \text{ in 1000 mg of water}}$$

TABLE 3
Partitioning: Alachlor in Lasso with Powdered Human Stratum Corneum

	[14C] Alachlor percent dose
Stratum corneum	90.3 ± 1.2
Lasso supernatant	5.1 ± 1.2
Water only wash of stratum corneum	4.6 ± 1.3
10% Soap and water wash	77.2 ± 5.7
50% Soap and water wash	90.0 ± 0.5

Note: [14C] Alachlor in Lasso® EC formulation (1:20 dilution) mixed with powdered human stratum corneum, let set for 30 min, then centrifuged. Stratum corneum wash with: (1) water only, (2) 10% soap and water, and (3) 50% soap and water.

From Bucks, D. A. W., Wester, R. C., Mobayen, M. M., Yang, D., Maibach, H. I., and Coleman, D. L., *Toxicol. Appl. Pharmacol.*, 100, 417, 1985. With permission.

standard static system *in vitro* was 0.12 ± 0.05% of applied dose through human adult abdominal skin. In contrast, *in vivo*, in man the absorption was 7.0 ± 2.8%. The discrepancy appeared to be due primarily to the insolubility of triclocarban in the small volume of saline used in the reservoir of the static system. By changing to a continuous flow system, in which the volume of saline was greatly increased, the solubility of triclocarban was no

TABLE 4
Percutaneous Absorption of Triclocarban *In Vitro* and
In Vivo

System	Dose absorbed (% ± SD)
Static system (23°C)	
Human adult abdominal skin (n = 8)	0.13 ± 0.05
Continuous-flow system (23°C)	
Human adult abdominal system (n = 12)	6.0 ± 2.0
Man, *in vivo* (n = 5)	7.0 ± 2.8

From Wester, R. C., Maibach, H. I., Surinchak, J., and Bucks, D. A. W.,
Percutaneous Penetration, Marcel Dekker, New York, 1985, 223. With
permission.

longer the limiting factor in absorption and the extent of absorption *in vitro* approached that
of absorption *in vivo*.[3]

The validity of using excised skin depends on three assumptions. The first is that no
living process affects the skin's impermeability. If the contribution of metabolism in maintaining cellular lipids is discounted, then for compounds that strictly follow Fick's law of
diffusion, this assumption would be valid. In contrast, the drug-metabolizing enzyme activity
in the epidermis is greatly dependent on tissue viability. (It should be emphasized that
determination of skin absorption by measurement of radioactivity in the collection vessel
when using radio-labeled compounds does not distinguish between the unchanged compound
and its metabolites.) It cannot be assumed that excised skin (usually stored) will retain full
enzymatic activity.

The second assumption is that the dermis does not affect penetration. The problem of
compatibility with hydrophobic compounds was discussed earlier with triclocarban. Additionally, Reifenrath[16] has compared the penetrability of different thicknesses of skin and has
shown the dermis to be the rate-limiting step in the penetration of DDT.

The third assumption is that skin surface conditions *in vitro* are similar to those in life.
However, the artificial retention of a quantity of liquid on the skin surface is clearly different
from the situation *in vivo*. With volatile compounds (mosquito repellents, perfume fragrances), the ratio of volatility to penetration is very dependent on surface conditions. Another
major unknown factor with surface conditions is the bacterial population and what role it
might play in percutaneous absorption.

Despite the concern regarding these assumptions, excised diffusion chambers are easy
to use and capable of producing rapid, reproducible results. However, once the results are
obtained, it may be necessary to check the findings *in vivo*.

IV. *IN VITRO:* INDIVIDUAL AND REGIONAL VARIATION

Table 5 presents the results *in vitro* percutaneous absorption of a test article in three
different human skin sources. The data were summarized over five different formulations
(each run in all three skin sources). What is of interest is the receptor fluid accumulation
in human skin source #2. What it shows is that the barrier properties of human skin source
#2 were such that no skin absorption occurred. A formulation comparison with only that
human skin source would have provided completely negative data.[4]

Table 6 gives the *in vitro* percutaneous absorption of pentadeconic acid from two formulations (A and B) in two human skin sources. Skin content shows that formulations were

TABLE 5
Individual Variation in *In Vitro* Percutaneous Absorption

Human skin source	Skin	Percent dose (mean ± SD)		Total recovery
		Surface wash	Receptor fluid	
#1	5.0 ± 2.4	85.7 ± 7.8	2.5 ± 4.5	93.2 ± 6.0
#2	3.2 ± 2.8	83.7 ± 9.5	0.3 ± 0.2	87.7 ± 8.1
#3	2.6 ± 1.0	72.1 ± 12.3	4.4 ± 5.0	79.0 ± 12.5

Note: Skin sources: #1 68-year-old white male abdominal
#2 69-year-old white male thigh
#3 33-year-old white male abdominal

From Wester, R. C. and Maibach, H. I., *In Vitro Percutaneous Absorption: Principles, Fundamentals, and Applications,* CRC Press, Boca Raton, FL, 1991, chap. 4. With permission.

TABLE 6
***In Vitro* Percutaneous Absorption of Pentadeconic Acid**

Skin source	Formulation	Skin	Percent dose (mean ± SD)		Total recovery
			Surface wash	Receptor fluid	
#1	A	5.4 ± 0.8	80.4 ± 8.9	0.05 ± 0.04	85.9 ± 9.5
	B	6.4 ± 1.8	85.6 ± 5.2	0.15 ± 0.03	92.1 ± 4.5
#2	A	5.8 ± 1.5[a]	86.2 ± 6.6	0.03 ± 0.03	92.1 ± 5.2
	B	14.2 ± 23[a]	80.0 ± 5.3	0.03 ± 0.01	94.2 ± 4.3

Note: Skin sources: #1 43-year-old white female thigh
#2 7-year-old Hispanic female thigh

[a] Statistically different ($p = 0.006$).

From Wester, R. C. and Maibach, H. I., *In Vitro Percutaneous Absorption: Principles, Fundamentals, and Applications,* CRC Press, Boca Raton, FL, 1991, chap. 4. With permission.

only distinguishable in one of the human skin samples. This suggests that decisions based on only one human skin source may be misleading.[4]

Table 7 gives the *in vitro* percutaneous absorption of taurocholic acid. The data are summarized over five formulations and individualized for each human skin source and repeated skin absorption for each of the skin sources. What is of interest is the receptor fluid content (24-h accumulation) and the totally different results in skin source #3 compared to the other two skin sources (#1 and #2). Again, individual variations can be significant, but this only becomes apparent when other skin sources are also used.[4]

Table 8 compares the *in vitro* percutaneous absorption of DDT and benzo(a)pyrene in two different human skin sources. Skin content differences should be noted for both individual skin sources and chemical penetrating into human skin.[4]

Table 9 compares the *in vitro* permeability of coumarin, griseofulvin, and propranolol across human abdominal skin and scalp skin. For coumarin and propranolol, the scalp showed higher permeability. *In vivo* in man, Feldmann and Maibach[5] identified the scalp as a higher absorbing area; so these *in vitro* results of Ritshcel et al. agree with the *in vivo* results.[6]

TABLE 7
***In Vitro* Percutaneous Absorption of
Taurocholic Acid**

Skin source	Skin content[a]	Receptor fluid content[b]
#1	0.5 ± 0.3	8.9 ± 12.9
#1 repeat	1.3 ± 1.3	8.2 ± 6.0*
#2	1.4 ± 1.8	16.5 ± 18.0
#2 repeat	3.4 ± 6.3	11.7 ± 14.3
#3	1.4 ± 1.5	0.2 ± 0.3
#3 repeat	0.8 ± 0.8	0.3 ± 0.7[c]

Note: Skin sources: #1   41-year-old white male thigh
                       #2   32-year-old white male thigh
                       #3   53-year-old white male thigh

[a]   Mean percent dose ± SD; n = 5.
[c]   Significant difference ($p < 0.02$).

From Wester, R. C. and Maibach, H. I., *In Vitro Percutaneous Absorption: Principles, Fundamentals, and Applications,* CRC Press, Boca Raton, FL, 1991, chap. 4. With permission.

TABLE 8
**Individual Human Skin Source Variation in
In Vitro Percutaneous Absorption**

Human skin source	Skin content[a]	
	DDT	Benzo(a)pyrene
#1	6.7 ± 1.5[b]	27.4 ± 10.9
#2	29.6 ± 8.1[b]	20.0 ± 8.6

Note: Skin sources: #1   41-year-old white male thigh
                       #2   21-year-old white female thigh

[a]   Mean percent dose ± SD (n = 3) of skin content following
      24 *in vitro* percutaneous absorption study.
[b]   Significant difference ($p = 0.005$).

From Wester, R. C. and Maibach, H. I., *In Vitro Percutaneous Absorption: Principles, Fundamentals, and Applications,* CRC Press, Boca Raton, FL, 1991, chap. 4. With permission.

Individual variation and regional variation exists for *in vivo* percutaneous absorption, and as shown in this chapter, the same variability exists for *in vitro* percutaneous absorption. For human skin this can be critical because of human skin availability and the tendency to perhaps only use one human skin source to conserve supply. We recommended, if possible, the use of multiple human skin sources.

V. *IN VITRO:* SHORT-TERM SKIN EXPOSURE

Skin exposure from drug therapy and from environmental chemical contaminants are usually of an extended period of a day or longer. However, some exposures will be of a

TABLE 9
In Vitro **Permeability across Human**
Abdominal and Scalp Skin

	Steady-state flux (mg/cm²/h)	
Drug	**Abdominal skin**	**Scalp skin**
Coumarin	130 ± 78[a]	172 ± 64[a]
Griseofulvin	10 ± 6	16 ± 8
Propranolol	30 ± 6[a]	42 ± 11[a]

[a] Significant difference ($p < 0.05$) abdominal vs. scalp skin.

From Ritschel, W. R., Sabouni, A., and Hussain, A. S., *Methods Fundam. Exp. Clin. Pharmacol.*, 11, 643, 1989. With permission.

considerably shorter duration in terms of minutes. In either case, the first steps of percutaneous absorption happen within the first 30 min. Rougier and co-workers[7-9] have shown that *in vivo* absorption for the longer term can be predicted from this first 30-min exposure. With binding to powdered human stratum corneum and with *in vitro* percutaneous absorption, our data reflect these early happenings. Our *in vivo* data confirm some of the predicting potential for longer exposure. The message is that even short-term chemical skin exposure can result in significant skin chemical content. This may be good for some drug therapy. This may not be good for some environmental exposure.

A study was done while human skin was exposed to cadmium in water for only 30 min, followed by skin surface wash with soap and water. One half of the replicates (nine each; three human skin sources and three replicates) were stopped at 30 min. Human skin content was 2.3 ± 3.3%, and no cadmium was detected in the plasma receptor fluid. The other half of the replicates were perfused for an additional 48 h. The skin content was 2.7 ± 2.2%, not significantly different ($p = 0.77$). However, some 0.6 ± 0.8% dose had diffused into the plasma receptor fluid ($p = 0.04$). Therefore, cadmium has the ability to bind to human skin during a short exposure in water, not be completely removed with soap and water wash, and subsequently be absorbed into the body (Table 10).

Table 11 summarizes short-term 30-min exposure for *p*-nitroaniline. *In vitro* absorption, *in vivo* absorption, and binding to powdered human stratum corneum are all in agreement, and all show the potential for short-term absorption. Table 12 gives additional data for DDT and benzo(a)pyrene. Short-term (25 min) exposure can be as extensive as 24-h exposure.[10]

VI. *IN VIVO* PERCUTANEOUS ABSORPTION METHODS

A. SKIN STRIPPING: SHORT-TERM EXPOSURE

The stripping method determines the concentration of chemical in the stratum corneum at the end of a short application period (30 min), and linear extrapolation predicts the percutaneous absorption of that chemical for longer application periods. The chemical is applied to skin of animals or humans and after the 30-min skin application time, the stratum corneum is removed by successive tape application and removal. The tape strippings are assayed for chemical content. Rougiér, Lotte, and co-workers have established a linear relationship between this stratum corneum reservoir content and percutaneous absorption using the standard urinary excretion method. The major advantages of this method are (1) the elimination of urinary (and fecal) excretion to determine absorption and (2) the appli-

TABLE 10
Exposure of Cadmium in Water to Human Skin for 30 min Followed by Skin Surface Wash with Soap and Water and then 48-h Perfusion with Human Plasma

| Treatment | Percent dose | |
	Skin content	Plasma receptor fluid
30-min exposure only	2.3 ± 3.3	0.0 ± 0.0
30-min exposure followed by 48-h perfusion	2.7 ± 2.2	0.6 ± 0.8
Statistics	$p = 0.77$	$p = 0.04$[a]

n = 9 (three human skin sources and three replicates each).

Note: This study simulates a 30-min exposure of human skin to cadmium in water (swim, bathe) followed by soap and water wash. Cadmium is able to bind to human skin in the 30-min exposure time and then be absorbed into the body during the remainder of the day.

[a] Statistically significant difference.

TABLE 11
In Vivo Percutaneous Absorption of *p*-Nitroaniline in the Rhesus Monkey Following 30-min Exposure to Surface Water: Comparisons to *In Vitro* Binding and Absorption

Phenomenon	Percent dose absorbed/bound
In vivo percutaneous absorption, Rhesus monkey	4.1 ± 2.3
In vitro percutaneous absorption, human skin	5.2 ± 1.6
In vitro binding, powdered human stratum corneum	2.5 ± 1.1

From Wester, R. C., Mobayen, M., and Maibach, H. I., *J. Toxicol. Environ. Health,* 21, 367, 1987. With permission.

TABLE 12
Skin Exposure Time and Skin Content during *In Vitro* Percutaneous Absorption

Chemical	Formulation	Percent dose 24-h exposure	Skin content 25-min exposure
DDT	Acetone	18.1 ± 13.4	16.7 ± 13.2
	Soil	1.0 ± 0.7	1.8 ± 1.4
Benzo(a)pyrene	Acetone	23.7 ± 9.7	5.1 ± 2.1
	Soil	1.4 ± 0.8	0.14 ± 0.13

Note: *In vivo* percutaneous absorption in the Rhesus monkey for 24-h application time was 18.9 ± 9.4% (acetone vehicle) and 3.3 ± 0.5% (soil) for DDT. For benzo(a)pyrene, 51.0 ± 2.0 (acetone vehicle) and 13.2 ± 3.4% (soil) was absorbed.

From Wester, R. C., Maibach, H. I., Bucks, D. A. W., Sedik, L., Melendres, J., Liao, C., and Di Zio, S., *Fundam. Appl. Toxicol.,* 15, 510, 1990. With permission.

cability to nonradio-labeled determination of percutaneous absorption because the skin strippings contain adequate chemical concentrations for nonlabeled assay methodology. This is an exciting new system for which more research is needed to establish limitations.[7-9]

VII. SKIN FLAPS

The methodology is to surgically isolate a section of skin on an animal such that the blood supply is singular and this singular source can be used to collect chemical in blood as the chemical absorbs through skin. The isolated skin section can be used for percutaneous absorption while intact on the *in vivo* animal, or the skin section with its intact blood vessels mounted in an *in vitro* perfusion system to study percutaneous absorption. The isolated perfused porcine skin flap (IPPSF) is surgically created on a pig, and then the viable flap with intact blood supply can be mounted in an *in vitro* perfusion system. The absorption of chemicals through skin and metabolism within the skin can be determined by assay of the blood vessel perfusate. The IPPSF model offers advantages in that it is an alternative in *in vitro* animal model and that metabolism of chemicals penetrating skin can be determined.[11,12] The skin sandwich flap (SSF) is an island flap that has split-thickness skin grafted to its subcutaneous surface directly under the superficial epigastric vasculature. In this setting, the dermis of the donor skin and subcutaneous tissue of the host flap grow together, sandwiching the vessels supplying the flap, the superficial epigastric vessels. Two additional steps allow this sandwich to be converted to an island sandwich flap, which is isolated on its vasculature and transferred to the rat's back by a series of surgical procedures. The juncture on the femoral vessels supplying and draining the flap can be readily visualized with an incision in the groin and is accomplished routinely. The exposed vein draining the flap tolerates multiple venopunctures. The SSF can be constructed with either human, pig, or rat skin as the donor skin.[13]

VIII. SYSTEMIC BIOAVAILABILITY (BLOOD AND EXCRETA)

Percutaneous absorption *in vivo* is usually determined by the indirect method of measuring radioactivity in excreta following topical application of the labeled compound. In human studies, plasma levels of the compound are extremely low following topical application, often below assay detection level, so it is necessary to use tracer methodology. The labeled compound, usually carbon-14 or tritium, is applied to the skin. The total amount of radioactivity excreted in urine (or urine plus feces) is then determined. The amount of radioactivity retained in the body or excreted by some route not assayed (CO_2, sweat) is corrected by determining the amount of radioactivity excreted following parenteral administration. This final amount of radioactivity is then expressed as the percent of the applied dose that was absorbed.

The equation used to determine percutaneous absorption is:

$$\text{Percent} = \frac{\text{Total radioactivity following topical administration}}{\text{Total radioactivity following parenteral administration}} \times 100$$

Determination of percutaneous absorption from urinary radioactivity does not account for metabolism by skin. The radioactivity in urine is a mixture of parent compound and metabolites. Plasma radioactivity can be measured and the percutaneous absorption determined by the ratio of the areas under the plasma concentration vs. time curves following topical and intravenous administration. Radioactivity in blood and excreta can include both the applied compound and metabolites. If the metabolism by skin is extensive and different

TABLE 13
Bioavailability of Topical Nitroglycerin
Determined from Plasma Nitroglycerin,
Plasma ^{14}C, and Urinary Excretion of ^{14}C

Method	Mean bioavailability (%)
Plasma nitroglycerin AUC[a]	56.6 ± 2.5
Plasma total radioactivity AUC[a]	77.2 ± 6.7
Urinary total radioactivity[b]	72.7 ± 5.8

[a] Absolute bioavailability of nitroglycerin and ^{14}C:

$$\text{Percent} = \frac{\text{AUC (ng·h/ml)/topical dose}}{\text{AUC (ng·h/ml)/i.v. dose}} \times 100$$

[b] Percent = (total ^{14}C excretion following topical administration)/total ^{14}C excretion following i.v. administration × 100.

From Wester, R. C., Noonan, P. K., Smeach, S., and Kosobud, L., *J. Pharm. Sci.,* 72, 745, 1983. With permission.

from that of other systemic tissues, then this method is not valid because the pharmacokinetics of the metabolites can be different from that of the parent compound. However, in practice, this method has given results similar to those obtained from urinary excretion.[14]

The only way to determine the absolute bioavailability of a topically applied compound is to measure the compound by specific assay in blood or urine following topical and intravenous administration. This is difficult to do since plasma concentrations after topical administration are often very low. However, as more sensitive assays are developed, estimates of absolute topical bioavailability will become a reality. A comparison of these methods was performed by using [^{14}C]-nitroglycerin in Rhesus monkeys (Table 13). The difference between the estimate of absolute bioavailability (56.6%) and that of ^{14}C (72.7 to 77.2%) is the percent of compound metabolized in the skin as the compound was being absorbed. For nitroglycerin, this is about 20%.[14]

IX. SURFACE DISAPPEARANCE

Another approach used to determine *in vivo* percutaneous absorption is to measure the loss of radioactive material from the surface as it penetrates the skin. Recovery of an ointment or solution following skin application is difficult because total recovery from the skin is never assured. With topical application of a transdermal delivery device, the total unit can be removed from the skin and the residual amount of drug in the device can be determined. The difference between the applied and the residual dose is assumed to be the amount of drug absorbed. One must be aware that the skin may act as a reservoir for unabsorbed material.

X. BIOLOGICAL RESPONSE

Another *in vivo* method of estimating absorption is to use a biological/pharmacological response.[15] Here, a biological assay is substituted for a chemical assay and absorption is estimated. An obvious disadvantage to the use of biological response is that it is only good for compounds that will elicit an easily measurable response. An example of a biological response would be the vasoconstrictor assay when the balancing effect of one compound is compared to that of a known compound. This method is perhaps more qualitative than quantitative.

Other qualitative methods of estimating *in vivo* percutaneous absorption include whole body autoradiography and fluorescence. Whole body autoradiography will give an overall picture of dermal absorption followed by the involvement of other body tissues with the absorbed compound.

REFERENCES

1. **Wester, R. C., Mobayen, M., and Maibach, H. I.,** *In vivo* and *in vitro* absorption and binding to powdered human stratum corneum as methods to evaluate skin absorption of environmental chemical contaminants from ground and surface water, *J. Toxicol. Environ. Health,* 21, 367, 1987.
2. **Bucks, D. A. W., Wester, R. C., Mobayen, M. M., Yang, D., Maibach, H. I., and Coleman, D. L.,** *In vitro* percutaneous absorption and stratum corneum binding of alachlor: effect of formulation dilution with water, *Toxicol. Appl. Pharmacol.,* 100, 417, 1985.
3. **Wester, R. C, Maibach, H. I., Surinchak, J., and Bucks, D. A. W.,** Predictability of *in vitro* diffusion systems: effect of skin types and ages on percutaneous absorption of triclocarban, in *Percutaneous Penetration,* Bronaugh, R. and Maibach, H., Eds., Marcel Dekker, New York, 1985, 223.
4. **Wester, R. C. and Maibach, H. I.,** Individual and regional variation with *in vitro* percutaneous absorption, in *In Vitro Percutaneous Absorption: Principles, Fundamentals, and Applications,* Bronaugh, R. and Maibach, H., Eds., CRC Press, Boca Raton, FL, 1991, chap. 4.
5. **Feldmann, R. J. and Maibach, H. I.,** Regional variation in percutaneous penetration of ^{14}C cortisone in man, *J. Invest. Dermatol.,* 48, 181, 1967.
6. **Ritschel, W. A., Sabouni, A., and Hussain, A. S.,** Percutaneous absorption of coumarin, griseofulvin and propranolol across human scalp and abdominal skin, *Methods Fundam. Exp. Clin. Pharmacol.,* 11, 643, 1989.
7. **Rougier, A., Dupuis, D., Lotte, C., Roguet, R., and Schaefer, H.,** Correlation between stratum corneum reservoir function and percutaneous absorption, *J. Invest. Dermatol.,* 81, 275, 1983.
8. **Dupuis, D., Rougier, A., Roguet, R., Lotte, C., and Kalopissis, G.,** *In vivo* relationship between horny layer reservoir effect and percutaneous absorption in human and rat, *J. Invest. Dermatol.,* 82, 353, 1984.
9. **Rougier, A., Dupuis, D., Lotte, C., Roguet, R., Wester, R. C., and Maibach, H. I.,** Regional variation in percutaneous absorption in man: measurement by the stripping method, *Arch. Dermatol. Res.,* 278, 465, 1986.
10. **Wester, R. C., Maibach, H. I., Bucks, D. A. W., Sedik, L., Melendres, J., Liao, C., and Di Zio, S.,** Percutaneous absorption of [^{14}C]-DDT and [^{14}C]-benzo(a)pyrene from soil, *Fundam. Appl. Toxicol.,* 15, 510, 1990.
11. **Riviere, J. E., Bowman, K. F., Monteiro-Riviere, N. A., Dix, L. P., and Carver, M. P.,** The isolated perfused porcine skin flap (IPPSF), *Fundam. Appl. Toxicol.,* 7, 444, 1986.
12. **Riviere, J. E., Bowman, K. F., and Monteiro-Riviere, N. A.,** On the definition of viability on isolated perfused skin preparations, *Br. J. Dermatol.,* 116, 739, 1987.
13. **Pershing, L. K. and Krueger, G. G.,** Human skin sandwich flap model for percutaneous absorption, in *Percutaneous Absorption,* 2nd ed., Bronaugh, R. and Maibach, H., Eds., Marcel Dekker, New York, 1989, 397.
14. **Wester, R. C., Noonan, P. K., Smeach, S., and Kosobud, L.,** Pharmacokinetics and bioavailability of intravenous and topical nitroglycerin in the Rhesus monkey: estimate of percutaneous first-pass metabolism, *J. Pharm. Sci.,* 72, 745, 1983.
15. **McKenzie, A. W. and Stoughton, R. B.,** Method for comparing percutaneous absorption of steroids, *Arch. Dermatol.,* 86, 608, 1962.
16. **Reifenrath,** personal communication.

Chapter 5

IN VITRO SKIN METABOLISM

Robert L. Bronaugh and Steven W. Collier

TABLE OF CONTENTS

I. INTRODUCTION

Investigators conducting percutaneous absorption studies have recently become aware of the potential for significant metabolism in skin during absorption of chemicals.[1,2] Although the diffusional barrier properties of skin are present under nonviable conditions, the chemical identity of compounds absorbed through skin may differ from that of the chemicals absorbed *in vivo* if biotransformations that occur in skin do not exist in the diffusion cell.

The *in vitro* diffusion cell system permits sampling of absorbed material directly below the skin surface. Not only does this system allow accurate measurements of percutaneous absorption, but it also permits measurements of skin metabolism that completely exclude systemic metabolic events. It is difficult to assess skin metabolism from *in vivo* studies. Even blood specimens obtained near the site of topical chemical application may be influenced by systemic metabolism in blood and other tissues.

II. METABOLISM IN SKIN

Frequently, the rates of cutaneous metabolism are lower than metabolic rates in the liver, and the identification of metabolites *in vivo* is sometimes precluded by the large quantity of blood that would need to be collected. Experimental techniques for the study of skin metabolism *in vitro* include the use of cytosolic and microsomal skin fractions or homogenates,[3-6] cultures of isolated skin cells,[7] reconstituted epidermis,[8] and organ cultures of explanted primary skin tissue.[1,9-12] *In vitro* skin perfusion models, such as the isolated perfused porcine skin flap, can be used for measurement of percutaneous absorption with concomitant metabolism.[13] The experimental design and data interpretation must be given important consideration when any of these techniques is used. For example, enzymatic assays with skin subcellular fractions or homogenates are usually optimized for the enzyme processes that are anticipated for the specific chemical being studied. Unexpected biotransformations that would result in unusual metabolites may therefore be overlooked. With each passage, cells in culture undergo a selection process that may change their metabolizing capacities. Moreover, the phenotype that the cells express is that of the cells that thrive under the culture conditions provided and not necessarily the phenotype of the primary tissue from which they were originally isolated. There are also special considerations for organ explant cultures, such as maintaining viability and preventing microbial overgrowth.

Microsomal and cytosolic fractions have been used in our laboratory to study the activities and kinetic parameters of some enzymatic processes and to compare activities between skin and liver.[2,14,15] A study was conducted to compare the activities of aryl hydrocarbon hydroxylase (AHH), 7-ethoxycoumarin deethylase (ED), a glutathione-*S*-transferase (GST) in rodent species and organs. AHH and ED activities were determined in microsomal fractions prepared from liver and skin.[14] Microsomal fractions were prepared according to a method derived from Mukhtar and Bickers,[16] and Rettie et al.[17]

Interspecies differences were apparent in AHH activity in both skin and liver. AHH activity in mouse skin was two to six times greater than AHH activity in the skin of rats. Cutaneous AHH activity in Sencar mice was somewhat greater than in BALB/c mice in these experiments, whereas hepatic AHH activity was equivalent in both strains. AHH activity in the livers of both strains of mice was equivalent and was approximately seven times greater than AHH activity in the livers of either rat strain. Interspecies differences were also apparent in ED activity of both skin and livers. Both mouse strains exhibited similar cutaneous activity, which was about 20 times greater than that of either rat strain. A similar pattern of ED activity exists in the livers.

Interspecies differences were also observed in GST activity of both skin and livers (Table 1). Because GST activity of cytosolic fractions was so much greater than either AHH or

TABLE 1
Kinetic Parameters of Cytosolic Glutathione-*S*-Transferase (GST) Activity in Two Strains of Mice and Rats

Strain of species	Parameter[a]	Liver	Skin
Fuzzy rat	V_{max}	1432	14
	K_m	0.17	0.13
Osborne-Mendel rat	V_{max}	963	10
	K_m	0.18	0.11
BALB/c mouse	V_{max}	218,165	
	K_m	0.36	0.51
Sencar mouse	V_{max}	1984	67
	K_m	0.24	0.33

Note: GST activity was determined in cytosolic fractions prepared from liver and skin according to a method slightly modified from Habig et al.[18] by using chlorodinitrobenzene (CDNB) as substrate. Liver assays contained 0.05 mg of protein, and skin assays contained approximately 2.0 mg of protein; both were incubated for 5 min at 37°C. Conjugation of CDNB with glutathione was determined at four concentrations ranging from 0.08 to 0.64 m*M*. Values from three to four specimens at each substrate concentration were averaged and kinetic parameters [(V_{max} and K_m (Michaelis constant)] were derived with Lineweaver-Burk plots.

[a] Units are as follows: V_{max} (nm/min/mg) and K_m (m*M*).

ED activity of microsomal fractions, it was possible to derive kinetic parameters for this enzyme. The maximum velocity (V_{max}) of liver in either strain of mice is about twice that found in rats, and the V_{max} of skin in either strain of mice is more than five times greater than it is in rats. No interspecies differences in the affinity of this enzyme for chlorodinitrobenzene was observed.

This study demonstrates a markedly greater metabolic capacity of mouse skin compared with that of rat skin. Activity of two Phase I enzymes, AHH and ED, is several times higher in BALB/c and Sencar mice than in Fuzzy and Osborne-Mendel rats. Also, the V_{max} of the Phase II enzyme GST was several times higher in both mouse strains than in either rat strain.

The overall metabolizing capacity of skin was less than that of liver by nearly two orders of magnitude. The AHH activity in skin is only 0.3% (BALB/c), 0.6% (Sencar), 0.8% (Osborne-Mendel), or 1.0% (Fuzzy) of its activity in liver. ED activity in skin is only 0.2% (BALB/c), 0.3% (Sencar), or 0.1% (Osborne-Mendel and Fuzzy) of its activity in liver. Similarly, the V_{max} of GST in skin is only 1% (rats) or 3% (mice) of its V_{max} in liver.

The microsomal metabolism of caffeine, butylated hydroxytoluene (BHT), acetylethyl tetramethyl tetralin (AETT), and salicyclic acid was also studied in the skin and livers of Fuzzy rats.[2] The results of these determinations and other values for skin and liver activities from published studies are given in Table 2. A similar pattern is seen in the relative ratios of skin and liver activities; that is, skin activity may be as much as one or two orders of magnitude less than liver activity.

Enzyme activity is lower in skin than in liver, possibly because microsomal protein concentration is tenfold lower in skin than in the livers of rats.[24] Even if a compound is metabolized extensively after systemic administration, as are some of the compounds listed in Table 2, it may still undergo little or no biotransformation during passage through the skin.

TABLE 2
Cutaneous and Systemic Metabolism in the
Rat

	Activity of microsomal enzymes	
Compound	Liver	Skin (pmol/min/mg protein)
Caffeine	4.1	N.D.
BHT	17,000	113
AETT	667	2.5
Salicylic acid	15.3	N.D.

Other compounds metabolized during percutaneous
absorption

Benzo(a)pyrene	1,670[19]	4.7[20]
Testosterone	4,200[21]	10[22]
Estradiol	16,840[23]	25[22]

Note: N.D. = Not detected.

III. DETERMINATION OF SKIN METABOLISM DURING *IN VITRO* PERCUTANEOUS ABSORPTION STUDIES

A. MAINTENANCE OF SKIN VIABILITY

The purpose of biological experimentation is to gain information on the functioning of a specimen by manipulating environmental variables and observing the responses. Because *in vitro* studies inevitably invite comparison with the situation *in vivo,* the cell or tissue being studied must function normally for valid conclusions. Logically, the first step in validating an *in vitro* system must be to determine the integrity and viability of the tissue being studied. The techniques for assessment of viability are applicable to tissues of most types and origins. From a medical standpoint, a tissue-specific viability assay for a skin section might be its ability to be successfully grafted.[25] Suggested parameters to determine the viability of skin in *in vitro* tests[26-29] include:

- Mitotic figures
- Cellular activity
- Transformation *in vitro*
- Vital dye exclusion
- "Take" and survival time
- Autoradiography
- pH changes *in vitro*
- Hair growth
- Pigment changes
- Ultrastructural changes
- Changes in histological tinctorial properties
- Respiration quotients
- Enzyme activity
- Glucose uptake and utilization

The scope of many of these methods overlap. Taylor suggested that a battery of tests be used to gather as many data as possible to make an assessment of tissue viability.[30] These types of tests can be broken down into three broad categories: (1) operations, (2) biochemistry, and (3) histology. These categories address, respectively, the overall function of skin as an organ, the specific biochemical processes in the cell, and the histopathological appearance of the skin specimen as compared with a fresh specimen.

Percutaneous absorption and metabolism studies are conducted for finite durations, and therefore active multiplication of the perfused tissue is not necessary for most investigations. The high concentrations of amino acids, vitamins, and lipids are occasionally sources of interferences in analytical assays for parent compound and metabolites. To develop a simplified medium to function as a receptor fluid in flow-through diffusion cell studies that would maintain skin viability and not interfere in most analytical procedures, we prepared a series of physiological buffer solutions. We then perfused the solutions under freshly obtained dermatome sections of Fuzzy rat skin in flow-through diffusion cells. Viability was determined by the ability of the skin sections to maintain aerobic and anaerobic glucose utilization at initial rates and to maintain metabolism of topically applied steroids. The histological appearance of the skin after 24 h of perfusion was also used to assess viability.[31] We tested minimal essential media (MEM) with: (1) a 10% fetal bovine serum (FBS) supplement, (2) a Dulbecco's modified phosphate-buffered saline (DMPBS), (3) a Hanks buffered saline solution (BSS) with 0.1% glucose (25 mM) buffered with N-hydroxyethyl-piperazine-N'-2-ethanesulfonic acid (Hepes) to yield a Hepes-buffered Hanks balanced salt solution (HHBSS), and (4) a phosphate-buffered saline solution with glucose (PBSG) for its ability to maintain skin section viability. MEM ($\pm$FBS), DMPBS, and HHBSS are able to sustain glucose utilization, steroid metabolism, and histological appearance for 24 h.

The receptor fluids MEM, DMPBS, and HHBSS are able to sustain viability of thin dermatome skin sections in flow-through diffusion cells.[31] Other studies have used physiological receptor fluids, such as medium 1640[32] or Tyrode's solution,[33] to maintain skin viability. The use of thick skin preparations is not only undesirable for the study of percutaneous diffusion, but also might affect viability by limiting diffusion of perfusate and dissolved oxygen to the viable epidermis. Skin uses most glucose anaerobically, producing lactate. Maintaining viability in static cells is complicated by the cumulative depletion of glucose from the receptor fluid and by the buffer, which must have the capacity to maintain physiological pH. The use of PBSG or distilled water does not maintain skin viability in flow-through diffusion cells and would not be expected to do so in static cells.

Serum use might be justified in the study of lipophilic compound penetration and metabolism as a means of providing a more lipophilic receptor fluid; however, the variability of serum and its intrinsic ability to metabolize some compounds is of concern. The use of bovine serum albumin for this purpose instead of whole serum may represent a less costly and better-defined alternative.

The determination of metabolism in skin is predicated on the assumption that viable tissue is used in the study. Investigators have used skin in organ cultures for measurements of metabolism during percutaneous absorption.[1,2,9-12,15] The condition of the skin during the experiment remains a preoccupation with investigators correlating *in vitro* skin metabolism results to the *in vivo* situation. Receptor fluids such as distilled water, normal saline, phosphate-buffered saline, and organic solvent solutions are not compatible with the continued metabolic functioning of freshly prepared skin. In static diffusion cells with physiological receptor solutions, skin loses viability quickly if the receptor fluid is not frequently replaced.[34] An incubation study of skin sections demonstrates how skin rapidly loses it glycolytic ability.

Dermatome sections of hairless guinea pig skin (200 μm thick, 14 mm in diameter) were floated in 3 ml of HHBSS or HHBSS with 1.5% of poloxamer 188 or poloxamer 338. The HHBSS solutions contained 1 g of dextrose per liter (5.6 mmol). Initial rates of glycolysis

TABLE 3

Percutaneous Absorption and Metabolism in Viable Fuzzy Rat Skin

	Receptor fluid		Skin	
Compound[a]	Radioactivity absorbed (%)[b]	% Metabolized[c]	Radioactivity absorbed (%)[b]	% Metabolized[c]
DDT (4)	0.59 ± 0.2	0	48.3 ± 3.0	0
AETT (8)	1.8 ± 0.2	15.0 ± 2.4	30.8 ± 3.9	1.1 ± 0.3
Caffeine (4)	30.9 ± 2.4	0	2.1 ± 0.2	0
BHT (4)	2.3 ± 0.1	26.8 ± 0.2	11.1 ± 0.9	2.4 ± 0.2
Salicylic acid (4)	12.2 ± 0.8	0	7.7 ± 1.0	0

[a]   Values are the mean ± SE of the number of determinations in parentheses.
[b]   Absorption is expressed as the percentage of the applied dose penetrating skin.
[c]   Metabolites were measured in ethyl acetate extracts of receptor fluid and skin homogenates. Values are expressed as the percentage of absorbed compound metabolized.

were unchanged when 1.5% surfactant was added. Lactate production ceased after 2.5 to 3 μmol of lactate accumulated in the medium (after approximately 12 h of incubation) for all receptor fluids. This volume of accumulated lactate represents a 9% utilization of the total available glucose in the incubation medium. For studies of long duration using viable skin in static diffusion cells, one must periodically replace diffusion cell contents. The use of flow-through diffusion cells eliminates this problem by continually replacing receptor fluid. The skin flap system described by Riviere[13] can be successfully configured to use a recirculating receptor fluid with a reservoir of suitable size and with proper monitoring and adjustment of receptor fluid glucose concentrations.

B. ABSORPTION AND METABOLISM STUDIES

We performed initial absorption and metabolism studies in our laboratory with Fuzzy rat skin, using a group of lipophilic compounds known to be extensively metabolized after systemic administration (Table 3).[2] Absorption values are presented for both receptor fluids and skin in Table 3 because, for water-insoluble compounds, absorbed material does not partition freely from skin into the receptor fluid. Total percutaneous absorption is therefore determined by summing absorbed compounds in both the skin and receptor fluid.

Measurable metabolism in skin was observed for only two of the test compounds, BHT and AETT. For both compounds, the relatively small amount of absorbed material found in the receptor fluid contained the highest percentage of metabolites. However, when the absorbed but unmetabolized material in skin was accounted for, only 1.9% of the absorbed AETT and 6.6% of the absorbed BHT were metabolized. Cutaneous and systemic microsomal enzyme activity were higher for these test compounds than for other compounds in the study (Table 2). It may be possible to predict the extent of skin metabolism during absorption if the liver microsomal activity of the compound is known.

Other lipophilic compounds metabolized by microsomal enzymes formed the basis for comparing animal and human skin absorption and metabolism.[15] Benzo(a)pyrene (BP) is extensively absorbed through skin in a 24-h period after application from an acetone vehicle. Absorption ranged from 30% in human skin to 40 to 60% absorption through mouse, rat, and guinea pig skin (Figure 1). Most of the absorbed material remained in the dermatome sections of skin assembled in the diffusion cells. Small amounts of more polar metabolites were formed during percutaneous absorption and readily diffused into the receptor fluid. Metabolites were divided into classes based on thin-layer chromatography (TLC) separations: aqueous, polar, diol, hydroxy, and quinone. Over 24 h, only about 6%, or 0.7 nmol of BP

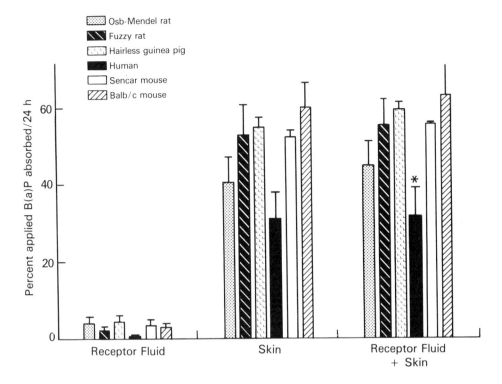

FIGURE 1. Percentage of applied benzo(a)pyrene (BP) absorbed in 24 h by skin in flow-through diffusion cells. Applied dose was 3 μg of BP per square centimeter of tissue in 10 μl of acetone. Values from two to five skin specimens from each experimental subject were averaged and constituted one experiment. Values are means ± SE of three to four experiments. Rat, guinea pig, and human skin were 200-μm dermatome sections; mouse skin was full thickness. The asterisk indicates a value significantly different from those for hairless guinea pig, Fuzzy rat, and BALB/c mouse ($p < 0.05$) and different from Sencar mouse ($p < 0.10$).

per square centimeter, was metabolized by full-thickness Sencar mouse skin and about 3%, or 0.3 nmol of BP per square centimeter, was metabolized by hairless guinea pig dermatomed skin (Figure 2). BP was less consistently metabolized in viable human skin in HHBSS receptor fluid than in skin in distilled water–10% FBS control solutions, although metabolism may have occurred at a rate below the limit of measurement of our TLC assay (about 6 pmol of BP per square centimeter metabolized in 24 h.

7-Ethoxycoumarin (7-EC) was also absorbed readily through human and rodent skin (Figure 3). The compound and its metabolite, 7-hydroxycoumarin, are so water soluble that essentially all of the absorbed material is found in the receptor fluid. When 24-h absorption values are compared, values are similar for human and rodent skin. A small percentage of the absorbed 7-EC was metabolized; it ranged from less than 0.05% (human dermatome skin) to about 1.3% of the absorbed dose in full-thickness Sencar mouse skin (Figure 3).

The percentage of nitroglycerin (GTN) metabolized during percutaneous absorption through rat skin[35] was greater than the percentage of substrates metabolized by microsomal enzyme activity reported earlier. GTN is metabolized by the soluble enzyme GST to 1,2-glyceryl dinitrate and 1,3-glyceryl dinitrate. Some decomposition of GTN occurred during absorption in diffusion cells through skin made nonviable with a phosphate-buffered saline receptor fluid. However, substantially more metabolites were formed when MEM was used as the receptor fluid to maintain viability of skin. Over the 24-h absorption period, 71% of the nitrates penetrating the MEM-perfused skin were present as GTN and 29% were present as dinitrate metabolites.

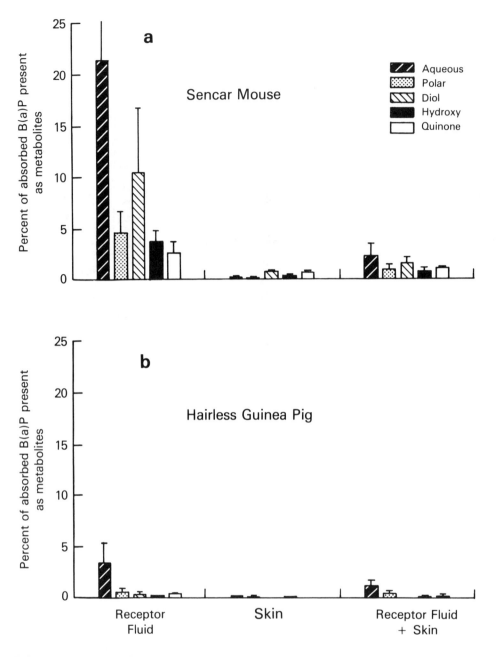

FIGURE 2.   Percnetage of absorbed benzo(a)pyrene (BP) metabolized by Sencar mouse (a) and hairless guinea pig (b) skin in 24 h in flow-through diffusion cells. Values are mean ± SE of four or five determinations. Treatment and skin specimens are as described in Figure 1.

Substantial acetyltransferase activity has also been reported in skin during percutaneous absorption in hairless guinea pig and human skin.[36] The absorption and metabolism of benzoic acid and two primary amine-containing derivatives, *p*-aminobenzoic acid (PABA) and its ethyl ester (benzocaine), were studied in viable skin *in vitro*. Dermatome sections of hairless guinea pig and human skin were placed in flow-through diffusion cells and kept viable (with HHBSS receptor fluid) or made nonviable (distilled water receptor fluid) for control studies.

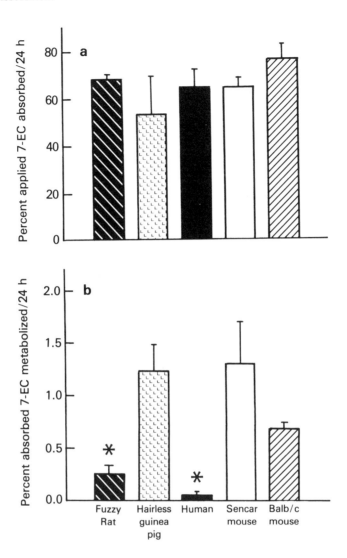

FIGURE 3.　Percentage of applied 7-ethoxycoumarin (7-EC) absorbed in 24 h (a) and percentage of absorbed 7-EC metabolized in 24 h (b) by skin in flow-through diffusion cells. Applied dose was 5 to 16 μg per square centimeter in 10 μl of acetone. Values from two to four skin specimens from each experimental subject were averaged and constituted one experiment. Values are means ± SE of two or three experiments. Rat, guinea pig, and human skin were 200-μm dermatome sections; mouse skin was full thickness. Asterisks indicate that values are significantly different from those for hairless guinea pig and Sencar mouse (LSD test, $p < 0.05$).

Benzocaine was absorbed rapidly through hairless guinea pig skin with most absorption occurring within the first 6 h after application in an ethanol vehicle (Figure 4). PABA absorption occurred slowly over a 48-h period. The two acidic compounds, benzoic acid and PABA, were absorbed at different rates through viable and nonviable skin. A decrease in pH of the nonviable skin would explain the more rapid absorption through this tissue. Total benzocaine absorption (expressed as percentage of the applied dose) was less in human skin (48%) than in hairless guinea pig skin (77%).

Benzoic acid that was absorbed by skin was partially converted to the glycine conjugate hippuric acid (Table 4). The receptor fluid contained essentially all the hippuric acid formed, which was equal to 6.9% of the absorbed radioactivity. The lack of hippuric acid formation

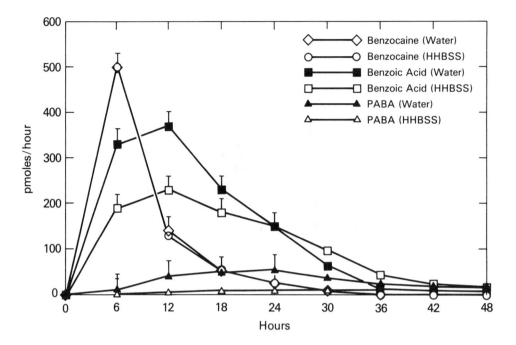

FIGURE 4. Percutaneous absorption rates in viable (Hepes-buffered Hanks balanced salt solution receptor fluid) and nonviable (water receptor fluid) skin. Values are the mean ± SE of four determinations in each of three animals.

in the distilled water control cells confirmed the enzymatic nature of the conjugation reaction. PABA was substantially biotransformed in skin to acetyl-PABA by acetylation of the primary amino group. Although the receptor contained only about 20% of the absorbed PABA, 60.7% of this material was identified as the acetylated metabolite (Table 5).

The metabolism of benzocaine was of particular interest because of two important potential sites for skin metabolism. Acetylation of the primary amine was expected to occur as was cleavage of the ethyl ester by esterases in the skin. Acetylation was the primary metabolic pathway for benzocaine in both hairless guinea pig and human skin (Table 6). Essentially all of the absorbed radioactivity was found in the receptor fluid; acetylbenzocaine made up 82% of the absorption products of hairless guinea pig skin and 57% of those of human skin. Much smaller quantities of the esterase products PABA and acetyl-PABA were present. Preliminary studies show that benzocaine is not a good substrate for skin esterase and that more lipophilic ethylbenzoate esters are hydrolyzed at faster rates.[37]

Recently, the percutaneous absorption and metabolism of phenanthrene have been studied in the hairless guinea pig.[38] Good agreement was obtained between absorption values from *in vivo* studies (80% applied dose absorbed) and values from viable skin in flow-through diffusion cells (89.7%). Phenanthrene was metabolized *in vitro* to phenanthrene 9,10-dihydrodiol, 3,4-dihydrodiol, and 1,2-dihydrodiol, and to traces of hydroxyphenanthrenes. No Bay-region 3,4-dihydrodiol epoxide or its rearranged product was found, which is consistent with the negative cancer bioassay data for phenanthrene reported in the literature. Based on the total radioactivity *in vitro,* the absorbed material contained 92% parent compound and 7% dihydrodiol metabolites.

TABLE 4
Benzoic Acid Metabolism in Hairless
Guinea Pig Skin[a]

Location and compound	HHBSS	Water (control)
Receptor fluid		
Benzoic acid	73.8 ± 8.0	92.3 ± 1.7
Hippuric acid	6.9 ± 3.4	0.1 ± 0.1
Polar	11.6 ± 4.1	2.5 ± 1.3
Skin		
Benzoic acid	83.2 ± 2.9	84.2 ± 7.2
Hippuric acid	0.6 ± 0.6	0.0 ± 0.0
Polar	3.8 ± 0.8	8.1 ± 1.7

[a] Percentage of absorbed dose. Values are the mean ± SE of four determinations in each of three animals.

TABLE 5
p-Aminobenzoic Acid (PABA)
Metabolism in the Hairless Guinea
Pig[a]

Location and compound	HHBSS	Water
Receptor fluid		
PABA	24.6 ± 8.1	93.0 ± 2.4
Acetyl-PABA	60.7 ± 7.7	1.6 ± 0.4
Skin		
PABA	86.0 ± 7.7	83.9 ± 2.2
Acetyl-PABA	6.7 ± 1.1	6.2 ± 0.6

[a] Values are the percentage of the absorbed dose and are the mean ± SE of four determinations in each of three animals.

IV. CONCLUSIONS

In vitro diffusion cell methods are valuable for skin absorption and metabolism studies. The ability to isolate viable skin in an *in vitro* system helps investigators quantify absorption and biotransformation products. This ability is particularly valuable for metabolism studies because metabolism in skin often occurs at a lower rate than in major metabolic organs, such as the liver. Results from *in vivo* skin metabolism studies are often difficult to interpret because of interference from systemic enzyme activity.

TABLE 6
Benzocaine Metabolism in Human and Hairless Guinea Pig Skin[a]

Location and compound	Hairless guinea pig			Human HHBSS
	HHBSS	Water (control)	HHBSS + inhibitor[b]	
Receptor fluid				
Benzocaine	4.4 ± 1.7	66.8 ± 5.2	89.4 ± 1.1	27.7 ± 12.3
Acetyl-benzocaine	82.4 ± 1.1	7.2 ± 1.2	0	56.7 ± 12.2
PABA	0.2 ± 0.1	5.4 ± 0.4	6.1 ± 1.1	1.5 ± 0.7
Acetyl-PABA	6.2 ± 0.7	7.7 ± 0.3	0	11.4 ± 2.9
Polar	2.5 ± 1.0	6.0 ± 0.5	2.0 ± 0.1	1.2 ± 0.3
Skin				
Benzocaine	25.8 ± 3.1	30.0 ± 1.4	62.1 ± 2.6	26.4 ± 3.0
Acetyl-benzocaine	31.7 ± 4.3	20.5 ± 3.3	0	47.7 ± 5.2
PABA	5.5 ± 0.7	12.7 ± 0.6	8.7 ± 3.1	2.3 ± 1.3
Acetyl-PABA	9.7 ± 1.1	8.1 ± 0.9	0	7.3 ± 1.9
Polar	14.6 ± 1.7	16.6 ± 0.8	7.5 ± 1.0	4.7 ± 0.5

[a] Values are the mean ± SE of four determinations in each of three animals. The inhibitor and human experiment represent four determinations in one subject.

[b] Skin in diffusion cells pretreated for 4 h with 0.1 M iodoacetamide in HHBSS; then HHBSS only used for absorption and metabolism measurements.

REFERENCES

1. **Kao, J., Patterson, F. K., and Hall, J.,** Skin penetration and metabolism of topically applied chemicals in six mammalian species, including man: an *in vitro* study with benzo[a]pyrene and testosterone, *Toxicol. Appl. Pharmacol.,* 81, 502, 1985.
2. **Bronaugh, R. L., Stewart, R. F., and Strom, J. E.,** Extent of cutaneous metabolism during percutaneous absorption of xenobiotics, *Toxicol. Appl. Pharmacol.,* 99, 534, 1989.
3. **Bickers, D. R., Dutta-Choudhury, T., and Mukhtar, H.,** Epidermis: a site of drug metabolism in neonatal rat skin. Studies on cytochrome P-450 content and mixed-function oxidase and epoxide hydrolase activity, *Mol. Pharmaocl.,* 21, 239, 1982.
4. **Andersson, P., Edsbäcker, S., Ryrfeldt, Å, and Von Bahr, C.,** *In vitro* biotransformation of glucocorticoids in liver and skin homogenate fraction from man, rat, and hairless mouse, *J. Steroid Biochem.,* 16, 787, 1982.
5. **Cheung, Y. W., Li Wan Po, A., and Irwin, W. J.,** Cutaneous biotransformation as a parameter in the modulation of the activity of topical corticosteroids, *Int. J. Pharm.,* 26, 175, 1985.
6. **Kulkarni, A. P., Nelson, J. L., and Radulovic, L. L.,** Partial purification and some biochemical properties of neonatal rat cutaneous glutathione S-transferases, *Comp. Biochem. Physiol.,* 87B, 1005, 1987.
7. **Coomes, M. W., Norling, A. H., Pohl, R. J., Müller, D., and Fouts, J. R.,** Foreign compound metabolism by isolated skin cells from the hairless mouse, *J. Exp. Pharmacol. Ther.,* 225, 770, 1983.
8. **Pham, M., Magdalou, J., Siest, G., Lenoir, M., Bernard, B. A., Jamoulle, J., and Shroot, B.,** Reconstituted epidermis: a novel model for the study of drug metabolism in human epidermis, *J. Invest. Dermatol.,* 94, 749, 1990.
9. **Holland, J. M., Kao, J. Y., and Whitaker, M. J.,** A multisample apparatus for kinetic evaluation of skin penetration *in vitro:* the influence of viability and metabolic status of the skin, *Toxicol. Appl. Pharmacol.,* 72, 272, 1984.
10. **Kao, J., Hall, J., and Holland, J. M.,** Quantitation of cutaneous toxicity: an *in vitro* approach using skin organ culture, *Toxicol. Appl. Pharmacol.,* 68, 206, 1983.
11. **Kao, J., Hall, J., Shugart, L. R., and Holland, J. M.,** An *in vitro* approach to studying cutaneous metabolism and disposition of topically applied xenobiotics, *Toxicol. Appl. Pharmacol.,* 75, 289, 1984.

12. **Kao, J. and Hall, J.**, Skin absorption and cutaneous first pass metabolism of topical steroids: *in vitro* studies with mouse skin in organ culture, *J. Pharmacol. Exp. Ther.*, 241, 482, 1987.

13. **Riviere, J. E.**, Biological factors in absorption and permeation, *Cosmet. Toiletries*, 105(10), 85, 1990.

14. **Storm, J. E., Stewart, R. F., and Bronaugh, R. L.**, Interstrain and species differences in xenobiotic metabolizing capacity in skin: evidence of enhanced activity in SENCAR mice, *Toxicologist*, 8, 126, 1988.

15. **Storm, J. E., Collier, S. W., Stewart, R. F., and Bronaugh, R. L.**, Metabolism of xenobiotics during percutaneous penetration: role of absorption rate and cutaneous enzyme activity, *Fundam. Appl. Toxicol.*, 15, 132, 1990.

16. **Mukhtar, H. and Bickers, D. R.**, Age related changes in benzo(a)pyrene metabolism and epoxide-metabolizing enzyme activities in rat skin, *Drug Metab. Dispos.*, 11, 562, 1983.

17. **Rettie, A. E., Williams, F. M., Rawlins, M. D., Mayer, R. T., and Burke, M. D.**, Major differences between lung, skin, and liver in the microsomal metabolism of homologous series of resorufin and coumarin ethers, *Biochem. Pharmacol.*, 35, 3495, 1986.

18. **Habig, W. H., Pabst, M. J., and Jakoby, W. B.**, Glutathione-S-transferase. The first enzymatic step in mercapturic acid formation, *J. Biol. Chem.*, 249, 7130, 1974.

19. **Van Cantfort, J., DeGraeve, J., and Gielen, J. E.**, Radioactive assay for aryl hydrocarbon hydroxylase. Improved method and biological significance, *Biochem. Biophys. Res. Commun.*, 79, 505, 1977.

20. **Bickers, D. R., Mukhtar, H., and Yang, S. K.**, Cutaneous metabolism of benzo(a)pyrene: comparative studies in C57BL/6N and DBA/2N mice and neonatal Sprague-Dawley rats, *Chem. Biol. Interact.*, 43, 263, 1983.

21. **Cheng, K. C. and Schenkman, J. B.**, Testosterone metabolism by cytochrome P-450 isozymes RLM$_3$ and RLM$_5$ and by microsomes, *J. Biol. Chem.*, 258, 11738, 1983.

22. **Davis, B. P., Rampini, E., and Hsia, S. L.**, 17β-Hydroxysteroid dehydrogenase of rat skin, *J. Biol. Chem.*, 247, 1407, 1972.

23. **Cheng, K. C. and Schenkman, J. B.**, Metabolism of progesterone and estradiol by microsomes and purified cytochrome P-450 RLM$_3$ and RLM$_5$, *Drug Metab. Dispos.*, 12, 222, 1984.

24. **Bickers, D. R.**, The skin as a site of drug and chemical metabolism, in *Current Concepts in Cutaneous Toxicity*, Drill, V. A. and Lazar, P., Eds., Academic Press, New York, 1980, 95.

25. **Merrell, S. W., Shelby, J., Saffle, J. R., Kruegar, C. G., and Navar, P. D.**, An *in vivo* test of viability for cryopreserved skin, *Curr. Surg.*, 43, 296, 1986.

26. **Perry, V. P.**, A review of skin preservation, *Cryobiology*, 3, 109, 1966.

27. **Malinin, T. I. and Perry, V. P.**, A review of tissue and organ viability assay, *Cryobiology*, 4, 104, 1967.

28. **May, S. R. and DeClement, F. A.**, Skin banking. III. Cadaveric allograft skin viability, *J. Burn Care Rehabil.*, 2, 128, 1981.

29. **Jensen, H. S.**, Skin viability studies *in vitro*, *Scand. J. Plast. Reconstr. Surg.*, 18, 55, 1984.

30. **Taylor, A. C.**, Cryopreservation of skin: discussion and comments, *Cryobiology*, 3, 192, 1966.

31. **Collier, S. W., Sheikh, N. M., Sakr, A., Lichtin, J. L., Stewart, R. F., and Bronaugh, R. L.**, Maintenance of skin viability during *in vitro* percutaneous absorption/metabolism studies, *Toxicol. Appl. Pharmacol.*, 99, 522, 1989.

32. **Hawkins, G. S. and Reifenrath, W. G.**, Influence of skin source, penetration cell fluid, and partition coefficient on *in vitro* skin penetration, *J. Pharm. Sci.*, 75, 378, 1986.

33. **Smith, L. H. and Holland, J. M.**, Interaction between benzo[a]pyrene and mouse skin in organ culture, *Toxicology*, 21, 47, 1981.

34. **Macpherson, S. E., Scott, R. C., and Williams, F. M.**, Metabolism of pesticides during percutaneous absorption *in vitro*, in *Prediction of Percutaneous Penetration: Methods, Measurements, Modelling*, Scott, R. C., Guy, R. H., and Hadgraft, J., Eds., IBC, London, 1990, 135.

35. **Storm, J. E., Bronaugh, R. L., Carlin, A. S., and Simmons, J. E.**, Cutaneous metabolism of nitroglycerin in viable rat skin *in vitro*, *Int. J. Pharm.*, 65, 265, 1990.

36. **Nathan, D., Sakr, A., Lichtin, J. L., and Bronaugh, R. L.**, *In vitro* skin absorption and metabolism of benzoic acid, *p*-aminobenzoic acid, and benzocaine in the hairless guinea pig, *Pharm. Res.*, 7, 1147, 1990.

37. **Wu, S. T., Shiu, G., Skelly, J., Stewart, R., and Bronaugh, R. L.**, Skin metabolism of xenobiotics: *in vitro* esterase hydrolysis of ethyl benzoate and its analogs, *Pharm. Res.*, 7 (Suppl.), S-263, 1990.

38. **Ng, K. M. E., Chu, I., Bronaugh, R. L., Franklin, C. A., and Somers, D. A.**, Percutaneous absorption/metabolism of phenanthrene — comparison of *in vitro* and *in vivo* results, *Fundam. Appl. Toxicol.*, 16, 517, 1991.

Chapter 6

ANIMAL MODELS FOR PERCUTANEOUS ABSORPTION*

Ronald C. Wester and Howard I. Maibach

TABLE OF CONTENTS

* Reported with permission from *Cutaneous Bioavailability.*

I. INTRODUCTION

The ideal way to determine the percutaneous absorption of a compound in man is to do the actual study in humans. Mechanism and parameters of percutaneous absorption elucidated *in vivo* with human skin are most relevant to the clinical situation. However, many compounds are potentially too toxic to test *in vivo* in humans, and so their percutaneous absorption must be tested in animals. Likewise, until more complete animal to human validation studies become available, not all investigators will have access to human volunteers. Mechanism studies and studies on factors affecting absorption must, therefore, be explored using animals and *in vitro* technqiues. We have two basic criteria to judge whether an animal model is good. First, the animal model gives the *same* percutaneous absorption as that in man. If this is not possible, then the animal model should be *consistently different* from that in man. Therefore, this chapter stresses studies where comparative studies between an animal model (or the skin from that animal) is compared directly to man.

A. NONHUMAN PRIMATES

Table 1 summarizes the comparative data on *in vivo* percutaneous absorption between the rhesus monkey and man. Wester and co-workers[1-8] first showed that percutaneous absorption for a variety of compounds is similar for the rhesus monkey and man. Other researchers[9,10] have made the same conclusion. The rhesus monkey has certain physical advantages in that the inner parts of the arms, legs, and trunk are relatively hairless; similar to that in man. Also, since percutaneous absorption differs from regions of the body,[11] the same anatomic site can be used in both the rhesus and man (i.e., ventral forearm). Finally, the rhesus monkey is sufficiently large such that serial blood samples can be obtained to determine chemical blood concentrations and pharmacokinetic parameters. This is the most relevant animal model for percutaneous absorption. An obvious disadvantage of the rhesus monkey is its availability and cost.

Table 2 gives the data by Bartek et al.[13,14] comparing percutaneous absorption in the squirrel monkey to man. For the limited number of compounds studied (three), absorption is consistently two to three times that in man. Advantages and disadvantages of the squirrel monkey would be similar to that of the rhesus monkey.

B. PIG

Percutaneous absorption studies with the pig usually mean the miniature pig. *In vivo* comparison between the pig and man was done by Bartek et al.[13,14] for several compounds (Table 3). Percutaneous absorption in the pig ranged from zero to four times that in man. Absorption in the pig distinguishes between high- and low-absorbing compounds in a fashion similar to man. In this study the back of the pig was used and compared to the ventral forearm of man (potential regional difference). The pig has a potential disadvantage of a concentrated subcutaneous body fat. It is conceivable that a lipid-soluble compound upon absorption would concentrate in the fat rather than the central compartment where biological fluid would be sampled (blood, urine).

One way to take advantage of the permeability characteristics of pig skin (relative to man) and avoid the systemic fat distribution probelm is to use pig skin in an *in vitro* situation. Roberts and Mueller[15] compared nitroglycerin transdermal flux *in vitro* in skin from Yucutan pig, hairless mouse, and human. Nitroglycerin delivery from Yucatan pig skin was closer to that in human skin than that from hairless rat.

Another method developed to take advantage of pig skin permeability characteristics is that of a skin flap. The methodology is to surgically isolate a portion of skin (pig) so that a singular blood supply is created to collect blood containing the chemical that has been

TABLE 1
In Vivo **Percutaneous Absorption in Rhesus Monkey and Man**

Compound	Percent dose absorbed		Difference
	Rhesus monkey	Man	
2,4-Dinitrochlorobenzene[a]	52 ± 4	54 ± 6	0
Nitrobenzene[a]	4 ± 1	2 ± 1	×2
Cortisone[a]	5 ± 3	3 ± 2	×2
Testosterone[a]	18 ± 10	13 ± 3	0
Hydrocortisone[a]	3 ± 1	2 ± 2	0
Benzoic acid[a]	60 ± 8	43 ± 16	0
Diethyl maleate[b]	68 ± 7	54 ± 7	0
DDT[c]	19 ± 9	10 ± 4	×2
Retinoic acid[d]	2 ± 1	1 ± 0.2	×2

[a]   Wester et al.[1-8]
[b]   Bronaugh et al.[9]
[c]   Wester et al.[12]
[d]   Franz and Lehman.[10]

TABLE 2
In Vivo **Percutaneous Absorption in Squirrel Monkey and Man**[13,14]

Compound	Percent dose absorbed		Difference
	Squirrel monkey	Man	
Lindane	16 ± 11	9 ± 4	×2
Parathion	30 ± 10	10 ± 6	×3
Malathion	19 ± 2	8 ± 3	×2

TABLE 3
In Vivo **Percutaneous Absorption in Pig and Man**[13,14]

Compound	Percent dose absorbed		Difference
	Pig	Man	
Haloprogin	20 ± 10	11 ± 4	×2
N-acetylcysteine	6 ± 0	2 ± 2	×3
Testosterone	29 ± 9	13 ± 3	×2
Cortisone	4 ± 2	3 ± 2	0
Caffeine	32 ± 5	47 ± 20	0
Butter yellow	42 ± 10	22 ± 5	×2
DDT	43 ± 8	10 ± 4	×4
Lindane	38 ± 3	9 ± 4	×4
Parathion	14 ± 1	10 ± 6	0
Malathion	15 ± 2	8 ± 3	×2

TABLE 4

Comparative Percutaneous Absorption of IPPSF *In Vitro* Model and Man *In Vivo*

	Percent dose absorbed				
	IPPSF model				
Compound	Skin	Perfusate	Skin + perfusate	Man *in vivo*	Difference[c]
Salicylic acid	7.1 ± 2.3	0.4 ± 0.4	7.5 ± 2.6	6.5 ± 5.0	0
Theophylline	11.2 ± 3.6	0.6 ± 0.8	11.8 ± 3.8	16.9 ± 11.3	×2
PABA[a]	5.2 ± 3.4	0.7 ± 0.5	5.9 ± 3.7	21.9 ± 17.6	×4
2,4-Damine[b]	2.4 ± 0.5	1.5 ± 0.2	3.8 ± 0.6	2.7 ± 1.0	0

[a]   Para amino benzoic acid.
[b]   2,4-Dimethylamine.
[c]   Difference of man and skin + perfusate in IPPSF.

absorbed through skin. The skin flap can be used to study percutaneous absorption *in vitro*. The absorption of chemicals through skin and metabolism within the skin can be determined by assay of the perfusate. The isolated perfused porcine skin flap (IPPSF) is an alternative *in vitro* animal model used by Rivier et al.[16,17] A study directly compared absorption in the IPPSF model and that in man. For each compound the same dose, vehicle, and skin concentration were used. The results are in Table 4. Compared to man *in vivo*, absorption in IPPSF perfusate was not predictive of that in man. When skin content is added to perfusate, the results range from zero to four times difference from that in man. This is the same range as that seen with pig *in vivo* (Table 3). The results suggest that partitioning of compound from vehicle/surface may be predictive of that in man, but that perfusate accumulation is limiting. Perfusate accumulation limitation is also prevalent with other types of *in vitro* diffusion.[18]

C. RAT

Percutaneous absorption is the primary route of exposure to pesticide users. Percutaneous absorption studies in man would be a logical procedure to determine the most relevant data. However, regulatory agencies traditionally desire animal data. The choice of experimental animal has become the rat. This choice was made predominantly because of availability of the species.[19] However, percutaneous absorption in the rat is often greater than in man, and the difference in percutaneous absorption between rat and man is not consistent from compound to compound.[20] Table 5 summarizes *in vivo* percutaneous absorption in rat and man.[13,14,18,21,23,24] The difference in absorption of rat ranges from zero to nine times that of man, and the difference is not consistent between the two species. Table 6 gives additional data on comparative absorption in rat and rhesus monkey (and man where the data exists for all three species). Percutaneous absorption in the rat is much higher than that in the rhesus monkey and not with a constant difference between the two species.[25,26]

D. RABBIT

Table 7 summarizes the data on comparative *in vivo* percutaneous absorption between rabbit and man.[13,14] For most compounds, absorption in the rabbit is much higher than in man, and the difference between the two species is not consistently different (zero-to-ten-times range).

E. GUINEA PIG

Table 8 compares the *in vivo* percutaneous absorption of cortisone in guinea pig with that in Rhesus monkey and man.[8] The data indicate that skin absorption, at least for cortisone,

TABLE 5
In Vivo Percutaneous Absorption in Rat and Man

Compound	Percent dose absorbed		Difference
	Rat	**Man**	
Haloprogin[a]	96 ± 14	11 ± 4	×9
N-Acetylcystein[a]	4 ± 4	2 ± 2	×2
Testosterone[a]	47 ± 3	13 ± 3	×4
Cortisone[a]	25 ± 4	4 ± 2	×6
Caffeine[a]	53 ± 12	48 ± 21	0
Butter yellow[a]	48 ± 2	22 ± 5	×2
Trichlorocarbanilide[b]	16 ± 9	7 ± 3	×2
Lindane[c]	31 ± 10	9 ± 4	×3
Parathion[d]	95 ± 3	46 ± 5	×2

[a]   Bartek and La Budde.[13,14]
[b]   Northroot et al.[21] and Wester et al.[18]
[c]   Moody and Ritter,[22] and Feldman and Maibach.[23]
[d]   Shah et al.[24] and Feldmann and Maibach.[23]

TABLE 6
In Vivo Percutaneous Absorption in Rat, Rhesus Monkey, and Man

Compound	Percent dose absorbed		
	Rat	**Rhesus**	**Man**
Lindane	31 ± 10	18 ± 4	9 ± 4
Dinoseb	86 ± 1	5 ± 3	—
Testosterone	47 ± 3	18 ± 10	13 ± 3
Fenitrothion[a]	84 ± 12	21 ± 10	—
Aminocarb[b]	88 ± 6	37 ± 14	—
DEET	36 ± 8	14 ± 5	—

[a]   Moody and Franklin.[25]
[b]   Moody et al.[26]

TABLE 7
In Vivo Percutaneous Absorption in Rabbit and Man[13,14]

Compound	Percent dose absorbed		Difference
	Rabbit	**Man**	
Haloprogin	113 ± 16	11 ± 4	×10
N-Acetylcysteine	2 ± 1	2 ± 2	0
Testosterone	70 ± 8	13 ± 3	×5
Cortisone	30 ± 9	3 ± 2	×10
Caffeine	69 ± 6	47 ± 20	0
Butter yellow	100 ± 8	22 ± 5	×5
DDT	46 ± 1	10 ± 4	×5
Lindane	51 ± 30	9 ± 4	×6
Parathion	98 ± 8	10 ± 6	×10
Malathion	65 ± 11	8 ± 3	×8

TABLE 8
Effect of Rubbing Cream on Dermal Absorption
of Cortisone: Comparison of Rhesus Monkey
and Guinea Pig with Man[8]

	Cortisone dose absorbed (% ± SD)	
	Rub	Spread
Man		3.4 ± 1.6[a]
Rhesus monkey	6.2 ± 2.7	5.3 ± 3.3[b]
Guinea pig	20.2 ± 3.7	20.1 ± 4.3[b]

[a]   Acetone vehicle.
[b]   Cream vehicle.

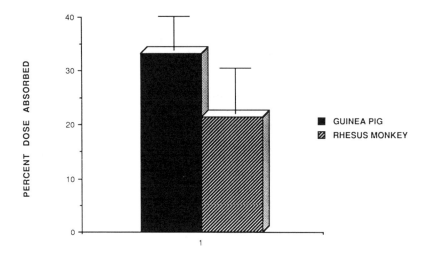

FIGURE 1. Percutaneous absorption of the PCB aroclor 1242 in the rhesus monkey and the guinea pig. Dose and vehicle (acetone) were the same for both species. (From Wester, R. C., Bucks, D. A. W., Maibach, H. I., and Anderson, J., *J. Toxical. Environ. Health,* 27, 65, 1989. With permission.)

is greater in the guinea pig. Absorption of the PCB aroclor 1242 is greater in guinea pig than in the rhesus monkey (Figure 1).[27] El Dareer et al.[28] compared *in vivo* percutaneous absorption study of 2-mercaptobenzothiazole (MBT) and 2-mercaptobenzothiazole disulfide (MBTS) in guinea pig and rat. The absorption of MBT was actually higher in guinea pig (33.4 ± 8.2%) than in rat (male 16.1 ± 2.8%; female 17.5 ± 1.9%). Absorption of MBTS was slightly higher in guinea pig (12.2 ± 2.2%) than rat (male 5.9 ± 1.8%; female 7.8 ± 0.7%).

These studies suggest percutaneous absorption in the guinea pig is more like that of rat, rather than like that of man.

F. HAIRLESS RAT, MOUSE, AND GUINEA PIG

Rougier et al.[29,30] compared the *in vivo* percutaneous absorption of some compounds in the hairless rat with that of man. The skin site of application was the back of the hairless rat and the arm, abdomen, postauricular, and forehead sites in man. Table 9 shows that

TABLE 9
In Vivo Percutaneous Absorption in Hairless Rat and Man[28,29]

	Man				Hairless Rat
Compound	Arm	Abdomen	Postauricular	Forehead	Back
Benzoic acid (sodium salt)	4 ± 1	8 ± 1	10 ± 1	12 ± 2	5 ± 1
Caffeine	12 ± 2	8 ± 1	12 ± 1	22 ± 2	8 ± 1
Benzoic acid	9 ± 1	15 ± 2	22 ± 5	27 ± 3	14 ± 1
Acetylsalicylic acid	17 ± 1	17 ± 3	29 ± 5	35 ± 3	16 ± 1

regional variation existed for absorption of benzoic acid, caffeine, and acetylsalicylic acid in man. The *in vivo* absorption from the back of the hairless rat was remarkably similar to that in man, especially from the arm (the traditional human site for skin absorption determination). If such information for other future compounds and other studies are the same, the hairless rat could be a relevant animal model. In an *in vitro* study, the percutaneous absorption of 5-iodo-2-deoxycytidine in hairless rat skin was sixfold greater than in human skin.[31] What effect *in vitro* methodology might have contributed to the difference between the two skin sources is not known. Therefore, guarded optimism must be used.

Bond and Barry[31,33,34] did a series of *in vitro* studies to compare percutaneous absorption in the hairless mouse. Their conclusions are that percutaneous absorption in hairless mouse skin is greater than in man, and that skin absorption enhancers will act differently in hairless mouse skin compared to man. Additionally, the stability of hairless mouse skin in water is different than that of man (important for *in vitro* diffusion studies). Some 50-fold differences are possible between hairless mouse skin and human skin. Therefore, hairless mouse does not appear to be a reliable animal model relevant to man.

The hairless guinea pig has also become available as an animal model. Moon et al.[35] determined the *in vivo* percutaneous absorption of hydrocortisone to be 17 ± 4% and benzoic acid to be 34 ± 9% in the hairless guinea pig. These absorption values would be high for hydrocortisone, but similar for benzoic acid when compared to published data for man (Table 1). More information is needed on this specie.

The hairless laboratory animals are attractive for use in skin studies because of the obvious lack of hair to interfere with chemical bioavailability. The situation is certainly analogous to man's lack of hair. Regarding relevance to man for percutaneous absorption, the hairless mouse seems like a poor choice for relevant interpretation for man. More information is needed for the hairless rat and the hairless guinea pig.

II. DOSE RESPONSE

In vivo percutaneous absorption can vary depending upon skin concentration. Therefore, a topical dose response can give added information about the relevance of an animal model. Figure 2 shows the dose response comparison between rhesus monkey and man for benzoic acid, testosterone, and hydrocortisone. The dose response is the same for both rhesus and man *in vivo*.[2] A similar dose response is shown for hairless rat and man in Figure 3.[29,30] Both species show increased linear absorption with increased dose. There is a difference in the absolute amount absorbed between the two species.

Table 13 gives the *in vivo* percutaneous absorption dose response for dinoseb in the rhesus monkey and rat.[24,37] Absorption of dinoseb in the rat is nearly total for all doses applied. In the rhesus monkey, absorption is much less and relatively consistent for the dose range. This table also illustrates the value of dose accountability with skin absorption studies.

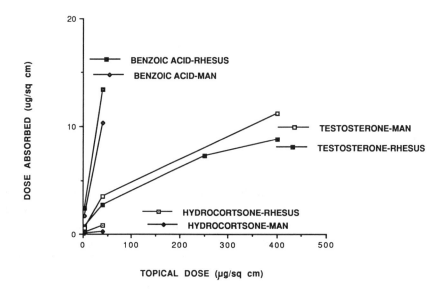

FIGURE 2. The topical dose response for *in vivo* percutaneous absorption of benzoic acid and hydrocortisone in rhesus monkey and man. (From Wester, R. C. and Maibach, H. I., *J. Invest. Dermatol.*, 67, 518, 1976. With permission.)

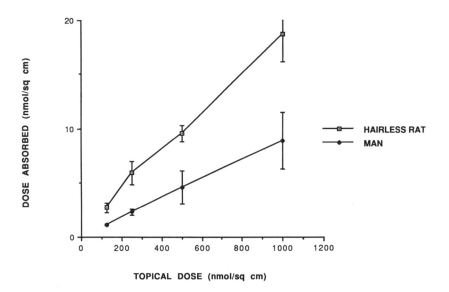

FIGURE 3. The topical dose response for *in vivo* percutaneous absorption of benzoic acid in hairless rat and man. (From Rougier, A., Dupuis, D., Lotte, C., Roguiet, R., Wester, R. C., and Maibach, H. I., *Arch. Dermatol. Res.*, 278, 465, 1986. With permission.)

TABLE 10
Effect of Body Site on Rat Skin
Permeability[36]

Compound	Permeability constant (cm/h × 10⁴)
Water	
Back	4.9 ± 0.4
Abdomen	13.1 ± 2.1
Urea	
Back	1.6 ± 0.5
Abdomen	18.8 ± 5.5
Cortisone	
Back	1.7 ± 0.4
Abdomen	12.2 ± 0.6

TABLE 11
Percutaneous Absorption of Fenitrothion,
Aminocarb, DEET, and Testosterone in
Rhesus Monkey and Rat[1,2,25,26]

Chemical	Specie	Applied dose absorbed (%) skin site		
		Forehead	Forearm	Back
Fenitrothion	Rhesus	49	21	
	Rat			84
Aminocarb	Rhesus	74	37	
	Rat			88
Testosterone	Rhesus	20	9	
	Rat			47
DEET	Rhesus	33	14	
	Rat			36

Most of the dinoseb dose remained on the skin surface and was recovered in the soap and water surface wash.

III. REGIONAL VARIATION IN ANIMALS

Percutaneous absorption data obtained in man is most relevant for human exposure. However, many estimates for humans are made from animal models. Therefore, regional variation in animals may affect prediction for humans. Also, if regional variation exists in an animal, that variation should be relative to humans.

Bronaugh et al.[36] reported the effect of body site (back vs. abdomen) on male rat skin permeability. Abdominal rat skin was more permeable to water, urea, and cortisone (Table 10). Skin thickness (stratum corneum, whole epidermis, whole skin) is less for the abdomen than for the back. With the hairless mouse, showed dorsal skin was found to be more permeable than abdominal skin (reverse that of the male rat). Hairless mouse abdominal skin is thicker than dorsal skin (also reverse that of the male rat).

Skin absorption in the rhesus monkey is considered to be relevant to that of humans. Table 11 shows the percutaneous absorption of testosterone,[1,2] fenitrothion, aminocarb, and diethyltoulamide (DEET)[25,26] in the rhesus monkey compared with the rat. Note that for the

TABLE 12
Percutaneous Absorption Ratio for Scalp and
Forehead to Forearm in Man and Rhesus
Monkey[11,25,26]

Chemical	Specie	Percutaneous absorption ratio	
		Scalp/forearm	Forehead/forearm
Hydrocortisone	Man	3.5	6.0
Benzoic acid	Man		2.9
Parathion	Man	3.7	4.2
Malathion	Man		3.4
Testosterone	Rhesus	2.3	
Fenitrothion	Rhesus		2.3
Aminocarb	Rhesus		2.0
DEET	Rhesus		2.4

TABLE 13
In Vivo **Percutaneous Absorption of**
Dinoseb in Rhesus Monkey and Rat[24,37]

Applied dose (μg/cm^2)	Skin penetration (%)	Dose accountability (%)
Rat		
51.5	86.4 ± 1.1	87.9 ± 1.8
128.8	90.5 ± 1.1	91.5 ± 0.6
643.5	93.2 ± 0.6	90.4 ± 0.7
Rhesus Monkey		
43.6	5.4 ± 2.9	86.0 ± 4.0
200.0	7.2 ± 6.4	81.2 ± 18.1
3620.0	4.9 ± 3.4	80.3 ± 5.2

rhesus there is regional variation between forehead (scalp) and forearm. If one determines the ratio of forehead (scalp)/forearm for the rhesus and compares the results with humans, the similarities are the same (Table 12). Therefore, the rhesus monkey probably can be a relevant animal model for human regional variation.[11]

IV. *IN VITRO* SPECIE COMPARISON

Comparative studies have been done *in vitro* between skin sources from various species. *In vitro* absorption adds another variable to the database, but comparative information seems to match that of the *in vivo* results. Table 14 compares results with the absorption of paraquat. The variability between specie skin sources are great.[8] This was confirmed by Scott and Corrigan[38] for diquat (Figure 4).

The advantage of *in vitro* absorption is the ability to rapidly obtain results. This is most desirable where formulations and enhancers need to be screened. A handy choice would be a ready supply of animal skin. Bond and Barry's work[32-34] suggest the hairless mouse is a bad choice because skin from that specie responds to formulation changes different from that of human skin. Catz and Friend[39] studied the effect of enhancers on levonorgestrel

TABLE 14
Absorption of Paraquat and Water through Human and Laboratory Animal Skin
(*In Vivo* and *In Vitro*)[8]

Species	Paraquat permeability rate (μg/cm^2)
Human (*in vivo*)[a]	0.03
Human (*in vitro*)[b]	0.5

Species (*in vitro*)[c]	Permeability constant (cm/h $\times$ 10^5)	
	Water	Paraquat
Human	93	0.7
Rat	103	27.2
Hairless rat	130	35.3
Nude rat	152	35.5[d]
Mouse	164	97.2[d]
Hairless mouse	254[d]	1,065.0[d]
Rabbit	253[d]	92.9[d]
Guinea pig	442[d]	196.0[d]

[d] Significantly different from human.

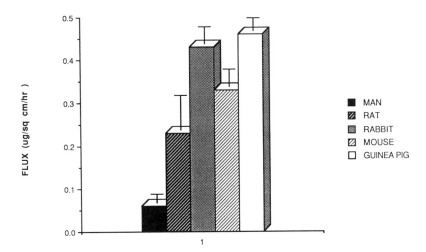

FIGURE 4. Comparative *in vitro* percutaneous absorption of diquat from skin of various species. (From Scott, R. C. and Corrigan, M. A., *Toxic In Vitro*, 4, 137, 1990. With permission.)

transdermal delivery. Figure 5 shows that other animal skin sources as well as hairless mouse skin will respond to formulation changes differently than that of human skin. The development of false positive results (and false hopes) with the use of animal skin can be detrimental to a research project.

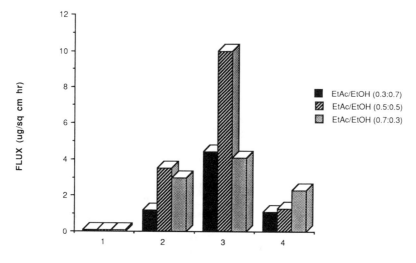

(1) HUMAN (2) RAT (3) HAIRLESS MOUSE (4) HAIRLESS GUINEA PIG

FIGURE 5.   Differences in levonorgestrel steady-state flux due to formulation changes. Skin from various animal species respond differently from man among themselves regarding enhancer formulations. (From Catz, P. and Friend, D. R., *Int. J. Pharmacol.*, 58, 93, 1990. With permission.)

V. OTHER SPECIES

Other species used for percutaneous absorption include goat, horse, cat, dog, chimpanzee,[8] marmoset,[40] and shed snake skin.[41] Additionally, human and animal can be grafted to other animals. Percutaneous absorption can be determined in the grafted skin on the living animal[42] or in a skin sandwich flap.[43] There is little comparative data with which to determine relevance to man.

VI. ANIMAL SKIN PHYSICAL AND CHEMICAL PARAMETERS

Monteiro-Riviere et al.[44] determined the interspecie and interregional differences of the comparative histological thickness and blood flow at five skin sites in nine species (monkey, pig, dog, cat, cow, horse, rabbit, rat, and mouse). Meyer and Neurand[45] reported the comparison skin pH values in domesticated and laboratory mammals (dog, cat, horse, pig, cattle, sheep, goat, rabbit, guinea pig, and rat). Oku et al.[46] reported the distribution of branch-chain fatty acid in the skin surface lipid of laboratory animals (rat, mouse, hamster, and rabbit). Bronaugh[36] has reported differences in percutaneous absorption with male and female rat skin. Thus there are many physical and biochemical distinctions between skin of a variety of animals, including man. Additionally, the skin of a haired animal will probably be shaved before use. A variety of these variables plus the dosing variables (concentration, surface area, formulation, time, etc.) makes the definitive animal model a challenge.

VII. SUMMARY

In reviewing studies comparing percutaneous absorption between animals and man, care must be taken to ascertain what influences the methodology may have had on the data.

Differences in results can be due to different techniques used in the studies. This becomes important when the data from an animal study are compared with published literature values on absorption in man. Subtle differences in technology may not be readily expressed in the printed methodology.

When comparing the percutaneous absorption of species, it becomes obvious that differences do exist. Some of these differences are due to the species themselves, and some of the differences are due to techniques used in the study. Various parameters affect percutaneous absorption.[20] It becomes important that in any species comparative study, the methods and techniques used be as close to each other as possible. Some of the variables, such as site of application, occlusion, dose concentration, surface area, and vehicle, can be controlled by the investigator. Some variables, such as skin metabolism, skin age, and skin condition, may in part be difficult for an investigator to control.

The perfect comparative study probably cannot be done; however, the data in the literature suggest that differences in percutaneous absorption exists between species. Compared with absorption in man, absorption in common laboratory animals (rat and rabbit) is high. Absorption in the pig and monkey (squirrel and rhesus) appears to be more predictive of that *in vivo*, whereas the comparative *in vitro* studies done with skin from different species favorably agree with the *in vivo* results. Information for the hairless animal species are, to date, somewhat mixed. Data with hairless rat are somewhat encouraging, while data with hairless mouse are very discouraging.

Thus, it appears that the animals models most predictive of percutaneous absorption in man are the weanling pig and monkey. It may be difficult to have access to monkeys and weanling pigs. This then does not mean that the investigator has to do meaningless studies *in vitro* or *in vivo* with rats and rabbits. What it means is that the results obtained must be carefully explained within the scope of the methods and species used. Correlations, predictions of results to humans, must be made with utmost care. For critical studies, percutaneous absorption in man remains the best option.

REFERENCES

1. **Wester, R. C. and Maibach, H. I.,** Percutaneous absorption in the rhesus monkey compared to man, *Toxicol. Appl. Pharmacol.,* 32, 394, 1975.
2. **Wester, R. C. and Maibach, H. I.,** Rhesus monkey as an animal model for percutaneous absorption, in *Maibach, Animal Models in Dermatology,* Churchill Livingstone, New York, 1975, 133.
3. **Wester, R. C. and Maibach, H. I.,** Relationship of topical dose and percutaneous absorption in rhesus monkey and man, *J. Invest. Dermatol.,* 67, 518, 1976.
4. **Wester, R. C. and Noonan, P. K.,** Relevance of animal models for percutaneous absorption, *Int. J. Pharmacol.,* 7, 99, 1980.
5. **Wester, R. C., Noonan, P. K., Cole, M. P., and Maibach, H. I.,** Percutaneous absorption of testosterone in the newborn rhesus monkey: comparison to the adult, *Pediatr. Res.,* 11, 737, 1977.
6. **Wester, R. C., Noonan, P. K., and Maibach, H. I.,** Recent advances in percutaneous absorption using the rhesus monkey model, *J. Soc. Cosmet. Chem.,* 30, 297, 1979.
7. **Wester, R. C., Noonan, P. K., and Maibach, H. I.,** Variations in percutaneous absorption of testosterone in the rhesus monkey due to anatomic site of application and frequency of application, *Arch. Dermatol.,* 267, 229, 1980.
8. **Wester, R. C. and Maibach, H. I.,** Animal models for percutaneous absorption, in *Models in Dermatology,* Maibach, H. and Lowe, N., Eds., S. Karger, Basel, 1985, 159.
9. **Bronaugh, R. L., Wester, R. C., Bucks, D., Maibach, H. I., and Sarason, R.,** *In vivo* percutaneous absorption of fragrance ingredients in rhesus monkey and man, *Fed. Chem. Toxicol.,* 28, 369, 1990.
10. **Franz, T. J. and Lehman, P.,** Percutaneous absorption of retinoic acid in monkey and man, in *Pharmacology of Retinoids in the Skin,* Reichert, U. and Shroot, B., Eds., S. Karger, Basel, 1989, 174.

11. **Wester, R. C. and Maibach, H. I.,** Regional variation in percutaneous absorption, in *Percutaneous Absorption,* Bronaugh, R. and Maibach, H., Eds., Marcel Dekker, New York, 1989, 111.

12. **Wester, R. C., Maibach, H. I., Bucks, D. A. W., Sedik, L., Melendres, J., Liao, C., and DiZio, S.,** Percutaneous absorption of DDT and benzo(a)pyrene from soil, *Fundam. Appl. Toxicol.,* 15, 510, 1990.

13. **Bartek, M. J., La Budde, J. A., and Maibach, H. I.,** Skin permeability *in vivo:* comparison in rat, rabbit, pig and man, *J. Invest. Dermatol.,* 58, 114, 1972.

14. **Bartek, M. J. and La Budde, J. A.,** Percutaneous absorption *in vitro,* in *Animal Models in Dermatology,* Maibach, H. Ed., Churchill Livingstone, New York, 1975, 103.

15. **Roberts, M. E. and Mueller, K. R.,** Comparison of *in vitro* nitroglycerin (TNG) flux across Yucatan pig, hairless mouse, and human skins, *Pharm. Res.,* 7, 673, 1990.

16. **Riviere, J. E., Bowman, K. F., Monteiro-Riviere, N. A., Dix, L. P., and Carver, M. P.,** The isolated perfused porcine skin flap (IPPSF), *Fundam. Appl. Toxicol.,* 7, 444, 1986.

17. **Riviere, J. E., Bowman, K. F., and Monteiro-Riviere, N. A.,** On the definition of viability on isolated perfused skin preparations, *Br. J. Dermatol.,* 116, 739, 1987.

18. **Wester, R. C., Maibach, H. I., Surinchak, J., and Bucks, D. A. W.,** Predictability of *in vitro* diffusion systems: effect of skin types and ages on percutaneous absorption of triclocarban, in *Percutaneous Absorption,* Bronaugh, R. and Maibach, H., Eds., Marcel Dekker, New York, 1985, 223.

19. **Zendzian, R. P.,** Skin penetration method suggested for Environmental Protection Agency requirements, *J. Am. Coll. Toxicol.,* 8, 829, 1989.

20. **Wester, R. C. and Maibach, H. I.,** Cutaneous pharmacokinetics: 10 steps to percutaneous absorption, *Drug Metab. Rev.,* 14, 169, 1983.

21. **North-Root, H., Corbin, N., and Demetrulias, J. L.,** Skin deposition and penetration of triclocarban, in *Percutaneous Absorption,* Bronaugh, R. and Maibach, H., Eds., Marcel Dekker, New York, 1985, 141.

22. **Moody, R. P. and Rittér, L.,** Dermal absorption of the insecticide lindane in rats and rhesus monkeys: effect of anatomical site, *J. Toxicol. Environ. Health,* 28, 161, 1989.

23. **Feldmann, R. J. and Maibach, H. I.,** Percutaneous penetration of some pesticides and herbicides in man, *Toxicol. Appl. Pharmacol.,* 28, 126, 1974.

24. **Shah, P. V., Fisher, H. L., Monroe, R. J., Chernoff, W., and Hall, L. L.,** Comparison of the penetration of 14 pesticides through the skin of young and adult rats, *J. Toxicol. Environ. Health,* 21, 353, 1987.

25. **Moody, R. P. and Franklin, C. A.,** Percutaneous absorption of the insecticides fenitrothion and aminocarb, *J. Toxicol. Environ. Health,* 20, 209, 1987.

26. **Moody, R. P., Benoit, F. M., Riedel, D., Rutter, L., and Franklin, C.,** Dermal absorption of the insect repellant Deet in rats and monkeys: effect of anatomic site and multiple exposure, personal communication.

27. **Wester, R. C., Bucks, D. A. W., Maibach, H. I., and Anderson, J.,** Polychlorinated biphenyls (PCBs). Dermal absorption, systemic elimination, and dermal wash efficiency, *J. Toxicol. Environ. Health,* 12, 511, 1983.

28. **El Career, S. M., Kalin, J. R., Tellery, K. F., and Hill, D. L.,** Disposition of 2-mercaptobenzothiazole and 2-mercaptobenzothiazole disulfide in rats dosed intravenously, orally, and topically and in guinea pigs dosed topically, *J. Toxicol. Environ. Health,* 27, 65, 1989.

29. **Rougier, A., Dupuis, D., Lotte, C., Roguet, R., and Schaefer, H.,** *In vivo* correlation between stratum corneum reservoir function and percutaneous absorption, *J. Invest. Dermatol.,* 81, 275, 1983.

30. **Rougier, A., Dupuis, D., Lotte, C., Roguiet, R., Wester, R. C., and Maibach, H. I.,** Regional variation in percutaneous absorption in man: measurement by the stripping method, *Arch. Dermatol. Res.,* 278, 465, 1986.

31. **Wepierre, J., Marty, J. P., Chanal, J. L., and Sincholle, D.,** Percutaneous absorption of 5-iodo-2-deoxycytidine in the hairless rat and in man, *Eur. J. Drug Metab. Pharmacokinet.,* 9, 79, 1984.

32. **Bond, J. R. and Barry, B. W.,** Hairless mouse skin is limited as a model for assessing the effects of penetration enhancers in human skin, *J. Invest. Dermatol.,* 90, 810, 1988.

33. **Bond, J. R. and Barry, B. W.,** Limitations of hairless mouse skin as a model for *in vitro* permeation studies through human skin: hydration damage, *J. Invest. Dermatol.,* 90, 486, 1988.

34. **Bond, J. R. and Barry, B. W.,** Damaging effect of acetone on the permeability barrier of hairless mouse skin compared with that of human skin, *Int. J. Pharm.,* 412, 91, 1988.

35. **Moon, K. C., Wester, R. C., and Maibach, H. I.,** Diseased skin models in the hairless guinea pig: *in vivo* percutaneous absorption, *Dermatologica,* 180, 8, 1990.

36. **Bronaugh, R. L., Stewart, R. F., and Congdon, E. R.,** Differences in permeability of rat skin related to sex and body site, *J. Soc. Cosmet. Chem.,* 34, 1237, 1983.

37. **Wester, R. C., McMaster, J., Bucks, D. A. W., Bellet, E. M., and Maibach, H. I.,** Percutaneous absorption in rhesus monkeys and estimation of human chemical exposure, in *Biological Monitoring for Pesticide Exposure: Measurement, Estimation, and Risk Reduction,* Wang, R., Franklin, C., Honeycutt, R. C., and Reinert, J. C., Eds., ACS Press, Washington, D.C., 1989, 152.

38. **Scott, R. C. and Corrigan, M. A.,** The *in vitro* percutaneous absorption of diquot: a species comparison, *Toxic. In Vitro,* 4, 137, 1990.

39. **Catz, P. and Friend, D. R.,** Transdermal delivery of levonorgestral. VIII. Effect of enhancers on rat skin, hairless mouse skin, hairless guinea pig skin, and human skin, *Int. J. Pharmacol.,* 58, 93, 1990.

40. **Scott, R. C., Corrigan, M. A., Smith, F., and Mason, H.,** The influence of skin structure on permeability: an intersite and interspecies comparison with hydrophilic penetrants, *J. Invest. Dermatol.,* 96, 921, 1991.

41. **Rigg, P. C. and Barry, B. W.,** Shed snake skin and hairless mouse skin as model membranes for human skin during permeation studies, *J. Invest. Dermatol.,* 94, 235, 1990.

42. **Wester, R. C. and Maibach, H. I.,** Advances in percutaneous absorption, *in Cutaneous Toxicity,* Dull, V. and Lazar, P., Eds., Raven Press, New York, 1984, 29.

43. **Kruger, G. G. and Shelby, J.,** Biology of human skin transplanted to the nude mouse. I. Response to agents which modify epidermal proliferation.

44. **Nonteiro-Riviere, N. A., Bristol, D. G., Manning, T. O., Rogers, R. A., and Rivier, J. E.,** Interspecies and interregional analysis of the comparative histologic thickness and laser Doppler blood flow measurements at five cutaneous sites in nine species, *J. Invest. Dermatol.,* 95, 582, 1990.

45. **Meyers, W. and Neurand, K.,** Comparison of skin pH in domesticated and laboratory mammals, *Arch. Dermatol. Res.,* 283, 16, 1991.

46. **Oku, H., Nakanishi, T., Kumamoto, K., and Chinen, I.,** Distribution of branch-chain fatty acid in the skin surface lipid of laboratory animals, *Comp. Biochem. Physiol.,* 96B, 475, 1990.

Chapter 7

A COMPARATIVE STUDY OF THE KINETICS AND BIOAVAILABILITY OF PURE AND SOIL-ADSORBED BENZENE, TOLUENE, AND *m*-XYLENE AFTER DERMAL EXPOSURE

Mohamed S. Abdel-Rahman, Gloria A. Skowronski, and Rita M. Turkall

TABLE OF CONTENTS

0-8493-7357-3/93/$0.00 + $.50
© 1993 by CRC Press, Inc.

I. INTRODUCTION

Over the past decade, soil contamination has been recognized as a significant public health concern. The range of contaminants in soil is now known to be very broad, including pertroleum, heavy metals, dioxins, pesticides, and organic solvents.[1] Furthermore, numerous sources of exposure to contaminated soil by workers and communities exist. These include industrial and transportation accidents, leaking storage tanks and pipelines, poor design and management of waste disposal sites, and habitation on or near a previously contaminated site.

Since human skin is the largest organ of the body (surface area approximately 18,000 cm^2), it has the potential to absorb significant amounts of contaminants into the body.[2] Until recently, assessment of potential health risks following dermal exposures to contaminated soil has relied almost exclusively on extrapolations from data derived with pure chemicals. Soil is a complex matrix, however, which can interact with the pollutants. Therefore, the percentage of the total chemical in the soil which enters the body, i.e., bioavailability, may differ from that seen following exposure to the contaminant alone. Chemical adsorption to soil and subsequent dermal bioavailability are governed by several factors: (1) physical and chemical parameters of soil, such as particle size, moisture content, percentage organic content, and percentage clay;[3] (2) characteristics of the chemical, for example, concentration, octanol-water partition coefficient, and volatility;[4] and (3) physiological and environmental parameters including pH, temperature, hydration, site of application, surface area exposed, and skin condition.[5]

Several dermal bioavailability studies have been performed on 2,3,7,8-tetrachlorodibenzo-*p*-dioxin (TCDD), a compound persistent in soil.[6-8] Less attention has been given to dermal bioavailability determinations of volatile pollutants in soil. Benzene together with toluene and xylene are among the most prevalent volatile contaminants identifed at dumpsites as well as being the major aromatic components of gasoline.[9,10] Benzene is used both as a solvent and as a starting material in chemical synthesis.[11,12] Because of its leukemia- and cancer-causing properties,[13,14] benzene is often considered to be more hazardous to health than toluene or xylene. Toluene is employed as a solvent for paints, plastics, varnishes, and resins, and in the manufacture of explosives, drugs, perfumes, and other synthetic chemicals.[15] Although toluene lacks benzene's chronic hematopoietic effects, animal experiments indicate that toluene is more acutely toxic than benzene. High concentrations of inhaled toluene cause symptoms of central nervous system (CNS) depression such as headache, dizziness, weakness, and loss of coordination. Furthermore, chronic exposure to toluene may cause permanent CNS damage.[11,15,16] Commercial xylene is a mixture of three isomers, ortho-, meta-, and para-xylene, with *m*-xylene predominating (usually 60 to 70%).[17] Xylene is extensively used as a solvent, a cleaning agent, a degreaser, and a starting material and intermediate in the chemical industry. Xylene is also present in a broad range of consumer products such as aerosols, paints, varnishes, shellacs, and rust preventatives.[17] Gerarde reported that xylene may be more acutely toxic than benzene or toluene.[18] When inhaled at high concentrations, symptoms included disturbed vision, dizziness, tremors, salivation, cardiac stress, CNS depression, confusion, and coma.[19] Xylene causes dehydration and defatting of the skin which may lead to erythema, blistering, or dermatitis.[18]

Previous oral bioavailability studies from this laboratory have demonstrated that adsorption to clay soil increased the bioavailability of benzene relative to the pure compound.[20] Although the plasma time course for the absorption of toluene administered by the oral route was altered by adsorption to sandy soil, the quantity of absorbed chemical was not changed.[21]

The aim of this study was to utilize pharmacokinetic techniques to evaluate the dermal bioavailability of a class of related chemicals (benzene, toluene, *m*-xylene) adsorbed to soil

TABLE 1
Soil Properties

	Atsion	Keyport
Sand (%)	90	50
Silt (%)	8	28
Clay (%)	2	22
Organic matter (%)	4.4	1.6
Particle size, mm (%)		
0.05–0.1	22.2	17.0
0.1–0.25	76.3	65.3
0.25–0.5	1.2	13.6
0.5–1.0	0.2	3.4
1.0–2.0	0.1	0.7
>2.0	0	0.1

vs. the pure compounds. Moreover, the effects of methyl functional group addition to the benzene nucleus were assessed. Thus, this research provides a means of identifying differences in bioavailability so that more informed predictions of health risks can be made following dermal exposure to contaminated soils.

II. METHODS

A. CHEMICALS

Uniformly labeled ^{14}C-benzene with a specific activity of 60 mCi/mmol and radiochemical purity of $\geq98\%$ was purchased from ICN Pharmaceuticals, Irvine, CA. (Ring-U-^{14}C)toluene (16.4 mCi/mmol, 95% pure) was purchased from Amersham Corporation, Arlington Heights, IL). Prior to use, the benzene and toluene radioisotopes were diluted to 133.3 μCi/ml, respectively, with nonradioactive high-performance liquid chromatography (HPLC)-grade benzene and toluene (Aldrich Chemical Company, Milwaukee, WI). *m*-(Ring-U^{14}C)xylene with a specific activity of 7.2 mCi/mmol and radiochemical purity of $\geq98\%$ was custom synthesized by E.I. DuPont de Nemours and Co., Inc., New England Nuclear Research Products (Boston, MA). The radioisotope was diluted to 88.8 μCi/ml with unlabeled anhydrous (>99% purity) *m*-xylene (Aldrich).

B. SOILS

Two soils were examined: (1) an Atsion sandy soil collected from an aquifer drill site of the Cohansey sand formation near Chatsworth in south central New Jersey and (2) a Keyport clay soil collected from 2 to 3 ft below the surface of the Woodbury formation near Moorestown in southwestern New Jersey. The Atsion soil is a deep, poorly drained soil (>150,000 acres) formed in Atlantic coastal plain sediments of New Jersey and New York. Warmer thermic equivalent Leon soil (>500,000 acres) is found extensively in Maryland, Virginia, North Carolina, South Carolina, Alabama, Georgia, and Florida.[22] The Keyport soil is a moderately well-drained soil formed on clay bed marine deposits of the Atlantic inner-coast plain. Moderate amounts (>70,000 acres) of Keyport soil are found in New Jersey, Delaware, Maryland, and Virginia with similar soils occurring from New Jersey to Texas.[23] Mechanical, sieve, and organic matter analyses were performed by the Cooperative Extension Service, Cook College, Rutgers University, New Brunswick, NJ. Physical characteristics of the soils are summarized in Table 1. Because of the Atsion soil's 90% sand content and the Keyport soil's 22% clay content, these soils will be referred to as sandy and clay, respectively. Gas chromatography/mass spectrometry (Hewlett Packard model

5995A) analysis of combined ether-hexane (1:1) and methylene chloride soil extracts using OV-17 and Tenax columns (Supelco, Bellefonte, PA) and by direct solid probe did not reveal any contamination by the chemicals.

C. ANIMALS

Male Sprague-Dawley rats (275 to 300 g) were purchased from Taconic Farms (Germantown, NY) and quarantined for at least 1 week prior to administration of the chemical. Animals were housed three per cage and were maintained on a 12-h light/dark cycle at constant temperature (23 to 25°C) and humidity (50 to 55%). Ralston Purina rodent chow (St. Louis, MO) and tap water were provided ad libitum.

D. TREATMENT

One half hour prior to the administration of radioisotope, a shallow glass cap (Q Glass Company, Towaca, NJ) circumscribing a 13-cm^2 area was tightly fixed with Lang's jet liquid acrylic and powder (Lang Dental Manufacturing Co., Inc., Chicago, IL) on the lightly shaved right costoabdominal region of each ether-anesthetized animal. Care was taken to prevent any abrasion of the skin during shaving and cap attachment. The glass cap was attached immediately after shaving the animal and remained on the rat for the duration of the study. Three hundred microliters of ^{14}C-benzene (40 µCi), 225 µl of ^{14}C-toluene (30 µCi), or 225 µl of m-(^{14}C)xylene (20 µCi) alone or after the addition of 750 mg of soil was introduced through a small opening in the cap which was immediately sealed. The weight of soil and volumes of the chemicals were selected so that a uniformly contaminated soil could be prepared without excess fluid. Minimum radioisotope doses of 40, 30, and 20 µCi, respectively, for benzene, toluene, and m-xylene were necessary to achieve detectable amounts of radioactivity in plasma. The applied chemical doses were 20 mg of benzene or 15 mg each of toluene or m-xylene per cm^2 skin. Rats were rotated from side to side so that soil-adsorbed chemical covered the entire circumscribed area.

E. ABSORPTION AND ELIMINATION

In the absorption and elimination studies, each treatment group contained six to ten rats. Heparinized blood samples were collected by cardiac puncture under light ether anesthesia at selected time points up to 48 h for toluene and up to 72 h for benzene or m-xylene. Blood samples were centrifuged at 1000 × g for 5 min to obtain plasma. A 100-µl aliquot of plasma was added to 10 ml of Aquasol-2 (DuPont) and radioactivity was measured in a Beckman LS-7500 liquid scintillation spectrometer. Sample quench was corrected by using the H-ratio method.

F. EXCRETION, TISSUE DISTRIBUTION, AND METABOLISM

In the excretion studies, groups of six rats each were administered chemical alone or chemical adsorbed to soil as described earlier. Animals were housed in all-glass metabolism chambers (Bio Serv Inc., Frenchtown, NJ) for the collection of expired air, urine, and fecal samples. Expired air was passed through activated charcoal tubes (SKC Inc., Eighty-Four, PA) for the collection of the parent compounds, then bubbled through traps filled with ethanolamine:ethylene glycol monomethyl ether (1:2 v/v) for the collection of ^{14}CO$_2$.[24] Charcoal tubes and trap mixtures were collected at 1, 2, 6, 12, 24, and 48 h after administration of compound. Urine samples were collected at 12, 24, and 48 h while fecal samples were collected at 24 and 48 h. Charcoal was extracted with glacial acetic acid, and fecal samples were homogenized in deionized water. Aliquots of the charcoal extracts, ethanolamine:ethylene glycol monomethyl ether mixture, fecal homogenates, and urine were then dispensed directly into Aquasol-2 for analysis.[20]

At the conclusion of the excretion studies, rats were sacrificed by an overdose of ether. Then, the surface of the treated skin area was washed with ethyl alcohol to determine the percentage of radioactivity remaining on skin application sites. Ethyl alcohol (1 to 3 ml) was introduced into the glass cap and the animals were rotated from side to side. Aliquots of ethanol wash (100 μl) were removed from the skin, added to scintillation vials containing Aquasol-2 and counted. Glass caps were removed from the rats and the following tissue specimens were collected: brain, thymus, thyroid, esophagus, stomach, duodenum, ileum, lung, pancreas, adrenal, testes, skin, fat, carcass, bone marrow, liver, kidney, spleen, heart, and whole blood. Radioactivity was determined in the skin application site and in the untreated skin of the rat (left side). Three areas of fat were examined, namely fat beneath treated and untreated skin and subrenal fat. Samples of liver, kidney, spleen, heart, and whole blood were solubilized by heating with 70% perchloric acid and 30% hydrogen peroxide. The remainder of the tissue specimens were processed in Protosol and radioactivity quantitated.[20]

The identity, quantity, and time for appearance of metabolites in urine were monitored by HPLC. Urine samples from the excretion studies were extracted with ethyl acetate, *n*-butanol, ether and amyl alcohol, respectively, for the determination of benzene, toluene, and *m*-xylene metabolites. Aliquots of the extracts were analyzed on a Beckman Model 334 chromatograph equipped with a Hitachi 100-40 spectrophotometric detector and a 24 × 0.46 cm I.D., C-18 column (Ultrasphere-ODS, Beckman/Altex, San Ramon, CA). Gradients of water-acetonitrile, 3.8% phosphoric acid-acetonitrile, and acidified water (adjusted to pH 3.0 with phosphoric acid)-methanol were utilized as mobile phases, respectively, for benzene, toluene, and *m*-xylene.[20,25,26] Nonradioactive standards were used to establish a time course for the appearance of metabolites in column eluant fractions. Eluants from radioactive samples were collected in liquid scintillation vials to which Aquasol-2 was added for analysis.

G. DATA ANALYSIS

Exploratory data analysis was utilized on replicate plasma time course data.[27,28] The curve-fitting procedure which was utilized is called smoothing. For these studies, a "4235EH" smoother (a statistical procedure for treating the data) was used.[28] Each replicate was smoothed over all time points, a median value was calculated for all smoothed replicates at each time point, and a second smooth was applied to these median values. The final smoothed data were used to calculate rate constants, half-lives of absorption and elimination from plasma by regression analysis, and the method of residuals.[29] The area under the plasma concentration curve (AUC) was calculated by the trapezoidal rule using individual replicates and is reported as the mean ± standard error of the mean (SEM). Comparison of slopes was determined by analysis of covariance. The values for excreta, tissue distribution, and metabolites were reported as the mean ± SEM. Statistical differences between treatment groups were determined by analysis of variance (ANOVA) followed by Scheffé's multiple range test or Newman-Kuells multiple range test. A *p* value less than 0.05 was considered statistically significant.

III. RESULTS

A. COMPARISON BETWEEN BENZENE, TOLUENE, AND *m*-XYLENE ABSORPTION AND ELIMINATION

The plasma absorption and elimination half-lives of dermally administered ^{14}C-benzene, toluene, and *m*-xylene alone or adsorbed to soil are presented in Table 2. Adsorption of toluene to sandy soil produced a statistically significant decrease in the half-life of absorption relative to pure compound; the absorption half-life of sandy soil-adsorbed toluene was similar

TABLE 2
Plasma Half-Lives of Radioactivity in Rats Treated Dermally with
^{14}C-Benzene, ^{14}C-Toluene, or ^{14}C-*m*-Xylene[a]

| | $t_{1/2}$ (h) | | | | | |
| | Absorption | | | Elimination | | |
Treatment	Benzene	Toluene	*m*-Xylene	Benzene	Toluene	*m*-Xylene
Chemical alone	3.1	6.4	0.9	23.0	10.3	15.1
+ Sandy soil	3.6	3.2[b]	0.6	24.5	10.8	16.5
+ Clay soil	4.4	5.6	2.4[b]	19.4	10.8	15.4

[a]   Values represent six to ten rats per group.
[b]   Significantly different from chemical alone by ANOVA followed by the Newman-Kuells multiple range test ($p < 0.05$).

TABLE 3
Area Under the Plasma Concentration Time
Curve (AUC) for Rats Treated Dermally with
^{14}C-Benzene, ^{14}C-Toluene, or ^{14}C-*m*-Xylene[a]

| | Percent initial dose/ml · h | | |
Treatment	Benzene	Toluene	*m*-Xylene
Chemical alone	0.41 ± 0.21	0.23 ± 0.10	0.23 ± 0.03
+ Sandy soil	0.22 ± 0.08[b]	0.22 ± 0.10	0.15 ± 0.03
+ Clay soil	0.17 ± 0.07[b]	0.25 ± 0.11	0.26 ± 0.02

[a]   Values represent the mean ± SEM of six to ten rats per group.
[b]   Significantly different from chemical alone by ANOVA followed by the Newman Kuells multiple range test ($p < 0.05$).

to that of sandy soil-adsorbed benzene. However, the absorption half-lives of toluene alone and toluene adsorbed to clay soil were longer than their benzene counterparts.

The half-life of pure *m*-xylene absorption was sevenfold shorter than pure toluene and threefold shorter than pure benzene. Sandy soil-adsorbed *m*-xylene had an absorption half-life which was five- to sixfold shorter than that of sandy soil-adsorbed benzene or toluene. On the other hand, clay soil adsorption significantly increased *m*-xylene absorption vs. chemical alone, resulting in an absorption half-life for clay soil-adsorbed *m*-xylene which was only twofold shorter than that of clay soil-adsorbed benzene or toluene.

Although neither of the soils altered the half-lives of elimination from plasma for any of the three chemicals compared to the pollutants alone, the elimination half-lives following all of the toluene treatments were about twofold shorter than all benzene half-lives and 1.5-fold shorter than all *m*-xylene half-lives (Table 2). Both soils significantly decreased the AUC for benzene (Table 3). On the contrary, the adsorption of toluene or *m*-xylene to sandy or clay soil did not produce any significant differences between AUCs.

B. EXCRETION

In all benzene treatment groups, the major route of excretion was via the urine (Table 4). During the 48-h collection period, 86% of the administered dose was excreted in the urine of the pure benzene group. Sandy soil and clay soil significantly decreased urinary excretion to 64 and 45% of the initial dose, respectively, during the same time period.

TABLE 4
Urinary Excretion of Radioactive Compounds from Rats Treated Dermally with ^{14}C-Benzene, ^{14}C-Toluene, or ^{14}C-*m*-Xylene[a]

Treatment	Percent initial dose		
	Benzene	**Toluene**	***m*-Xylene**
Chemical alone	86.2 ± 2.1	75.5 ± 14.2	42.7 ± 6.5
+ Sandy soil	64.0 ± 2.8[b]	84.8 ± 10.4	41.8 ± 1.7
+ Clay soil	45.5 ± 4.8[b]	80.1 ± 1.2	46.1 ± 4.7

[a] Values represent the mean ± SEM of six rates per group 48 h after dermal treatment.

[b] Significantly different from chemical alone by ANOVA followed by Scheffé's multiple range test ($p < 0.05$).

TABLE 5
Expired Air Excretion of ^{14}C-Benzene, ^{14}C-Toluene, or ^{14}C-*m*-Xylene from Dermally Treated Rats[a]

Treatment	Percent initial dose		
	Benzene	**Toluene**	***m*-Xylene**
Chemical alone	12.8 ± 1.1	3.8 ± 1.2	61.9 ± 7.3
+ Sandy soil	5.9 ± 1.3[b]	5.4 ± 1.3	66.9 ± 11.0
+ Clay soil	10.1 ± 1.4	0.9 ± 0.4	52.8 ± 7.3

[a] Values represent the mean ± SEM of six rats per group 48 h after dermal treatment.

[b] Significantly different from chemical alone by ANOVA followed by Scheffé's multiple range test ($p < 0.05$).

Toluene-derived radioactivity was also mainly eliminated in the urine following all treatments (Table 4). Unlike benzene and toluene, urine was a secondary route of excretion for the pure and sandy soil-adsorbed *m*-xylene groups (43 and 42%, respectively) (Table 4). However, in the clay soil-adsorbed *m*-xylene group, both urine (Table 4) and expired air (Table 5) were nearly equal routes of excretion (46 to 53%). Excretion of radioactivity after percutaneous absorption of *m*-xylene occurred primarily through the lungs for the pure and sandy soil groups (62 and 67% of the initial dose, respectively) (Table 5). Rats receiving benzene or toluene excreted less of the dose in expired air than *m*-xylene with the percentage of radioactivity in the expired air of the sandy soil-adsorbed benzene group being significantly lower than the control (Table 5). Carbon dioxide comprised less than 1% of the total radioactivity in the expired air of all treatment groups for the three pollutants (data not shown). Negligible amounts of radioactivity (less than 0.5% of the dose) were also recovered in the feces of the same groups (data not shown).

C. TISSUE DISTRIBUTION

Soil-related differences were found in tissue distribution following treatment with each of the chemicals. Postadministration of soil-adsorbed benzene (48 h), the highest concentration of radioactivity was found in the treated skin, followed by the kidney and liver in both soil groups. In the pure benzene group, the kidney contained the largest amount of

TABLE 6
Urinary Metabolites of ^{14}C-Benzene, ^{14}C-Toluene, or ^{14}C-*m*-Xylene from Dermally Treated Rats[a]

	Treatment		
Metabolite	Chemical alone	+ Sandy soil	+ Clay soil
Benzene			
Phenol	37.7 ± 3.1	44.2 ± 4.3	45.5 ± 2.7
Hydroquinone	19.0 ± 1.6	10.2 ± 1.3[b]	24.7 ± 0.1
Catechol	13.1 ± 1.1	14.4 ± 4.9	11.6 ± 0.8
Benzenetriol	5.4 ± 2.2	17.3 ± 6.8	13.1 ± 2.1
Toluene			
Hippuric acid	71.8 ± 4.6	64.5 ± 2.5	63.3 ± 2.3
Undetermined	19.4 ± 1.6	25.4 ± 0.5	27.4 ± 1.2
m-Xylene			
Methylhippuric acid	69.8 ± 6.6	86.5 ± 2.3	82.6 ± 2.2
Xylenol	2.6 ± 0.4	16.3 ± 2.3[b]	9.9 ± 3.1
Parent compound	0.4 ± 0.1	2.3 ± 0.2	1.4 ± 0.8

[a] Values represent percentage of total radioactivity (mean ± SEM)
    present in 0 to 12-h collection period for six rats per group treated
    dermally with chemical alone or adsorbed to soil.
[b] Significantly different from chemical alone by ANOVA followed by
    Scheffé's multiple range test ($p < 0.05$).

radioactivity, followed by liver and treated skin. Clay soil treatment significantly increased radioactivity in treated skin, while significantly decreasing radioactivity in treated fat compared to benzene alone.[30]

As with benzene, the skin application sites contained the greatest amount of radioactivity after treatment with toluene or *m*-xylene. A statistically significant decrease in ^{14}C-activity occurred in the treated skin of the clay soil-adsorbed toluene group and in the untreated skin of both soil-adsorbed toluene groups.[25] Tissue analysis revealed an increase in radioactivity of treated skin for both *m*-xylene soil groups.[26] However, with the addition of methyl groups to the benzene ring, greater concentrations of radioactivity appeared in treated, untreated, and subrenal fat than in all other tissues in the toluene and *m*-xylene studies.[25,26] Moreover, a significant increase in radioactivity in the fat beneath the treated skin of the *m*-xylene clay soil-adsorbed group was observed.[26] Ethanol washes of treated sites at necropsy established that ≤0.5% of the applied dose was loosely retained on the skin treatment sites in all studies.

D. METABOLISM

The urinary metabolites of ^{14}C-benzene, toluene, and *m*-xylene are presented in Table 6. Phenol was the major metabolite (38 to 46%) in the urine of all benzene groups. Smaller quantities of hydroquinone, catechol, and benzenetriol were also detected with hydroquinone significantly decreased in the 0 to 12-h collection period for the sandy soil group. In all toluene treatment groups, hippuric acid was the primary metabolite (63 to 72%), followed by a smaller amount of an undetermined metabolite (19 to 27%). No statistical differences were found between toluene treatments. Methylhippuric acid was identified as the major urinary metabolite (70 to 86%) in all *m*-xylene groups. Minor amounts of xylenol and the parent compound were also found. A statistically significant increase in xylenol occurred in the sandy soil group 12 h after dosing. Similar metabolite percentages were detected in the 12 to 24 and 24 to 48 h urine of all treatment groups for all three chemicals.

IV. DISCUSSION

The results of this study indicate that adsorption to either a sandy or clay soil decreased the bioavailability of benzene when male rats were treated topically. A significant decrease was observed in the AUC of either soil-adsorbed treatment group. However, benzene demonstrated a relatively greater affinity for clay soil than for sandy soil, since a significant increase of radioactivity in the skin application site and decrease of radioactivity in the fat below the treated site occurred in the clay treatment group.

In previous studies, Franz measured the percutaneous absorption of benzene alone in animals and man.[31] Franz applied ^{14}C-benzene either to the backs of animals, allowing the chemical to flow over the skin seeking its own area, or to the forearms of human subjects. Following the application of a thin layer of benzene (5 to 10 μl/cm^2) absorption was found to average less than 0.2% of the applied dose in all species studied. Maibach and Anjo also reported that only about 0.17% of a single dose of benzene was absorbed when applied to the forearms of monkeys.[32]

Under the conditions of the previous studies, the applied compound quickly volatilized and was lost to the atmosphere. Our research utilized a glass cap to minimize volatilization losses of compound and loss of soil. Susten et al., using a stainless-steel skin depot to minimize volatilization losses of benzene, applied a rubber solvent mixture containing 0.5% (v/v) benzene to the skin of hairless mice for 4 h.[33] Data derived from their study, together with observations made during tire-building operations, suggest that a worker could absorb 4 to 8 mg of benzene daily through the skin. Hanke et al. showed that benzene penetrated the skin of human forearm at a rate of 0.24 to 0.48 mg/cm^2/h after 1.25 h of continuous occluded exposure.[34] Longer skin contact (72 h) may have been a factor in increasing the percentage of absorption in our investigation. Although occlusive coverings have been known to enhance the percutaneous absorption of a compound by increasing the temperature and hydration of the skin,[5] it should be kept in mind that individuals living in the community or waste disposal site workers could be exposed to contaminated soil during seasons of high heat and humidity. Use of the occlusive covering would mimic these conditions.

Although the chemical structures of benzene and toluene differ by only a methyl group, toluene exhibited a lower affinity for soil than observed for benzene. Sandy soil-adsorbed toluene was absorbed into the blood faster as evidenced by the shorter half-life of absorption, but the amount of radioactivity that was absorbed in all toluene groups was the same as reflected in their AUCs and excretion recoveries (>78% in all cases).

Various data have been published on the rate of dermal absorption of toluene. Dutkiewicz and Tyras[35] reported that the absorption rate of toluene through human skin is very high (14 to 23 mg/cm^2/h compared to 1.2 mg/cm^2/h (calculated from the half-life of absorption) in our study. The absorption rates of Dutkiewiez and Tyras were based on the difference between the doses of applied and remaining toluene on skin after immersion of both hands in liquid toluene. However, Sato and Nakajima showed that toluene is poorly absorbed through the skin.[36] Because of the low absorption of toluene in their work, Sato and Nakajima proposed that the rate of toluene absorption reported by Dutkiewicz and Tyras may not be the absorption rate, but the rate of penetration of toluene into the stratum corneum. Toluene may be confined to the stratum corneum rather than penetrating further into the dermis and on to the systemic circulation. Since Dutkiewicz and Tyras only measured toluene on skin, no data are available on the amount of parent compound and/or metabolites in the remaining tissues and excreta; hence, it is difficult to ascertain whether toluene is confined to the stratum corneum in their work.

Relative to benzene and toluene, the absorption of radioactivity into plasma for all *m*-xylene treatment groups was faster. The rate of pure *m*-xylene absorption in the present study (8.3 mg/cm^2/h) most closely resembled that of Dutkiewicz and Tyras (4.5 to 9.6 mg/cm^2/h)[35] and was much faster than the absorption rate (120 μg/cm^2/h) reported by Engström et al.[37] and Lauwerys et al.[38] As in their toluene work, Dutkiewicz and Tyras based their *m*-xylene calculation on the difference between the dose applied and remaining on skin. On the other hand, Engström et al. and Lauwerys et al. calculated the absorption rate of *m*-xylene based on the amount of methylhippuric acid appearing in urine after exposing human volunteers to *m*-xylene.

It is apparent from the present *m*-xylene excretion and tissue distribution studies that 100% of the topically administered compound was absorbed in all treatments. *m*-Xylene has a higher octanol-water partition coefficient (3.20) than benzene (2.13) and toluene (2.69), therefore, it was more readily bioavailable.[39-41] McDougal et al. quantitated the dermal penetration of benzene, toluene, and xylene vapors after a 4-h exposure in rats and found that the permeability (centimeter per hour) for the three compounds correlated linearly with fat/air and muscle/air partition coefficients in the order xylene > toluene > benzene.[42]

After dermal application of *m*-xylene alone or adsorption to sandy soil, the majority of radioactivity was exhaled. The data suggest that the rapid absorption of *m*-xylene (50% of the dose in less than 1 h) may be saturating metabolic pathways, thus permitting the compound to be expired unmetabolized in these groups. The tighter adsorption of *m*-xylene to the clay soil resulted in an increase in the half-life of absorption and a shift in excretion pathways. Less activity was excreted through the lungs as parent compound, while more radioactivity appeared in urine following metabolism when *m*-xylene was adsorbed to clay soil. Moreover, fat beneath the treated skin of the clay soil group contained a significant increase in radioactivity compared to the other *m*-xylene groups. The fat may be acting as a reservoir for future release of compound in the clay soil group. The rapid absorption of radioactivity in the pure and sandy soil groups may provide less opportunity for bioaccumulation in fat.

In vivo studies by Bartek et al. indicate that skin permeability in rat is different from that in humans, depending on the lipid solubility of the compound.[43] Whether or not this factor translates into pharmacokinetic differences between rat and man is difficult to assess, however, since human volunteers for this type of research are difficult to recruit and comprehensive examination of bioavailability and tissue distribution in humans is impossible due to the unavailability of necessary tissue samples.

V. CONCLUSIONS

Under the conditions of these studies, the degree of dermal bioavailability observed for the benzene-toluene-xylene series after adsorption to soil was dependent on the complex interaction between the clay content of the soils, the volatility of the chemicals from the soils, and the lipid solubility of the compounds. For benzene and *m*-xylene, it appears that a higher percentage of clay, more so than the percentage of organic matter in the soil, was governing the retention of these pollutants in soil and, therefore, altering dermal penetration. Particle size of the two soils was essentially equivalent and does not appear to be a factor in these studies. Because volatilization losses were lowest for *m*-xylene, the amount of *m*-xylene available for absorption was greater than that of benzene or toluene. With the addition of methyl groups to the benzene ring, the value of the lipid/water partition coefficients for the benzene derivative series was significantly increased. *m*-Xylene, having the highest partition coefficient, was absorbed the quickest and with some accumulation of radioactivity in adipose tissue after an acute treatment with *m*-xylene adsorbed to clay soil. The bioaccumulation effect may become more pronounced during chronic exposures. Thus, clay soil adsorbed *m*-xylene was more bioavailable than *m*-xylene alone, sandy soil adsorbed *m*-xylene, or soil-adsorbed benzene and toluene. Minimal differences between soil-adsorbed toluene and toluene alone imply that dermal exposures to soil-adsorbed toluene have the potential for producing health risks similar to the pure compound, while the health risk to soil-adsorbed benzene exposure is less than from exposure to the contaminant alone.

VI. ACKNOWLEDGMENT

This research was supported as a project of the National Science Foundation/Industry/ University Cooperative Center for Research in Hazardous and Toxic Substances at New Jersey Institute of Technology, an Advanced Technology Center of the New Jersey Commission on Science and Technology.

REFERENCES

1. **Kostecki, P. T. and Calabrese, E. J.,** Council for Health and Environmental Safety of Soils (CHESS): a coalition to standardize approaches to soil contamination problems, in *Petroleum Contaminated Soils,* Vol. 2, Calabrese, E. J. and Kostecki, P. T., Eds., Lewis Publishers, Chelsea, MI, 1989, 485.
2. **Wester, R. C. and Maibach, H. I.,** Human skin binding and absorption of contaminants from ground and surface water during swimming and bathing, *J. Am. Coll. Toxicol.,* 8, 853, 1989.
3. **Duffus, J. H.,** *Environmental Toxicology,* John Wiley & Sons, New York, 1980, 101.
4. **Ibbotson, B. G., Gorber, D. M., Reades, D. W., Smyth, D., Munro, I., Willes, R. F., Jones, M. G., Granville, G. C., Carter, H. J., and Hailes, C. E.,** A site-specific approach for the development of soil cleanup guidelines for trace organic compounds, in *Petroleum Contaminated Soils,* Vol. 1, Kostecki, P. T. and Calabrese, E. J., Eds., Lewis Publishers, Chelsea, MI, 1989, 321.
5. **Wester, R. C. and Maibach, H. I.,** *In vivo* percutaneous absorption, in *Dermatotoxicology,* 3rd ed., Marzulli, F. N. and Maibach, H. I., Eds., Hemisphere, Washington, D.C., 1987, 135.
6. **Poiger, H. and Schlatter, C. H.,** Influence of solvents and adsorbents on dermal and intestinal absorption of TCDD, *Food Cosmet. Toxicol.,* 18, 477, 1980.
7. **Shu, H., Teitelbaum, P., Webb, A. S., Marple, L., Brunck, B., Dei Rossi, D., Murray, F. J., and Paustenbach, D.,** Bioavailability of soil-bound TCDD: dermal bioavailability in the rat, *Fundam. Appl. Toxicol.,* 10, 335, 1988.
8. **Roy, T. A., Yang, J. J., Krueger, A. J., and Mackerer, C. R.,** Percutaneous absorption of neat 2,3,7,8-tetrachlorodibenzo-p-dioxin (TCDD) and TCDD sorbed on soils, *Toxicologist,* 10, 308, 1990.
9. United States Environmental Protection Agency (USEPA), National Priority List for Cleanup, Washington, D.C., 1985.
10. **Korte, F. and Boeldefeld, E.,** Ecotoxicological review of global impact of petroleum industry and its products, *Ecotoxicol. Environ. Saf.,* 2, 55, 1978.
11. **Sandmeyer, E. E.,** Aromatic hydrocarbons, in *Patty's Industrial Hygiene and Toxicology,* Vol. 2B, Clayton, G. D. and Clayton, F. E., Eds., John Wiley & Sons, New York, 1981, 3253.
12. **Marcus, W. L.,** Chemical of current interest — benzene, *Toxicol. Ind. Health,* 3, 205, 1987.
13. **Cronkite, E. P., Bullis, J. E., Inoue, T., and Drew, R. T.,** Benzene inhalation produces leukemia in mice, *Toxicol. Appl. Pharmacol.,* 75, 358, 1984.
14. **Maltoni, C., Conti, B., and Cotti, G.,** Benzene: a multipotential carcinogen. Results of long-term bioassays performed at the Bologna Institute on Oncology, *Am. J. Ind. Med.,* 4, 589, 1983.
15. **Von Burg, R.,** Toxicology update, toluene, *J. Appl. Toxicol.,* 1, 140, 1981.
16. **Bergman, K.,** Whole-body autoradiography and allied tracer techiques in distribution and elimination studies of some organic solvents, *Scand. J. Work Environ. Health,* Suppl. 1, 5, 1979.
17. **Fishbein, L.,** Xylenes: uses, occurrence and exposure, *IARC Sci. Publ.,* 85, 109, 1988.
18. **Gerarde, H. W.,** *Toxicology and Biochemistry of Aromatic Hydrocarbons,* Elsevier, London, 1960, 171.
19. **Gitelson, S., Aladjemoff, L., Ben-Hador, S., and Katznelson, R.,** Poisoning by a malathion-xylene mixture, *J. Am. Med. Assoc.,* 197, 165, 1966.
20. **Turkall, R. M., Skowronski, G., Gerges, S., Von Hagen, S., and Abdel-Rahman, M. S.,** Soil adsorption alters kinetics and bioavailability of benzene in orally exposed male rats, *Arch. Environ. Contam. Toxicol.,* 17, 159, 1988.

21. **Turkall, R. M., Skowronski, G. A., and Abdel-Rahman, M. S.,** Differences in kinetics of pure and soil-adsorbed toluene in orally exposed male rats, *Arch. Environ. Contam. Toxicol.,* 20, 155, 1991.

22. National Cooperative Soil Survey: Official Series Description — Atsion Series, Soil Conservation Service, U.S. Department of Agriculture, Washington, D.C., 1977.

23. National Cooperative Soil Survey: Official Series Description — Keyport Series, Soil Conservation Service, U.S. Department of Agriculture, Washington, D.C., 1972.

24. **Jeffay, H. and Alvarez, J.,** Liquid scintillation counting of carbon-14; use of ethanolamine-ethylene glycol monomethyl ether-toluene, *Anal. Chem.,* 33, 612, 1961.

25. **Skowronski, G. A., Turkall, R. M., and Abdel-Rahman, M. S.,** Effects of soil on percutaneous absorption of toluene in male rats, *J. Toxicol. Environ. Health,* 26, 373, 1989.

26. **Skowronski, G. A., Turkall, R. M., Kadry, A. M., and Abdel-Rahman, M. S.,** Effects of soil on the dermal bioavailability of m-xylene in male rats, *Environ. Res.,* 51, 182, 1990.

27. **Tukey, J. W.,** *Exploratory Data Analysis,* Addison-Wesley, Reading, MA, 1977, 205.

28. **Velleman, P. F. and Hoaglin, D. C.,** *Applications, Basics, and Computing of Exploratory Data Analysis (ABC's of EDA),* Duxbury Press, Boston, 1981, 159.

29. **Gibaldi, M. and Perrier, D.,** *Pharmacokinetics,* Marcel Dekker, Inc., New York, 1975, 281.

30. **Skowronski, G. A., Turkall, R. M.,and Abdel-Rahman, M. S.,** Soil adsorption alters bioavailabiltiy of benzene in dermally exposed male rats, *Am. Ind. Hyg. Assoc. J.,* 49, 506, 1988.

31. **Franz, T. J.,** Percutaneous absorption of benzene, in *Advances in Modern Environmental Toxicology, Applied Toxicology of Petroleum Hydrocarbons,* Vol. 6, MacFarland, H. N., Holdsworth, C. E., Mac-Gregor, J. A., Call, R. W., and Lane, M. L., Eds., Princeton Scientific Publishers, Inc., Princeton, NJ, 1984, 61.

32. **Maibach, H. I. and Anjo, D. M.,** Percutaneous penetration of benzene and benzene contained in solvents used in the rubber industry, *Arch. Environ. Health,* 36, 256, 1981.

33. **Susten, A. S., Dames, B. L., Burg, J. R., and Niemeier, R. W.,** Percutaneous penetration of benzene in hairless mice: an estimate of dermal absorption during tire-building operations, *Am. J. Ind. Med.,* 7, 323, 1985.

34. **Hanke, J., Dutriewicz, T., and Piotrowski, J.,** Absorption of benzene through the skin in man, *Med. Prog.,* 12, 413, 1961.

35. **Dutkiewicz, T. and Tyras, H.,** Skin absorption of toluene, styrene, and xylene by man, *Br. J. Ind. Med.,* 25, 243, 1968.

36. **Sato, A. and Nakajima, T.,** Differences following skin or inhalation exposure in the absorption and excretion kinetics of trichloroethylene and toluene, *Br. J. Ind. Med.,* 35, 43, 1978.

37. **Engström, K., Husman, K., and Riihimäki, V.,** Percutaneous absorption of m-xylene in man, *Int. Arch. Occup. Environ. Health,* 39, 181, 1977.

38. **Lauwerys, R. R., Dath, T., Lachapelle, J. M., Buchet, J. P., and Roels, H.,** The influence of two barrier creams on the percutaneous absorption of m-xylene in man, *J. Occup. Med.,* 20, 17, 1978.

39. **Leo, A., Hansch, D., and Elkins, D.,** Partition coefficients and their uses, *Chem. Rev.,* 71, 525, 1971.

40. **Grisham, J. W., Ed.,** Factors influencing human exposure, in *Health Aspects of the Disposal of Waste Chemicals,* Pergamon Press, New York, 1986, 40.

41. **Freed, V. H., Chiou, C. T., and Hague, R.,** Chemodynamics: transport and behavior of chemicals in the environment — a problem in environmental health, *Environ. Health Perspect.,* 20, 55, 1977.

42. **McDougal, J. N., Jepson, G. W., Clewell, J. H., III, and Andersen, M. E.,** Pharmacokinetics of organic vapor absorption, in *Pharmacology and the Skin,* Vol. 1, Shroot, B. and Shaefer, H., Eds., S. Karger, Basel, 1987, 245.

43. **Bartek, M. J., LaBudde, J. A., and Maibach, H. I.,** Skin permeability *in vivo:* comparison in rat, rabbit, pig and man, *J. Invest. Dermatol.,* 58, 114, 1972.

Chapter 8

PREDICTION OF HUMAN PERCUTANEOUS ABSORPTION WITH PHYSICOCHEMICAL DATA

Robert L. Bronaugh and Curtis N. Barton

TABLE OF CONTENTS

I. INTRODUCTION

During skin exposure studies, it would be useful for scientists to be able to estimate the skin absorption of a chemical. Numerous studies have examined the relationship of various properties of chemicals that might aid in this endeavor. Reasonable correlations have been obtained between percutaneous absorption and lipid/water partition coefficients for certain homologous series of compounds: phenols,[1] alcohols,[2] steroids,[3] and hair dyes.[4] In these studies, percutaneous absorption was expressed in terms of a permeability constant (K_p) measured from an aqueous vehicle. In this way, one could extrapolate between doses, because a K_p value by definition is the flux normalized for concentration. The octanol/water partition coefficient (K_{ow}) is now most commonly used as a measure of lipophilicity, because numerous published K_{ow} values are available and because methods are available for K_{ow} estimations based on chemical structure.

It is generally accepted, however, that no reliable correlation could be expected between K_p and K_{ow} with a group of chemicals of diverse molecular structure.[4] Other factors in addition to the oil/water partition coefficient determine rates of percutaneous absorption. Recent studies have shown correlations between percutaneous absorption and molecular volume[5] and molecular weight.[5,6] Inclusion of a molecular weight term (along with K_{ow}) in a regression equation improved the prediction of K_p values for a group of phenols and steroid compounds.[7]

II. COLLECTION OF DATA

We examined the correlation of previously published human percutaneous absorption values with physicochemical properties, using a multiple regression analysis. We thought that use of water solubility [water] values in addition to values for K_{ow} and molecular weight might help us predict skin penetration, in that chemicals with similar K_{ow} values can have vastly different solubilities in water. In addition, it seemed that separating chemicals into three groups on the basis of chemical structure (monocyclic, polycyclic, and aliphatic) might further improve these correlations, because earlier studies correlating small homologous groups of compounds were successful.

Many of the values were obtained from a data base prepared by the U.S. Environmental Protection agency (EPA) for dermal exposure assessment (the individual references cited are provided).[7a] Other values were obtained from Flynn[8] and other sources indicated in Tables 1 to 3. Physical-chemical values were obtained from various sources, as indicated in the tables. The water solubility (also designated as [water]) of compounds listed as being infinitely soluble was expressed numerically as 1000 g/l.

III. CORRELATION ANALYSIS

We performed the correlation analyses using SAS (SAS/STAT Release 6.03, SAS Institute Inc., Cary, NC).

When log K_{ow} was plotted against log K_p, a poor correlation was obtained (Figure 1A). The R^2 value for all compounds combined was 0.22 (Table 4). When the compounds were divided into three groups on the basis of similarity in chemical structure, an improved R^2 correlation was obtained for each group (Table 4; polycyclic = 0.43, monocyclic = 0.63, and aliphatic = 0.60). The correlation of molecular weight with percutaneous absorption was found to be poor in the combined group ($R^2 = 0.15$) and even worse when compounds were separated into the smaller groups (Figure 2 and Table 4).

However, when a multiple regression analysis was conducted combining the predictive effects of log K_{ow} and molecular weight on log K_p, the improved correlation (as indicated

TABLE 1
Monocyclic Aromatic Compounds: Physicochemical and Absorption Values

Compound	log K_{ow}	log Water solubility (g/l)	Molecular weight	Measured log K_p[a]	Estimated log K_p ±95% C.I.	Ref.
Amobarbital	1.96	0.769	226	−2.64	−3.11 ± 1.56	8
Aniline	0.9	1.54	93.1	−1.39	−2.24 ± 1.46	9
Barbital	0.65	0.74	184	−3.95	−3.71 ± 1.51	8
Benzene	2.13	−0.15	78.1	−0.95	−0.94 ± 1.48	10
Benzyl alcohol	1.10	1.60	108	−2.22	−2.27 ± 1.44	8
p-Bromophenol	2.65	2.16	173	−1.44	−1.77 ± 1.46	1
Butabarbital	1.65	—	212	−3.71	−3.20 ± 1.53	8
2-Chlorophenol	2.15	1.45	129	−1.48	−1.61 ± 1.43	1
4-Chlorophenol	2.39	1.43	129	−1.44	−1.39 ± 1.44	1
4-Chloro-*m*-phenylenediamine	0.84	0.45[b]	142	−2.68	−2.97 ± 1.45	4
Chloroxylenol	3.39	−0.48	157	−1.28	−0.89 ± 1.48	1
p-Cresol	1.94	1.30	108	−1.76	−1.51 ± 1.44	1
o-Cresol	1.95	1.40	108	−1.80	−1.51 ± 1.44	1
m-Cresol	1.96	1.40	108	−1.82	−1.52 ± 1.44	1
2,4-Diclorophenol	2.92	0.653	163	−1.22	−1.39 ± 1.46	1
1,4-Dioxane	−0.42	3.0	88.1	−3.37	−3.36 ± 1.53	11
Ephedrine	1.03	1.70[c]	165	−2.22	−3.11 ± 1.46	8
Ethylbenzene	3.15	−0.68	106	0.14	−0.41 ± 1.49	12
4-Ethylphenol	2.26	0.68[d]	106	−1.46	−1.20 ± 1.45	1
3-Nitrophenol	2.0	1.13	139	−2.25	−1.89 ± 1.43	1
4-Nitrophenol	1.91	1.20	139	−2.25	−1.97 ± 1.43	1
p-Phenylenediamine	−0.7	1	108	−3.62	−3.88 ± 1.54	4
o-Phenylenediamine	0.15	1.62	108	−3.35	−3.12 ± 1.48	4
2-Nitro-*p*-phenylenediamine	0.53	0.26[b]	153	−3.30	−3.39 ± 1.47	4
Phenol	1.46	1.82	94.1	−2.09	−1.75 ± 1.45	1
Resorcinol	0.8	3.16	110	−3.62	−2.56 ± 1.45	1
Styrene	2.95	−0.60[e]	104	−0.17	−0.56 ± 1.48	13
Thymol	3.3	0.0	150	−1.28	−0.88 ± 1.47	1
Toluene	2.73	−0.14	92.1	0.00	−0.59 ± 1.48	13
2,4,6-Trichlorophenol	3.69	−0.046	198	−1.23	−1.17 ± 1.54	1

Note: K_{ow}, octanol/water partition coefficient; K_p, permeability constant; C. I., confidence interval.

[a] K_p is expressed as centimeters per hour.
[b]   Division of Colors and Cosmetics, Food and Drug Administration, Washington, D.C.
[c]   *U.S. Pharmacopeia*, 23rd Rev., U. S. Pharmacopeial Convention, Rockville, MD, 1990, 1851.
[d]   Valvani, S. C. and Yalkowsky, S. H., Solubility and partitioning in drug design, in *Physical Chemical Properties of Drugs,* Yalkowski, S. H., Sinkula, A. A., and Valvani, S. C., Eds., Marcell Dekker, New York, 1980, 201–229.
[e]   *Handbook of Chemistry and Physics*, CRC Press, Boca Raton, FL, 1990, 16–27.

by the R^2 values) increased much more than the sum of the values obtained when the two variable were examined by themselves (Figures 3 to 6 and Table 4). For the monocyclic compounds (the best case), only 15% of the uncertainty in the estimation of log K_p was unaccounted for if the log K_{ow} and molecular weight were known. This remaining uncertainty in the estimation could easily be explained by the error in obtaining the human absorption data used in these analyses. It is known that the permeability properties of human skin can vary five- to tenfold. Variability in the data for the many compounds tabulated for this comparison is probably due in part to the use of various specimens of human skin. The studies that provided absorption data were conducted in many laboratories, which also contributed to the observed variability.

TABLE 2
Aliphatic Compounds: Physicochemical and Absorption Values

Compound	log K_{ow}	log Water solubility (g/l)	Molecular weight	Measured log K_p[a]	Estimated log K_p ±95% C.I.	Ref.
2,3-Butanediol	−0.92	—	90.1	−4.30	−3.72 ± 1.52	14
Butanoic acid	0.79	3.0	88.1	−3.00	−2.45 ± 1.44	8
n-Butanol	0.88	1.96	74.1	−2.60	−2.22 ± 1.45	15
2-Butanone	0.28	2.40	72.1	−2.35	−2.64 ± 1.45	14
n-Decanol	4.00	−1.50[b]	158	−1.10	−0.91 ± 1.54	15
Ethanol	−0.31	3.0	46.1	−1.10	−2.77 ± 1.48	15
2-Ethoxyethanol	−0.10	3.0	90.1	−3.52	−3.13 ± 1.46	14
Ethyl ether	0.89	1.78	74.1	−1.77	−2.22 ± 1.45	14
Heptanoic acid	2.50	0.38	130	−1.70	−1.68 ± 1.46	8
n-Heptanol	2.72	0	116	−1.49	−1.36 ± 1.47	14
Hexanoic acid	1.90	1.03	116	−1.85	−1.96 ± 1.44	8
n-Hexanol	2.03	0.80[b]	102	−1.56	−1.71 ± 1.45	16
Methanol	−0.77	3.0	32.0	−2.89	−2.94 ± 1.51	17
Nitroglycerine	2.0	0.097	227	−1.96	−3.17 ± 1.74	8
N-Nitrosodiethanolamine	−1.18	3.0	134	−5.22	−4.42 ± 1.66	18
n-Nonanol	3.62	−0.84[b]	144	−1.22	−1.03 ± 1.51	15
Octanoic acid	3.00	−0.17	144	−1.60	−1.48 ± 1.48	8
n-Octanol	2.97	−0.52	130	−1.21	−1.34 ± 1.48	17
Pentanoic acid	1.30	1.52	102	−2.70	−2.24 ± 1.44	8
n-Pentanol	1.56	1.43	88.2	−2.22	−1.89 ± 1.45	15
n-Propanol	0.25	3.0	60.1	−2.85	−2.52 ± 1.46	14
Water	−1.38	3.0	18.0	−2.82	−3.23 ± 1.55	19

Note: K_{ow}, octanol/water partition coefficient; K_p, permeability constant; C. I., confidence interval.

[a] K_p is expressed as centimeters per hour.
[b] Small, D. M., *The Physical Chemistry of Lipids*, Plenum Press, New York, 1986, 233–284.

The importance of knowing [water] to predict percutaneous absorption was also examined. Multiple regression analyses were performed on the compounds in Tables 1 to 3 when [water] data were available. As shown in Table 4, no improvement was obtained in predicting log K_p by including [water] in the regression equation. The solubility properties of the molecules as they relate to percutaneous absorption are adequately represented by the K_{ow}.

Statistical analysis was performed to determine whether membership in a chemical structure group was useful for estimating log K_p. A multiple regression model with distinct intercepts and slopes for the three chemical structure groups was compared to a model with a common intercept and slopes for all compounds (combined group). The model with distinct intercepts and slopes for the three chemical structure groups fit significantly better ($F_{6,85}$ = 4.76, P = 0.0003), a result that indicates group membership was useful for estimating log K_p. The model with separate slopes and intercepts for the groups yielded an R^2 of 0.72 for the entire set of compounds, whereas the model (combined group) in which group membership was ignored yielded an R^2 of 0.63. The regression equations for the three chemical structure groups and the combined group are

Aliphatic	$\log K_p = -2.01 + 0.731 \log K_{ow} - 0.0115 \text{ MW}$
Monocyclic	$\log K_p = -1.77 + 0.895 \log K_{ow} - 0.0137 \text{ MV}$
Polycyclic	$\log K_p = -3.93 + 0.637 \log K_{ow} - 0.0026 \text{ MV}$

TABLE 3
Polycyclic Aromatic Compounds: Physicochemical and Absorption Values

Compound	log K_{ow}	log Water solubility (g/l)	Molecular weight	Measured log K_p[a]	Estimated log K_p ±95% C.I.	Ref.
Aldosterone	1.08	—	360	−5.52	−4.18 ± 1.44	8
Atropine	1.79	0.34	289	−5.05	−3.54 ± 1.43	8
Caffeine	−0.07	1.3	194	−3.25	−4.48 ± 1.49	20
Codeine	0.89	0.92	299	−4.31	−4.14 ± 1.44	8
Cortexolone	2.52	—	346	−4.13	−3.23 ± 1.42	8
Corticosterone	1.97	−0.99	346	−4.22	−3.58 ± 1.42	8
Cortisone	1.64	−0.55	360	−5.00	−3.82 ± 1.43	8
Deoxycorticosterone	2.88	−0.93[b]	330	−3.35	−2.95 ± 1.42	8
Digitoxin	1.73	−2.0	765	−4.89	−4.82 ± 1.63	8
Estradiol	2.69	−2.82	272	−2.28	−2.93 ± 1.43	8
Estriol	2.47	—	288	−4.40	−3.11 ± 1.43	8
Estrone	2.76	−1.52	270	−2.44	−2.88 ± 1.43	8
Etorphine	1.86	—	412	−2.44	−3.82 ± 1.43	8
Fentanyl	4.37	−1.29	337	−2.25	−2.02 ± 1.46	8
Fluocinonide	3.19	—	495	−2.77	−3.19 ± 1.45	8
Hydrocortisone (HC)	1.61	−0.55	362	−5.52	−3.85 ± 1.43	8
HC *N,N*-dimethyl succinamate	2.03	—	490	−4.17	−3.91 ± 1.45	8
HC hemipimelate	3.26	—	505	−2.75	−3.17 ± 1.45	8
HC hexanoate	4.48	—	461	−1.75	−2.28 ± 1.47	8
HC hydroxy hexanoate	2.79	—	477	−3.04	−3.39 ± 1.44	8
HC methyl pimelate	3.70	—	519	−2.27	−2.92 ± 1.46	8
HC methyl succinate	2.58	—	477	−2.27	−3.53 ± 1.44	8
HC octanoate	5.49	—	489	−1.21	−1.70 ± 1.51	8
HC pimelamate	2.31	—	504	−3.05	−3.77 ± 1.45	8
HC proprionate	3.00	—	419	−2.47	−3.11 ± 1.43	8
HC succinamate	1.43	—	462	−4.59	−4.22 ± 1.45	8
Hydromorphone	1.25	0.35	285	−4.82	−3.88 ± 1.43	8
Hydroxypregnenolone	3.00	—	330	−3.22	−2.88 ± 1.43	8
17a-Hydroxy-progesterone	2.74	−2.19[c]	330	−3.22	−3.04 ± 1.42	8
Isoquinoline	2.03	0.66[d]	129	−1.78	−2.97 ± 1.48	8
Meperidine	2.72	0.82	247	−2.43	−2.84 ± 1.44	8
Morphine	0.62	−0.70	285	−5.03	−4.28 ± 1.45	8
2-Naphthol	2.84	0.0	144	−1.55	−2.50 ± 1.48	1
Naproxen	3.18	—	230	−3.40	−2.50 ± 1.45	8
Nicotine	1.17	3.0	162	−1.71	−3.61 ± 1.47	8
Phenobarbital	1.47	0.0	232	−3.34	−3.60 ± 1.44	8
Pregnenolone	3.13	—	316	−2.82	−2.76 ± 1.43	8
Progesterone	3.87	−2.14	314	−2.82	−2.28 ± 1.45	8
Scopolamine	1.23	2.02	303	−4.30	−3.94 ± 1.43	8
Sucrose	−2.25	3.3	342	−5.28	−6.25 ± 1.61	8
Sufentanyl	4.59	−1.47	388	−1.92	−2.02 ± 1.47	8
Testosterone	3.32	−2.0	288	−3.40	−2.57 ± 1.44	8

Note: K_{ow}, octanol/water partition coefficient; K_p, permeability constant; C. I., confidence interval.

[a] K_p is expressed as centimeters per hour.
[b] Valvani, S. C. and Yalkowsky, S. H., Solubility and partitioning in drug design, in *Physical Chemical Properties of Drugs,* Yalkowsky, S. H., Sinkula, A. A., and Valvani, S. C., Eds., Marcell Dekker, New York, 1980, 201–229.
[c] Lundberg, B., Temperature effect on the water solubility and water-octanol partition of some steroids, *Acta Pharm. Suec.,* 16, 151, 1979.
[d] Pfleiderer, W., The solubility of heterocyclic compounds, in *Physical Methods in Heterocyclic Chemistry,* Vol. 1, Academic Press, New York, 1963, 177–188.

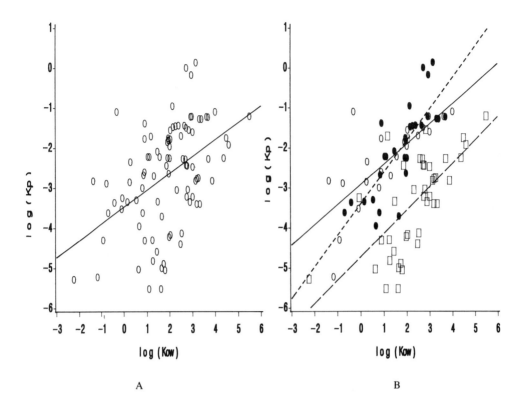

FIGURE 1.   Linear regression of the log of the permeability constant (K_p) and the log of the octanol/water partition coefficient (K_{ow}). (A) Combined plot of all compounds; (B) separate plots for three different groups: aliphatic ($\circ$), monocyclic ($\bullet$), and polycyclic ($\square$). The linear regression is shown for each of the following groups: aliphatic (solid line), monocyclic (short dashed line), and polycyclic (long dashed line).

Combined $\log K_p = -2.57 + 0.680 \log K_{ow} - 0.0062 \text{ MV}$

where MW is molecular weight.

The accuracy of estimation for the regression equations for the chemical structure groups was evaluated by examining the average size of the estimation errors for log K_p. The mean absolute error, which is the mean of the absolute values of the differences between the observed and estimated values, was computed for each group. The mean absolute errors for log K_p were 0.41 for the aliphatic compounds, 0.34 for the monocyclic compounds, and 0.71 for the polycyclic compounds. These errors for log K_p correspond to multiplicative factors for K_p of 2.57 for the aliphatic compounds, 2.19 for the monocyclic compounds, and 5.13 for the polycyclic compounds.

The log K_p values for all compounds were estimated with the appropriate equation, and 95% confidence intervals were calculated to give a measure of worst case error for each compound (Tables 1 to 3). Possible errors in estimation were similar for all compounds and averaged about ±1.5 orders of magnitude.

IV. CONCLUSIONS

The log K_p values for 94 compounds previously determined in human skin have been reasonably accurately predicted by using a multiple correlation analysis with log K_{ow} and

TABLE 4
Correlation Analyses Data

Compounds	Variable	R^2 value data set[a] Complete	[Water]
Polycyclic	log K_{ow}	0.43	
	MW	0.01	
	log [water]		0.02
	log K_{ow}, MW	0.49	0.60
	log K_{pw}, MW, log [water]		0.62
Monocyclic	log K_{ow}	0.63	
	MW	0.08	
	log [water]		0.33
	log K_{ow}, MW	0.85	0.85
	log K_{ow}, MW, log [water]		0.86
Aliphatic	log K_{ow}	0.60	
	MW	0.07	
	log [water]		0.49
	log K_{ow}, MW	0.72	0.67
	log K_{ow}, MW, log [water]		0.69
All compounds	log K_{ow}	0.22	
	MW	0.15	
	log [water]		0.01
	log K_{ow}, MW	0.63	0.73
	log K_{ow}, MW, log [water]		0.74

Note: K_{ow}, octanol/water partition coefficient; MW, molecular weight.

[a] Data set used was either all appropriate compounds tabulated in Tables 1 to 3 (complete) or compounds with water solubility [water] data listed in the tables.

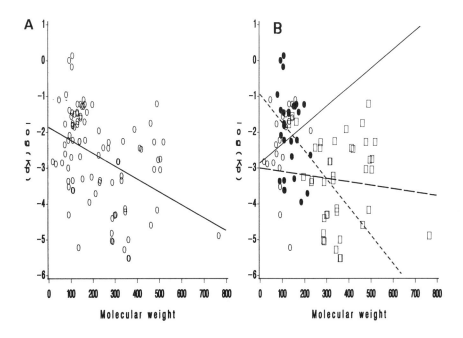

FIGURE 2. Linear regression of the log of the permeability constant (K_p) and molecular weight. (A) Combined plot of all compounds; (B) separate plots for three different groups: aliphatic (○), monocyclic (●), and polycyclic (■). The linear regression is shown for each of the following groups: aliphatic (solid line), monocyclic (short dashed line), and polycyclic (long dashed line).

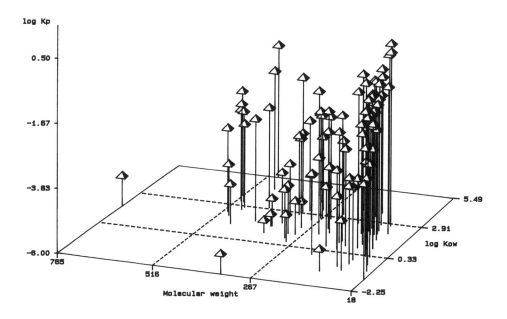

FIGURE 3. Three-dimensional plot of the log of the permeability constant (K_p), the log of the octanol/water partition coefficient (K_{ow}), and the molecular weight for all compounds.

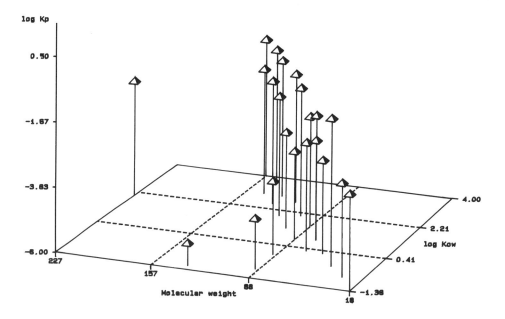

FIGURE 4. Three-dimensional plot of the log of the permeability constant (K_p), the log of the octanol/water partition coefficient (K_{ow}), and molecular weight for aliphatic compounds.

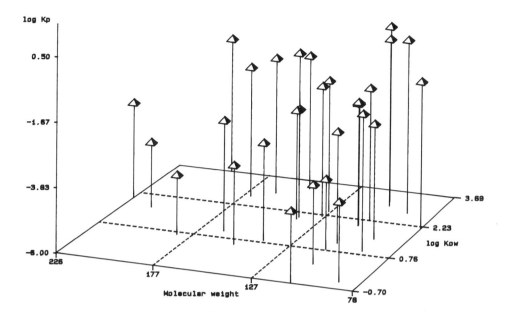

FIGURE 5. Three-dimensional plot of the log of the permeability constant (K_p), the log of the octanol/water partition coefficient (K_{ow}), and molecular weight for monocyclic compounds.

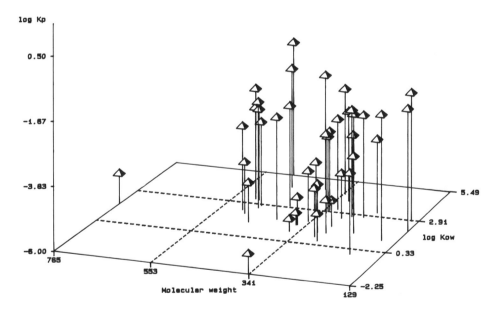

FIGURE 6. Three-dimensional plot of the log of the permeability constant (K_p), the log of the octanol/water partion (K_{ow}), and molecular weight for polycyclic compounds.

molecular weight as independent variables. A significant reduction in error is obtained by dividing the compounds into three broad groups on the basis of similar chemical structure before performing the regression analysis. Although the error in estimation of log K_p may be as large as ± 1.7 orders of magnitude, the permeability values may be useful when no experimental values can be obtained.

REFERENCES

1. **Roberts, M. S., Anderson, R. A., and Swarbrick, J.,** Permeability of human epidermis to phenolic compounds, *J. Pharm. Pharmacol.,* 29, 677, 1977.
2. **Scheuplein, R. J.,** Mechanism of percutaneous absorption. I. Routes of penetration and influence of solubility, *J. Invest. Dermatol.,* 45, 334, 1965.
3. **Scheuplein, R. J., Blank, I. H., Brauner, G. J., and MacFarlane, D. J.,** Percutaneous absorption of steroids, *J. Invest. Dermatol.,* 52, 63, 1969.
4. **Bronaugh, R. L. and Congdon, E. R.,** Percutaneous absorption of hair dyes: Correlation with partition coefficients, *J. Invest. Dermatol.,* 83, 124, 1984.
5. **Kasting, G. B., Smith, R. L., and Cooper, E. R.,** Effect of lipid solubility and molecular size on percutaneous absorption, *Pharmacol. Skin,* 1, 138, 1987.
6. **Roy, S. D. and Flynn, G. L.,** Transdermal delivery of narcotic analgesics: Comparative permeabilities of narcotic analgesics through human cadaver skin, *Pharm. Res.,* 6, 825, 1989.
7. **Anderson, B. A. and Raykar, P. V.,** Solute structure-permeability relationships in human stratum corneum, *J. Invest. Dermatol.,* 93, 280, 1989.
7a. **U. S. Environmental Protection Agency,** Dermal exposure assessment: principles and applications, Interim Rept. No. EPA/600/8-91/011B, Washington, D. C., 1992.
8. **Flynn, G. L.,** Physicochemical determinants of skin absorption, in *Principles of Route-to-Route Extrapolation for Risk Assessment,* Gerrity, T. R., and Henry, C. J., Eds., Elsevier Science Publishing Co., New York, 1990, 93.
9. **Baranowska-Dutkiewicz, B.,** Skin absorption of aniline from aqueous solutions in man, *Toxicol. Lett.,* 10, 367, 1982.
10. **Blank, I. H. and McAuliffe, D. J.,** Penetration of benzene through human skin, *J. Invest. Dermatol.,* 85, 522, 1985.
11. **Bronaugh, R. L.,** Percutaneous absorption of cosmetic ingredients, in *Principles of Cosmetics for the Dermatologist,* Frost, P. and Horowitz, S. N., Eds., C. V. Mosby Co., St. Louis, MO, 1982, 244.
12. **Dutkiewicz, T. and Tyras, H.,** A study of the skin absorption of ethylbenzene in man, *Br. J. Ind. Med.,* 24, 330, 1967.
13. **Dutkiewicz, T. and Tyras, H.,** Skin absorption of toluene, styrene, and xylene by man, *Br. J. Ind. Med.,* 25, 243, 1968.
14. **Blank, I. H., Scheuplein, R. J., and MacFarlane, D. J.,** Mechanism of percutaneous absorption. III. The effect of temperature on the transport of non-electrolytes across the skin, *J. Invest. Dermatol.,* 49, 582, 1967.
15. **Scheuplein, R. J. and Blank, I. H.,** Mechanism of percutaneous absorption. IV. Penetration of non-electrolytes (alcohols) from aqueous solutions and from pure liquids, *J. Invest. Dermatol.,* 60, 286, 1973.
16. **Bond, J. and Barry, B.,** Limitations of hairless mouse for *in vitro* permeation studies through human skin, *J. Invest. Dermatol.,* 90, 486, 1988.
17. **Southwell, D., Barry, B., and Woodford, R.,** Variations in permeability of human skin within and between specimens, *Int. J. Pharm.,* 18, 299, 1984.
18. **Bronaugh, R. L., Congdon, E. R., and Scheuplein, R. J.,** The effect of cosmetic vehicles on the penetration of *N*-nitrosodiethanolamine through excised human skin, *J. Invest. Dermatol.,* 76, 94, 1981.
19. **Bronaugh, R. L., Stewart, R. F., and Simon, M.,** Methods for *in vitro* percutaneous absorption studies. VII. Use of excised human skin, *J. Pharm. Sci.,* 75, 1094, 1986.
20. **Bronaugh, R. L. and Franz, T. J.,** Vehicle effects on percutaneous absorption: *in vitro* and *in vitro* comparisons with human skin, *Br. J. Dermatol.,* 115, 1, 1986.

Chapter 9

DERMAL ABSORPTION OF TCDD: EFFECT OF AGE

Yolanda Banks Anderson and Linda S. Birnbaum*

TABLE OF CONTENTS

---

* This document has been reviewed in accordance with U.S. Environmental Protection Agency policy and approved for publication. Mention of tradenames or commercial products does not constitute endorsement or recommendation for use.

I. INTRODUCTION

Absorption through the skin is a major route for direct entry of chemicals into systemic circulation. Although the skin functions as a barrier following exposure, it is selectively permeable and does not completely prevent penetration of many chemicals. The principles governing percutaneous absorption of therapeutic agents and endogenous substances as well as the factors modulating absorption after intentional application have been well studied.[1] Changes in absorption which occur during development and aging of skin have also been investigated.[2] These principles and factors are also applicable to absorption which occurs after dermal exposure to environmental chemicals. However this group of chemicals has been less well examined.

TCDD, 2,3,7,8-tetrachlorodibenzo-*p*-dioxin, is an environmental contaminant which has been detected in every segment of the environment. Several human populations have received acute and chronic exposure to this chemical through ingestion, inhalation, and dermal contact. Observed toxic responses include chloracne, hyperpigmentation of the nails, gastrointestinal disorders, weight loss, anorexia, nausea, joint pain, and fatigue.[3] Chloracne, the most diagnostic indicator of acute human exposure to TCDD, has been reported to persist for 30 years.[4,5]

This chapter will briefly review literature reports on the effect of age on absorption of topically applied substances. The focus of this chapter, however, will be recent *in vivo* studies of the effect of age on percutaneous absorption of TCDD.

II. CHANGES IN PERCUTANEOUS ABSORPTION WITH AGE

A. HUMAN STUDIES

Clinical reports of systemic toxicity after dermal application of therapeutic agents in very young age groups suggest that the skin of persons in these age groups is much more permeable to some chemicals than that of adults.[6,7] Reports of studies regarding absorption in children at other ages are extremely limited. Skin barrier function in infants and children has been measured primarily in terms of ability to limit water loss.[8,9]

Differences in dermal absorption between adult and aged persons have also been reported. Results from *in vivo* absorption studies conducted by Christophers and Kligman[10] indicate decreased percutaneous absorption of topically applied chemicals in aged subjects and decreased dermal clearance after intradermal injection. They also reported increased *in vitro* penetration of hydrophilic fluorescent dyes through epidermis from elderly persons, resolving the difference between the *in vivo* and *in vitro* results in terms of the effect of blood flow on the absorption process.[10] Tagami[11] observed decreased penetration of fluorescent dyes through the stratum corneum and prolonged dermal clearance of radio-labeled sodium chloride in aged persons. In a more recent study, Rougier et al.[12] and Roskos et al.[13,14] reported decreased absorption of benzoic acid in aged subjects. Roskos et al.[13,14] also reported an age-related decrease in penetration of hydrocortisone. However, no age-related changes in penetration of testosterone and estradiol were observed.

Physiologic changes in the skin at different ages have been examined in an attempt to explain differences in absorption. Holbrook[15] observed that the epidermis of the full-term infant was similar to that of the adult. Evans and Rutter[16] also reported that human skin is structurally mature at birth although it continues to thicken during the first few months of life. The increased thickness is believed due to formation of characteristic ridges in the interface between the epidermis and dermis. Fisher,[17] however, suggested that infant stratum corneum is thicker than adult stratum corneum and that this indicated structural immaturity. No differences in stratum corneum thickness were observed between adult and aged human skin, although the epidermis of aged persons is decreased in thickness and the epidermal-dermal interface becomes flattened.[10,18,19] Vogel et al.[20] reported that the total thickness of human skin increases during maturation, reaches a maximum thickness at 27 years of age, and decreases during senescence. Leveque et al.[21] observed that the decrease in skin thickness begins around 45 years of age. Changes in the collagen and elastin content and structure with age have been reported.[18,19,22]

Other structural changes have also been reported to influence the changes in absorption observed in various age groups. McCormack et al.[23] suggested that the higher skin lipid content of infants was responsible for the observed increased penetration of fatty acids, but not nonvolatile alcohols. Other investigators have reported differences in skin lipid composition and content in various age groups, yet have not suggested a relationship to absorption changes.[24–27] Regression of the capillaries perfusing the papillary dermis of aged skin has also been reported.[28–30]

B. OTHER SPECIES

Results of studies comparing dermal absorption in young and adult animals indicate that observed changes may be chemical- and species-specific. Wester et al.[31] examined percutaneous absorption of testosterone in newborn and adult monkeys and reported no differences. These workers cautioned that the threefold increase in the ratio of dose to total body surface area in the younger group could lead to systemic effects. Shah et al.[32] examined dermal absorption of 14 pesticides in young and adult rats, reporting that differences in absorption between the two groups were compound- and dose-specific. In a separate study of developmental changes in absorption of the hydrophilic pesticide, carbofuran, absorption at all doses was increased in the younger age group.[33] Observed differences in penetration of 2,4,5,2′,4′,5′-hexachlorobiphenyl through young compared with adult rat skin were time-dependent.[34] Total absorption was greater in the younger age group 72 h after application, but no differences were observed at 120 h. Behl and co-workers observed differences in absorption of testosterone and a series of alkanols throughout the life span of the hairless mouse. They concluded that the observed differences were due to changes in skin thickness during stages of the single hair cycle.[35,36]

Few studies have investigated physiologic changes which can affect absorption in animal skin. Vogel[20] reported fluctuations in the thickness of rat skin, however no observations of changes associated with individual skin layers were reported. In studies of the ultrastructure of stratum corneum from neonatal and adult guinea pigs and rats, Singer et al.[37] reported finding no differences between the two age groups. Monteiro-Riviere et al.[38] examined skin thickness of Fischer 344 rats ranging in age from 1 to 24 months. Epidermal thickness (not including stratum corneum) decreased between 1 and 3 months of age and remained constant from 3 to 24 months. Dermal thickness increased sharply between 1 and 2 months of age and then remained unchanged. Skin blood flow is also reported to vary with age. Monteiro-Riviere et al.[38] reported that blood flow through the interscapular region of the skin was unchanged from 1 to 2 months, decreased between 2 and 3 months of age, and remained constant from 3 to 24 months. Yates and Hiley[39] also reported an age-related decrease in blood flow through the skin, but the decrease noted in this study occurred in middle-aged rats (11 to 12 months old) as compared to young adult (3 to 4 months old) Wistar rats. The discrepancies in these reports may be due to methodological differences. Yates and Hiley[39] measured entrapment of radio-labeled microspheres in skin capillaries, whereas Monteiro-Riviere et al.[38] determined blood flow using laser doppler velocimetry, which measures the movement of blood cells. Therefore, although age has been shown to affect dermal absorption, there is no consensus on the basis for these changes.

III. DERMAL ABSORPTION OF TCDD

TCDD is the most potent isomer in the larger class of toxic polychlorinated dibenzo-*p*-dioxins and dibenzofurans. Due to its pervasiveness and persistence, there is great potential for dermal contact with this chemical. Its half-life in soil is reported to range from 1 to 2 years in surface layers to 10 to 12 years when located in deeper layers.[40,41] TCDD has also been detected in commonly used paper products, including cosmetic tissue and newspaper.[42] Its presence has been primarily attributed to chlorine bleaching.[43,44]

Earlier studies evaluated the content of TCDD in the liver, the major depot for this chemical, as an indicator of absorption following dermal application.[45,46] Poiger and Schlatter[45] examined penetration of TCDD after application in methanol, polyethylene glycol, vaseline, and soil. Approximately 15% of the dose was detected in the liver 24 h after application of TCDD in methanol; dermal application of TCDD in the other vehicles resulted in a dramatic decrease in the percent of the dose detected in the liver compared to application in methanol. In the same report, the authors noted that increasing the contact time between the soil and TCDD prior to application resulted in a decrease in absorption. This change was due to increased adsorption of TCDD onto the soil particles. Shu et al.[46] also examined dermal absorption of TCDD after application in soil and observed that peak levels of TCDD in the liver were observed 48 h after application. Absorption was not affected by dose nor was absorption altered when crankcase oil was added to the TCDD-soil mixture. The bioavailability of TCDD after application in soil relative to oral administration in corn oil was estimated at 1.3% in haired and hairless rats and was independent of the soil preparation method (laboratory-contaminated vs. environmentally contaminated).

A. EFFECT OF AGING

Recently, the question of the effect of aging on dermal absorption of TCDD was addressed in male Fischer 344 rats. This species was chosen in order to compare results from this study with results obtained after administration via other routes. Age groups were selected to represent various stages of aging: early adulthood (10-week-old rats), middle age (36-week-old rats), and senescence (96-week-old rats). Because of the approximately twofold difference in body weight between 10-week-old and 36- or 96-week-old rats, two groups of 10-week-old rats were included in this study. This allowed comparison of age effects while maintaining a constant dose per surface area as well as while applying a constant dose per body weight. In contrast to previously mentioned studies, the extent of percutaneous absorption was determined by measuring tissue distribution and elimination of TCDD following topical application. The methods used in this study have been reported elsewhere.[47]

Cumulative absorption of TCDD 72 h after dermal application to the dorsal interscapular region is shown in Table 1. Absorption efficiency, measured as percent of administered dose, was less in 36- and 96-week-old rats than in either group of 10-week-old rats. There was an approximately 2.5- to threefold difference in relative absorption of TCDD between 10-week-old and 36-week-old rats. Absorption in senescent (96-week-old) rats was unchanged compared to rats from the middle-age (36-week-old) group. Also shown in Table 1 are data for the mass of TCDD absorbed in each age group. The majority of the unabsorbed dose of TCDD was located on the surface or within the stratum corneum of the application site.[48] Other studies in our laboratory have shown that absorption of a lower dose of TCDD is essentially complete within 72 h after application.[49] It is not known whether this is true for this higher dose or whether the TCDD absorption time course at this dose would be limited by the solubility of TCDD in stratum corneum lipids.

Distribution of absorbed TCDD to major tissue depots was independent of the age of the animal, as shown in Table 2. Liver, adipose tissue, skin, and muscle accounted for the

TABLE 1

Cumulative Dermal Absorption of TCDD 72 h after Dermal Application to 1.8 cm^2

Age (weeks)	Dose nmol/kg	Dose nmol/rat	Absorption[a] % Administered dose	Absorption[a] Mass (nmol)
10	100	20	17.7 ± 2.6[b]	3.7 ± 0.4[b]
10	200	40	13.4 ± 1.9[b]	5.4 ± 0.8[b]
36	100	40	5.6 ± 2.5	2.2 ± 1.0
96	100	40	4.5 ± 0.7	1.8 ± 0.3

[a] Values are mean ± SD of three to four rats.
[b] Significantly greater than 36- and 96-week-old rats ($p < 0.05$).

TABLE 2

Tissue Distribution and Elimination of Absorbed Dose of TCDD 72 h after Dermal Application to 1.8 cm^2

Age (weeks)	Dose[a]	Liver	Adipose tissue	Skin	Muscle	Feces	Urine
10	100 (20)	55.8 ± 3.5[b]	17.0 ± 3.3	5.6 ± 1.7	10.8 ± 2.6[c]	10.3 ± 3.3	1.0 ± 0.5[d]
10	200 (40)	58.6 ± 2.5	10.8 ± 1.8	3.9 ± 0.9	3.8 ± 1.1	15.0 ± 2.0	3.0 ± 0.7[d]
36	100 (40)	56.0 ± 9.0	10.4 ± 2.1	8.1 ± 3.6	8.1 ± 3.6	10.5 ± 4.4	6.7 ± 2.0
96	100 (40)	56.0 ± 7.6	14.7 ± 4.8	9.6 ± 4.3	4.3 ± 2.8	8.8 ± 3.0	7.0 ± 2.3

[a] nmol/kg (nmol/rat).
[b] Values represent percentage of absorbed dose (mean ± SD) of three to four rats.
[c] Significantly different from 96-week-old rats ($p < 0.05$).
[d] Significantly different from 36- and 96-week-old rats ($p < 0.05$).

majority of absorbed chemical detected in the tissues 72 h after treatment. The ratio of the relative amount of TCDD in liver and fat (approximately four to one) was consistent with previously reported results.[50] On the basis of results from earlier studies,[51,52] it was assumed that the absorbed dose was parent chemical and not metabolite. Elimination data supported this assumption. In older age groups, elimination of TCDD via the urine was increased. This presumably reflected age-dependent alteration in renal function such as glomerular sclerosis, hyperfiltration, increased glomerular permeability to large molecules such as albumin, and progressive loss of renal function.[53] Since albumin transports TCDD in the blood,[54] these changes would result in increased elimination of TCDD in the urine. A similar, although isomer-specific, increase in urinary elimination has been reported for other polyhalogenated aromatic hydrocarbons.[55]

B. DEVELOPMENTAL CHANGES

Developmental changes in percutaneous absorption of a low dose of TCDD (200 pmol) were recently examined by comparing absorption in weanling (3-week-old), juvenile (5-week-old), pubescent (8-week-old), and young adult (10-week-old) rats.[56] This dose (100- to 200-fold lower than the dose used in the aging study) was selected to preclude induction of systemic effects in the younger age groups. Middle-aged (36-week-old) rats were included in the developmental study in order to overlap the aging study. The methods used in this study were identical to those used to examine the effect of senescence on dermal absorption of this chemical.[47]

Preliminary results indicate that absorption of TCDD decreased during maturation. These changes are in addition to those observed during aging. In each age group, the majority of unabsorbed chemical remaining at the site of application could be removed by swabbing with acetone. Observed differences in distribution of absorbed TCDD to major depots were related to age-related variation in body mass of these tissues, however changes in tissue concentration were also observed. Elimination of the absorbed dose was independent of the age of the animal.

IV. CONCLUSIONS

These studies show that percutaneous absorption of TCDD decreases with increasing age. A major decline occurs prior to middle age, and preliminary results suggest that this is preceded by another age-related decrease which occurs during maturation. These changes are consistent with previously mentioned reports of increased percutaneous absorption in younger age groups compared to adults.[32-34] They are also consistent with reports of differences in absorption between young adult and aged humans.[10,12] Previous studies have not established whether there is a gradual change in absorption of chemicals prior to adulthood or whether there is a more specific period during which the skin becomes less permeable. The results presented here suggest that decrements in dermal absorption of TCDD are not gradual, but occur during distinct periods of development and aging. By using TCDD as a model for other lipophilic dioxins and furans, the results of these studies can be extrapolated to this class of chemicals. In fact, a similar decrease in absorption during senescence has been observed for 2,3,4,7,8-pentachlorodibenzofuran, another congener in this class.[47] Therefore, decreased absorption with increasing age may result in a decreased risk of systemic exposure and potential systemic effects following dermal exposure to TCDD and other dioxins and furans.

REFERENCES

1. **Barry, B. W.,** Properties that influence percutaneous absorption, in *Dermatological Formulations: Percutaneous Absorption,* Barry, B. W., Ed., Marcel Dekker, New York, 1983, 127.
2. **Ramussen, J. E.,** Percutaneous absorption in children, in *Year Book of Dermatology,* Dobson, R. L., Ed., Year Book Medical Publishers, Chicago, 1979, 25.
3. **Skene, S. A., Dewhurst, I. C., and Greenberg, M.,** Polychlorinated dibenzo-*p*-dioxins and polychlorinated dibenzofurans: the risks to human health. A review, *Hum. Toxicol.,* 8, 173, 1989.
4. **Suskind, R. R. and Hertzberg, V. S.,** Human health effects of 2,4,5-T and its toxic contaminents, *J. Am. Med. Assoc.,* 251, 2372, 1984.
5. **Caputo, R., Monti, M., Ermacora, E., Carminati, G., Gelmetti, C., Gianotti, R., Gianni, E., and Pucinessi, V.,** Cutaneous manifestations of tetrachlorodibenzo-*p*-dioxin in children and adolescents, *J. Am. Acad. Dermatol.,* 19, 812, 1988.
6. **West, D. P., Worobec, S., and Solomon, L. M.,** Pharmacology and toxicology of infant skin, *J. Invest. Dermatol.,* 76, 147, 1981.
7. **Wester, R. C. and Maibach, H. I.,** Comparative percutaneous absorption, in *Neonatal Skin,* Maibach, H. and Boisits, E. K., Eds., Marcel Dekker, New York, 1982, 137.
8. **Wildnauer, R. H. and Kennedy, R.,** Transepidermal water loss of human newborns, *J. Invest. Dermatol.,* 54, 483, 1970.
9. **Wilson, D. R. and Maibach, H.,** An *in vivo* comparison of skin barrier function, in *Neonatal Skin,* Maibach, H. and Boisits, E. K., Eds., Marcel Dekker, New York, 1982, 101.
10. **Christophers, E. and Kligman, A. M.,** Percutaneous absorption in aged skin, in *Advances in the Biology of the Skin,* Montagna, W., Ed., Pergammon Press, Oxford, 1965, 163.
11. **Tagami, H.,** Functional characteristics of aged skin, *Acta Dermatal. Kyoto Engl. Ed.,* 66–67, 19, 1971.

12. **Rougier, A., Lotte, C., Corcuff, P., and Maibach, H. I.,** Relationship between skin permeability and corneocyte size according to anatomic site, age, and sex in man, *J. Soc. Cosmet. Chem.,* 39, 15, 1988.
13. **Roskos, K. V., Guy, R. H., and Maibach, H. I.,** Percutaneous absorption in the aged, *Dermatol. Clin.,* 4, 455, 1986.
14. **Roskos, K. V., Maibach, H. I., and Guy, R. H.,** The effect of aging on percutaneous absorption in man, *J. Pharmacokinet. Biopharm.,* 17, 617, 1989.
15. **Holbrook, K. A.,** A histological comparison of infant and adult skin, in *Neonatal Skin,* Maibach, H. and Boisits, E. K., Eds., Marcel Dekker, New York, 1982, 3.
16. **Evans, N. J. and Rutter, N.,** Development of the epidermis in the newborn, *Biol. Neonate,* 49, 74, 1986.
17. **Fisher, L. B.,** *In vitro* studies on the permeability of infant skin, in *Percutaneous Absorption,* Bronaugh, R. L. and Maibach, H. I., Eds., Marcel Dekker, New York, 1985, 213.
18. **Lavker, R. M.,** Structural alterations in exposed and unexposed aged skin, *J. Invest. Dermatol.,* 73, 59, 1979.
19. **Lavker, R. M., Zheng, P., and Dong, G.,** Aged skin: a study by light, transmission electron, and scanning electron microscopy, *J. Invest. Dermatol.,* 88, 44s, 1987.
20. **Vogel, H. G.,** Effects of age on the biomechanical and biochemical properties of rat and human skin, *J. Soc. Cosmet. Chem.,* 34, 453, 1983.
21. **Leveque, J. L., Corcuff, P., de Rigal, J., and Agache, O.,** *In vivo* studies of the evolution of physical properties of the human skin with age, *Int. J. Dermatol.,* 23, 322, 1984.
22. **Shuster, S., Black, M. M., and McVitie, E.,** The influence of age and sex on skin thickness, skin collagen and density, *Br. J. Dermatol.,* 93, 639, 1975.
23. **McCormack, J. J., Boisits, E. K., and Fisher, L. B.,** An *in vitro* comparison of the permeability of adult versus neonatal skin, in *Neonatal Skin,* Maibach, H. I. and Boisits, E. K., Eds., Marcel Dekker, New York, 1982, 149.
24. **Ramasastry, P., Downing, D. T., Pochi, P. E., and Strauss, J. S.,** Chemical composition of human skin surface lipids from birth to puberty, *J. Invest. Dermatol.,* 54, 139, 1970.
25. **Nazzaro-Porro, M., Passi, S., Boniforti, L., and Belsito, F.,** Effects of aging on fatty acids in skin surface lipids, *J. Invest. Dermatol.,* 73, 112, 1979.
26. **Saint Leger, D., Francois, A. M., Leveque, J. L., Stoudemayer, T. J., Grove, G. L., and Kligman, A. M.,** Age-associated changes in stratum corneum lipids and their relation to dryness, *Dermatologica,* 177, 159, 1988.
27. **Stewart, M. E., Steele, W. A., and Downing, D. T.,** Changes in the relative amounts of endogenous and exogenous fatty acids in sebaceous lipids during early adolescence, *J. Invest. Dermatol.,* 92, 371, 1989.
28. **Montagna, W. and Carlisle, K.,** Structural changes in aging human skin, *J. Invest. Dermatol.,* 73, 47, 1979.
29. **Kligman, A. M.,** Perspectives and problems in cutaneous gerontology, *J. Invest. Dermatol.,* 73, 39, 1979.
30. **Fenske, N. A. and Lober, C. W.,** Structural and functional changes of normal aging skin, *J. Am. Acad. Dermatol.,* 15, 571, 1986.
31. **Wester, R. C., Noonan, P. K., Cole, M. P., and Maibach, H. I.,** Percutaneous absorption of testosterone in the newborn Rhesus monkey: comparison to the adult, *Ped. Res.,* 11, 737, 1977.
32. **Shah, P. V., Fisher, H. L., Sumler, M. R., Monroe, R. J., Chernoff, N., and Hall, L. L.,** Comparison of the penetration of 14 pesticides through the skin of young and adult rats, *J. Toxicol. Environ. Health,* 21, 353, 1987.
33. **Shah, P. V., Fisher, H. L., Month, N. R., Sumler, M. R., and Hall, L. L.,** Dermal penetration of carbofuran in young and adult Fischer 344 rats, *J. Toxicol. Environ. Health,* 22, 207, 1987.
34. **Fisher, H. L., Shah, P. V., Sumler, M. R., and Hall, L. L.,** *In vivo* and *in vitro* dermal penetration of 2,4,5,2′,4′,5′-hexachlorobiphenyl in young and adult rats, *Environ. Res.,* 50, 120, 1989.
35. **Behl, C. R., Flynn, G. L., Kurihara, T., Smith, W. M., Bellatone, N. H., Gatmaitan, O., Higuchi, W. I., and Ho, N. F. H.,** Age and anatomical site influences on alkanol permeation of skin of the male hairless mouse, *J. Soc. Cosmet. Chem.,* 35, 237, 1984.
36. **Behl, C. R., Flynn, G. L., Linn, E. E., and Smith, W. M.,** Percutaneous absorption of corticosteroids: age, site and skin-sectioning influences on rates of permeation of hairless mouse skin by hydrocortisone, *J. Pharma. Sci.,* 73, 1287, 1984.
37. **Singer, E. J., Wegmann, P. C., Lehman, M. D., Christensen, M. S., and Vinson, L. J.,** Barrier development, ultrastructure, and sulfhydryl content of the fetal epidermis, *J. Soc. Cosmet. Chem.,* 22, 119, 1971.
38. **Monteiro-Riviere, N. A., Banks, Y. B., and Birnbaum, L. S.,** Laser doppler measurements of cutaneous blood flow in ageing mice and rats, *Toxicol. Lett.,* 57, 329, 1991.
39. **Yates, M. S. and Hiley, C. R.,** The effect of age on cardiac output and its distribution in the rat, *Experentia,* 35, 78, 1979.

40. U.S. Environmental Protection Agency, Health assessment document for polychlorinated dibenzo-*p*-dioxins, EPA/600/8–84/014F, Washington, D.C., 1985.

41. Agency for Toxic Substances and Disease Registry, Toxicological profile for 2,3,7,8-tetrachlorodibenzo-*p*-dioxin, ATSDR/TP–88/23, Atlanta, GA, 1989.

42. **Beck, H., Eckart, K., Mathar, W., and Wittkowski, E.,** Occurrence of PCDD and PCDF in different kinds of paper, *Chemosphere,* 17, 51, 1988.

43. **Amendola, G., Barna, D., Blosser, R., LaFleur, L., McBride, A., Thomas, F., Tiernan, T., and Whittemore, R.,** The occurrence and fate of PCDDs and PCDFs in five bleached kraft pulp and paper mills, *Chemosphere,* 18, 1181, 1989.

44. **Clement, R. E., Tashiro, C., Suter, S., Reiner, E., and Hollinger, D.,** Chlorinated dibenzo-*p*-dioxins (CDDs) and dibenzofurans (CDFs) in effluents and sludges from pulp and paper mills, *Chemosphere,* 18, 1189, 1989.

45. **Poiger, H. and Schlatter, C.,** Influence of solvents and adsorbents on dermal and intestinal absorption of TCDD, *Food Cosmet. Toxicol.,* 18, 477, 1980.

46. **Shu, H., Teitelbaum, P., Webb, A. S., Marple, L., Brunck, B., Dei Rossi, D., Murray, F. J., and Paustenbach, D.,** Bioavailability of soil-bound TCDD: dermal bioavailability in the rat, *Fundam. Appl. Toxicol.,* 10, 335, 1988.

47. **Banks, Y. B., Brewster, D. W., and Birnbaum, L. S.,** Age-related changes in dermal absorption of 2,3,7,8-tetrachlorodibenzo-*p*-dioxin and 2,3,4,7,8-pentachlorodibenzofuran, *Fundam. Appl. Toxicol.,* 15, 163, 1990.

48. **Banks, Y. B.,** unpublished data, 1990.

49. **Banks, Y. B. and Birnbaum, L. S.,** Absorption of 2,3,7,8-tetrachlorodibenzo-*p*-dioxin (TCDD) after low dose dermal exposure, *Toxicol. Appl. Pharmacol.,* 107, 302, 1991.

50. **Brewster, D. W., Banks, Y. B., and Birnbaum, L. S.,** Comparative dermal absorption of 2,3,7,8-tetrachlorodibenzo-*p*-dioxin and three polychlorinated dibenzofurans, *Toxicol. Appl. Pharmacol.,* 97, 156, 1989.

51. **Rose, J. Q., Ramsey, J. C., Wentzler, T. H., Hummel, R. A., and Gehring, P. J.,** The fate of 2,3,7,8-tetrachlorodibenzo-*p*-dioxin following single and repeated oral doses to the rat, *Toxicol. Appl. Pharmacol.,* 36, 209, 1976.

52. **Birnbaum, L. S.,** The role of structure in the disposition of halogenated aromatic xenobiotics, *Environ. Health Perspect.,* 6, 11, 1985.

53. **Goldstein, R. S., Tarloff, J. B., and Hook, J. B.,** Age-related nephropathy in laboratory rats, *FASEB J.,* 2, 2241, 1988.

54. **Henderson, L. O. and Patterson, D. G.,** Distribution of 2,3,7,8-tetrachlorodibenzo-*p*-dioxin in human whole blood and its association with, and extractability from, lipoproteins, *Bull. Environ. Contam. Toxicol.,* 40, 604, 1988.

55. **Birnbaum, L. S.,** Changes in the disposition of two hexachlorobiphenyls in senescent rats, in *Liver and Aging,* Kitani, K., Ed., Elsevier, Amsterdam, 1982, 99.

56. **Jackson, J. A., Banks, Y. B., and Birnbaum, L. S.,** Maximal dermal absorption of TCDD occurs in weanling rats, *Toxicologist,* 10, 309, 1990.

Chapter 10

PERCUTANEOUS ABSORPTION OF CHEMICALS FROM WATER DURING SWIMMING AND BATHING

Ronald C. Wester and Howard I. Maibach

TABLE OF CONTENTS

0-8493-7357-3/93/$0.00 + $.50
© 1993 by CRC Press, Inc.

I. INTRODUCTION

Contamination of ground and surface water and the transfer of contaminants from water into the human body are major concerns. The obvious first effect is that contaminants are ingested by drinking water. However, when the large surface area of skin is exposed to contaminated water by daily bathing and swimming, skin absorption may also be significant. Brown et al.[1] suggested that skin absorption of contaminants in drinking water has been underestimated and that ingestion may not constitute the sole, or even the primary, route of exposure.

The evolution of skin resulted in a tissue that protects precious body fluids and constituents from excessive uptake of water and contaminants in the external environment. The outermost surface of the skin that emerged for humans is the stratum corneum, which restricts, but does not prevent, penetration of water and other molecules. This is a complex lipid-protein structure, which is exposed to contaminants during bathing and swimming.

Industrial growth has resulted in the production of organic chemicals and toxic metals whose disposal results in contamination of water supplies. As a man or woman settles into a tub or pool, the skin, the largest organ of the body (surface area approximately 18,000 cm^2) acts as a lipid sink (stratum corneum) for lipid-soluble contaminants. Skin also serves as a transfer membrane for water and whatever contaminants may be dissolved in it.

Percutaneous absorption implies transfer of agents through the epidermis, the deeper layers of skin, and the general circulation *in vivo*, which includes both transport through the skin and local clearance. The first part of the process is accomplished by simple diffusion across the skin. Both *in vitro* and *in vivo* methodology can be used.[2,3]

II. BINDING TO HUMAN STRATUM CORNEUM

Table 1 demonstrates the binding and partitioning of various chemical contaminants between water and powdered human stratum corneum following contact for 30 min (simulating a short swim or soak in tub).[4] Lipid-soluble chemicals such as alachlor and PCBs readily partition out of water and bind to stratum corneum. Benzene partitions between air, water, and skin. Nitroaniline, a more water-soluble chemical, shows some linear binding over a tenfold concentration range. Thus, a variety of contaminating chemicals in water will be attracted/bound to skin (Figure 1). The other factor involved in bathing is the use of soap. Figure 2 shows that alachlor bound to powdered human stratum corneum will partition back into a water matrix if that water contains soap.[5]

III. *IN VITRO* PERCUTANEOUS ABSORPTION

Table 2 shows *in vitro* human skin absorption and distribution of nitroaniline (tenfold dose-response) and 54% PCBs. The amounts in the skin or systemic absorption for either the more water-soluble nitroaniline or the more lipid-soluble 54% PCBs are considerable (5 to 12%) after only a 30-min exposure. Since the percent doses of nitroaniline in skin and absorbed through skin are the same for the tenfold dose range, we must assume no dose dependency for absorption of nitroaniline from water over this dose range (Figure 3). Figure 4 shows the same type of *in vitro* data for benzene and includes comparisons for single vs. 3× water exposures. Not only is benzene absorbed out of water by human skin, but multiple exposure (3×) is at least cumulative and more likely multiple dosing results in enhanced absorption. With multiple dosing, more benzene is kept in contact with skin (evaporation continues, but benzene available for absorption is replaced with the multiple dosing). Water exposure may be longer than 30 min, e.g., children at the swimming pool, training for the Olympics. Table 3 shows the absorption of alachlor from water. Over time, the amount

TABLE 1
Partitioning of Chemical Contaminants Between
Water and Powdered Human Stratum Corneum

Chemical	Percentage dose partitioned to skin
Benzene	16.6 ± 1.4
54% PCBs	95.7 ± 0.6
Nitroaniline (0.49 μg/ml)	3.9 ± 1.5
Nitroaniline (1.6 μg/ml)	3.7 ± 0.7
Nitroaniline (4.9 μg/ml)	2.5 ± 1.1
Alachlor	91.8 ± 0.1
Water wash #1	90.1 ± 0.08
Water wash #2	88.7 ± 0.05
Water wash #3	87.7 ± 0.02

PARTITION FROM WATER TO POWDERED HUMAN STRATUM CORNEUM IN 30 MINUTES

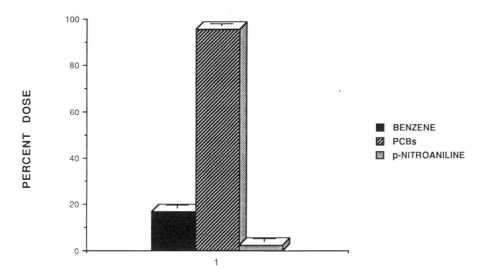

FIGURE 1. Chemicals will partition out of water and bind to powdered human stratum corneum. This partitioning varies for different chemicals.

entering the skin and absorbing into plasma increases (recovery in water surface decreases: accountability remains constant).[5]

IV. *IN VIVO* ABSORPTION

In vitro absorption and stratum corneum binding can be confirmed by *in vivo* studies. While human percutaneous absorption studies *in vivo* are the preferred method, some *in vivo* animal models, such as the rhesus monkey, are suitable substitutes.[3] Table 4 shows the *in vivo* absorption of nitroaniline (30-min exposure to water solution). The percentage of dose absorbed is similar to that from *in vitro* and binding studies. Thus, there is some validity to the surrogate systems, at least for nitroaniline. We should not overgeneralize the interpretation of this result; however, more validation is required.

ALACHLOR PARTITION FROM POWDERED HUMAN STRATUM CORNEUM

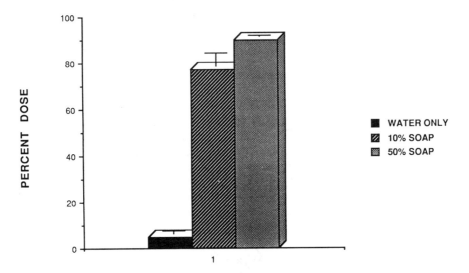

FIGURE 2.   Alachlor will bind to human powdered stratum corneum and not be washed away with water. Soap in water will reverse the binding.

TABLE 2
In Vitro Percutaneous Absorption and Human Skin Distribution (%) of Nitroaniline and 54% PCBs from Water

Parameter[a]	Nitroaniline 4.9 μg/ml	Nitroaniline 1.6μg/ml	Nitroaniline 0.49 μg/ml	54% PCBs (1.6 μg/ml)
Percutaneous absorption (systemic)	5.2 ± 1.6	4.4 ± 3.9	5.0 ± 0.80	0.03 ± 0.00
Surface bound/stratum corneum	0.28 ± 0.17	0.31 ± 0.04	0.21 ± 0.03	6.8 ± 1.0
Epidermis and dermis	0.61 ± 0.25	0.4 ± 0.7	2.3 ± 0.6	5.5 ± 0.7
Total (skin/systemic)	6.1	5.1	7.5	12.3
Skin wash/residual	92.4 ± 0.8	88.3 ± 2.3	85.2 ± 1.1	71.2 ± 5.5
Apparatus wash	0.47 ± 0.03	0.02 ± 0.004	0.08 ± 0.01	0.25 ± 0.2
Total (accountability)	99.0	93.4	92.8	83.3

V. WATER DILUTION AND PERCUTANEOUS ABSORPTION

Some marketed formulations of pesticides are diluted with water during use and then further diluted when they enter the environment (ground and surface water). Changes in formulation/dilution may influence percutaneous absorption.

The first example is dinoseb formulated with Premerge-3 only, and with Premerge-3 diluted 1:40 with water (at two concentrations). Neither concentration change nor dilution with water changed *in vivo* skin absorption in the rhesus monkey (Table 5).

Table 6 shows *in vitro* absorption of alachlor. Lasso® emulsifiable concentrate diluted with water 1:20, 1:40, or 1:80 was placed on human skin at two concentrations (0.5 or 1.0 ml over 5.7 cm² skin surface area). The data in Table 7 show that with increasing water dilution, the percentage of dose absorbed increased significantly ($p \leq 0.01$). If the mass (microgram) of alachlor diffusing into the plasma receptor fluid is compared for theoretical (dilution reduces mass) and actual levels, the skin absorption increases by many times

p-NITROANILINE SKIN ABSORPTION FROM WATER: 30 MINUTE EXPOSURE

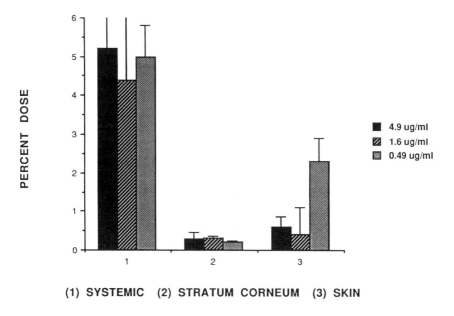

FIGURE 3. The *in vitro* human skin absorption of *p*-nitroaniline from water is dose-independent over a tenfold dose range.

BENZENE IN VITRO SKIN ABSORPTION FROM WATER: SINGLE AND MULTIPLE DOSES

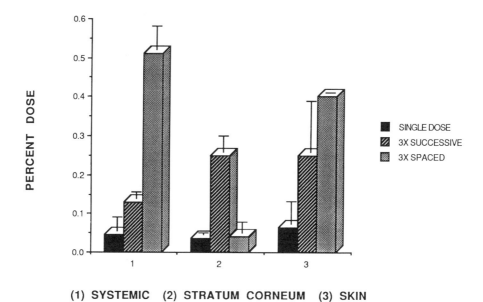

FIGURE 4. Benzene absorption into and through human skin from water increases with multiple skin applications, probably due to increasing skin contact of benzene.

TABLE 3
In Vitro Human Skin Absorption and
Distribution of Alachlor from Water
Over Time

	Percentage applied dose		
Parameter	0.5 h	4 h	8 h
Receptor fluid (plasma)	0.01	0.28	0.82
Skin digest	19.6	30.5	33.1
Skin wash	77.0	61.4	43.4
Apparatus	0.3	1.0	9.0
Total accountability	96.91	93.2	86.2

TABLE 4
In Vivo Percutaneous Absorption of *p*-Nitroaniline in the
Rhesus Monkey Following 30-Min Exposure to Surface
Water: Comparisons to *In Vitro* Binding and Absorption

Phenomenon	Percent dose absorbed/bound
In vivo percutaneous absorption (rhesus monkey)	4.1 ± 2.3
In vitro percutaneous absorption (human skin)	5.2 ± 1.6
In vitro binding (powdered human stratum corneum)	2.5 ± 1.1

TABLE 5
Percutaneous Absorption of Dinoseb in the Rhesus
Monkey: Formulation Diluted with Water

Dosage formulation	Percent dose absorbed (n = 4; 24-h exposure)
High (3.6 mg/cm^2) Premerge-3	5 ± 3
Mid (0.2 mg/cm^2) Premerge-3:water, 1:40	7 ± 6
Low (0.04 mg/cm^2) Premerge-3:water, 1:40	5 ± 3

(although overall percentage efficiency remains low compared to total dose applied). It appears that the thermodynamic forces determining alachlor skin absorption change with water dilution (at least in an *in vitro* system using human skin).[4] However, recent *in vivo* studies with alachlor failed to confirm increased absorption with increased dilution. The *in vitro* results are thus probably due to the *in vitro* methodology.[6]

VI. PERCUTANEOUS ABSORPTION OF CADMIUM FROM WATER

Cadmium-109 as the chloride salt was dissolved in water and applied to the surface of human skin suspended in a diffusion cell with human plasma as the receptor fluid. Percutaneous absorption was determined following 16 h of exposure. Table 7 gives the cadmium content of plasma, skin, surface skin wash with soap and water, and total accountability for four skin sources. Cadmium binds to human skin (Figure 5). Perhaps more disturbing is that some of the cadmium in skin partitioned into plasma (Figure 6). Thus, the heavy metal cadmium has an affinity for human skin, and the percutaneous absorption of cadmium does occur.

TABLE 6
In Vitro **Percutaneous Absorption of Human Skin:**
Effect of Dilution on Alachlor Absorption Over 8 h

Volume (ml/ 5.7 cm²)	Dilution[a]	Percent dose plasma receptor fluid	Statistics 1:20	1:40
0.5	1:20	0.8 ± 0.1	—	—
	1:40	2.1 ± 0.4	N.S.[b]	—
	1:80	3.7 ± 0.7	$p < 0.01$	—
1.0	1:20	0.5 ± 0.1	—	—
	1:40	1.8 ± 0.1	$p < 0.01$	—
	1:80	3.9 ± 0.3	$p < 0.01$	$p < 0.01$

Actual and Theoretical Absorption with Dilution

		Mass (µg) plasma receptor fluid	
		Actual	Theoretical
0.5	1:20	90	—
	1:40	130	45.0
	1:80	110	22.5
1.0	1:20	110	—
	1:40	210	55.0
	1:80	230	27.5

TABLE 7
In Vitro **Percutaneous Absorption of Cadmium**
Through Human Skin

	Percent applied dose			
Skin source	Water 5 µl/cm²			
	Plasma	Skin	Wash	Total
#1	0.5 ± 0.2	8.8 ± 0.6	93.3 ± 3	103 ± 3
#2	0.6 ± 0.6	12.7 ± 11.7	74 ± 1	88 ± 20
	2.5 µl/cm²			
#1	0.2 ± 0.2	2.4 ± 1.6	86 ± 3	89 ± 2
#4	0.1 ± 0.04	3.2 ± 4.4	88 ± 13	92 ± 12

VII. SOAP-AND-WATER WASH

Bathing (as opposed to swimming) with soap may affect contaminant binding and skin absorption. Figure 2 showed the partitioning of alachlor with powdered human stratum corneum. Alachlor binds to stratum corneum and will not come off with water-only wash. However, 10% Ivory Liquid soap in water wash removed 77% of the alachlor, and 50% soap in water removed 90% of the alachlor. This may be useful news for the pesticide applicator/farmer because washing with soap and water probably removes a large portion of this and other pesticides bound to skin.[4]

Figure 7 shows the *in vivo* skin decontamination of araclor 1242 (42% PCBs) with soap and water wash. PCBs in mineral oil and trichlorbenzene (vehicle constituents of trans-

CADMIUM IN VITRO HUMAN SKIN ABSORPTION FROM WATER

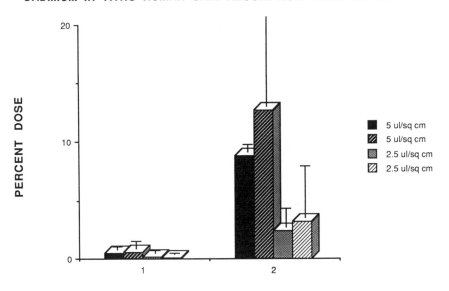

(1) PLASMA RECEPTOR FLUID   (2) SKIN CONTENT

FIGURE 5.   Cadmium will absorb into and through human skin.

CADMIUM IN HUMAN PLASMA RECEPTOR FLUID

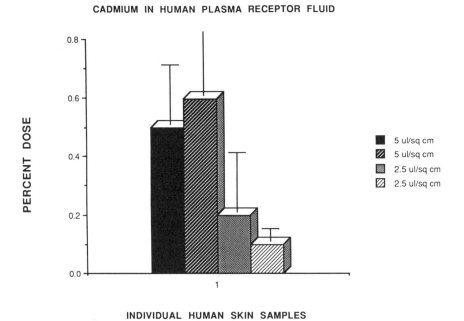

INDIVIDUAL HUMAN SKIN SAMPLES

FIGURE 6.   Cadmium in human plasma receptor fluid varies with individual human skin source.

formers) was applied to skin. Washing the skin with soap and water removed 70% of the surface-bound PCBs (0.0 time) and this efficiency of removal lasted for 3 h. At 6 h, removal decreased to 50%, and at 24 h, only 30% was removed. With time, skin absorption of PCBs limited the amount that could be removed by washing the skin with soap and water.[8]

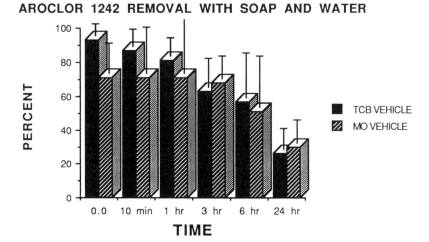

FIGURE 7. Aroclor 1242 (42% PCBs) removal from skin with soap and water wash is time-dependent.

VIII. DISCUSSION

We are just beginning to determine the potential health hazards associated with skin absorption of contaminants from ground and surface water during swimming and bathing. Human skin surface area is large, and human skin has the potential to attract, bind, and absorb significant amounts of contaminants from water into the body. We have proposed methods (binding to human stratum corneum; *in vitro* and *in vivo* percutaneous absorption), and we have reported some findings (linearity: cumulative potential; influence of soap). Additional information is needed for risk assessment.

REFERENCES

1. **Brown, H. S., Bishop, D. R., and Rowan, C. A.,** The role of skin absorption as a route of exposure for volatile organic compounds (VOCs) in drinking water, *Am. J. Public Health,* 74, 479, 1984.
2. **Skelly, J. P., Shah, V. P., Maibach, H. I., Guy, R. H., Wester, R. C., Flynn, G., and Yacobi, A.,** FDA and AAPS reports of the workshop on principles and practices of *in vitro* percutaneous penetration studies: relevance to bioavailability and bioequivalence, *Pharmacol. Res.,* 4, 265, 1987.
3. **Wester, R. C. and Maibach, H. I.,** *In vivo* animal models for percutaneous absorption, in *Percutaneous Absorption,* Bronaugh, R. and Maibach, H., Eds., Marcel Dekker, New York, 1985, 251.
4. **Wester, R. C., Mobayen, M., and Maibach, H. I.,** *In vivo* and *in vitro* absorption and binding to powdered stratum corneum as methods to evaluate skin absorption of environmental chemical contaminants from ground and surface water, *J. Toxicol. Environ. Health,* 21, 367, 1987.
5. **Bucks, D. A. W., Wester, R. C., Mobayen, M. M., Yange, D., Maibach, H. I., and Coleman, D. L.,** *In vitro* percutaneous absorption and stratum corneum binding of alachlor: effect of formulation dilution with water, *Toxicol. Appl. Pharmacol.,* 100, 917, 1987.
6. **Wester, R. C. and Maibach, H. I.,** Human skin binding and absorption of contaminants from ground and surface water during swimming and bathing, *J. Am. Coll. Toxicol.,* 8, 803, 1989.
7. **Wester, R. C., Maibach, H. I., Sedek, L., Melendres, J., DiFio, S., Jamall, I., and Wade, I.,** *In vitro* percutaneous absorption of cadmium from water and soil into human skin, *Fund. Appl. Toxicol.,* 1, 19, 1992.
8. **Wester, R. C., Maibach, H. I., Bucks, D. A. W., McMaster, M., Mobayen, M., Sarason, R., and Moore, A.,** Percutaneous absorption and skin decontamination of PCBs. *In vitro* studies with human skin and *in vitro* studies in the Rhesus monkey *J. Toxicol. Environ. Health,* in press.

Chapter 11

PERCUTANEOUS ABSORPTION OF CONTAMINANTS FROM SOIL

Ronald C. Wester, Daniel A. W. Bucks, and Howard I. Maibach

TABLE OF CONTENTS

0-8493-7357-3/93/$0.00 + $.50
© 1993 by CRC Press, Inc.

I. INTRODUCTION

Hazardous substances can cause adverse effects in humans only if exposure occurs. Soil has recently been recognized as a potentially important medium of exposure. Skin absorption is now recognized as a port of entry into the human body for environmental hazardous substances. The soil is a medium with which human skin has constant contact. This can be work-related (farming, waste hazard disposal), recreational (gardening), or a child's delight (beach, sand box).

A major dilemma in establishing regulatory limits for environmental pollutants is the establishment of environmental standards or limits for chemical concentrations in soil at industrial and residential sites. Factors and assumptions used to predict the bioavailability of a chemical from soil significantly affect the establishment of a virtually safe dose or acceptable daily exposure level of a compound in soil. The two major concerns in setting relevant contamination levels are public safety and cost/feasibility of cleanup. Public safety depends upon the inherent toxicity of the hazardous chemical and the bioavailability (rate and extent of systemic absorption). The cost of remediation varies dramaticallly with the level to which soil must be decontaminated and excessive remediation means that limited resources will be spent without providing additional protection of public health. For example, Paustenbach et al.[1] reproted that the cost of removing and disposing of soil containing more than 1 ppb 2,3,7,8-tetrachlorodibenzo-p-dioxin (TCDD) at the Castlewood site in Missouri to be \$17 million. However, if the level for cleanup at that site were set at 10 ppb, the cost would drop to \$6 million and minimal action would be required at 100 ppb TCDD. The authors state other sites, such as Times Beach, MO, show an even more dramatic relationship between the cost of cleanup and the degree of remediation.

II. DDT

DDT residues can still be detected in California soils some 13 years after the last use of DDT. Every one of 99 sites sampled in 32 counties showed detectable levels ranging from trace amounts to 15 ppm.[26] DDT can be absorbed through human skin.[2]

The soil used in this study has been designated as Yolo County soil sample 65-California 57-8 (26% sand, 26% clay, 48% silt). DDT-contaminated soil was prepared at levels of 10 ppm of compound to soil and 0.04 g soil per squared centimeter of skin area. Table 1 and Figure 1 give the *in vitro* percutaneous absorption of DDT from soil and acetone vehicle into human skin. DDT readily penetrated human skin when applied in acetone vehicle. Significantly less DDT penetrated human skin from soil. DDT would not readily partition from human skin into plasma in the receptor phase. Skin surface wash removed most of the remaining chemical. Total chemical accountability was greater than 80%.[3]

Figure 1 shows the *in vivo* percutaneous absorption of DDT in the rhesus monkey. An average 18.9 ± 9.4% DDT was absorbed from acetone vehicle and a significantly less ($p = 0.04$) 3.3 ± 0.5% was absorbed from soil.[3] Absorption of DDT in man from acetone vehicle is included.[2]

Hawkins and Reifenrath[4] studied DDT skin absorption *in vitro* using pig skin. DDT readily penetrated into skin (47 ± 7%), but not into receptor fluid (0.2 ± 0.2%). Bronaugh et al.,[5] using rat skin, reported similar proportions of 48 ± 3% in skin and 0.6 ± 0.2% in the receptor fluid. Wester et al.[3] using human skin found less DDT (18 ± 13%) in human skin, but agree that little DDT will partition into aqueous receptor fluids. *In vivo* percutaneous absorption of DDT reported as percent of applied dose was 46.3 ± 1.4% (rabbit),[6] 43.4 ± 7.9% (pig)[6] 1.5 ± 2.0% (squirrel monkey),[6] and 10.4 ± 3.6% (man).[2] Data for absorption of DDT in rhesus monkey was not significantly different ($p > 0.05$) from man for the acetone formulation application.

TABLE 1
In Vitro **Percutaneous Absorption of DDT from Soil into Human Skin**

	Percent applied dose			
	Skin	**Plasma receptor fluid**	**Surface wash**	**Total**
Soil vehicle	1.0 ± 0.7	0.04 ± 0.01	95.6 ± 18.2	96.6 ± 18.4
Control acetone vehicle	18.1 ± 13.4	0.08 ± 0.02	63.7 ± 12.8	82.0 ± 9.8

Note: 24-h skin application time.

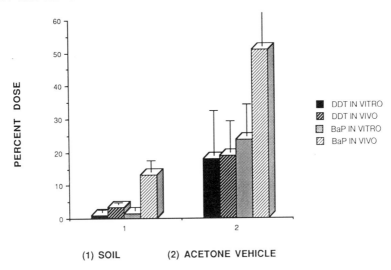

FIGURE 1. *In vitro* and *in vivo* percutaneous absorption of DDT and benzo(a)pyrene (BaP) from soil on human skin.

The general conclusion for absorption of DDT from soil is that it will occur, but that the absolute amount will be less than if the DDT chemical only was deposited on skin. The best estimate for DDT chemical absorption in human skin is 10 to 20% for 24-h exposure.

III. BENZO(A)PYRENE

The first, and historically most important, carcinogen isolated from coal tar was benzo(a)pyrene. It is the smallest and therefore most easily synthesized unsubstituted polycyclic aromatic hydrocarbon that will produce tumors. The epoxide diol of benzo(a)pyrene is the ultimate carcinogen responsible for the adducts of benzo(a)pyrene with DNA.[7] Benzo(a)pyrene in contact with skin is absorbed into the metabolically active epidermis and converted to these metabolites which produce the tumors.[8-10]

Table 2 and Figure 1 give the *in vitro* percutaneous absorption of benzo(a)pyrene from soil and actone vehicle into human skin. Benzo(a)pyrene readily penetrated human skin when applied in acetone vehicle. Significantly less benzo(a)pyrene penetrated human skin from soil. Benzo(a)pyrene would not readily partition from human skin into plasma in the receptor phase. Skin surface wash removed most of the remaining chemical. Total chemical accountability was 77 to 93%. Figure 1 gives the *in vivo* percutaneous absorption of

TABLE 2

In Vitro **Percutaneous Absorption of Benzo(a)pyrene from Soil into Human Skin**

| | Percent applied dose | | | |
| | | Plasma receptor | | |
Human skin source	Skin	fluid	Surface wash	Total
Soil vehicle	1.4 ± 0.9	0.01 ± 0.004	91.2 ± 13.2	92.7 ± 13.1
Control acetone vehicle	23.1 ± 9.7	0.09 ± 0.02	53.0 ± 17.1	76.89 ± 14.7

Note: 24-h skin application time.

benzo(a)pyrene in the rhesus monkey. An average $51.0 \pm 22.0\%$ benzo(a)pyrene was absorbed from acetone vehicle, and a significantly less ($p = 0.015$) $13.2 \pm 3.4\%$ was absorbed from soil.[3]

Bronaugh and Stewart[11] reported the *in vitro* percutaneous absorption of benzo(a)pyrene through rat skin to be 3.7% using normal saline as the receptor phase and 56.0% when the surfactant solubilizer 6% PEG-20 oleyl ether was added to the receptor phase. Yang et al.[12] determined benzo(a)pyrene percutaneous absorption *in vitro* through rat skin (using PEG-20 oleyl ether) to be 38.1%. Wester et al.[3] reported *in vitro* percutaneous absorption of benzo(a)pyrene through human skin with human plasma as receptor fluid (and no surfactant solubilizer) was 0.09%. Kao[13] reported benzo(a)pyrene *in vitro* percutaneous absorption to be 1.9% in rat skin and 2.7% in human skin. Thus, the amount of 23.7% in human skin[3] represents that amount which the surfactant solubilizer placed in the receptor fluid. *In vivo*, benzo(a)pyrene percutaneous absorption was 48.3% and 35.3% in the rat[11,12] compared to 51% in the rhesus monkey.[3]

Yang et al.[12] reported that percutaneous absorption of benzo(a)pyrene from soil matrix *in vitro* was 8.4%, giving a reduction ratio of 8.4:38.1% = 0.2. *In vivo* the ratio was reduced to 9.2:35.3% = 0.26. In Wester et al.,[3] soil reduction ratio *in vitro* was 1.4:23.7% = 0.06, and *in vivo*, it was 13.2:51.0% = 0.26. Therefore, both studies suggest that percutaneous absorption of benzo(a)pyrene from soil is about 25% of that from other more conventional vehicles.

IV. CHLORDANE

Chlordane is an insecticide for which the commercial product is a mixture containing 60 to 75% of the pure compound and 25 to 40% of related compounds. The EPA has cancelled registration of pesticides containing this compound with the exception of its use through subsurface ground insertion for termite control and the dipping of roots or top of nonfood plants. Therefore, knowledge of potential chlordane skin absorption from soil is important. Table 3 gives the percutaneous absorption of pure chlordane from soil and from an acetone solution (acetone quickly evaporated after application to skin). During *in vitro* absorption through human skin, a larger quantity of chlordane adhered to skin from direct deposit in acetone solution; however, quantities reaching the human plasma receptor fluid were the same. *In vivo* percutaneous absorption in the rhesus monkey showed that an equal quantity of chlordane was absorbed through skin from soil ($4.2 \pm 1.8\%$) and acetone solution ($6.0 \pm 2.0\%$). The affinity of chlordane for skin is further illustrated in the *in vivo* decontamination of rhesus monkey skin (Figure 2). It took several soap and water washes to remove chlordane from skin.[14]

TABLE 3
Percutaneous Absorption of Chlordane

| | Percent applied dose[a] | | |
| | *In vitro* | | |
Vehicle	Plasma receptor fluid	Skin	*In vivo* Urine
Soil	0.04 ± 0.05[b]	0.34 ± 0.31[c]	4.2 ± 1.8[b]
Acetone	0.07 ± 0.06[b]	10.8 ± 8.2[c]	6.0 ± 2.0[b]

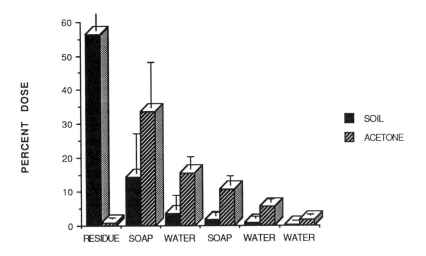

FIGURE 2. Chlordane skin decontamination from soil and acetone applied *in vivo* to rhesus monkey. Soap and water wash sequence was done following 24-h skin application.

V. LEAD

Roels et al.[15] conducted an environmental study concerning blood lead (Pb) levels among 11-year-old children attending schools situated less than 1 and 2.5 km from a primary lead smelter. Age-matched control children from a rural area and an urban area were contemporaneously examined. Samples analyzed for lead were hand rinses, blood, air, and soil. The hand rinses were collected by slowly pouring 500 ml of 0.1 N NHO_3 over the palm of one hand while the fingers were slightly spread. As expected, lead exposure based upon blood lead levels decreased with distance from the smelter. Inhalation of airborne lead was the major source of slightly increased lead levels in adults, whereas children demonstrated an increased blood lead level resulting from soiled hands. The authors conclude that in a lead smelter area, the enforcement of a permissible limit for airborne lead alone may not necessarily prevent an excessive exposure of children to environmental lead since past emission of lead and possible transport of lead-containing dirt (e.g., through road transport) will maintain a high level of lead in soil-dust and dirt irrespective of the current concentration of atmospheric lead. Using the data published by Roels et al.,[15] we estimate the mass of soil per area of skin adhering to the children as 2.3 ± 0.3 mg soil per squared centimeter hand (assuming an area of 70 cm² for the child's palm and fingers and the rinsing procedure completely removes the dirt) (Table 4). This value agrees with Driver et al.,[16] who reported soil adherence to be 0.2 to 2 mg soil per squared centimeter skin depending upon soil type and particle size.

TABLE 4
Mass of Soil per Unit of Area Adhering to Children

Location	Soil (pb μg/g)	Hand (μg pb)	mg Soil/cm^2 hand
Rural	114	17	2.1
Urban	112	20.4	2.6
2.5 km	466	62.2	1.9
<1 km	2560	436	2.4

Note: mean = 2.3; SD = 0.3.

Adapted from Roels, H., Buchet, J. P., Lauwerys, R. R., Bruaux, P., Claeys-Thoreau, F., Lafontaine, A., and Verduyn, G., *Environ. Res.*, 22, 81, 1980. With permission.

VI. CADMIUM

Heavy metals are of toxicological concern and they have concentrated in water and soil where humans come in contact with them in the environment. The following was a study to determine potential skin bioavailability of cadmium. Soil (Yolo County 65-California-57-8) was passed through 10-, 20-, and 48-mesh sieves. Soil retained by 80 mesh was mixed with radioactive cadmium-109 at 13 ppb. Water solutions of cadmium-109 at 116 ppb were prepared for comparative analysis. Human cadaver skin was dermatomed to 500 μm and used in glass diffusion cells with human plasma as the receptor fluid (3 ml/h flow rate) for a 16-h skin application time. Cadmium in water (5 μl/cm^2) penetrated skin to concentrations of 8.8 $\pm$ 0.6% and 12.7 $\pm$ 11.7% of applied dose for two human skin sources. Percent doses absorbed into plasma were 0.5 $\pm$ 0.2% and 0.6 $\pm$ 0.6%, respectively. Cadmium from soil (0.04 g soil per squared centimeter) penetrated skin at concentrations of 0.06 $\pm$ 0.01% and 0.13 $\pm$ 0.05% for the two human skin sources. Amounts absorbed into plasma were 0.02 $\pm$ 0.01% and 0.07 $\pm$ 0.03%. Most of the nonabsorbed cadmium was recovered in the soap and water skin surface wash. Cadmium from water (2.5 μl/cm^2) and from soil (0.02 g soil per squared centimeter) was applied to two additional human skin sources at the indicated reduced amounts. Results were similar to the first tests. Cadmium will penetrate human skin and be absorbed into plasma (Table 5). Absorption from soil is less than absorption from water, but still significant (Figure 3).[17]

VII. TCDD (DIOXIN)

Poiger and Schlatter[18] studied the effect of solvents and adsorbents on the *in vivo* percutaneous penetration of TCDD using the hairless rat animal model (sex and site of application not published). The percutaneous penetration of TCDD from soil, methanol, activated carbon-water paste, petrolatum, polyethylene glycol 1500, or polyethylene glycol 1500-water was examined under aluminum foil-occluded conditions. TCDD topical exposure levels ranged from 6.5 to 43 ng/cm^2. Soil was sieved to <160 μm, then ground in a mortar prior to formulation with TCDD. Each formulation was applied over 3 to 4 cm^2 for 24 h. Percutaneous penetration was assessed by measurement of radioactivity in the liver 24 h postapplication of compound. The relative dermal bioavailability of TCDD from these for-mulations was assessed by comparison of liver TCDD levels at 24 h from dosing. This type of analysis reported by these authors indicated that adsorption of TCDD to activated carbon completely prevented its percutaneous penetration. Our statistical analysis of their data

TABLE 5
In Vitro Percutaneous Absorption of Cadmium through Human Skin

Percent applied dose

Skin source	Soil				Water			
	0.04 g/cm²				5 µl/cm²			
	Plasma	Skin	Wash	Total	Plasma	Skin	Wash	Total
#1	0.02 ± 0.01	0.06 ± 0.02	102 ± 17	102 ± 17	0.5 ± 0.2	8.8 ± 0.6	93 ± 3	103 ± 3
#2	0.07 ± 0.03	0.13 ± 0.05	82 ± 33	83 ± 33	0.6 ± 0.6	12.7 ± 11.7	74 ± 11	88 ± 20
	0.02 g/cm²				2.5 µl/cm²			
#3	0.02 ± 0.02	0.08 ± 0.06	106 ± 2	106 ± 2	0.2 ± 0.2	2.4 ± 1.6	86.3 ± 3	89 ± 2
#4	0.02 ± 0.02	0.08 ± 0.06	88 ± 13	92 ± 12	0.1 ± 0.04	3.2 ± 4.4	88 ± 13	92 ± 12

CADMIUM IN VITRO HUMAN SKIN ABSORPTION FROM SOIL

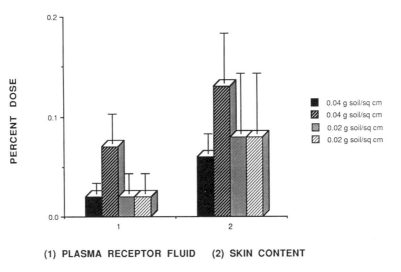

(1) PLASMA RECEPTOR FLUID   (2) SKIN CONTENT

FIGURE 3.   Cadmium skin and plasma receptor fluid content following 16-h *in vitro* absorption from soil through human skin.

indicated a significant difference ($p < 0.01$, ANOVA) in TCDD penetration from the MeOH, petrolatum, and soil formulations. TCDD penetration from MeOH was significantly greater ($p < 0.05$, Newman-Keuls test) than from petrolatum or soil formulations. There was no significant difference ($p > 0.05$, Newman-Keuls test) between soil and petrolatum formulations. In addition, a significant difference ($p < 0.01$, ANOVA) in TCDD penetration from the polyethylene glycol 1500, polyethylene glycol-15% water, and soil formulations was observed. Newman-Keuls test results indicated that TCDD penetration from soil was significantly less ($p < 0.05$) than from polyethylene glycol 1500 or polyethylene glycol-15% water formulations, but no significant difference ($p < 0.05$) in TCDD penetration from the polyethylene glycol 1500 and polyethylene glycol-15% water formulations. The authors conclude that TCDD penetration is highly dependent on the formulation in which it is applied and that mixing TCDD with soil or activated carbon results in reduced compound.

Shu et al.[19] determined TCDD bioavailability from Times Beach, MO, soil in the presence and absence of used crankcase oil. Tritiated TCDD was added to the soil and mixed by passage through the 40-mesh screen three times. In studies employing crankcase oil, the used oil was first added to the soil to achieve 0.5 or 2.0% (w/w) prior to [^{3}H]-TCDD addition. *In vivo* dermal bioavailability of soil-bound TCDD was estimated using male Sprague-Dawley and hairless (naked ex-Backcross and Holtzman strain) rats. A 12-cm^2 area of the back was the site of application. Haired rats were lightly clipper-shaven prior to dosing. The TCDD-contaminated soil was applied by gentle rubbing. The site of application was covered with a nonocclusive perforated aluminum eye protector whose edges were covered with foam, backed with adhesive, and affixed in place with masking tape. The relative dermal bioavailability of TCDD from these formulations was assessed by comparison of liver TCDD levels at 24 h from dosing, therefore the rate or total extent of TCDD percutaneous penetration is not known. However, relative differences in percutaneous penetration between the various treatment modalities should not be affected. The data indicated that variations in TCDD dose (10 compared to 100 ppb TCDD in soil) and oil concentration (0, 0.5, and 2.0%) did not affect the relative percent applied dose penetrating skin. A 4-h exposure resulted in about 60% of an applied dose penetrating from a 24-h exposure. There was no significant difference in TCDD percutaneous penetration from soil between lightly clipper-shaven rats and hairless rats. Unfortunately the authors did not do the control study involving TCDD penetration in the absence of soil.

Using the data from Poiger and Schlatter,[18] Kimbrough et al.[20] of the Centers for Disease Control (CDC) have stated: "1 ppb of 2,3,7,8-TCDD in soil is a reasonable level at which to begin consideration of action to limit human exposure for contaminated soil." Subsequently, Kissel and McAvoy[21] reexamined the CDC risk assessment using a fugacity-based dermal exposure model incorporated in a physiologically based pharmacokinetic model. Their results suggest that CDC's dermal risk assessment is not conservative and predict dermal bioavailabilities well in excess of that assumed by the CDC.

Brewster et al.[22] have studied the *in vivo* dermal absorption of ^{3}H-TCDD using male F344 rats. Radio-labeled TCDD was applied at six dose levels (0.00015, 0.001, 0.01, 0.1, 0.5, and 1.0 µmol/kg) from an acetone solution to the intracapular region of the back (total surface area of application not reported). The site of application was covered with a stainless steel perforated cap. The percentage of applied dose absorbed (defined as the difference between the administered dose and the amount in the application site) declined with increasing dose while the absolute total mass (micrograms per kilogram) absorbed increased nonlinearly with dose. Absorption of TCDD ranged from approximately 40% of the applied dose when applied at 0.001 and 0.00015 µg/kg to approximately 18% at the 0.1, 0.5, and 1.0 µg/kg dose levels. Major tissue depots included liver, adipose tissue, skin, and muscle. The investigators conclude "[Our] results indicate that the dermal absorption of these compounds is incomplete and that systemic toxicity following acute dermal exposure to levels found in the environment is unlikely. Although the potential for systemic toxicity after acute environmental exposure to these chemicals is low, chronic low-level dermal exposure to these compounds ... could result in bioaccumulation of [TCDD] body burdens sufficient to induce toxicity."

Evident from the statements from the several laboratories cited is the lack of consensus on the dangers of topical TCDD exposure. Further research to determine a potential safe level of TCDD exposure from soil is warranted.

VIII. BENZENE, TOLUENE, AND XYLENE

Skowronski et al.[23–25] studied the *in vivo* percutaneous penetration of benzene and toluene when applied as pure chemical and when incorporated with sandy or clay soils. Male Sprague-Dawley rats were dosed on 13 cm^2 of lightly shaved costoabdominal skin. The application site was occluded with a glass cap. Chemicals were applied for 48 h. Percutaneous penetration was assessed by urinary excretion of radio-label.

The authors showed that the percutaneous penetration of benzene was significantly higher ($p < 0.05$) when applied as the pure chemical (86%) compared to application with either soil and penetration was significantly higher ($p < 0.05$) from sandy soil compared to clay soil. The skin absorptions of toluene and xylene were not affected by either soil. The authors conclude that percutaneous penetration of a chemical from soil can approach levels observed when applied as pure compound.

IX. DISCUSSION

Exposure assessment of an area with chemical contaminated soil is linked to considerations of excess risks of developing specific adverse health effects as a result of the total cumulative dose an individual receives. The total cumulative dose is a function of several factors, including:

1.   Concentration(s) of contaminant(s) in the soil
2.   Location of and access to contaminated areas
3.   Type(s) of activity(ies) in contaminated areas
4.   Duration of exposure
5.   Specific exposure mechanisms
6   Soil type and moisture content
7.   Amount of soil adhering to skin
8.   Skin site and total area of exposure
9   Time of exposure (acute and chronic)

Accumulation of compound in plants and the half-life of the compound in the soil are additional important factors in the determination of relevant cleanup levels and methodology to be employed. Dose-rate may be an important factor in acute exposure assessment. The potential of increased risk from receiving high doses at susceptible life stages may be offset or exceeded by repair mechanisms operative at times of lesser dose. Clearly, these mandate further study.

In general, these investigations have revolved around a specific site with investigators employing many different techniques, methodologies (*in vitro* or *in vivo*), surface concentrations, and volumes of application to animals and/or humans. Of concern to all is the extrapolation of results obtained using animal models to estimate human body burden following topical exposure. Animal skin is usually more permeable than human. When rank ordered, rat skin is more permeable than monkey or man, and monkey skin closely approximates the permeability of man.[10] Therefore, in studies employing rodents, one expects the result to overestimate percutaneous penetration in man. Furthermore, we believe that results obtained using *in vitro* systems must be substantiated *in vivo* before use in establishing regulatory guidelines because of the wide range in results one can obtain employing different *in vitro* methodologies.

In many cases, one must realize that experimental conditions employed in some of these investigations do not mimic actual environmental exposure situations, and therefore the results reported may not be relevant to estimate potential human body burden following topical exposure. Results to date indicate that soil does not enhance chemical percutaneous penetration. However, under certain experimental conditions, levels of chemical absorption from soil may approach those following topical application of pure chemical.

Hopefully this chapter will stimulate discussion between the regulatory committees, industrial personnel, and academicians. A consensus on guidelines should be followed in conducting relevant studies to answer the questions raised.

REFERENCES

1. **Paustenbach, D. J., Shu, H. P., and Murray, F. J.,** A critical examination of assumptions used in risk assessments of dioxin contaminated soil, *Reg. Toxicol. Pharm.,* 6, 284, 1986.
2. **Feldmann, R. J. and Maibach, H. I.,** Percutaneous penetration of some pesticides and herbicides in man, *Toxicol. Appl. Pharmacol.,* 28, 126, 1974.
3. **Wester, R. C., Maibach, H. I., Bucks, D. A. W., Sadik, L., Melendres, J., Liao, C., and DiZio, S.,** Percutaneous absorption of [14C]-DDT and [14C]-benzo(o)pyrene from soil, *Fundam. Appl. Toxicol.,* 1990.
4. **Hawkins, G. S., Jr. and Reifenrath, W. G.,** Development of an *in vitro* model for determining the fate of chemicals applied to skin, *Fundam. Appl. Toxicol.,* 4, 5133, 1984.
5. **Bronaugh, R. L., Stewart, R. F. and Storm, J. E.,** Extent of cutaneous metabolism during percutaneous absorption of xenobiotics, *Toxicol. Appl. Pharmacol.,* 99, 534, 1989.

6. **Bartek, M. J. and LaBudde, J. A.,** Percutaneous absorption, in *Animal Models in Dermatology,* Maibach, H., Churchill Livingstone, New York, 1975, 103.

7. **Scribner, J. D.,** Chemical carcinogenesis, in *Environmental Pathology,* Mottett, N. K., Ed., Oxford University Press, New York, 1985, 17.

8. **Noonan, P. K. and Wester, R. C.,** Cutaneous metabolism of xenobiotics, in *Percutaneous Absorption,* Bronaugh, R. and Maibach, H., Eds., Marcel Dekker, Inc., New York, 1989, 53.

9. **Noonan, P. K. and Wester, R. C.,** Cutaneous biotransformations and some pharmacological and toxicological implications, in *Dermatotoxicology,* Marzulli, F. and Maibach, H., Eds., Hemisphere, Washington, D.C., 1987, 71.

10. **Wester, R. C. and Maibach, H. I.,** Dermatotoxicology, in *Environmental Pathology,* Mottet, N. K., Ed., Oxford University Press, New York, 1985, 181.

11. **Bronaugh, R. L. and Stewart, R. F.,** Methods for *in vitro* percutaneous absorption studies. VI. Preparation of the barrier layer, *J. Pharm. Sci.,* 75, 487, 1986.

12. **Yang, J. J., Roy, T. A., Krueger, A. J., Neil, W., and Mackerer, C. R.,** *In vitro* and *in vivo* percutaneous absorption of benzo[a]pyrene from petroleum crude-fortified soil in the rat, *Bull. Environ. Contam. Toxicol.,* 43, 207, 1989.

13. **Kao, J.,** The influence of metabolism on percutaneous absorption, in *Percutaneous Absorption,* Bronaugh, R. and Maibach, H., Eds., Marcel Dekker, Inc., New York, 1989, 259.

14. **Wester, R. C., Maibach, H. I., Sadik, L., Melendres, J., Liao, C., and DiZio, S.,** Percutaneous absorption of chlordane from soil, 1990.

15. **Roels, H., Buchet, J. P., Lauwerys, R. R., Bruaux, P., Claeys-Thoreau, F., Lafontaine, A., and Verduyn, G.,** Exposure to lead by the oral and the pulmonary routes of children living in the vicinity of a primary lead smelter, *Environ. Res.,* 22, 81, 1980.

16. **Driver, J. H., Konz, J. J., and Whitmyre, G. K.,** Soil adherence to human skin, *Bull. Environ. Contam. Toxicol.,* 43, 814, 1989.

17. **Wester, R. C. Maibach, H. I., Sedik, L., Melendres, J., DiZio, S., Jamall, I., and Wade, M.,** Percutaneous absorption of cadmium from water and soil, *Toxicologist,* 11, 289, 1990.

18. **Poiger, H. and Schlatter, C. H.,** Influence of solvent and adsorbents on dermal and interstitial absorption of TCD, *Fundam. Comment. Toxicol.,* 18, 477, 1980.

19. **Shu, H., Teitelbaum, P., Webb, A. S., Marple, L., Brunck, B., Del Rossi, D., Murray, F. J., and Paustenbach, D.,** Bioavailability of soil-bond TCDD: dermal bioavailability in the rat, *Fundam. Appl. Toxicol.,* 10, 335, 1988.

20. **Kimbrough, R. D., Falk, H., Stehr, P., and Fries, G.,** Health implications of 2,3,7,8-tetrachlorodibenzodioxion (TCDD) contamination of residential soil, *J. Toxicol. Environ. Health,* 14, 47, 1984.

21. **Kissel, J. C. and McAvoy, D. R.,** Reevaluation of the dermal bioavailability of 2,3,7,8-TCDD in soil, *Hazardous Waste Hazardous Mat.,* 6, 231, 1989.

22. **Brewster, D. W., Banks, Y. B., Clark, A.-M., and Birnbaum, L.,** Comparative dermal absorption of 2,3,7,8-teytrachlorodibenzo-p-dioxin and three polychlorinated dibenzofurans, *Toxicol. Appl. Pharmacol.,* 97, 156, 1989.

23. **Skrowronski, G. A., Turkall, R. M., and Abdel-Rahman, M. S.,** Soil adsorption alters bioavailability of benzene in dermally exposed rats, *Am. Ind. Hyg. Assoc. J.,* 49, 506, 1988.

24. **Skrowronski, G. A., Turkall, R. M., Abdel-Rahman, M. S.,** Effects of soil on percutaneous absorption of toluene in male rats, *J. Toxicol. Environ. Health,* 26, 373, 1989.

25. **Skrowronski, G. A., Turkall, R. M., Abdel-Rahman, M. S.,** Effect of soil on the dermal bioavailability of *m*-xylene in male rats, *Environ. Res.,* 51, 182, 1990.

26. State of California, 1985.

Route of Exposure and Toxicological Response

Chapter 12

GENERAL OVERVIEW ON TOXICOLOGICAL RESPONSES AND ROUTES OF CHEMICAL EXPOSURE*

Rhoda G. M. Wang

TABLE OF CONTENTS

---

* The opinions expressed in this chapter are those of the author and not necessarily those of the California
  Environmental Protection Agency Department of Pesticide Regulation.

I. HISTORICAL BACKGROUND: INTENTIONAL VS. NONINTENTIONAL CHEMICAL EXPOSURE

Categorically, with respect to the route of exposure which ultimately affects the amount of dose received there is a distinct difference between pharmacological consideration and involuntary chemical exposure consideration. For therapeutic purpose, a chosen drug delivery system is applied in achieving the desirable plasma concentration within the shortest time interval. Unlike drug treatment which offers beneficial therapeutic effects, chemical exposure is unintentional and undesirable, and relatively speaking, the route of exposure is random and may include inadvertent ingestion via inhalation or dermal routes, or a combination of the above.

II. A BLACK BOX NAMED DOSE

A. DEFINITION OF DOSE

Exposure dose refers to the amount of chemical through contacts by humans in the environment, in the air, water, soil, or on the skin. External dose refers to calibrated dosages given to animals, e.g., through gavage or mixed in a diet, etc. Loosely speaking, internal dose refers to the dose that has been absorbed by the biological system, usually expressed as a percent of an applied dose. However, simply calculating the recovered dose from the excreta and tissues/organs and exhaled air as the fraction of absorbed dose is inadequate to express the dynamic nature of chemical disposition in a biological system.

B. STAGES OF AWARENESS ONE TO THREE — DOSE CONCEPT DEVELOPMENT

Stage One — Is the period of denying any consideration of route of exposure. Typically, it was assumed that when exposed to a toxicant, uptake by the biological system is complete. This assumption, though physiologically implausible, however, given the circumstance that extrapolation of animal dose to human dose was under similar route of exposure conditions, the interspecies incompatibility between animals and humans may be greatly reduced, such as extrapolation of dietary exposure in animals to similar exposure in humans. There is, nonetheless, a much greater disparity when extrapolating animal oral data to human exposure conditions for other than oral ingestion. For instance, in estimating dermal dose it is typical, as a general practice, to consider that upon skin contact with a toxicant, complete absorption by the skin is achieved. Thus, the tendency is to equate exposure dose with the external dose which invariably introduces large dosimetry errors.

Stage Two — This is the period of applying a dermal absorption rate as a factor to modify exposure dose. Efforts have been made by the scientific community to improve dose estimate methodology regarding route-to-route extrapolation. Typically, an absorption factor is applied in estimating the dermal dose. A 24-h dermal absorption rate observed in one or more test animal species is used to adjust the exposure dose and arrive at a loosely defined "absorbed dose". It is obvious this strategy has many loopholes. First, the disposition of a xenobiotic in a biological system by a chosen route of exposure impacts the absorption process and determines the kinetic of that chemical being introduced to the blood stream, which is followed by tissue distribution, biotransformation, excretion, etc. This is a dynamic process and cannot be represented by a simple, finite number. Secondly, there are known, significant differences in the percutaneous absorption processes between species. It appears that the rat is on the higher end (absorption rate is high both *in vivo* and *in vitro*) and that humans are on the lower end, with rabbits and pigs in between.[1,2] It is also known that rhesus monkeys tested with a variety of chemicals display an order of dermal absorption closer to humans.[3] Unfortunately, most dermal absorption data are derived from rats.

Stage Three — The period of developing internal dose. Physiologically based pharmacokinetic (PBPK) models describing the fate of a toxicant upon exposure are playing increasingly important roles in health risk assessment. One of the attributes of PBPK models is their ability to more accurately predict tissue dose, which reflects the internal dose. With the knowledge of internal dose the applications are limitless, one of which is more realistically predicting human risk. Abundant information can be found in the literature regarding multicompartment model development for volatile organic compounds through inhalation exposure,[4–7] ingestion,[5,6] multiple route of exposure,[5] and dermal exposure.[8]

III. IMPACT OF ROUTE OF EXPOSURE ON TOXICOLOGICAL RESPONSE

A. CARCINOGENICITY

A pending question confronting the scientific community is the correlation of the test results generated from animal bioassay. Traditionally, this involves feeding animals with a test material mixed in a diet, in drinking water, or through intrastomach intubation for the lifetime of the animal, then correlating those results to that of the general human population, or to occupational workers exposed to toxicants through dermal and/or inhalation exposure in a discontinuous fashion for but a fraction of their lifetimes.

Generally speaking, the route of exposure plays an essential role in carcinogenesis. It determines not only quantitatively the incidence but, more importantly, the histogenetic nature and site of the induced tumors. For instance, when 1,3-dichloropropene (1,3-D) was fed by gavage to rats and mice, malignant tumors of various histogenic types were observed, including hepatocellular carcinomas, some of which had a high grade of malignancy and distant metastasis. Also observed were transitional cell carcinomas of the bladder.[9] However, through inhalation exposure in rats and mice, no statistically significant increase of tumor incidence was observed. An increased incidence of bronchioloalveolar adenomas (benign lung tumors) was observed in male mice of the highest dose group.[10] In a chronic dermal bioassay with 1,3-D in mice, repeated administration of this chemical three times a week to the shaved back of the mice for up to 85 weeks did not result in developing systemic tumors.[11] In other cases, such as with ethylene dibromide (EDB), there were only quantitative differences of tumor incidence between oral and inhalation exposure. Thus, the critical question is again the differential tissue dose achieved under various routes of exposure which dictates tissue response and is highly chemical dependent.

Due to the high cost for lifetime dermal carcinogenicity bioassay in rodents, especially when confronted with the experimental difficulties in restraining test animals from ingestion through grooming or licking, fewer bioassays under dermal route of administration are available for analysis.

Dermal oncogenicity studies are usually done on industrial chemicals that may have potential dermal exposure.[12,13] This type of testing usually results in developing local skin tumors. Historically, dermal application of a carcinogen has been used to study the mechanism of carcinogenicity for decades.[14] In order to simulate the route of exposure to toxicants in the natural environment of humans, NTP has made efforts to increase the number of chronic bioassays using inhalation or dermal route of exposure in test animals. Currently, out of nine completed and fully evaluated skin painting studies, two were tested positive through the dermal route. It is generally recognized that the dermal route of exposure to carcinogens is less likely to produce positive systemic carcinogenicity than the oral route for the simple reason that tissue dose is comparatively lower.

B. DEVELOPMENTAL TOXICITY

More recently the U.S. Environmental Protection Agency (EPA) sponsored several workshops on the acceptability and interpretation of dermal developmental toxicity studies. In these workshops, details of experimental procedures in conducting dermal teratology studies were discussed.[15,16] A general consensus was reached concerning the absolute necessity of making available dermal absorption and pharmacokinetic data for the evaluation of systemic blood concentration of a toxicant prior to conducting dermal developmental toxicity studies. By applying various potency factors, Wang and Schwetz developed a ranking system to analyze animal test results.[17] Using actual exposure scenario and this ranking system, Wang evaluated various developmental toxicants.[18]

The route of exposure of a toxicant governs the target organ dose. When the blood/placenta critical tissue concentration has been achieved, a developmental toxicant will exert its adverse effect in pregnant women. This is especially critical during the first trimester of pregnancy. Exposure to a developmental toxicant during this period of pregnancy at an internal dose equal to or exceeding the threshold dose is most likely to result in birth defects of the unborn. In test animals, the same principle governs; that is, the critical blood/placenta concentration achieved during organogenesis determines the presence, absence, or nature of the malformation.

A few examples are cited here to illustrate the essential role of exposure route. To compare the absorption profile via oral and dermal routes on dinocap (Karathane®), a fungicide/miticide, numerous studies were performed using C^{14} labeled dinocap on rabbits and rhesus monkeys.[19] Practically all of a given oral dose of C^{14} dinocap (0.5 to 25mg/kg) was eliminated in the excreta of rabbits within 48 h. C^{14} dinocap when given orally is absorbed rapidly, reaching peak plasma concentration within 1 to 3 h. In contrast, dermal application resulted in overall low absorption rates, e.g., in a rabbit multiple dermal dose study only 6% of 13 daily dermal applications of C^{14} dinocap (6 h/d) was absorbed. Comparatively, peak plasma concentration through the dermal route was not only much lower than oral, but the rate of achieving such concentration was much more sluggish. Most of the administered dose can be recovered by skin wiping, indicating non-absorption. Percutaneous studies in monkeys have indicated comparable or lower rates of absorption than those found in rabbits.

The dermal absorption profile obtained in rabbits explains the different outcomes of developmental toxicity via oral or dermal routes of exposure. When administered by gavage, dinocap induced frank teratogenicity manifested as hydrocephalus and malformation of the neural tube and skull with an NOEL as low as 0.5 mg/kg/d.[20] Such observation was not found in a dermal developmental toxicity testing in the same species in which the highest dose tested was 100 mg/kg/d. Clearly, the effect of exposure route is demonstrated here, and the reason for negative finding was due to the lack of achieving critical tissue dose.

In recent years, increasing attention has been placed on adverse effects in males due to toxicant exposure. Although inhalation exposure in animal studies can be successfully conducted in male animals to demonstrate the presence or absence of an adverse reproductive effect,[21] such studies are very difficult if not impossible to perform under the dermal route of exposure. Technically, it is not feasible to conduct multiple-generation reproduction studies through the dermal route of exposure due to the difficulties in constraining animals after skin painting and also conducting mating. Sound epidemiologic studies in humans offer more reliable results from this type of potential adverse effect.

Even with relevant route of administration, e.g., oral gavage and dietary exposure, distinctly different potency of developmental toxicity on similar toxicants may be observed. Such is true in the case of benomyl, a herbicide. When dosed by gavage to animals, it demonstrates more potent developmental toxic effects at lower dosage than when administered to animals in a diet mix during organogenesis.[22] Few agents, however, demonstrate a strength of developmental toxicity similar to that of oral administration when administered dermally. One of the exceptions is a potent animal teratogen, nitrofen. When given dermally it was equally teratogenic compared to oral dosing.[23] This is due to the fact that nitrofen is readily and rapidly absorbed through the skin. The above examples simply point out that the route of administration along with other essential biological parameters must be taken into consideration when evaluating human health risks.

C. NEUROTOXICITY

Conventionally, testing of neurotoxicity is done through oral administration of a test compound at high dosages to fully atropinized hens, followed with the observation of delayed neurotoxic effects. The delivered dose through oral and dermal routes to the systemic circulation may differ due to bioavailability differences and other effects, such as GI (gastrointestinal) hydrolysis and metabolism. Using DEF (*S,S,S*-tributylphosphorotrithioate) as an example, at high dosages when given orally, DEF produced delayed neurotoxicity in Leghorn hens. Interestingly, at similarly high dosages when given dermally, DEF demonstrated more intense neurotoxicity. This is due to the hydrolysis of DEF in the GI tract generating a mercaptan compound that caused delayed acute toxicity, whereby the active chemical species that caused delayed neurotoxicity was substantially reduced due to hydrolysis in the stomach.[24] Through hydrolytic alteration of the effective chemical species of DEF in the GI, which resulted in demonstrably lesser neurotoxic effect upon oral administration, a reverse of commonly known sequence of oral vs. dermal effect is seen here. Thus, the route of exposure not only governs the quantitative toxic response, it may alter altogether or modify the nature of the response.

IV. ROUTE OF EXPOSURE, PBPK MODELING, AND RISK ESTIMATE

Route of exposure can determine both the quantitative and qualitative outcomes of a toxicological response. Extrapolation from one route to the other is often difficult if not impossible. One of the applications of PBPK modeling is to facilitate the interroute extrapolation. Reitz et al. applied PBPK modeling with methyl chloroform under various routes of exposure and explicitly demonstrated dose route extrapolation relationship,[25] and studies of parenteral and oral route extrapolation on various chemicals also provide insight of route-to-route extrapolation.[26–29]

One essential aspect of PBPK modeling is the inclusion of the vital physiological and anatomical considerations, such as blood flow and ventilation rates, metabolic constants of essential metabolic pathways, and tissue solubilities and bindings, etc. The incorporation of these physiological parameters facilitates the derivation of tissue and target organ dose, which is then jointly used with dose scaling and extrapolation techniques between any species for dose response estimation.

Progress has been made in recent years to incorporate pharmacokinetic data in low dose extrapolation, and in developing quantitative risk assessment that incorporates the intrinsic factors of pharmacokinetic considerations.[30,31] A recent study reported the diversified yet convergent opinions of scientific experts on formaldehyde cancer risk assessment using pharmacokinetic approaches.[32] This approach to resolving uncertainties offers a promising avenue for risk characterization and may define where research is needed to fill the gap and improve risk analysis.

The effort put forth to compare toxicological responses on similar toxicants through various routes of exposure is considered very primitive and must be improved as more data become available. Instead of comparing the different outcomes from toxicity testing, a primitive way to perform a task, we should concentrate on understanding the relationship between tissue dose and tissue response. If a chemical cannot achieve certain concentration at a target organ, e.g., the placenta during organogenesis, it is safe to assume that a teratogenic effect cannot be provoked. Thus, concentrating on tissue/organ dosimetry is of utmost importance.

In the beginning of the 1990s we will see more cancer risk assessment using the tissue dose/tissue response approach. Using PBPK and multistage modeling, Bois et al.[33] performed risk assessment on tetrachloroethylne, a probable human carcinogen. Animal data were used to obtain parameters such as V_{max} and K_m scaling coefficient, and the rate of metabolic formation of an epoxide (as the ultimate carcinogen) was computed through PBPK modeling for risk estimate.

The authors discuss potential extrapolation uncertainties and unknown factors such as lacking of the precise mechanism of carcinogenesis in humans. Undoubtedly, much more work is needed to refine this approach. However, the development of tissue dose concept through PBPK modeling offers an improved method for realistic dose and risk estimation, the beginning of a new chapter in human risk assessment.

V. DISCUSSION

There is a growing need to investigate the differential risks due to various exposure routes of toxicants in humans. The route of exposure to toxic chemicals of humans in the environment could be very different from that in laboratory-tested animals, which are predominantly through gavage or feeding of a test chemical through intrastomach intubation or in the diet. Human exposure to toxic chemicals n the environment is predominantly through inhalation and dermal exposure, especially in occupationally related situations.

Given the black box named dose in many situations, it is impractical to expect reasonable extrapolation from animal data to humans under dissimilar route of exposure conditions, especially when the response is a threshold response. Physiologically, it is known that when a dose is modified to a fraction of that dose which usually produces a pharmacological or toxicological response, then the new dose when introduced to the biological system may indeed evoke no response. Thus, the traditional practice of applying safety factors on something unknown would result in a greater ''unknown'' and does not fit into the physiological basis. The true sense of identifying the problem, making interspecies comparisons, and ensuring human exposure to environmental toxicants that does not cause unacceptable risks is though researching the tissue dose and toxicological effects evoked and making comparisons and decisions accordingly.

REFERENCES

1. **Bartek, M. J., La Budde, J. A., and Maibach, H. I.,** Skin permeability in vivo: Comparison in rat, rabbit, pig and man. *J. Invest. Dermatol.*, 58, 114, 1975.
2. **Bartek, M. J. and La Budde, J. A.,** Percutaneous absorption *in vitro*, in *Animals Models in Dermatology*, Maibach, A. B. Ed., Churchill Livingstone, New York, 1975, 102.
3. **Wester, R. C., McMaster, J., Bucks, D. A. W., Bellet, E. M., and Maibach, H. I.,** Percutaneous Absorption in Rhesus Monkeys and Estimation of Human Chemical Exposure, in *Biological Monitoring for Pesticide Exposure* (ACS Symposium Series 381), Wang, R. G., Franklin, C. A., Honeycutt, R. C., and Reinert, J. C., Eds., Washington, D.C., 1989, Chapter 12.
4. **Bogan, J. T. and McKone, T. E.,** Linking indoor air and pharmacokinetic models to assess tetrachloroethylene, *Risk Analysis*, 8, 509, 1988.

5. **Reitz, R. H., McDougal, J. N., Himmelstein, M. W., Nolan, T. J., and Schumann, A. M.,** Physiologically based pharmacokinetics modeling with methylchloroform: implication for interspecies, high dose/ low dose, and dose route extrapolations, *Toxicol. Appl. Pharmacol.,* 87, 185, 1987.

6. **Andersen, M. E., Clewell, H. J., III, Gargar, M. L., Smith, F. A., and Reitz, R. H.,** Physiologically based pharmacokinetics and the risk assessment process for methylene chlorine, *Toxicol. Appl. Pharmacol.,* 87, 185, 1987.

7. **Pausteinbach, D. J., Clewell, H. J., III, Gargas, M. L., and Andersen, M. E.,** A physiologically based pharmacokinetic model for inhaled carbon tetrachloride, *Toxicol. Appl. Pharmacol.,* 85, 191, 1988.

8. **McDougal, J. N., Jepson, G. W., Clewell, H. J., III, MacNaughton, M. G., and Andersen, M. E,** A physiological pharmacokinetic model for dermal absorption of vapors in the rat, *Toxicol. Appl. Pharmacol.,* 85, 286, 1986.

9. **Wang, G. M.,** Evaluation of pesticides which pose carcinogenicity potential in animal testing, I. Developing a tumor data evaluation system, *Reg. Toxicol. Pharmacol.,* 4, 355, 1984.

10. **Lomax, L. G., Scott, W. T., Johnson, K. A., Calhoun, L. L., Yano, B. L., and Quast, J. F.,** The chronic toxicity and oncogenicity of inhaled technical grade 1,3-dichloropropene in rats and mice, *Fundam. Appl. Toxicol.,* 12, 418, 1989.

11. **VanDuuren, G. L., Goldschmidt, B. M., Loewengart, G., Smith, A. C., Melchionne, S., Seldman, I., and Roth, D.,** Carcinogenicity of halogenated olefinic and aliphatic hydrocarbon in mice, *J. Nat. Cancer Inst.,* 63(6), 1433, 1979.

12. **McKee, R. H., Stubblefield, W. A., Lewis, S. C., Scala, R. A., Simon, G. S., and Depass, L. R.,** Evaluation of the dermal carcinogenic potential of tar sands bitumen-derived liquids, *Fundam. Appl. Toxicol.,* 7, 228, 1986.

13. **Dapass, L. R., Fowler, E. H., and Weil, C. S.,** Dermal oncogenicity studies on various ethyleneamines in male C3H mice, *Fundam. Appl. Toxicol.,* 9, 807, 1987.

14. **Mukhtar, H., Merk, H. F.,and Achsr, M.,** Skin chemical carcinogenesis, *Clin. Dermatol.,* 7, 1, 1989.

15. **Kimmel, C. A. and Francis, E. L.,** Proceedings of the Workshop on the Acceptability and Interpretation of Dermal Developmental Toxicity Studies, U.S. Environmental Protection Agency, Washington, D.C., 1988.

16. **Patton, D. E.,** Risk Assessment Forum Workshop Report on Dermal Developmental Toxicity Studies, U.S. Environmental Protection Agency, Washington, D.C., 1988.

17. **Wang, G. M. and Schwetz, B. A.,** An evaluation system for ranking chemicals with teratogenic potential, *Teratogenesis, Carcinogenesis, and Mutagenesis,* 7, 133, 1987.

18. **Wang, G. M.,** Regulatory decision making and the needs for and the use of exposure data on pesticide determined to be teratogenic in test animals, *Teratogen. Carcinogen. Mutagen.,* 8, 117, 1988.

19. **Longacre, S. L., DiDonato, L. J., Wester, R. C., Maibach, H. I., Hurt, S. S., and Costlow, R. D.,** Dinocap dermal absorption in female rabbits and rhesus monkeys, in *Biological Monitoring for Pesticide Exposure,* Wang, R. G., Franklin, C. A., Honeycutt, R. C., and Reinert, J. C., Eds., ACS Symposium Series 382, Washington, D.C., 1989, 11.

20. **Costlow, R. D., Lutz, M. F., Kane, W. W., Hurt, S. S., and O'Hara, G. P.,** Dinocap: developmental toxicity studies in rabbits, *Toxicologist,* 6, 85, 1986.

21. **Wang, G. M., Kier, L. D., and Pounds, G. W.,** Male fertility study on N,N-Dimethylacetamide administered by the inhalation route to Sprague Dawley rats, *J. Toxicol. Environ. Health,* 27, 297, 1989.

22. **Kovlock, R. J., Chernoff, N., Gray, L. E., Jr., Gray, J. A., and Whitehouse, D.,** Teratogenic effects of Benomyl in the Wistar rat and CD-1 mouse, with emphasis on the route of administration, *Toxicol. Appl. Pharmacol.,* 62, 44, 1982.

23. **Hirsekorn, J. M., Kane, W. W., and Black, D. L.,** Tok E-25 Percutaneous Teratology study with Post Partum Evaluation, Rohm and Haas Rep. 81, R-0154, 1982.

24. **Abou-Donia, M. B., Graham, D. G., Timmons, P. R., and Redchert, B. L.,** Delayed neurotoxic and late acute effect of S,S,S-Tributyl Phosphorotrithioate on the hen: effect of route of administration, *Neurotoxicity,* 1, 425, 1979.

25. **Reitz, R. H., McDougal, J. N., Himmelstein, M. W., Nolan, R. J., and Schumann, A. M.,** Physiologically based pharmacokinetic modeling with Methylchloroform: implications for interspecies, high dose/ low dose and dose route extrapolations, *Toxicol. Appl. Pharmacol.,* 95, 185, 1988.

26. **Fisher, J. W., Whittaker, T. A., Taylor, D. H., Clewell, H. J., III, and Andersen, M. E.,** Physiologically based pharmacokinetic modeling of the lactating rat and nursing pup: a multiroute exposure mode for Trichloroacetic acid, *Toxicol. Appl. Pharmacol.,* 102, 497, 1990.

27. **Gilette, J. R.,** Dose, species, and route extrapolation: general aspects in workshop proceedings, in *Pharmacokinetics in Risk Assessment,* Vol. 8, National Academy Press, Washington, D.C., 1987, 96.

28. **McDougal, J. N., Jepson, G. W., Clewell, H. J., III, MacNaughton, M. G., and Andersen, M. E.,** A physiological pharmacokinetic model for dermal absorption of vapors in the rat, *Toxicol. Appl. Pharmacol.,* 85, 286, 1986.

29. **Frederick, C. B.,** Contact Site Carcinogenicity — Estimation of and Upper Limit for Risk of Dermal Carcinogenicity Based on Oral Dosing Site Tumors, ILSI/EPA Workshop on Principles of Route-to-Route Extrapolation, Hilton Head, South Carolina, March, 1990.
30. **Whittemore, A. S., Grosser, S. C., and Silver, A.,** Pharmacokinetics in low dose extrapolation using animal cancer data, *Fundam. Appl. Toxicol.,* 7, 183, 1986.
31. **Thorslund, T. W., Brown, C. C., and Charnley, G.,** Biologically motivated cancer risk models, *Risk Anal.,* 7, 109, 1987.
32. **Hawkins, N. C. and Graham, J. D.,** Expert scientific judgement and cancer risk assessment: a pilot study of pharmacokinetic data, *Risk Anal.,* 8, 614, 1988.
33. **Bois, F. Y., Zeise, L., and Tozer, T. N.,** Precision and sensitivity of pharmacokinetic models for cancer risk assessment: Tetrachloroethylene in mice, rats, and humans, *Toxicol. Pharmacol.,* 102, 300, 1990.

Chapter 13

ACUTE TOXICITY TESTING BY THE DERMAL ROUTE

Roy C. Myers and Linval R. DePass

TABLE OF CONTENTS

0-8493-7357-3/93/$0.00 + $.50
© 1993 by CRC Press, Inc.

I. INTRODUCTION

The potential for dermal exposure to chemicals in today's world seems endless. Intentionally or unintentionally, we may be exposed to a myriad of chemical substances such as pharmaceuticals, agricultural products, industrial chemicals, consumer products, food contaminants, and pollutants (natural and man-made) of the air, soil, and water. Our homes, workplaces, and the great outdoors are potential sources of contamination.

A traditional and basic concept of toxicology is that while any chemical may be tolerated in sufficiently small amounts, every chemical may be harmful in sufficiently large amounts.[1,2] To help protect ourselves from the undesirable effects of chemical substances, we must define the nature and severity of their possible hazards to health. Although we may be exposed to chemicals by any number of routes — oral, dermal, inhalation, intravenous (pharmaceuticals or substances of abuse), ocular, nasal, anal, vaginal, etc. — in this chapter we will consider exposure by one of the most common routes, dermal contact.

Although human subjects are sometimes used for evaluating the safety of chemicals, especially pharmaceutical products, acute toxicity testing is first conducted in appropriate animal models. Generally, relatively high doses are administered in acute toxicity tests to examine the potential for gross biological effects up to and including death. One purpose of acute toxicity testing is to mimic a single "accidental" human exposure. Beyond this, however, important data may be collected on specific target organs for the compounds, as well as dermal bioavailability and irritancy.

Frequently, because of constraints of time and resources, acute testing (defined as a single dose or multiple doses given within 24 h[3]) may be the only toxicity study performed on a given substance. Thus, data collected in such studies may be very valuable.

As will be seen in subsequent sections, establishing a specific course of action for animal testing is subject to a number of important considerations, including the nature of the product in question or the likelihood of dermal exposure. Other factors are cost of testing, amount of handling, method of transportation, and regulatory requirements.

II. STUDY DESIGN AND PURPOSE

Selection of the specific design of a dermal toxicity test involves several factors such as animal model (species), age, sex, duration of skin exposure, etc. While a repeated-dose dermal study may be conducted over a period of months or years to meet specific goals, acute testing is particularly useful for evaluation of the possible hazards from a single exposure to a compound, setting dose levels for longer-term studies, identification of potential target organs, fulfilling regulatory requirements, and/or establishing toxicity classifications.

A. ANIMAL MODEL
1. Species

Several species have been used in acute dermal toxicity testing, including the rat, rabbit, guinea pig, pig, and monkey. The choice is often based on practical considerations such as cost, ease of housing, the amount of available test substance, availability of disease-free animals, and ease of animal handling. Ideally, an animal model should closely resemble the human in sensitivity to dermally applied toxic compounds and permeability of the skin. One may also employ a species with a relatively high degree of dermal absorption to provide "worst-case" toxicity data for extrapolation to humans.

Variability in toxicity from species to species (and often between strains of the same species) is quite striking. The rabbit is usually considerably more sensitive to dermally applied chemicals than is the rat. For example, in a recent dermal toxicity study with methyldimethoxysilane,[4] the LD_{50}s for male and female New Zealand White rabbits were 1.47 and 0.73 ml/kg, respectively. Death occurred in as little as 5 min. However, doses of 16.0 ml/kg did not kill any Sprague Dawley rats after 24 h of occluded contact.

Numerous investigators have evaluated species differences in dermal toxicity, as well as differences in skin permeability, based on *in vivo* and *in vitro* test methods. Some examples are presented in Table 1. Generally, rabbit skin has been found to be the most permeable, followed by the rat, guinea pig, pig, monkey, and man. The monkey[10,11] or pig[11] may best predict dermal absorption in man. The guinea pig may be similar to the rabbit (a common species for acute toxicity testing) as a model for dermal absorption.[12,13] However, it should be remembered that toxicity from dermal contact may often be determined by factors other than degree of absorption.

Many regulatory guidelines recommend the rabbit for use in the acute dermal toxicity test, presumably because of its sensitivity to dermally applied chemicals, large dermal surface, docility, availability, and an ever-increasing database. The rat or guinea pig is frequently accepted as a test model, however, especially if minimal sample (or funding) is available or if a second species for comparative dermal toxicity is required. The most widely used rabbit strain is the New Zealand White; common rat strains include the Sprague Dawley, Wistar, Fischer (344), and Long-Evans (Hooded), and the most common guinea pig strain is the Hartley.

2. Age

Results of dermal toxicity testing depend not only on the species/strain used, but also on the age. Very young animals are generally (but not always) more sensitive to the toxic effects of a chemical by various routes of exposure.[14] Younger animals have limited detoxification capabilities due to incomplete enzyme system development.[15] Moreover, younger animals may have different dermal penetration characteristics than older ones.[16] As an animal ages, enzyme systems may again become less efficient and higher sensitivity to dermally applied chemicals may again develop. In humans, there is frequently a lessening of permeability in aging skin,[17] probably from changes in lipid content, hydration, and structure.[18]

Dermatology: Clinical and Basic Science

TABLE 1
Examples of Species Differences in the Dermal Toxicity and Penetration of Various Substances

Substance	Ref. 5 Organophosphorus compound	Ref. 5 Organophosphorus compound	Ref. 6 Cortisone	Ref. 6 Haloprogin	Ref. 7 Testosterone	Ref. 8 2,4-D Amine	Ref. 9 Water
Measurement	*In vitro* penetration	*In vivo* toxicity	*In vivo* penetration	*In vivo* penetration	*In vivo* penetration	*In vivo* penetration	*In vitro* penetration
Ranking[a]							
Mouse	—	4	—	—	—	—	—
Rat	1	6	2	2	2	2	3
Guinea pig	2	—	—	—	3	—	3
Rabbit	1	1	1	1	1	3	—
Dog	4	2	—	—	4	—	1
Pig	5	5	3	3	5	—	—
Monkey	3	3	—	—	6	—	2
Man	—	—	4	4	5	1	—
Max. ratio[b]	31	17	9	10	5	5	4

a Species rankings in order of highest (1) to lowest toxicity or penetration.
b Ratio of highest to lowest skin penetration or toxicity in the series.

In the rat, dermal absorption of TCDD decreased from 17% of the administered dose to 5% as the animal aged from 3 to 9 months.[19]

Generally, a young adult animal is employed for acute dermal toxicity testing. This is required by regulatory agencies, although it might be useful to conduct tests with younger or older animals if they are more relevant to the intended human exposure (to test baby lotions, for example). A rabbit weighing 2.0 to 3.0 kg (3 to 5 months of age) or a rat weighing 200 to 300 g (roughly 1.5 to 3 months old, depending on strain) is frequently used for acute dermal toxicity testing. Guinea pigs, if used, should be approximately 350 to 450 g (one to two months old).

3. Sex

Toxicity, as might be expected, may also be sex-dependent, but this is not always so. DePass et al.[20] evaluated 28 chemicals applied to intact skin for rabbit dermal toxicity (LD$_{50}$ determination) and obtained mean LD$_{50}$ values of 3.15 g/kg for males and 2.81 g/kg for females. Although these values were similar numerically, they were statistically different. In 17 dermal studies in which compounds were applied to abraded rabbit skin, the mean LD$_{50}$s for males and females were essentially identical. Subsequent experience in the same laboratory has confirmed that the female is generally more sensitive to the toxic effects of dermally applied chemical substances, although the absolute difference is generally minimal.

Hormonal differences could explain many of the instances of differing toxicity between male and female animals, but another consideration could be variability in dermal penetration. It has been reported that permeability constants of several materials were two to three times greater for female rats than for males and that the female rat stratum corneum was one half as thick as that of the male.[21]

In spite of infrequent large differences in toxic response between sexes or the more common slight sex-related differences, acute toxicity testing in only one sex is often sufficient, especially if one uses the more sensitive sex (usually the female). This is permitted (and encouraged) by many of the more recent testing guidelines (discussed later in this chapter). However, it is still a rather common procedure to use both sexes in acute dermal toxicity testing to make certain that there are no unexpected sex-related effects. Sometimes, one sex is used for a definitive study following preliminary testing in both sexes, or a few animals of the opposite sex may be treated concurrently for comparison.

4. Other Animal Considerations

Several additional factors should be considered in choosing the animal model for acute dermal toxicity evaluation. Important among these is the nutritional status of the animal. Deficiency of essential fatty acids, proteins, or vitamins, excess of sugar or fat, or presence of enzyme inducers in the diet may affect the degree of toxic response to a chemical.[22] Also critical to a dermal toxicity study is the disease status of the test animal. Diseases of the liver or kidney may have a significant impact on metabolism of a toxicant, for example. A change in the immune system may also be expected to alter an animal's response to a toxic chemical.

In dermal toxicity testing, it is most crucial that the dose site be free from injury or disease. These conditions will almost certainly affect the degree of penetration, but the increase (or decrease) of permeability will vary from chemical to chemical.[23] Unless a study is designed to include specific skin abnormalities, standard test animals from reputable vendors should be used. Proper care and nutrition must be provided and only healthy animals with clear skin are dosed. Care must also be taken to avoid the use of animals with unusually thick or irregular fur patterns as these conditions may also affect dermal absorption.

B. TEST CONDITIONS

1. Duration of Contact

Theoretically, any length of contact time could be chosen for an acute dermal toxicity study. Depending on specific circumstances (anticipated human exposure, regulatory concerns, practical limitations, species tested, research goals, etc.), contact could last for minutes, several hours, or for days. Common contact periods for animal tests are 4, 6, and 24 h. For humans, dermal contact of less than 8 h would seem most likely, but exposure for as long as 24 h is conceivable. Increase in contact time should permit greater absorption and, therefore, greater potential for toxic effect. In one study,[24] monocrotophos had percutaneous LD_{50}s (in rats) of 868 and 467 mg/kg for a single 1- and 4-h contact period, respectively. Wester and Maibach (quoting earlier work by Feldmann and Maibach)[25] reported that occluded skin contact with malathion for periods of 1, 4, and 24 h resulted in 12.7, 24.2 and 62.8% absorption, respectively.

Increases in toxicity do not always follow extended dermal exposure times, however. In a comparative study using 30 chemicals,[26] 4- and 24-h rat dermal LD_{50}s were quite similar (correlation coefficient = 0.90) as were 4- and 24-h rabbit dermal LD_{50}s (correlation coefficient = 0.82). (Interestingly, the rabbit LD_{50}s were generally well below the rat LD_{50}s even though the contact times were much shorter for the rabbit — further evidence of the greater sensitivity of the rabbit as discussed earlier). Another example was the percutaneous toxicity evaluation of bis[2-(dimethylamino)-ethyl]ether.[27] The 4-h LD_{50}s were 0.41 ml/kg and 0.63 ml/kg for male and female rabbits, respectively; 24-h tests resulted in an LD_{50} of 0.37 ml/kg for both sexes. These values were judged to be statistically equivalent. Several explanations may be offered as to the occasional lack of correlation between contact period and toxicity (or absorption).[28,29] Dermal penetration may be rapid (and relatively complete), the sample may evaporate after a short time, the saturation of metabolic pathways may limit the degree of conversion to a more toxic product, the skin may become less penetrable due to local changes, or there may be some absorptive limits in the skin (such as saturation of the dermal reservoir) or limits in general circulation.

Probably the most common contact period is 24 h, as prescribed by numerous regulatory guidelines. This allows adequate time for significant (worst-case) absorption to occur. The 4-h test is especially useful for smaller test animals, for which there may be difficulties in maintaining prolonged contact. If the animal must be restrained during the contact period, the 4-h test would be more appropriate since longer periods of restraint may result in considerable stress.

2. Choice of Dose Site

There is no doubt that the degree of absorption through the skin depends upon the specific location of dose application. This is amply demonstrated by the data in Table 2. Generally, the skin of the forehead, hand (dorsal surface), and abdomen appear to be more permeable than that of the chest, forearm, or back. The palm of the hand and ball of the foot are even more resistant to dermal penetration.

Detailed discussion of the reasons for regional variability is beyond the scope of this chapter. However, it is tempting to conclude that there is an inverse relationship between the thickness of the stratum corneum and the degree of penetration.[34] There is also evidence that permeability increases with the density of hair follicles.[30] While the follicles and sweat ducts may facilitate penetration, their effect is probably minimal due to the relatively small area they occupy.[35,36] Other factors that may contribute to regional permeability include lipid composition, morphologic structure (cell arrangement) of the stratum corneum,[34,37] and dermal vasculature.[36,38] Each of these factors may vary in importance depending upon the structure of the chemical. Of course, variability in dermal penetration will greatly influence the local and systemic toxicity of a chemical.

TABLE 2
Examples of Regional Differences in the Dermal Penetration of Various Substances

	Ref. 30	Ref. 31	Ref. 32	Ref. 8	Ref. 33	Ref. 33
Substance	Parathion	Hydrocortisone	Benzoic acid	2,4-D Amine	Urea	Caffeine
Species	Man	Man	Man	Monkey	Rat	Rat
Measurement	Excretion, % of dose	Excretion, % of dose	Excretion, per cm^2	Excretion, % of dose	Permeability constant	Permeability constant
Ranking[a]						
Forehead	1	1	1	1	—	—
Chest	—	—	3	—	—	—
Forearm	5	3	4	2	—	—
Hand	2	4	—	—	—	—
Abdomen	3	—	2	—	1	1
Back	—	2	5	—	2	2
Foot	4	5	—	—	—	—
Factor[b]	4.2	43	3.2	5.2	11.8	2.3

[a]   Location rankings in order of highest (1) to lowest penetration.
[b]   Ratio of highest to lowest in the series.

Although most regulatory guidelines do not clearly specify the exact location of application, it is a generally accepted practice to use the dorsal trunk area of the test animal. The dorsum provides a large, relatively smooth surface for dosing. It is easy to access and to maintain chemical contact. Moreover, application to the skin of the back permits minimal opportunity for ingestion, sample removal (rubbing against the cage, cleaning with the feet, etc.), and is relatively easy to cover. Standardization of dermal toxicity data can be achieved by using a single dose region.

3. Occlusion of the Dose Site

During the period of contact with a test chemical, a treated site may be left uncovered, be partially covered, or may be fully covered. Coverings may consist of semi-impermeable to completely occlusive materials. The degree of occlusion of the dose site may have a marked effect on the outcome of a dermal toxicity study.

In humans, absorption of a topical dose of hydrocortisone increased from 0.456% of applied dose to 4.48% after occlusion of the dose site.[39] The absorption of cinnamic acid through monkey skin, expressed as percentage of applied dose, rose from 38.6 to 83.9% when the dermal surface was occluded.[40] There was a similar increase from 17.8 to 60.8% in absorption through human skin (*in vitro*). In the rabbit (commonly used for dermal toxicity testing), one study[41] showed that 1.0 g/kg of hydroxylamine sulfate was not lethal to animals when the treated skin was covered with only gauze dressing. However, occlusion with plastic resulted in death of 90% of rabbits receiving one half of the this dose. It should be noted that absorption may not be enhanced by occlusion for every chemical.[42]

It has generally been accepted that occlusion alters skin permeability (and/or potential for toxic effect) by: (1) increasing hydration of the skin, (2) increasing temperature of the skin, (3) maintaining adequate contact of test substance with the skin, (4) limiting evaporation of volatile materials, and (5) preventing accidental removal of the test substance (from animal movement, grooming, or scratching).[7,29,42] While the loss of sample through evaporation[28] or some physical means is not to be ignored, hydration may be the most important factor.[38] The stratum corneum may show an increase of water content from 5 to 15% to 50% under occlusive conditions. Generally (but not always) such a change will significantly enhance dermal absorption.

Occlusion is probably applicable to human exposure, in many cases, as a human may receive inadvertent doses underneath clothing, gloves, boots, or other types of covering. Thus it is probably quite appropriate to test chemicals under the most severe contact scenario. On the other hand, many human exposures could take place on exposed (unoccluded) dermal surfaces and, therefore, toxicity information from similar exposures in animals would be quite useful.

The "standard" procedure for acute dermal toxicity studies defined by most regulatory agencies includes occlusive covering of the dose site. Usually this takes the form of thin plastic or rubber sheeting which is secured over the dosed area. The sheeting may be protected from removal by numerous methods which will be discussed in a later section.

Occlusion can be accomplished through the use of a variety of materials such as adhesive tapes, plastic or rubber sheeting, aluminum foil, as well as glass, metal, or plastic chambers.[24,39-43] For relatively small doses, a useful occlusive device is the HILLTOP CHAMBER®, which consists of adhesive tape, a plastic chamber, a WEBRIL® pad, and release paper.[44]

4. Abrasion of the Dose Site

The skin consists of two major layers, the epidermis (outer layer) and dermis (inner layer). The epidermis, in turn, is composed of several cell layers, the outermost of which is the stratum corneum (or "horny layer").[45] The stratum corneum (which is made up of rows of flattened keratinized cells) is essentially (but not entirely) responsible for the barrier properties of the skin.[9,35,46] Through this barrier, most exogenous materials are kept out and moisture is locked in. Any disturbance of this stratum corneum (through injury, disease, or aging) would be expected to have a significant effect on penetration (and, therefore, toxic effect) of various chemical substances. The underlying layer of the skin, the dermis, consists primarily of connective tissue with a network of hair follicles, blood vessels, sebaceous glands, sweat glands, and nerve endings. It is this skin layer that permits absorption and distribution of chemical substances.

In one *in vitro* study using human skin,[47] tri-*n*-propyl phosphate penetrated unstripped epidermis at the rate of 4.4 $\mu g/cm^2/min$. The rate increased to 14.4 $\mu g/cm^2/min$ when the stratum corneum was stripped away. Similarly, penetration of hydrocortisone (measured through urinary excretion of the labeled compound) rose from 4.48% of dose to 14.91% of dose by stripping the outer skin layers (with cellophane tape).[39] In the rat, penetration of formic acid (0.3 *M*) was measured at 0.94 $\mu mol/cm^2/h$ for whole skin, but increased to 17.5 $\mu mol/cm^2/h$ for the isolated dermis.[9]

Results from comparative *in vivo* toxicity tests using abraded and unabraded skin are not quite as dramatic (presumably because abrasion removes only portions, but not substantial areas, of the stratum corneum). One study involving eight chemicals indicated that dermal LD_{50}s were essentially similar for guinea pigs with intact and abraded skin.[12] In fact, most samples had *higher* LD_{50}s (lower toxicity) when applied to abraded animals. Skin abrasion was found to have a moderate, but statistically significant, decrease on male rabbit dermal LD_{50}s (mean LD_{50} for intact skin = 1.35 g/kg; mean LD_{50} for abraded skin = 0.91 g/kg) as reported by DePass et al.[20] Female rabbits demonstrated only a slight difference in mean LD_{50}s (1.21 g/kg for intact skin; 1.04 g/kg for abraded skin), which was not statistically significant. These studies used 16 chemicals (males) and 20 chemicals (females). It has been our continued experience that abrasion does not substantially alter the toxicity of a chemical. Most regulatory guidelines do not require abrasion, probably for this reason.

Certainly, skin abrasion is a part of the human experience. Hand, arms, or legs might be scratched, scraped, or otherwise injured before (or during) contact with a chemical substance. Thus, an animal toxicity test might include abrasion of the skin on at least some subjects. Indeed some regulatory guidelines (as discussed in a later section) still make this a requirement.

A serious problem associated with testing on abraded skin has been the choice of a consistent method for abrasion. Regulatory agencies give little guidance in this regard. It is generally suggested, however, that only the epidermis be abraded and that no bleeding occur. In our laboratory, the back edge of a scalpel is used to make several parallel strokes (2 to 3 cm apart) along the cutaneous surface to be treated. Some labs use the edge of a hypodermic needle and other procedures have included sandpaper, files, chemical erosion (solvents or known irritants), contact with dry ice, or stripping with adhesive tape.

5. Use of Vehicles/Chemical Penetration

It would be simple if all test substances could be applied neat (undiluted), but numerous factors prevent this. Only liquids can routinely be administered without moistening, dilution, or other preparation. Test substances may be viscous liquids, powders, bulky solids, fabrics, chips, granules, or pellets. In addition, very toxic liquids may require dilution to permit accurate measurement of small volumes and adequate skin coverage. Still other materials may require mixing with a vehicle to mimic use or transportation conditions of the chemical.

Absorption through the skin barrier (stratum corneum) is generally by passive diffusion.[21,23,35] Briefly, the absorption process follows several steps,[35,42] including adsorption of the molecule at the stratum corneal surface, diffusion through the stratum corneum (the rate limiting factor), release of the molecule into the viable epidermis, diffusion into the dermis, and then into the capillaries and circulating blood. (Complicating this process are the biotransformations that may occur in the skin or in other tissues; this in turn may affect absorption and/or toxicity.)

In considering the physical and chemical nature of a chemical, there are a few basic "rules" to skin penetration:[48] (1) substances with high lipid solubility penetrate relatively well, (2) inorganic electrolytes (ionized materials) penetrate poorly, (3) gases (if maintained in dermal contact) penetrate easily, and (4) very large molecules penetrate very poorly.

Obviously, use of a vehicle may greatly affect a substance's toxicity by influencing its rate of absorption. In the rabbit, the percutaneous LD_{50} for 1,3,5-triacryloylhexa-hydro-*s*-triazine when dosed as a 10% aqueous suspension was 0.80 g/kg.[49] When administered as an aqueous paste, the percutaneous LD_{50} was 0.57 g/kg. A 10% solution of this material in dichloromethane had a percutaneous LD_{50} of 0.07 g/kg.

A dramatic example of the effect of vehicle on dermal toxicity was reported by Brown and Muir.[24] They found that rat dermal LD_{50}s for the pesticide dicrotophos were approximately 20, 100, and 130 mg/kg when the respective vehicles were *n*-octanol, acetone, and isopropanol. These authors also demonstrated the effect of concentration on the toxicity of another pesticide (monocrotophos). The LD_{50}s (based on active ingredient) were roughly 55 mg/kg with a 5% solution (in hexylene glycol), 95 mg/kg with a 10% solution, and 145 mg/kg with a 58% solution.

Of considerable interest is the apparent increase of penetration of many substances when dimethylsulfoxide (DMSO) is used as a vehicle. DMSO readily penetrates the skin and then disappears.[29] *In vivo* testing with human subjects demonstrated a 25-fold increase in percutaneous absorption of hexopyrronium bromide (quaternary) through the use of a DMSO solvent.[50] Also reported was an increase in penetration of hydrocortisone through human skin, with 2.25% penetration using water and 15.1% using DMSO.

The absorption of a test substance from a vehicle depends on several factors including the partition of the substance between the vehicle and the skin, the solubility of substance in the vehicle, the concentration of the substance, and direct effect of the vehicle on the barrier.[7] Many organic solvents alter the stratum corneum through removal of its lipid constituents, thereby affecting (often increasing) penetration of other materials.[35] This condition may reverse in 2 or 3 d, but some solvents (for example, mixtures of polar and nonpolar components) may destroy the barrier structure itself.[9] Again, permeability will be substantially changed. It is believed that DMSO does not act as a "carrier" for a solute, but does greatly alter the permeability of the barrier (possibly by increasing hydration,[38] removing lipids, or producing structural damage).[35] Some oily or greasy vehicles may restrict water loss (increase hydration) and thus affect permeability.[51] Although soaps and surfactants do not appear to penetrate the skin very well, they have been shown to alter its barrier functions.[52]

For a detailed description of vehicles used in toxicology, the reader is referred to the review by Gad and Chengelis.[53] As a rule, a vehicle is chosen on the basis of its: (1) low reactivity with the test substance, (2) low toxicity, (3) ability to form a stable and homogeneous solution or suspension, (4) limited binding characteristics (which permits release of the test substance for absorption), and (5) minimal disturbance of the skin barrier.

The solvent of first choice is water (or saline), since it will usually have minimal influence on the test substance's toxicity. Unfortunately, water alone is not always effective in producing the homogeneous mixture required in toxicity testing. There are a number of useful suspending/wetting agents, including methyl cellulose, Tergitols®, Tweens®, or agar. These all can be used to make aqueous mixtures. However, if unsuitable mixtures result or, very importantly, if the substance is unstable (reactive) in water, other vehicles must be considered.

Another common and often successful solvent is vegetable oil, especially corn oil. Many hydrophobic chemicals dissolve or suspend well in this vehicle. Similarly, high molecular weight glycols (polyethylene, polypropylene, propylene glycol), low toxicity oils (mineral, peanut), or petrolatum gels are sometimes effective. Organic solvents such as acetone, methyl ethyl ketone, ethanol, or DMSO may be very good vehicles. However, their inherent toxicity and/or efficiency in facilitating the penetration of other chemicals across the skin barrier can confound study results. Human exposure is rarely associated with such solvents.

Warming is sometimes helpful in dosing viscous materials. Sometimes, simply heating the test substance without dilution will produce a more fluid (and thus a more dosable) material. Heat cannot be used when a substance is unstable at elevated temperatures. Moreover, excessively hot samples must not be applied to the skin.

Solids, if not already in powder form, should be mechanically ground (using a ball mill, mortar and pestle, or other methods) to produce a more suitable state for dosing (to give maximum dermal contact). Then a suitable vehicle can be added to permit good contact with the skin. Application of a dry powder will generally result in much lower toxicity than application of a moistened powder. Chlorfenvinphos, as a dry formulation, had rat dermal LD_{50}s (based on active ingredient) of 400 mg/kg to greater than 800 mg/kg. When applied wet, this material had an LD_{50} of 32 mg/kg.[24] Typically, regulatory guidelines recommend adding just enough vehicle to form a paste that can be easily spread on the dermal surface. As noted earlier, water should be the first choice, but other materials may be more suitable if adequate wetting does not occur. We typically add 0.5 to 2.0 ml of vehicle per gram of solid to produce an adequate paste. Some test substances may require special preparation as appropriate. Paper, fabrics, soft plastics, etc. may be cut into fine strips or squares before moistening.

6. Other Considerations

Several factors beyond those discussed earlier should be considered when performing an acute dermal toxicity test. One of these is restraint or immobilization of the animal during the contact period. As will be discussed later, the dose site must be protected during the dermal test and immobilization is one way to accomplish this. Indeed, early regulations and accepted test guidelines provided for complete animal restraint.[54] The standard range-finding test of Smyth et al. (with which the authors had considerable experience) also made use of immobilization techniques.[55]

Restraint of an animal for an extended period of dermal contact increases stress and prevents normal feeding. In theory, these factors may affect the outcome of a toxicity test. While this may sometimes occur, recent comparisons of LD_{50} data from several chemicals revealed little difference between data from restraint and nonrestraint procedures.[56] We recommend the use of nonimmobilization techniques, whenever possible, for humane reasons.

Although extreme ranges of temperature and humidity would not be expected in the laboratory, their effect on dermal toxicity cannot be ignored. In one *in vitro* (human skin) study, an increase of temperature from 10 to 40°C (at 50% humidity) enhanced penetration of acetylsalicylic acid by 15-fold.[57] At a constant temperature of 40°C, an increase of relative humidity from 50 to 100% resulted in a tenfold rise in penetration of this same material. Temperature increases in intact living systems would be expected to increase dermal blood flow, thereby enhancing penetration and, potentially, toxicity. Control of the testing environment can limit the effect of these factors.

C. REGULATORY CONSIDERATIONS

One primary consideration in establishment of a dermal toxicity test procedure is the specific requirements of regulatory agencies. If test results are to be submitted to one or more such bodies, or if there is potential for a given agency to regulate the manufacture, shipping, or use of a chemical, it is clear that testing must fit that specification. Thus, regulatory requirements frequently become the driving force in selecting test methods. There are numerous detailed reviews of appropriate regulatory requirements.[58-62] The reader should refer to these or, better still, to the original regulations for guidance in the regulatory process. A brief discussion of the function and relevance of the various regulatory guidelines is presented here. Specific test requirements appear in a later section.

1. Environmental Protection Agency (EPA)

The EPA has become one of the most influential forces in regulatory toxicology in recent years. Nearly every industrial, commercial, or agricultural chemical is controlled to some extent by this body. Exceptions are household products, pharmaceuticals, food additives, cosmetics, and radioactive products, though components of these may fall within EPA jurisdiction.

The EPA's regulations apply to two classes of chemicals — general industrial chemicals and biocides (pesticides). The former are regulated by the Toxic Substances Control Act (TSCA), while the latter fall under the Federal Insecticide, Fungicide, and Rodenticide Act (FIFRA). Most nonpesticide chemicals manufactured, processed, distributed, or used in the United States that may become an environmental problem or a health risk, and that are not specifically regulated by other agencies, must be tested as specified by the TSCA.[63] Any agent intended for poisoning or control of insects, animal pests, plants, bacteria, or fungi must be tested under FIFRA guidelines. Toxicity testing must comply with EPA Good Laboratory Practices (GLP).[64,65] Under the GLP, the EPA has the authority to conduct inspections of the testing facility and verify the validity of study data.

2. Consumer Product Safety Commission (CPSC)

The Federal Hazardous Substances Act (FHSA) was passed to regulate consumer products (used in the home, school, or recreational areas) that may cause injury or illness.[66] The act provides testing methods for toxicity evaluation of such products. The FHSA requires appropriate labeling of all home products, including hazard classification by various routes of exposure. In the absence of human exposure data, wording on the labels depends heavily on the results of animal testing. Thus, the consumer receives guidance as to the anticipated dangers in product use (and the producer may avoid consumer complaints and lawsuits).

3. Department of Transportation (DOT)

The U.S. DOT, under the Hazardous Materials Transportation Act, has established criteria for acute toxicity testing that allow classification of chemicals by degree of toxicity. Depending on the classification, specific labels, container construction, and modes of transportation are required. The intent is to prevent undue risk of injury or illness as a material is moved by road, rail, water, or air. Frequently, but not always, testing by the various other guidelines will allow for DOT classification.

4. Organization for Economic Cooperation and Development (OECD)

The OECD is an international organization dealing with numerous issues including toxicity testing. Members include European, Asian, and American countries. These nations have reached an agreement on general toxicity testing procedures so that test data may be accepted worldwide.

Basically, OECD procedures are very similar to EPA methods, but OECD does not have the authority to enforce compliance. This remains the function of individual national agencies. The OECD has essentially taken the lead in reviewing new test methods and promoting specific guidelines among the member nations. Once a new method is accepted by the OECD, member nations' agencies tend to adopt it as well. The acute dermal toxicity testing methods accepted by the OECD are generally adequate for the EPA.

5. Food and Drug Administration (FDA)

The Food, Drug, and Cosmetic Act of 1938 is one of the oldest pieces of legislation in the country relating to control of chemical hazards.[67] It empowers the FDA to regulate the use of foods (human and animal), drugs (human and animal), medical devices, and cosmetics. It has since been amended to cover food additives, pesticide residues, animal drug residues, food packaging, and artificial coloring. While early FDA guidelines apparently accepted the dermal testing method of Draize et al.,[54] recent requirements are not clear (more emphasis is placed on oral toxicity testing).

6. International Transportation Regulations

There are several agencies that regulate the international transportation of chemical substances by air or by water. Notable among these are the International Air Transport Association (IATA) Dangerous Good Regulations[68] and the International Maritime Organization (IMO) International Maritime Dangerous Goods Code.[69] These documents provide only brief test procedure information and are primarily aimed at establishing toxicity classification.

III. TEST METHODS

A. GENERAL PROCEDURES

Early dermal test procedures were patterned after the techniques of Draize et al.[54] Many basic aspects of the test have remained unchanged since then, but there has been considerable refinement over the years.

1. Dose Calculation

Doses are usually calculated on the basis of weight (or volume) of sample per unit of animal weight (in kilograms), such as milligrams per kilogram, grams per kilogram, or milliliters per kilogram. However, some investigators feel that the more difficult expression of sample weight per unit of surface area (grams per square centimeter) might be more appropriate. Solids are usually dosed as pastes; therefore, individual doses are weighed to give the appropriate sample weight per body weight. Liquids are generally administered undiluted and may be measured by weight or volume.

To administer several dose levels (for an LD_{50} determination, for example), doses may be varied by changes in dose volumes, keeping the concentration constant. Alternatively, the dose volume may remain constant while the concentration is changed. This permits consistency in the surface area covered by the dose, but is more time-consuming and eliminates the possibility of using only undiluted sample (as is recommended by regulatory guidelines).

2. Dose Application

As a first step in the application process, the fur is removed from the area of the animal to be dosed. This is accomplished in most instances through use of an electric clipper, although a razor could be used. Typically, the dorsum between the scapular area and the rump is clipped as closely as possible at least 24 h before dosing. This allows minor abrasions to heal before sample application. Care must be taken during clipping not to cut, nick, or otherwise injure the skin. The permeability of the skin could be altered by such an injury. We prefer to remove the fur from the entire trunk area (back, flanks, and belly) to be sure that the chemical is maintained in full contact with the clipped surface even if the chemical "migrates" to nondorsal regions. Moreover, this allows for more effective occlusion later.

The chemical to be dosed is weighed (and mixed with an appropriate vehicle) or measured with a suitably sized syringe and then applied to the dorsal surface of the test animal. The chemical is spread over as large an area as possible (with the syringe or other instrument) to make sure that the substance is freely available for absorption (and not "piled up" on the skin). Depending on the specific guideline requirements, the area of coverage is measured. Some regulatory guidelines request that the dose cover approximately 10% of the surface area (approximately 4 × 5 cm for rats, 12 × 14 cm for rabbits, and 7 × 10 cm for guinea pigs[60]). However, it is frequently quite difficult to fulfill this 10% coverage requirement. If a liquid is applied neat (undiluted) at a small dose, it will not be possible to cover 10% of the skin surface. On the other hand, large doses will cover substantially more than 10%.

When the test substance is very fluid and/or the dose volume is large, retention on the dorsal surface may be rather difficult. In such cases, all wrapping materials should be added first, and then the dose is introduced under the wraps (with a syringe). Alternatively, a moistened solid may be added to the gauze sheeting before being placed in contact with the skin.

3. Maintaining Dermal Contact

The dosed skin surface may be covered with a layer of gauze sheeting (or gauze may be added to the animal skin before dosing), depending on dose volume and on the specification of the guideline being followed. There are a number of techniques to assure that the gauze and occlusive (plastic) bandaging remain intact and undisturbed. Heavy cloth bandaging, adhesive tape, or elastic binding is sometimes applied to protect the layers of wrapping. Some investigators prefer to place a restrictive "Elizabethan" collar on the animal to prevent sample ingestion and removal of coverings. Other restrictive devices (including complete immobilization) are relatively common in dermal testing. We found that such

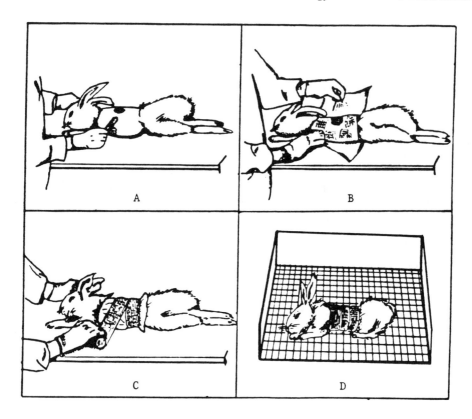

FIGURE 1. Dose application and occlusion in the rabbit. The test substance is applied to the clipped dorsal surface (A); the trunk of the animal is wrapped with gauze and plastic sheeting (B); Vetrap® Bandaging Tape is added to protect the occlusive (plastic) sheeting (C); and the animal is returned to its cage for the contact period (D). (Illustrations by S. M. Christopher.)

methods were unnecessary in the 24-h contact test. Undue animal stress, perturbation of the dose site, dermal injury, or limits to animal movement (and feeding) were frequently observed.

A more successful method has been to wrap the dosed animal (over the plastic sheeting) with Vetrap® Bandaging Tape (Animal Care Products/3M, St. Paul, MN).[56] This material is highly elastic and self-adhesive, but contains no sticky chemical substances. As needed, the Vetrap® is secured with strips of flexible plastic sheeting or adhesive tape. After dosing, the animal is returned to its cage with full access to feed and water. Stress is reduced and the dose site remains undisturbed for the full 24 h. This procedure has been highly successful with hundreds of studies involving rats and rabbits. Figure 1 shows the dose application and wrapping techniques as conducted with the rabbit. The sample remains in contact with the skin for the desired period of time, after which all covering and sample residue are removed. The residue may be removed with tissue, paper towels, or cotton which may be moistened (with warm water). Sometimes soap and water may be required.

Frequently during the contact period and for 14 d thereafter (or until death), the animal is closely observed for clinical signs. Weights are generally recorded just before dosing and weekly thereafter. Pathologic evaluation is conducted at death or sacrifice.

4. DOSE SELECTION (LD$_{50}$ TEST)

If dermal toxicity of a test substance is suspected to be relatively low or is completely unknown, a single limit test (usually at 2 g/kg) should first be conducted. If animals survive, no further testing should be necessary.

For an acute dermal toxicity test in which determination of the LD_{50} is required, toxicity information available on the test substance or on a chemically similar material is obtained. Using these data, one can then estimate a "ballpark" mortality dose level. With this rough lethal dose in mind, a series of range-finding doses is administered over a few (three to four) widely spaced levels, using one or two animals for each. We use a factor of four between dose levels. The goal is to quickly identify lethal and nonlethal doses. It may be necessary to repeat the process if this goal is not met. The preliminary dose selection process is diagramed in Figure 2.

During the range-finding phase, it is important to note major signs of toxicity and time to onset. This information, along with mortality data, becomes very useful in selecting doses for definitive LD_{50} determination. At a dose that appears to be minimally lethal, a full complement of animals is treated (frequently five per sex). Examination of the results from this first full dose level should indicate which subsequent dose levels are needed to complete the LD_{50} determination. Ideally, the investigator will have obtained levels with greater than 50% mortality (at or near 100%), approximately 50% mortality, and less than 50% mortality (at or near 0%) following treatment of three or four dose groups. Testing may continue at lower dose levels if a no-effect dose is required. Note that different doses may be chosen for males and females if a sex-related difference in toxicity is indicated.

The dose selection process can be accelerated by treating more animals on a larger number of dose levels (five or more) early in the study. This may result in excessive animal use (beyond the minimum requirements for the LD_{50} determination) and is, therefore, not recommended. It may not even be necessary to include all the animals that would be needed to determine a statistically "accurate" LD_{50}. Most regulatory agencies are satisfied with an approximate lethal dose or estimated LD_{50} value.

B. REGULATORY GUIDELINES

As mentioned earlier, the specific regulatory guideline is frequently the driving force in conducting acute dermal toxicity testing. Most guidelines require a dosing procedure similar to that described earlier. Beyond this, the various guidelines have unique requirements as outlined later. The original guidelines should be consulted for detailed information and several summaries are also available.[61,62]

1. EPA (TSCA and FIFRA)

By present EPA guidelines,[3,70,71] a single "limit dose" of 2000 mg/kg may be administered to five male and five female rabbits (2.0 to 3.0 kg). Alternatively, rats (200 to 300 g) or guinea pigs (350 to 450 g) may be used, but the rabbit is preferred. If all (or most) animals survive this dose level, no further dosing is required. If rabbits succumb, then LD_{50} determinations are recommended. Originally, the EPA required separate LD_{50}s for intact and abraded animals of both sexes.[72] Abrasion was later dropped as a required test condition. Recent EPA policy[73] allows for testing of only one sex (the more sensitive, usually female) with possible confirmatory doses in a few animals of the opposite sex. This change reflects the EPA's response to the pressures for reducing animal use in toxicology. A minimum of three dose levels are included in the LD_{50} test and these are spaced apart sufficiently to allow a range of toxic effects and mortalities. Concurrent controls are not needed unless the toxicity of the vehicle (if used) has not been evaluated. In this case, a separate LD_{50} test is conducted on the vehicle and the results are compared. Each dose group includes at least five animals (of either or both sexes), and dose levels may be varied through changes in dose volume (although concentration changes may sometimes be appropriate). Observations (discussed in a later section) are made daily through 14 d or, if signs persist, possibly longer. All animals on the test (unless excessively autolyzed or cannibalized) are subjected to gross pathologic examination.

	Dose Level 1/4 X	Dose Level X	Dose Level 4 X	Next Action
Result 1	No signs	No signs	No signs	Dose 1 at 16X
Result 2	No signs	No signs	Signs	Moderate signs at 4X: dose 1 at 16X Severe signs at 4X: dose group at 8X
Result 3	No signs	Signs	Signs	Moderate signs at 4X: dose 1 at 16X Severe signs at 4X: dose group at 8X
Result 4	Signs	Signs	Signs	Moderate signs at 4X: dose 1 at 16X Severe signs at 4X: dose group at 8X
Result 5	No signs	No signs	Death	Dose group at 2X
Result 6	No signs	Signs	Death	Dose group at 2X (possibly complete X)
Result 7	Signs	Signs	Death	Dose group at 2X (possibly complete X)
Result 8	No signs	Death	Death	Dose group at 1/2 X
Result 9	Signs	Death	Death	Dose group at 1/2 X (possibly complete 1/4 X)
Result 10	Death	Death	Death	Dose 1 at 1/16 X

FIGURE 2. Possible dose-related outcomes following initial range finding testing (with one animal dosed on each of three levels) and next action required toward definitive LD50 determination. Note that X is the initial dose, estimated to be the minimally lethal level. Signs of toxicity represent definite (readily visible) effects other than minor response (such as slight discharge, lethargy, gait change, etc.). Findings such as severe signs, delayed recovery from signs, delayed death, or unclear dose-response should also be considered when choosing follow-up action.

2. CPSC (under the FHSA)

Strictly speaking, the FHSA may require only fixed limit doses to permit toxicity classification. These limits are 2000 and 200 mg/kg. Frequently, however, enough doses are included to allow for an LD_{50} determination. While the dosing procedure is presented in some detail (and is similar to that described earlier),[74] little guidance is given as to observations required (other than lethality data). Ten rabbits per dose level (unspecified sex) are subjected to 24-h occluded contact with the test substance. Animal weight is 2.3 to 3.0 kg. Half of the test animals are abraded before doses are given. The FHSA guideline states that the rabbit should be immobilized during the 24-h contact period. However, this is considered to be excessively stressful to the animal. Instead, as previously discussed, the occlusive sheeting can be otherwise protected from removal or damage. Rabbits are observed for 14 d.

3. DOT

Until recently, the purpose of testing for the DOT was the determination of whether the chemical was a Class B poison (lethal at 200 mg/kg). New regulations require the assignment of test substances to "packing groups" (like those of international transportation regulations),[75] based on 50% or more lethality at 40 mg/kg (Packing Group I), at 40 to 200 mg/kg (Packing Group II), or at 200 to 1000 mg/kg (Packing Group III). A "statistically valid" number of rabbits (unspecified sex and age) are dosed for 24 h, and then survivors are observed for 14 d. There is no guidance for sample preparation. The animals may be sacrificed without weight determination or pathologic examination.

4. OECD

The original OECD requirements for acute dermal toxicity testing[76] are essentially the same as those for the EPA. In 1987, the guidelines were revised somewhat to require LD_{50} determination in only one sex (with confirmatory doses in a few animals of the other sex).[77] Moreover, animals showing severe distress may be humanely killed before the end of the study. The rat (200 to 300 g), rabbit (2 to 3 kg), or guinea pig (350 to 450 g) may be used. A limit test using a dose of 2000 mg/kg and five animals per sex may first be conducted. If no deaths result, further testing may not be necessary. However, whenever deaths occur in the limit test, an LD_{50} test is conducted with five animals on each of at least three dose levels. As noted previously, the more sensitive sex (usually the female) should be used. Doses should be spaced widely enough to give a range of toxicity and lethality. The OECD recommends that the dose be applied to approximately 10% of the body surface area, followed by appropriate occlusion. Observations are made over a 14-d period, but duration should not be fixed rigidly. Gross pathologic examination should be carried out for all animals.

IV. OBSERVATIONS

It is desirable to gather as much information as possible from an acute dermal toxicity study. For many substances, little or no investigation is conducted beyond initial acute toxicity testing. Most regulatory guidelines require the recording of local irritation, clinical signs, body weights, and gross pathology data. There are fewer stipulations as to histopathology, clinical chemistry, or hematology, although these are often suggested as options.

A. LOCAL IRRITATION

While there is considerable interest in the systemic effects that may result from acute dermal toxicity tests, one should not overlook the local toxicity that can also be evaluated. As a part of the acute dermal toxicity tests described earlier, one can easily evaluate the

TABLE 3
Draize Scoring System for Skin Irritation

Erythema and eschar formation	Value
No erythema	0
Very slight erythema (barely perceptible)	1
Well-defined erythema	2
Moderate to severe erythema	3
Severe erythema (beet redness) to slight eschar formation (injuries in depth)	4
Maximum possible	4

Edema formation	Value
No edema	0
Very slight edema (barely perceptible)	1
Slight edema (edges of area well defined by definite raising)	2
Moderate edema (raised approximately 1 mm)	3
Severe edema (raised more than 1 mm and extending beyond area of exposure)	4
Maximum possible	4

dose site and surrounding areas for irritation and/or corrosion. Irritation usually represents an inflammatory response characterized by redness (erythema), swelling (edema), and heat. Corrosion, on the other hand, is a form of irreversible tissue destruction such as necrosis (cell death) or ulceration. Other local reactions that might be observed include staining, hemorrhage, ecchymoses, fissuring, desquamation (sloughing of skin layers), leathery or crusty texture, and alopecia (fur or hair loss). Local reactions should be recorded either in descriptive terms or as formal scores such as those developed by Draize et al. (Table 3).[54]

Examination for dermal irritation should be made approximately 1 h after the end of the contact period, with subsequent examinations at least once a week. As specified by most guidelines for dermal irritation tests, it may be advisable to make daily observations for the first 3 d after dermal application, especially if no separate dermal irritation testing is planned. In fact, if relatively little irritation is observed after the acute dermal toxicity test (using doses of 0.1 ml or more), then a separate dermal irritancy test may not be necessary.

B. CLINICAL SIGNS

The observation of an animal's behavior and physical condition provides important information about the toxicity of a substance. For example, Smyth and Nair[1] reported many years ago the development of a kangaroo-like motion in rats dosed with cyclopropane derivatives. Closer examination revealed complete paralysis of the forelimbs and atrophy of the pectoral muscles. More recently, Ballantyne et al.[78] reported that 3-methyl-2-benzothia-zolinone hydrazone (MBTH) had relatively low dermal toxicity (based on lethality) in the rabbit, but produced delayed intermittent tremors and convulsions.

In an acute toxicity test, the following organs/systems are evaluated for clinical changes:

1. Skin and fur
2. Eyes and mucous membranes
3. Respiratory system
4. Circulatory system
5. Autonomic and central nervous system
6. Somatomotor activity
7. Behavior pattern
8. Appearance of tremors, convulsions, diarrhea, salivation, lethargy, sleep, and coma
9. Weight change

A list of common clinical observations, summarized from published sources (cited earlier) as well as our own experience, is presented in Table 4.

Examinations are made frequently during the contact period and at least once (preferably twice) daily thereafter. Close inspection (including handling) is made of each animal. All senses (touch, sight, auditory, and olfactory) should be fully utilized to detect any toxic effect. Each animal is weighed shortly before dose application, at least once each week after dosing, and at death or sacrifice.

Records must be made of time to onset of each clinical sign, its duration, and the time to resolution (or, of course, to death). Although each animal should be closely observed, special attention may be paid to certain signs that are indicative of anticipated toxicity (for example, administration of a carbamate pesticide may typically be followed by salivation, lacrimation, diarrhea, tremors, and convulsions; therefore, an investigator must watch carefully for these signs if a new pesticide is being studied).

C. CLINICAL CHEMISTRY AND HEMATOLOGY

For more extensive evaluation of acute dermal toxicity, useful data might be obtained from clinical chemistry (including urinalysis) and hematology determinations. Testing guidelines for acute dermal toxicity do not specifically require clinical pathology, but valuable information as to the target organs affected and the mechanism of the toxicologic action may be gained from this additional work.

A common evaluation, worth mentioning here, is the determination of cholinesterase levels in the blood, plasma, and brain in acute dermal studies involving potential cholinesterase inhibitors (such as organophosphorus or carbamate pesticides). Reduction of cholinesterase levels can give clues of poisoning before clinical signs develop, confirming also that dermal penetration has occurred. One simple clinical chemistry test that can provide insight into a chemical's toxicologic action is the use of HEMASTIX® Reagent Strips (Ames) to test urine for the presence of blood. We have found that there is good correlation between highly positive tests by HEMASTIX® and kidney and/or bladder damage apparent upon subsequent microscopic evaluation.

D. GROSS PATHOLOGY

Most acute dermal toxicity test guidelines recommend that all animals be examined for gross pathologic changes. It has been suggested that significant gross pathologic findings in acute toxicity tests are infrequent and not indicative of a specific effect.[79] Rather than performing a necropsy only at death or the usual sacrifice period (14 d), a better procedure may be to sacrifice some treated animals after 1 to 4 d for gross examination and to evaluate organ weights.[80]

In our experience, there are benefits to performing a necropsy on any animal that dies (unless there is severe autolysis) or that is sacrificed. During gross pathologic examinations, we have found evidence of hemorrhage, tissue necrosis (direct chemical corrosion), presence of test substance (crystals or staining) in the tissues, organ enlargement or atrophy, abnormal fluid accumulation, visceral pallor, gastrointestinal blockage (which may suggest that a death was not treatment-related), and broken backs or limbs (possibly suggesting injury as the cause of paralysis rather than treatment). Thus, systemic effects, target organs, or even cause of death may be better evaluated through the necropsy procedure than by clinical observations alone.

During the gross pathologic examination, external and internal organs are evaluated for size, color, texture, shape, firmness, position, and integrity. Attention is given to any fluids (clear, red, purulent) in the abdominal cavity, thoracic cavity, or pericardial sac. Organs with significant treatment-related lesions may be preserved in formalin for possible histo-

TABLE 4
Common Signs Observed During Acute Dermal Toxicity Testing

System or function involved	Common signs of toxicity
Behavioral	Hyperactivity or lethargy
	Hypersensitivity or hyposensitivity to stimuli (sound, light, handling)
	Aggression or irritability
	Confused, bizzare, or stereotypic activity
	Excessive grooming, scratching, or rubbing
	Excessive phonation/vocalization
Postural	Hunched position (kyphosis)
	Low carriage or high carriage
	Limbs splayed or rotated
	Drooping or raised-head position
Skin and fur	Staining, residue from discharge or diarrhea
	Skin irritation, inflammation, or corrosion (erythema, edema, necrosis, etc.)
	Piloerection
	Alopecia
	Texture change (rough, oily, dry, etc.)
	Cyanosis
	Pallor
	Abnormal temperature (cool or warm to the touch)
Ocular	Discharge (clear or colored)
	Opacity
	Iritis
	Periocular irritation
	Drooping eyelids (ptosis)
	Protrusion (exophthalmos)
	Abnormal movement (nystagmus)
	Abnormal pupil size (miosis or mydriasis)
	Involuntary blinking (blepharospasm)
Locomotor/muscular	Abnormal gait
	Ataxia (coordination loss)
	Prostration
	Hindlimb and/or forelimb weakness
	Hindlimb and/or forelimb paralysis
	Diminished surface (or air) righting reflex
	Diminished grip strength
	Diminished blink or pinch response
	Diminished muscle tone
Nervous	Tremors, twitches, spasms
	Convulsions, seizures (tonic, clonic)
	Somnolence, narcosis
	Catalepsy
Respiratory	Dyspnea (labored breathing: gasping, deep abdominal breathing)
	Apnea (brief cessation of breathing)
	Rapid breathing (tachypnea) or slow breathing (bradypnea)
	Audible breathing (rales, wheezing)
	Nasal discharge (clear, red, frothy)
Cardiovascular	Rapid heart rate (tachycardia) or slow heart rate (bradycardia)
	Irregular heart rate (arrhythmia)
	Pallor or redness of extremities

TABLE 4 (continued)
Common Signs Observed During Acute Dermal Toxicity Testing

System or function involved	Common signs of toxicity
Gastrointestinal, urinary	Salivation
	Emesis (retching or vomiting)
	Diarrhea, loose or watery feces
	Constipation, hardened feces
	Abdominal distention
	Blood in feces or urine
	Excessive or decreased urination
	(polyuria or anuria)

pathology. Organ weight may be an indicator of specific organ effect and might be considered at sacrifice, especially if organ weights are obtained for control animals to permit statistical comparison.

Gross pathologic examination data should be compared with other observational parameters, especially clinical signs. Special attention should be given to those organs that appear to have been affected based on clinical observations. As examples, an animal exhibiting frank hematuria might be examined for gross lesions (such as hemorrhage) in the bladder or kidney. Animals displaying severe weight loss should be evaluated for changes in the gastrointestinal tract. An animal with neurotoxic signs should be examined for lesions in the brain, spinal cord, or peripheral nerves.

E. HISTOPATHOLOGY

Most test guidelines for acute toxicologic evaluation do not give clear, detailed instruction for inclusion of histopathology. They generally recommend that it be considered if gross lesions are observed in animals surviving for longer than 24 h. Histopathology could be added to the protocol as an option to be considered upon review of clinical signs and gross lesions. A "pathology decision tree" has been proposed by Gad et al.[80] for the appropriate use of histopathology in acute studies. In order to gain maximum information from treated animals, some researchers routinely incorporate histopathology of certain common target organs (liver, heart, kidney, lung, gastrointestinal system, spleen, and brain) in their acute dermal testing protocol.

We have found that a number of industrial intermediates produce substantial kidney lesions that are visible microscopically, but not grossly. Histopathologic examination may also aid in pinpointing the cause of gross pathologic effects. In one study,[81] we found that trimethoxysilyldecane produced delayed hindlimb paralysis when administered by the percutaneous route. Microscopic examination of peripheral nerves revealed that severe, irreversible structural damage had occurred.

V. INTERPRETATION OF RESULTS

After the acute dermal toxicity test is completed, the data are collected, reviewed, and tabulated. Appropriate calculations are made to allow for interpretation and reporting.

A. LD$_{50}$ DETERMINATION

A major goal of an acute dermal toxicity study may be the determination of an LD$_{50}$ (a calculated dose, extrapolated from lethality data, that would be expected to kill 50% of treated animals). Although the value of the LD$_{50}$ has been questioned, several regulators and health officials still require it. A number of reviews of the LD$_{50}$ determination are available including those by Gad,[61] Gad and Weil,[82,83] and more recently, by DePass.[84]

The "traditional" methods of LD_{50} calculations include the probit analysis method[85,86] and the graphic method of Litchfield and Wilcoxon.[87] The former method requires relatively large numbers of animals on several dose levels that produce partial mortality ratios (0 and 100% mortality are not used). It is designed to provide accurate LD_{50} values, as well as 95% confidence limits and the slope of the dose-lethality curve. The latter method may not have some of the restrictions of the probit method (0 and 100% mortalities may be used), but relatively large numbers of animals are needed and substantial time is required. These methods have become more automated in recent years, however.

There are a number of methods for LD_{50} calculation that require relatively few animals. Included among these is the moving average method of Thompson.[88] To make this method more convenient, Thompson and Weil developed a series of table values that reduced the complexity of the calculations.[89,90] All-kill or no-kill dose levels can be used with this method. As few as two animals may be used per level (if four or more levels are included on study); two dose levels may be sufficient if four or more animals are used for each. It should be noted that this method requires a constant geometric factor between dose levels and that the number of animals per level must also be constant. The 95% confidence limits are included in the moving average calculation and, more recently, a successful method for slope determination has been developed.[91]

Weil et al.[92] reviewed a number of acute (LD_{50}) studies and compared : (1) rat oral LD_{50}s (probit method) using ten animals per level vs. LD_{50}s (moving average method) using five animals per level, (2) dermal LD_{50}s (moving average) using four vs. five rabbits per level, and (3) rat oral LD_{50}s (moving average) using factors of 5.0 2.0, 1.28 or 1.26 between dose levels. They concluded that LD_{50} values based on four or five animals per level were comparable to those based on ten animals each. Doses separated by a factor of 2.0 or less were recommended. Also, values obtained by probit analysis were highly correlated with those calculated by moving averages. DePass et al.[20] compared oral LD_{50}s using five animals per group with those using ten animals per group (on 11 chemicals) and calculated a correlation coefficient of 0.98. Comparison of LD_{50}s based on two to three dose levels vs. those with three to seven dose levels yielded a correlation coefficient of 0.99. When LD_{50}s calculated by the probit method were compared to those determined by moving averages, the correlation coefficient was 0.99 (as it also was for the slope comparison).

There are several other methods that may be used to calculate LD_{50}s, many of which are designed to use minimal numbers of animals. These have received varying degrees of acceptance by toxicologists and regulators. One method is the Up-And-Down Procedure.[93,94] In this method, animals are dosed one at a time. If the animal survives, a second animal receives a higher dose. If the first animal dies, a second animal receives a lower dose. This process continues until several lethal and nonlethal levels, separated by a relatively small factor, are obtained. Then, a series of calculations is used to determine an LD_{50} value. In a method published by Lorke,[95] three animals are administered each of three widely spaced doses (10, 100, and 1000 mg/kg). The results of this phase of the test are used to establish a second set of doses (usually three or four) which are more closely spaced. One animal may be sufficient for each of the second set of dose levels. The LD_{50} may then be estimated by standard calculations. Schütz and Fuchs[96] have proposed a method using a small number of animals (two or three per sex per group), but with extended observation times and multiple probe doses.

A number of investigators have presented methods to determine an approximate lethal dose (ALD) rather than the traditional LD_{50}. These methods basically estimate the lowest dose that would be expected to kill some (but not necessarily 50%) of animals tested. Early work in this area was reported by Deichmann and LeBlanc,[97] who administered a series of doses using one animal at each dose level. A factor of 1.5 was established between levels and dosing continued until a lethal and nonlethal dose were identified. The lowest lethal dose was considered to be the ALD. Kennedy et al.[98] used a similar method more recently to evaluate a number of chemicals (by the oral route). Using starting doses of 670 and 2250 mg/kg, doses were increased and/or decreased by a constant factor until the ALD was determined. When the ALDs were compared to LD_{50} values, the toxicity classification was the same by either method.

It should be pointed out that many of the proposed LD_{50} or ALD methods have been tested in the rat by the oral route. In theory, the methods should also apply to dermal studies. In our experience, care must be taken in interpreting data from very limited dermal testing since lethality results can be relatively inconsistent (especially in the rabbit). Variations in dermal absorption may account for this finding. Confidence limits and slopes can not be determined with many of these proposed methods. Also, with fewer animals there may be less likelihood of observing some significant clinical signs or tissue changes. As noted earlier, regulators may not yet routinely accept some of these methods as alternatives to the LD_{50}.

B. TARGET ORGAN IDENTIFICATION

One potential benefit from conducting an acute dermal toxicity test, beyond establishing lethal levels and gross clinical effects, is the determination of target organs (those organs or organ systems that are affected by the chemical). It could be said that target organs are infrequently identified in acute dermal toxicity tests, but that organs affected in single-dose tests exhibit agonal changes. Moreover, target organs of longer-term studies may not be the same ones as in acute studies. Gad and Chengelis[79] reviewed the results of studies for which the acute test pathology identified no target organs, whereas subchronic studies on the same compounds clearly indicated which organs were affected. Conversely, some target organs apparent in acute studies were not affected in subchronic studies with the same compounds. These findings would indeed indicate that cumulative effects from smaller doses may differ from effects of single administration of large doses. Nevertheless, it may be quite important to identify the organ injured by a single dose when evaluating acute hazards and developing therapeutic measures.

In our experience, there have been a number of instances of correlation between acute and repeated dose findings. For trimethoxysilyldecane, discussed earlier, lesions of the kidney and nerves seen in acute oral and dermal studies were confirmed during a 20-exposure inhalation study.[81] In fact, the latter study was conducted only after review of the single dose results.

C. DOSE-RESPONSE EVALUATION

With or without a goal of LD_{50} determination, one should examine the relationship of toxic effect (clinical signs, body and/or organ weight changes, lethality, clinical chemistries, pathologic changes, etc.) obtained from an acute dermal toxicity study and the amount of sample applied. When LD_{50}s are calculated as part of an acute dermal toxicity test, a number of associated values can also be calculated. Among these are the 95% confidence limits. As the name implies, these limits give us an idea of how acceptable the LD_{50} value may be. If we calculate an LD_{50} of 75 mg/kg and the 95% confidence limits are reasonably narrow, such as 50 to 100 mg/kg, we can judge the LD_{50} to be acceptable and indicative of a consistent dose response. On the other hand, wider 95% confidence limits of 15 to 135 mg/kg would suggest a high degree of uncertainty in our data and questionable dose response. It is also useful to calculate the slope to see how sensitive the test system is to changes in dose. In one recent study (unpublished), it was necessary to decrease a dose for females from 100 mg/kg (where all animals died) to 25 mg/kg to obtain a 0% lethality level. The slope of 5.60 reflected the relatively shallow dose response. In males treated with the same chemical, doses of 100 and 71 mg/kg gave 80 and 0% mortality, respectively, with the much steeper slope of 11.2.

To determine if an apparent dose response is significant or a random occurrence, statistical comparison may be made between dose groups and control groups or among all dose groups (with or without controls). In one study (unpublished), an industrial chemical was applied to rabbit skin (15 rabbits per dose group) for a 24-h period. Lungs (an apparent target organ for this material) were evaluated microscopically, and the following results were seen:

Dose (ml/kg)	Incidence of bronchopneumonia (%)
1.0	53*
0.5	40**
0.25	0
0.0	0

* Significant at $p < 0.01$.
** Significant at $p < 0.05$

This example illustrates the clear-cut dose response that often, but not always, occurs.

The reader is encouraged to refer to a number of sources for statistical methods and interpretations including several by Gad[99] or Gad and Weil,[82,83] one by Snedecor and Cochran,[100] and by Sokal and Rohlf.[101]

D. TOXICITY CLASSIFICATION

There are several published toxicity classification systems. One such system, not related to regulatory requirements, is based on an early scheme of Smyth et al.[104] This system, several regulatory classifications, and examples of chemicals that fit into the various categories are presented in Table 5. A classification system is quite useful in putting toxicity findings into words. These are frequently based on mortality data (such as LD_{50} values). However there is movement toward the use of nondeath endpoints (development of clinical signs) as has been described by the British Toxicology Society (BTS) in 1984.[111] The BTS method has been subjected to a series of validation studies and final conclusions have recently been reported.[112] Although these studies were conducted with oral doses to rats, the concepts should apply to any route of administration. Classifications are made as shown in Table 6.

The use of toxicity classification allows us to summarize relative toxicity (especially in comparing numerous test substances). Additionally, it makes regulatory categorization an easy task. However, toxicity classification may be of relatively little value to the toxicologist who is more concerned with dose-response relationships, time to onset of signs, duration of ill effects, reversibility of ill effects, target organ involvement, mechanism of toxic action, and perhaps the course of treatment required to counter the toxic action of a substance.

E. COMPARISON OF THE DERMAL ROUTE TO OTHER ROUTES OF ADMINISTRATION

We have seen previously that dermal toxicity is greatly influenced by species, site of application, condition of the skin, and so on. Likewise, toxicity from dermal administration may vary tremendously from that by other routes of dosing. Certainly, the skin barrier plays an important role here, as do differences in metabolic pathways, availability of chemical target organs, and elimination processes. Generally, the least severe and the slowest toxic response will be produced by dermal administration, compared to any other common route of administration (the usual order of acute toxicity is intravenous > inhalation > intraperitoneal > subcutaneous > oral > dermal).[113,114]

TABLE 5
Classification Systems for Acute Rabbit Dermal Toxicity (Based on $LD_{50}S$)

LD_{50} (mg/kg)	Equivalent dose, man (approx.)	Example	Classification			
			Smyth et al.[104]	EPA/FIFRA[110]	FHSA[74]	IATA/IMO/DOT[65,69,75]
>20,000	>1.5 qt	Polyethylene glycol[102]		Category IV (no precaution)		
2000—20,000	0.3 pt—1.5 qt	2-Heptanone[103]		Category III (harmful)		
>15,000	>1.2 qt	Ethylbenzene[103]	Relatively harmless			
5000—15,000	0.8 pt—1.2 qt	Carbon tetrachloride[104]	Practically nontoxic			
500—5000	1.3 oz—0.8 pt	n-Hexanol[105]	Slightly toxic			
200—2000	1 tbs—0.3 pt	Cyclohexanone[106]		Category II (possibly fatal)	Toxic	
200—1000	1 tbs—2.5 oz	Ethylenediamine[105]				Packing group III
50—500	1 tsp—1.3 oz	Methylisocyanate[106]	Moderately toxic			Packing group II
≤200	≤1 tbs	Isopropylmorpholine[106]				
40—200	1 tsp—1 tbs	4-Thiocyanoanilene[107]		Category I (fatal)	Highly toxic	
1—50	1 drop—1 tsp	Mevinphos[108]	Highly toxic			
≤40	1 tsp	Isodrin—rat[109]				Packing group I
<1	<1 drop	[a]	Extremely toxic			

[a] The most toxic example found was Thimet,[109] which had $LD_{50}S$ (in rats) of 2.5 to 6.2 mg/kg.

TABLE 6
BTS Fixed-Dose Procedure Classification

5 mg/kg produces mortality	Very toxic
5 mg/kg produces signs	Toxic
5 mg/kg produces no effect and 50 mg/kg produces mortality	Toxic
50 mg/kg produces signs	Harmful
50 mg/kg produces no effect and 500 mg/kg produces mortality	Harmful
500 mg/kg produces signs	No significant risk
500 mg/kg produces no effect and 2000 mg/kg produces mortality	Harmful
2000 mg/kg produces no mortality	No significant risk

There is ample evidence that animals are frequently much more sensitive to an oral dose than to a topical one. In the MBTH study cited earlier,[78] the dermal LD_{50} for female rats was more than 107 times greater than the oral LD_{50} for female rats. In male rabbits, the dermal LD_{50} was over 90 times greater than the oral LD_{50}. In many of the toxicity lists referenced previously and in our continuing experience, this trend prevails, although not always to the extent of the MBTH example. It should be remembered, however, that oral toxicity data are frequently obtained in the rat while dermal toxicity data are commonly obtained in the rabbit. Comparisons of routes of administration thus become confounded with species variations.

Examples of greater dermal toxicity relative to oral toxicity are not rare. Many pesticides appear to be just as toxic or more toxic by the dermal route than by the oral route.[26] This may be especially true for some organophosphates, although no consistent pattern can be discerned.[24] Smyth et al.[1] presented a number of examples including fluoroallyl alcohol (dermal toxicity = 43 times oral toxicity), di-2-*n*-hexylether (30 times), 5-indanol (eight times), formaldehyde (seven times), and tridecyl acrylate (seven times). These authors noted that curare is lethal on contact with abraded skin at a concentration much lower than that which is harmless by mouth. They concluded that such substances must be readily absorbed through the skin into the bloodstream, allowing contact with target tissues before detoxification by the liver (or other metabolizing organs). Moreover, the low pH of the stomach may break down toxic chemicals or inhibit their absorption. Of lesser importance, but not to be discounted, is the potential for metabolic action by the skin (resulting in more toxic metabolites).

When a significant (unexpected) effect is observed, it may be wise to conduct additional tests to determine whether that effect is route-specific or species-specific. In the MDMS study cited earlier,[4] rabbits receiving dermal doses of chemical showed a much more severe response (rapid death) than rats receiving higher oral doses. Therefore, rats were then given dermal doses and rabbits were administered oral doses. The rats were not affected, but rabbits had LD_{50}s similar to those obtained by dermal exposure. Thus the rabbit was confirmed to be the more sensitive species. Moreover, subcutaneous injections given to rats at dose levels that killed rabbits cutaneously produced no lethality. This suggested that differences in species sensitivity could not be explained by the effectiveness of the dermal barrier alone.

VI. ALTERNATIVE TEST METHODS

There has been continuing demand to reduce or eliminate certain acute toxicity tests because they are perceived as excessively stressful to the animals. The following sections describe briefly some of the potential alternative methods for acute dermal toxicity testing. It must be remembered that many alternative procedures have not been fully validated, accepted by regulators, or supported by the toxicology community. These conditions must be met before they can be used routinely. Considerations of alternatives to animal use generally follow the "3R" philosophy of Russel and Burch[115] — reduction of animal use, replacement of animal models, and refinement (improvement) of animal test procedures. For those desiring further review of alternative methods, several interesting papers are available.[116-118] Additionally, there are a number of newsletters and books published by A. M. Goldberg, The Johns Hopkins Center for Alternatives to Animals Testing, Baltimore, MD.

A. LIMITING ANIMAL NUMBERS/STRESS

Reduction of the number of animals used in acute dermal toxicity studies is feasible, as previously discussed. We have already seen that an LD_{50} can be determined with relatively few animals and only one sex may be required. The LD_{50} is not dramatically changed when such reductions are employed. Using only a few animals per dose group and limited numbers of groups has been gaining regulatory and scientific support. Depending on the proposed use of the data acquired and on whether more sophisticated, longer-term testing is planned, still further limitations may be in order.

The usefulness of the LD_{50} itself has also been seriously questioned. In an often-quoted paper by Zbinden and Flury-Roversi,[119] it is made clear that the LD_{50} does not represent a meaningful biologic constant, it is not relevant to human exposure, and that there is little benefit in attempting to obtain an "accurate" LD_{50} value. In our experience, LD_{50} values can vary for a given material by a factor of two times or more in separate studies that are conducted under the same conditions. Factors such as animal variability, seasonal effects, diurnal cycles, etc. may play a considerable role.

As noted previously, minimal lethal doses or approximate lethal doses may be quite adequate, if the test is conducted carefully. With these methods, administration of higher doses for the sole purpose of increasing numbers of deaths is not necessary. Instead, increasing the amount of biologic data (histopathology, clinical chemistry, etc.) from nonlethal doses may be more desirable.[80,120] If we eliminate death as an endpoint altogether, both animal numbers and animal stress can be reduced. Methods such as the British Toxicology Society fixed-dose procedure (discussed previously) have been developed to accomplish this goal and are gaining in acceptance.

Most regulatory agencies recommend a limit dose above which no further testing is necessary. This limit for acute dermal toxicity testing is 2000 mg/kg. Frequently, administration of test substance at this one level constitutes the complete acute dermal study. There may be some drawbacks to this approach, however. Critical effects may be missed. In the MBTH study discussed earlier,[78] delayed convulsions appeared in rabbits receiving single dermal doses of 4 g/kg and higher. This finding led to recommended changes in handling this chemical. It is possible (but not always true) that a single large dose may give us clues as to what might result upon repeated doses with smaller amounts of material. It is conceivable that cumulative toxicity might occur with frequent exposure and, in the absence of longer-term studies, acute studies may provide the only opportunity to evaluate potential toxic effects. Thus, it may be advisable to increase doses, within practical limits, to near lethal levels in a few animals.

Reducing animal stress can take many forms — early (humane) sacrifice of animals found to be moribund or suffering excessively, the use of analgesics to reduce local and systemic discomfort, or perhaps anesthesia of test animals during periods of extreme discomfort. Of course, there is legitimate concern as to the effect these may have on the outcome of the study. Will the reversal of clinical signs be missed by early sacrifice? Will the administration of pain-killing drugs interact with the chemical? Will signs fail to develop if an animal is anesthetized? These are all possibilities that have to be weighed against the concern for animal welfare. Perhaps these measures can be taken with only the most severely affected animals, accepting the fact that a certain level of stress may be unavoidable in the others.

B. USING NONMAMMALIAN AND *IN VITRO* METHODS

"Topical" doses can be given to nearly any species. Some suggested animals have been the earthworm, brine shrimp, the hydra, and a variety of insects. *Daphnia* may be a suitable model for predicting toxicity, and results may correlate with those from rat LD_{50} tests.[61] Monocellular organisms (protozoa, bacteria, yeast), tissue slices, embryos, and individual cells of a test species may play an increasingly important role in preliminary toxicity testing. Indeed, *in vitro* methods have been rapidly gaining acceptance. We have already seen that *in vitro* dermal penetration evaluation can be useful in providing relative absorption data for chemical substances. Skin penetration may provide clues to potential toxicity, but penetration does not equal toxicity.

Reviews of these and other procedures are available.[116,121] Cell cultures or tissue cultures from mice, rats, hamsters, chickens, monkeys, and man have been used to evaluate morphologic, functional, and biochemical alterations. These tissues have come from lungs, kidneys, livers, nerves, brain, blood, muscle, intestine, pancreas, spleen, and the immune system. Especially interesting are cells or tissue cultures from humans which provide the chance to look for specific target organ effects in humans. However, these tests are most likely to predict cytotoxicity, rather than systemic effects, as contrasted to *in vivo* studies in which various organ systems interact in the overall absorption, distribution, elimination, and metabolic processes.

C. STRUCTURE-ACTIVITY RELATIONSHIPS

Substances with similar molecular structures should produce similar toxic effects in living systems. In practice this is frequently true, but unfortunately there are often exceptions to this rule. Certainly, examination of the structure is an important first step in preparing to evaluate a chemical's potential for acute dermal toxicity.

Sometimes, minor alteration in molecular structure can have a rather striking effect on toxicity. In the acute rabbit dermal toxicity study (discussed previously) on methyldimethoxysilane,[4] the LD_{50}s were determined to be 0.73 ml/kg (females) and 1.47 ml/kg (males), with rapid death. For structurally similar trimethoxysilane, the rabbit dermal LD_{50}s were 6.73 and 7.46 ml/kg (females and males, respectively). Most deaths occurred after several days. Similarly, Loomis[122] gave an example of an organophosphate which, by changing one chlorine atom from the ortho to the meta position on an aromatic ring, became a neurotoxin.

Still, structure-activity relationship (SAR) modeling has received considerable attention in the last several years. Enslein has reported that his system can correctly predict a number of endpoints 85 to 95% of the time.[123] This system uses a database from animal test results, careful chemical structure modeling, complex statistical techniques, and advanced computer technology. Structures are keyed into the system and (hopefully) a reasonable prediction of biological effect is produced. Although predictions are made for carcinogenicity, mutagenicity, acute oral toxicity, and eye irritancy, acute dermal toxicity is not yet included in Enslein's program.

While SARs can be quite useful for quick screening of a large number of molecules (such as during early product development), the final toxicologic evaluation of a given product will almost certainly still require some animal testing.

VII. CONCLUSIONS

Because of the potential for human skin contact with chemicals in the home, at work, and in the environment, it is most appropriate that the toxic effects be studied using all the tools at our disposal, including animal testing. The acute dermal toxicity test is one such tool, which although quite straightforward, requires numerous decisions before conduct. Once the test has been completed and results have been obtained, we can attempt extrapolation to the human situation. Some products may be abandoned if deemed excessively toxic. Others may require special warnings, labels, and exposure controls. Some may be determined to be safe for direct application. Additional studies (longer-term) may be required in both animals and in man.

As much as we possibly can, we should consider the various alternative methods to traditional animal tests as long as scientific integrity is not compromised. It is becoming increasingly important to limit animal use and animal stress, but the protection of the human population must not be forgotten. With the integrated use of literature data, computer analyses, and animal testing, the hazards of human dermal exposure can be adequately evaluated.

ACKNOWLEDGMENTS

The authors wish to express their gratitude to Linda Farren for typing the manuscript and to Marilyn Thurn for her help in obtaining the reference materials.

REFERENCES

1. **Smyth, H. F., Jr. and Nair, J. H., III,** Experimental toxicology — discovering the unexpected, *Ind. Eng. Chem.,* 51, 75A, 1959.
2. **Loomis, T. A.,** *Essentials of Toxicology,* 3rd ed., Lea & Febiger, Philadelphia, 1978, 1.
3. Environmental Protection Agency (EPA), Health effects testing guidelines, *Fed. Regist.,* 50, 39398, 1985.
4. **Myers, R. C., Christopher, S. M., Fowler, E. H., and Ballantyne, B.,** Evaluation of methyldimethoxy-silane (MDMS) for acute toxicity and primary irritancy, *Toxicologist,* 11, 144, 1991.
5. **McCreesh, A. H.,** Percutaneous toxicity, *Toxicol. Appl. Pharmacol.,* 7, 20, 1965.
6. **Bartek, M. J., LaBudde, J. A., and Maibach, H. I.,** Skin permeability *in vivo*: comparison in rat, rabbit, pig and man, *J. Invest. Dermatol.,* 58, 114, 1972.
7. **Wester, R. C. and Noona, P. K.,** Relevance of animal models for percutaneous absorption, *Int. J. Pharm.,* 7, 99, 1980.
8. **Moody, R. P., Franklin, C. A., Ritter, L., and Maibach, H. I.,** Dermal absorption of the phenoxy herbicides 2,4-D, 2,4-D amine, 2,4-D isoocytyl, and 2,4,5-T in rabbits, rats, Rhesus monkeys, and humans: a cross species comparison, *J. Toxicol. Environ. Health,* 29, 237, 1990.
9. **Vinson, L. J., Singer, E. J., Koehler, W. R., Lehman, M. D., and Masurat, T.,** The nature of the epidermal barrier and some factors influencing skin permeability, *Toxicol. Appl. Pharmacol.,* 7, 7, 1965.
10. **Wester, R. C. and Maibach, H. I.,** Percutaneous absorption in the Rhesus monkey compared to man, *Toxicol. Appl. Pharmacol.,* 32, 394, 1975.
11. **Wester, R. C. and Maibach, H. I.,** Animal models for percutaneous absorption, in *Models in Dermatology,* Vol. 2, Maibach, H. I. and Lowe, Eds., S. Karger, Basel, 1985, 159.
12. **Roudabush, R. L., Terhaar, C. J., Fasset, D. W., and Dziuba, S. P.,** Comparative acute effects of some chemicals on the skin of rabbits and guinea pigs, *Toxicol. Appl. Pharmacol.,* 7, 559, 1965.
13. **Andersen, K. E., Maibach, H. I. and Anjo, M. D.,** The guinea pig: an animal model for human skin absorption of hydrocortisone, testosterone and benzoic acid, *Br. J. Dermatol.,* 102, 447, 1980.
14. **Goldenthal, E. I.,** A compilation of LD50 values in newborn and adult animals, *Toxicol. Appl. Pharmacol.,* 18, 185, 1971.

15. **Sipes, I. G. and Gandolfi, A. J.**, Biotransformation of toxicants, in *Casarett and Doull's Toxicology: The Basic Science of Poisons,* 3rd ed., Klaassen, C. D., Amdur, M. O., and Doull, J., Eds., Macmillan, New York, 1986, chap. 4.
16. **Fisher, L. B.**, Studies on the permeability of infant skin, in *Percutaneous Absorption,* 2nd ed., Bronaugh, R. L. and Maibach, H. I., Eds., Marcel Dekker, Inc., New York, 1989, chap. 9.
17. **Roskos, K. V., Guy, R. H., and Maibach, H. I.**, Percutaneous absorption in the aged, *Dermatol. Clin.,* 4, 455, 1986.
18. **Montagne, W. and Carlisle, K.**, Structural changes in aging human skin, *J. Invest. Dermatol.,* 73, 47, 1979.
19. **Banks, Y. B., Brewster, D. W., and Birnbaum, L. S.**, Age-related changes in dermal absorption of TCDD and 2,3,4,7,8-pentachlorodibenzofuran, *Toxicologist,* 9, 118, 1989.
20. **DePass, L. R., Myers, R. C., Weaver, E. V., and Weil, C. S.**, An assessment of the importance of number of dosage levels, number of animals per dosage level, sex, and method of LD50 and slope calculation in acute toxicity studies, in *Acute Toxicity Testing: Alternative Approaches,* Goldberg, A. M., Ed., Mary Ann Liebert, Inc., New York, 1984, chap. 9.
21. **Bronaugh, R. L. and Maibach, H. I.**, *In vitro* percutaneous absorption, in *Dermatotoxicology,* 3rd ed., Marzulli, F. N. and Maibach, H. I., Eds., Hemisphere, Washington, D. C., 1987, chap. 4.
22. **Lu, F. C.**, Modifying factors of toxic effects, in *Basic Toxicology: Fundamentals, Target Organs and Risk Assessment,* Hemisphere, Washington, D.C., 1985, chap. 5.
23. **Dugard, P. H.**, Skin permeability theory in relation to measurements of percutaneous absorption in toxicology, in *Dermatotoxicology,* 3rd ed., Marzulli, F. N. and Maibach, H. I., Eds., Hemisphere, Washington, D.C., 1987, chap. 3.
24. **Brown, V. K. H. and Muir, C. M. C.**, Some factors affecting the acute toxicity of pesticides to mammals when absorbed through skin and eyes, *Int. Pest Control,* 13, 16, 1971.
25. **Wester, R. C. and Maibach, H. I.**, Interrelationships in the dose-response of percutaneous absorption, in *Percutaneous Absorption,* 2nd ed., Bronaugh, R. L. and Maibach, H. I., Eds., Marcel Dekker, Inc., New York, 1989, chap. 21.
26. **Weil, C. S., Condra, N. I., and Carpenter, C. P.**, Correlation of 4-hour vs. 24-hour contact skin penetration toxicity in the rat and rabbit and use of the former for predictions of relative hazard of pesticide formulations, *Toxicol. Appl. Pharmacol.,* 18, 734, 1971.
27. **Ballantyne, B., DePass, L. R., Fowler, E. H., Garman, R. H., Myers, R. C., Troup, C. M., and Coleman, J. H.**, Dermatotoxicology of bis[2-(dimethylamino)ethyl]ether, *J. Toxicol. Cut. Ocular Toxicol.,* 5, 3, 1986.
28. **Reifenroth, W. G. and Spencer, T. S.**, Evaporation and penetration from skin, in *Percutaneous Absorption,* 2nd ed., Bronaugh, R. L. and Maibach, H. I., Eds., Marcel Dekker, Inc., New York, 1989, chap. 19.
29. **Wester, R. C. and Maibach, H. I.**, *In vivo* percutaneous absorption, in *Dermatotoxicology,* 3rd ed., Marzulli, F. N. and Maibach, H. I., Eds., Hemisphere, Washington, D.C., 1987, chap. 5.
30. **Maibach, H. I., Feldmann, R. J., Milby, T. H., and Serat, W. F.**, Regional variation in percutaneous penetration in man: pesticides, *Arch. Environ. Health,* 23, 208, 1971.
31. **Feldmann, R. J. and Maibach, H. I.**, Regional variation in percutaneous penetration of ^{14}C cortisol in man, *J. Invest. Dermatol.,* 48, 181, 1967.
32. **Rougier, A., Dupuis, D., Lotte, C., Roguet, R., Wester, R. C., and Maibach, H. I.**, Regional variation in percutaneous absorption in man: measurement by the stripping method, *Arch. Dermatol. Res.,* 278, 465, 1986.
33. **Bronaugh, R. L., Stewart, R. F., and Congdon, E. R.**, Differences in permeability of rat skin related to sex and body site, *J. Soc. Cosmet. Chem.,* 34, 127, 1983.
34. **Holbrook, K. A. and Odland, G. F.**, Regional differences in the thickness (cell layers) of the human stratum corneum: an ultrastructural analysis, *J. Invest. Dermatol.,* 62, 415, 1974.
35. **Scheuplein, R. J.**, Permeability of the skin: a review of major concepts, *Curr. Probl. Dermatol.,* 7, 172, 1978.
36. **Scheuplein, R. J.**, Mechanism of percutaneous absorption. II. Transient diffusion and the relative importance of various routes of skin penetration, *J. Invest. Dermatol.,* 48, 79, 1967.
37. **Elias, P. M. and Brown, B. E.**, The mammalian cutaneous permeability barrier: defective barrier function in essential fatty acid deficiency correlates with abnormal intercellular lipid deposition, *Lab. Invest.,* 6, 574, 1978.
38. **Idson, B.**, Percutaneous absorption (review article), *J. Pharma. Sci.,* 64, 901, 1975.
39. **Feldmann, R. J. and Maibach, H. I.**, Penetration of ^{14}C hydrocortisone through normal skin: the effect of stripping and occlusion, *Arch. Dermatol.,* 91, 661, 1965.
40. **Bronaugh, R. L., Stewart, R. F., Wester, R. C., Bucks, D., Maibach, H. I., and Anderson, J.**, Comparison of percutaneous absorption of fragrances by humans and monkeys, *Fd. Chem. Toxic.,* 23, 111, 1985.

41. **Derelanko, M. J., Gad, S. C., Gavigan, F. A., Babich, P. C., and Rinehart, W. E.,** Toxicity of hydroxylamine sulfate following dermal exposure: variability with exposure method and species, *Fundam. Appl. Toxicol.,* 8, 583, 1987.

42. **Bucks, D. A. W., Maibach, H. I., and Guy, R. H.,** Occlusion does not uniformly enhance penetration *in vivo* in *Percutaneous Absorption,* 2nd ed., Bronaugh, R. L. and Maibach, H. I., Eds., Marcel Dekker, Inc., New York, 1989, chap. 5.

43. **Sulzberger, M. B. and Witten, V. H.,** Thin pliable plastic films in topical dermatologic therapy, *Arch. Dermatol.,* 84, 1027, 1961.

44. **Quisno, R. A. and Doyle, R. L.,** A new occlusive patch test system with a plastic chamber, *J. Soc. Cosmet. Chem.,* 34, 13, 1983.

45. **Rongone, E. L.,** Skin structure, function and biochemistry, in *Dermatotoxicology,* 3rd ed., Marzulli, F. N. and Maibach, H. I., Eds., Hemisphere, Washington, D.C., 1987, chap. 1.

46. **Scheuplein, R. J.,** Mechanism of percutaneous adsorption. I. Routes of penetration and the influence of solubility, *J. Invest. Dermatol.,* 45, 334, 1965.

47. **Marzulli, F. N.,** Barriers to skin penetration, *J. Invest. Dermatol.,* 39, 387, 1962.

48. **Stoughton, R. B.,** Percutaneous absorption, *Toxicol. Appl. Pharmacol.,* 7, 1, 1965.

49. **Slesinski, R. S., Guzzie, P. J., Hengler, W. C., Myers, R. C., and Ballantyne, B.,** Studies on the acute toxicity, primary irritancy and genotoxic potential of 1,3,5-triacryloylhexahydro-s-triazine, *Toxicology,* 40, 145, 1986.

50. **Stoughton, R. B. and Fritsch, W.,** Influence of dimethylsulfoxide in human percutaneous absorption, *Arch. Dermatol.,* 90, 512, 1964.

51. **Barrett, C. W., Hadgraft, J. W., and Sarkang, I.,** The influence of vehicles on skin penetration, *J. Pharm. Pharmacol.,* Suppl. 16, 104T, 1964.

52. **Scala, J., McOsher, D. E., and Reller, H. H.,** The percutaneous absorption of ionic surfactants, *J. Invest. Dermatol.,* 50, 371, 1968.

53. **Gad, S. C. and Chengelis, C. P.,** Vehicles, in *Acute Toxicology Testing Perspectives and Horizons,* The Telford Press, Caldwell, NJ, 1988, appendix C.

54. **Draize, J. H., Woodard, G., and Calvery, H. D.,** Methods for the study of irritation and toxicity of substances applied topically to the skin and mucous membranes, *J. Pharmacol. Exp. Ther.,* 82, 377, 1944.

55. **Smyth, H. F., Jr., Carpenter, C. P., Weil, C. S., Pozzani, U. C., and Striegel, J. A.,** Range-finding toxicity data: list VI, *Am. Ind. Hyg. Assoc. J.,* 23, 95, 1962.

56. **Myers, R. C., Hufford, K. R., and Bellich, N. S.,** Using Vetrap® bandaging during rabbit dermal toxicity testing, *Lab. Anim.,* 18, 44, 1989.

57. **Fritsch, W. C. and Stoughton, R. B.,** The effect of temperature and humidity on the penetration of C^{14} acetylsalicyclic acid in excised human skin, *J. Invest. Dermatol.,* 41, 307, 1963.

58. **Merrill, R. A.** Regulatory toxicology, in *Casarett and Doull's Toxicology: The Basic Science of Poisons,* 3rd ed., Klaassen, C. D., Amdur, M. O., and Doull, J., Eds., Macmillan, New York, 1986, chap. 30.

59. **Beck, B. D., Calabrese, E. J., and Anderson, P. D.,** The use of toxicology in the regulatory process, in *Principles and Methods of Toxicology,* 2nd ed., Hayes, A. W., Ed., Raven Press, New York, 1989, chap. 1.

60. **Chan, P. K. and Hayes, A. W.,** Principles and methods for acute toxicity and eye irritancy, in *Principles and Methods of Toxicology,* 2nd ed., Hayes, A. W., Ed., Raven Press, New York, 1989, chap. 6.

61. **Gad, S. C. and Chengelis, C. P.,** Lethality testing, in *Acute Toxicology Testing Perspectives and Horizons,* The Telford Press, Caldwell, NJ, 1988, chap. 7.

62. **Ritter, L. and Franklin, C. A.,** Dermal toxicity testing: exposure and absorption, in *Handbook of In Vivo Toxicity Testing,* Arnold, D. L., Grice, H. C., and Krewski, D. R., Eds., Academic Press, San Diego, 1990, chap. 11.

63. Toxic Substances Control Act, 15 U.S.C., Section 2601, 1976.

64. Environmental Protection Agency, 40 CFR 792, Toxic Substances Control Act (TSCA); good laboratory practice standards, *Fed. Reg.,* 54, 34034, 1989.

65. Environmental Protection Agency, 40 CFR 160, Federal Insecticide, Fungicide and Rodenticide Act (FIFRA); good laboratory practice standards, *Fed. Reg.,* 54, 34067, 1989.

66. Federal Hazardous Substances Act, 15 U.S.C., Section 1261, 1976.

67. Federal Food, Drug, and Cosmetic Act, 21 U.S.C., Section 321, 1938.

68. International Air Transport Association, *Dangerous Goods Regulations,* 31st ed., 3.6-Class 6, Poisonous (Toxic) and Infectious Substances, 1990, 52.

69. International Maritime Organization, *International Maritime Dangerous Goods Code,* 30th ed., Class 6.1, Poisons, 1989.

70. Environmental Protection Agency, Pesticide Assessment Guidelines, Subdivision F, Hazard Evaluation: Human and Domestic Animals, No. 540/9–82–025, Section 81–2, 1982.

71. Environmental Protection Agency, New and Revised Health Effects Test Guidelines, No. 560/6–84–002, Section HG-Acute-Dermal, 1984.
72. Environmental Protection Agency, Proposed rules, *Fed. Reg.*, 43, 37356, 1978.
73. Environmental Protection Agency, Alternative methodology for acute toxicity testing (revised policy statement), 1988.
74. Consumer Product Safety Commission, Regulations under FHSA, Title 16, *Code of Federal Regulations*, Ch. II, Commercial Practices, Section 1500, 1989.
75. Department of Transportation, Rules and regulations, *Fed. Reg.*, 55, 52637, 1990.
76. Organization for Economic Cooperation and Development, Guidelines for testing of chemicals, Section 402, 1981.
77. Organization for Economic Cooperation and Development, Guidelines for testing of chemicals, Section 402, 1987.
78. **Ballantyne, B., Slesinski, R. S., and Myers, R. C.,** The acute toxicity and mutagenic potential of 3-methyl-2-benzothiazolinore hydrazone, *Toxicol. Ind. Health*, 4, 23, 1988.
79. **Gad, S. C. and Chengelis, C. P.,** Acute systemic toxicity tests, in *Acute Toxicity Testing Perspectives and Horizons*, The Telford Press, Caldwell, NJ, 1988, chap. 9.
80. **Gad, S. C., Smith, A. L., Cramp, A. L., Gavigan, F. A., and Derelanko, M. J.,** Innovative designs and practices for acute systemic toxicity studies, *Drug Chem. Toxicol.*, 7, 423, 1984.
81. **Ballantyne, B., Myers, R. C., Gill, M. W., Burleigh-Flayer, H. D., Fowler, E. H., and Garman, R. H.,** 1,10-bis-Trimethoxysilyldecane (TMSD): neurotoxic and nephrotoxic effects, *Toxicologist*, 11, 119, 1991.
82. **Gad, S. C. and Weil, C. S.,** Statistics for toxicologists, in *Principles and Methods of Toxicology*, 2nd ed., Hayes, A. W., Ed., Raven Press, New York, 1989, chap. 15.
83. **Gad, S. C. and Weil, C. S.,** *Statistics and Experimental Design for Toxicologists*, The Telford Press, Caldwell, NJ, 1986.
84. **DePass, L. R.,** Alternative approaches in median lethality (LD_{50}) and acute toxicity testing, *Toxicol. Lett.*, 49, 159, 1989.
85. **Bliss, C.,** The calculation of the dosage-mortality curve, *Anal. Appl. Biol.*, 22, 134, 1935.
86. **Finney, D. J.,** *Probit Analysis*, 3rd ed., Cambridge University Press, Cambridge, 1971.
87. **Litchfield, J. T., Jr. and Wilcoxon, T.,** A simplified method of evaluating dose-effect experiments, *J. Pharmacol. Exp. Ther.*, 96, 99, 1949.
88. **Thompson, W. R.,** Use of moving averages and interpolation to estimate median effective dose, *Bacteriol. Rev.*, 11, 115, 1947.
89. **Thompson, W. R. and Weil, C. S.,** On the construction of tables for moving average interpolation, *Biometrics*, 8, 51, 1952.
90. **Weil, C. S.,** Tables for convenient calculation of median-effective dose (LD_{50} or ED50) and instructions in their use, *Biometrics*, 8, 249, 1952.
91. **Weil, C. S.,** Economical LD_{50} and slope determination, *Drug Chem. Toxicol.*, 6, 595, 1983.
92. **Weil, C. S., Carpenter, C. P., and Smyth, H. F.,** Specifications for calculating the median effective dose, *Am. Ind. Hyg. Assoc. Q.*, 14, 200, 1953.
93. **Brownlee, K. A., Hodges, J. L., Jr., and Rosenblatt, M.,** The up-and-down method with small samples, *J. Sm. Stat. Assoc.*, 48, 262, 1953.
94. **Bruce, R.,** An up-and-down procedure for acute toxicity testing, *Fundam. Appl. Toxicol.*, 5, 151, 1985.
95. **Lorke, D.,** A new approach to practical acute toxicity testing, *Arch. Toxicol.*, 54, 275, 1983.
96. **Schütz, E. and Fuchs, H.,** A new approach to minimizing the number of animals used in acute toxicity testing and optimizing the information of test results, *Arch. Toxicol.*, 51, 197, 1982.
97. **Deichmann, W. B. and LeBlanc, T. J.,** Determination of the approximate lethal dose with about six animals, *J. Ind. Hyg. Toxicol.*, 25, 415, 1943.
98. **Kennedy, G. L., Jr., Ferenz, R. L., and Burgess, B. A.,** Estimation of acute oral toxicity in rats by determination of the approximate lethal dose other than the LD_{50}, *J. Appl. Toxicol.*, 6, 145, 1986.
99. **Gad, S. C. and Chengelis, C. P.,** Statistical analysis of acute toxicology and safety studies, in *Acute Toxicology Testing Perspectives and Horizons*, The Telford Press, Caldwell, NJ, 1988, chap. 12.
100. **Snedecor, G. W. and Cochran, W. G.,** *Statistical Methods*, 7th ed., Iowa State University Press, Ames, 1980.
101. **Sokal, R. R. and Rohlf, F. J.,** *Biometry*, 2nd ed., W. H. Freeman, San Francisco, 1981.
102. **Smyth, H. F., Jr., Carpenter, C. P., and Shaffer, C. B.,** The subacute toxicity and irritation of polyethylene glycols of approximate molecular weights of 200, 300, and 400, *J. Am. Pharm. Assoc.*, 34, 172, 1945.
103. **Smyth, H. F., Jr., Carpenter, C. P., Weil, C. S., Pozzani, U. C., and Striegel, J. A.,** Range-finding toxicity data: list VI, *Am. Ind. Hyg. Assoc. J.*, 23, 95, 1962.

104. **Hodge, H. C. and Sterner, J. H. (table by Smyth, H. F., Jr., Carpenter, C. P., and Weil, C. S.),** Tabulation of toxicity classes, *Ind. Hyg. Q.,* 10, 93, 1949.

105. **Smyth, H. F., Jr., Carpenter, C. P., and Weil, C. S.,** Range-finding toxicity data: list IV, *AMA Arch. Ind. Hyg. Occup. Med.,* 4, 119, 1951.

106. **Smyth, H. F., Jr., Carpenter, C. P., Weil, C. S., Pozzani, U. C., Striegel, J. A., and Nycum, J. S.,** Range-finding toxicity data: list VII, *Am. Ind. Hyg. Assoc. J.,* 30, 470, 1969.

107. **Carpenter, C. P., Weil, C. S., and Smyth, H. F., Jr.,** Range-finding toxicity data: list VIII, *Toxicol. Appl. Pharmacol.,* 28, 313, 1974.

108. World Health Organization, *Organophosphorus Insecticides: A General Introduction,* Environmental Health Criteria 63, World Health Organization, Geneva, 1986.

109. **Gaines, T. B.,** The acute toxicity of pesticides to rats, *Toxicol. Appl. Pharmacol.,* 2, 88, 1960.

110. Environmental Protection Agency, Code of Federal Regulations, 40, Section 162.10, 1984.

111. British Toxicology Society Working Party on Toxicity, A new approach to the classification of substances and preparations on the basis of their acute toxicity, *Hum. Toxicol.,* 3, 85, 1984.

112. **van den Heuvel, M. J., Clark, D. G., Fielder, R. J., Koundakjian, P. P., Oliver, G. J. A., Pelling, D., Tomlinson, N. J., and Walker, A. P.,** The international validation of a fixed-dose procedure as an alternative to the classical LD_{50} test, *Fd. Chem. Toxic.,* 28, 469, 1990.

113. **Klaassen, C. D.,** Principles of toxicology, in *Casarett and Doull's Toxicology: the Basic Science of Poisons,* 3rd ed., Klaassen, C. D., Amdur, M. O., and Doull, J., Eds., Macmillan, New York, 1986, chap. 2.

114. **Gad, S. C. and Chengelis, C. P.,** Route, formulations and vehicles, in *Acute Toxicology Testing Perspectives and Horizons,* The Telford Press, Caldwell, NJ, 1988, chap. 10.

115. **Russel, W. M. S. and Burch, R. L.,** *The Principles of Humane Experimental Technique,* Methuen & Co., London, 1959.

116. **Prince, H. N., Prince, D. L., Prince, R. N., Farber, T., Griffith, J. F., and Spira, H.,** Alternative to animal testing, *Chem. Times Trends,* July 7, 1987.

117. **Rack, L. and Spira, H.,** Animal rights and modern toxicology, *Toxicol. Ind. Health,* 5, 133, 1989.

118. **Zbinden, G.,** Reduction and replacement of laboratory animals in toxicological testing and research, *Biomed. Environ. Sci.,* 1, 90, 1988.

119. **Zbinden, G. and Flury-Roversi, M.,** Significance of the LD_{50} test for the toxicological evaluation of chemical substances, *Arch. Toxicol.,* 47, 77, 1981.

120. **Dayan, A.,** Complete programme for acute toxicity testing — not only LD_{50} determination, *Acta. Pharmacol. Toxicol.,* 52 (Suppl. 2), 31, 1983.

121. **Phillips, J. C., Gibson, W. B., Yam, J., Alden, C. L., and Hard, G. C.,** Survey of the QSAR and *in vitro* approaches for developing non-animal methods to supersede the *in vivo* LD_{50} test, *Fd. Chem. Toxic.* 28, 375, 1990.

122. **Loomis, T. A.,** Principles of biological tests for toxicity, in *Essentials of Toxicology,* 3rd ed., Lea & Febiger, Philadelphia, 1978, chap. 12.

123. **Enslein, K.,** An overview of structure-activity relationships as an alternative to testing in animals for carcinogenicity, mutagenicity, dermal and eye irritation, and acute oral toxicity, *Toxicol. Ind. Health,* 4, 479, 1988.

Chapter 14

SUBCHRONIC DERMAL EXPOSURE STUDIES WITH INDUSTRIAL CHEMICALS

David W. Hobson

TABLE OF CONTENTS

0-8493-7357-3/93/$0.00 + $.50
© 1993 by CRC Press, Inc.

I. INTRODUCTION

Subchronic toxicity studies may be conducted using a variety of different exposure routes (e.g., oral, inhalation, intravenous, dermal, etc.), but this chapter will focus on dermal studies. For the most industrial chemicals, subchronic toxicity studies are likely to be included as part of an overall toxicologic evaluation. Such evaluations are most often conducted in accordance with established guidelines to collect toxicologic data required to evaluate chemical or chemical formulations for possible adverse health effects. For industrial chemicals, subchronic toxicity data is required in order to establish procedures to protect workers from toxic injury, to notify government agencies of their toxicity prior to their entry into commerce, and to provide information to the consumer concerning any significant use related to toxic hazards. Like the acute dermal toxicity studies discussed in the previous chapter, subchronic dermal studies are currently considered to be essential in the assessment of risk for industrial chemicals or chemical mixtures with a potential for dermal exposure. In order to appreciate the need for conducting subchronic dermal toxicity with industrial chemicals, it is first necessary to have some understanding as to what constitutes subchronic toxicity and how it can be distinguished from acute or chronic toxicity.

The distinction between subchronic toxicity and acute or chronic toxicity is somewhat indistinct and can be confusing. Acute toxicity generally refers to toxic effects resulting a short time (e.g., less than a week) following a single dose or multiple doses administered within a 24-h period. Chronic toxicity generally involves adverse effects which are the result of continuous exposure or repeated exposures to a toxicant over a significant portion (e.g., greater than 10%) of the lifespan of a given species. Thus, subchronic toxicity falls somewhere between acute and chronic toxicity and distinctions between these two extremes are sometimes subject to debate. To make matters even worse, the term "subacute toxicity" was often used in older toxicology literature to refer to what we now term "subchronic toxicity." In order to help to eliminate some of this confusion, some regulatory bodies have established working definitions of subchronic exposure and subchronic toxicity. For example, the National Academy of Sciences (NAS)[1] defines subchronic exposure as occurring from a few days up to 6 months and the Organization for Economic Cooperation and Development (OECD)[2] characterizes subchronic toxicity as adverse effects resulting from repeated daily exposures to a chemical for part (not exceeding 10%) of the life span. In actual practice, most subchronic studies with industrial chemicals are conducted using toxicant exposures of between 2 weeks and 3 months (i.e., 14, 21, 28, or 90 d) in order to comply with test guidelines for the conduct of such studies established by a particular regulatory agency or scientific organization. Translated in human exposure terms, subchronic exposures represent persistent exposures which might occur as a result of process-related occupational exposures, frequent use of a particular product or formulation over a short period, wearing of a required or favorite article of clothing, lot-related exposures of a chronically used substance, military operations, etc.

Subchronic toxicity studies may be conducted with a variety of chemicals in different physical forms and via any one of several possible administration routes. In most cases, the purpose for conducting such studies is usually to provide information of use in health-related decision making. From a toxicologic perspective, subchronic studies are performed generally for three reasons:

1. To provide data on the short-term toxicity, accumulation, and disposition of a chemical or chemical mixture.
2. To obtain an estimate of the maximal exposure which produces no observable adverse effects (NOAEL).

3.   To aid in establishing doses and principal endpoints to be evaluated in chronic (e.g., 2-year rodent studies) or carcinogenicity studies.

Subchronic studies represent the first opportunity in the toxicologic evaluation of a chemical or mixture to determine whether or not accumulation is occurring at low doses and whether or not this accumulation produces an adverse effect in any detectable systemic effect. These effects may include any alteration which can be differentiated from "normal" using a myriad of available technologies (pathology, hematology, clinical chemistry, neurobehavioral changes, physiologic function tests, clinical observations, genetic abnormalities, etc.).

Subchronic studies are most commonly performed using laboratory animal models exposed via the oral (e.g., dosed feed, closed water, and gavage) or inhalation (e.g., whole animal and nose only) exposure routes. The dermal exposure route is not used as often. Perhaps this is due to a number of factors including the requirement for intensive technical effort, difficulty in controlling dermally delivered dosages, the occurrence of fewer chemicals which pose dermal exposure concerns, the perception that toxicity findings from another exposure route also apply for the dermal route, etc. This situation may change however, due to a recognition of the skin as a potential route of exposure for an increasing number of industrial chemicals and an increasing desire to perform toxicity studies via the anticipated routes of human exposure.[3-5]

The likehood of the skin being exposed to chemicals is relatively high. Such exposures may occur in an occupational setting, while using household chemicals, from cosmetic or toiletry formulations, and from dermally applied pharmaceutical agents in a variety of vehicles. In some instances, such subchronic dermal exposures may have rather insidious consequences.[6,7] This is especially true with industrial chemicals or chemical mixtures used in manual operations such as cleaning, dyeing, inking, painting, weaving sewing, etc. where repetitive exposures to the skin may occur over prolonged periods. It is not always possible, or desireable, in some industrial settings to require or guarantee the protection of workers from dermal exposure to chemicals (especially in operations requiring a high degree of manual dexterity). As a result, there is a need to conduct subchronic toxicity studies using skin as the route of exposure to certain types of industrial chemicals and mixtures.

Various types of industrial chemicals might be considered as candidates for subchronic dermal toxicity testing. The principal criterion for selection of a chemical for such testing should be the likelihood that the chemical will come into repetitive contact with the skin, either intentionally or accidentally, over prolonged periods. Examples of such materials might be solvents, cleaning agents, dyes, inks, fabric protectants, lubricants, fuels, etc.

Subchronic dermal toxicity studies, as well as subchronic studies by other exposure routes, are typically conducted using laboratory animals and only rarely with human subjects. Unlike subchronic studies using other exposure routes, dermal studies have inherent practical problems such as a requirement for the individual handling of each animal for chemical treatment, the need to prevent cross-contamination of animals, and routine hair-clipping that make them costly to perform. For these reasons fewer subchronic skin exposure studies have been conducted and protocols for such studies are not as well-defined as are the protocols for the more frequently performed dosed-feed, gavage, and dosed-water studies. In addition, these studies are conducted in different laboratories throughout the world and as a result the conditions for conducting the studies can be quite variable. This issue was the topic of an international committee of scientists convened in 1979 to discuss flaws in long-term study design and data interpretation and to propose guidelines for the design, conduct, analysis, and reporting that would improve the quality and comparability of the results from such studies.[8]

The terms "skin exposure," "skin painting," and "dermal studies" are often used interchangeably in the literature to describe studies where skin penetration is the exposure route. While "dermal studies" is semantically inappropriate, it is ingrained in the literature and will be used in the following discussion to mean subchronic skin exposure studies. This chapter will present some important considerations for designing and conducting subchronic toxicity studies using skin as a route of exposure. For the purpose of this chapter, it will be assumed that: (1) application of a chemical or chemical mixture to the skin does not necessarily infer that penetration through the skin and entry into the systemic circulation will occur, (2) "local" toxic effects are those that occur in or near the region skin at or most proximal to the application site, (3) "systemic toxicity" refers to adverse effects which occur in the bodily organ systems including the skin if affected in a generalized fashion, and (4) a chemical or chemical mixture must penetrate the skin and enter the systemic circulation before systemic toxicity arising from a dermal exposure can occur.

II. STUDY DESIGN

Actually, there is no singularly correct way to design a subchronic dermal toxicity study. Any dermal study which involves exposure to the test material that occurs within a subchronic time frame (e.g., 10% or less of the expected life span of the test species) for the test species can be legitimately considered to be subchronic. There are, however, some established regulatory guidelines for the conduct of dermal subchronic toxicity studies using laboratory animals.[1,2,9] Such guidelines have been found to be acceptable for the routine assessment of subchronic toxicity by various agencies including the OECD, the EPA, the FDA, the Health and Safety Commission of the United Kingdom, and the Japanese Ministry of Health and Welfare. Even though these regulatory guidelines have been promulgated to generally describe routine procedures to be followed when designing and conducting a subchronic study using rodents, variability exists between the recommendations included in one guideline and those of another. The various similarities and differences between these recommendations are discussed in detail elsewhere.[10] Because this chapter is concerned with subchronic toxicity studies conducted with industrial chemicals using the dermal exposure route, provisions of various guidelines recommended by the EPA[9] and the OECD[2] for the conduct of such studies will be used in reference to the design of studies involving rodents.

Most subchronic toxicity studies utilizing exposure routes other than the dermal route (i.e., intravenous, subcutaneous, intramuscular, oral, and inhalation) are principally designed to examine the systemic effects of chemical exposures and assume that all or most of the chemical enters the body. Because test substances are applied externally in dermal toxicity studies, there is generally more recognition of the fact that adverse effects arising from test substance exposure may be local as well as systemic. The rates of dermal penetration and metabolism are also becoming generally recognized as potentially significant design factors.[19,12] Thus, subchronic dermal studies have some unique design considerations in addition to design factors common to all types of subchronic studies. These considerations are especially evident in the selection of the test species, dosing procedure, dose formulation, endpoints to be evaluated, and quality control procedures.

A. TEST SPECIES

The species used most frequently in subchronic dermal toxicity studies with industrial chemicals are mice, rats, hamsters, guinea pigs, and rabbits. On occasion, studies are conducted with dogs, pigs, and nonhuman primates. Human studies are rare.[13] Species selection is usually based on practical considerations such as comparatively short life span, small body size, and availability.[3] Another consideration is species acceptability. Mice and rats are used extensively in toxicology and carcinogenicity studies and large databases are available for these species containing information on the normal histology, clinical chemistry values, spontaneous pathology, normal physiology, etc. which are of great value in the interpretation of toxicity study findings. Because one of the reasons given for conducting subchronic studies is to establish dose levels to be used in chronic toxicity or carcinogenicity tests, it follows that the selection of rats or mice as test species is generally acceptable. Rabbits and guinea pigs are also generally acceptable due to their routine use in dermal irritation and sensitization studies. Although often deemed acceptable due to the likelihood of producing findings of greater relevance to man than studies with rabbits or rodents, experiments using larger animal species (i.e., pigs, nonhuman primates, etc.) require more justification due to their expense and as to the relevance and need.

The nature and severity of a chemical's toxicity response is not only a function of the sensitivity of the animal species used in the study; it also may be related to the strain used within a given species. In selecting an appropriate animal strain, genetics is usually an important consideration (i.e., whether the animals are inbred, hybrid, or outbred). many laboratories have experience with outbred strains because they are more readily available and these strains are commonly more disease-resistant than inbred animals. However, randomly bred strains are subject to low genetic stability marked by genetic drift and this can produce considerable variation in response. Inbred strains, on the other hand, will reduce variation, but they are single genotypes and may not be representative of the species. F1 hybrids are a uniform genotype, but they have a level of heterozygosity more closely resembling the outbred animals and are often chosen for use in toxicity studies.[3] Examples of strains from species often used in subchronic dermal toxicity studies incude mice (B6C3F1, CD-1, C3H), rats (F344, Osborne-Mendel), guinea pigs (Hartley), and rabbits (New Zealand White).

The age of animals to be used in subchronic dermal studies is also important. This is because the studies are designed to represent an exposure relative to a significant portion of the animal's life span. Thus, it is possible to design the study to examine the effects of subchronic exposure during early development just as easily as for late in life. It is possible to design a subchronic dermal toxicity study using animals of almost any age. However, due to the need to standardize and to avoid potential age-related response variation, most regulatory guidelines recommend using young adult animals (e.g., rats, 200 to 300 g; guinea pigs, 350 to 450 g; rabbits, 2.0 to 3.0 kg).[2]

Another important consideration for selection of animals for dermal studies might also be the inherent excitability of the species. For most species used for subchronic dermal studies, the site of application will need to be clipped regularly to ensure consistent application of chemical. This procedure may cause stress to the animal. If unavoidable, consideration should be given to acclimating the animals to the routine handling required by the study for a week or two before study initiation.

More than one species and more than one sex per species may be required for the evaluation of a given chemical or mixture. This is due to the general recommendations that toxicity studies be performed with at least two mammalian species (one rodent and one nonrodent) to increase the extrapolative power of the results.[14,15] When possible, exposure groups consisting of both sexes (OECD guidelines[2] for rodent studies recommend ten and other guidelines[9,16] recommend 20 animals per sex per treatment group) are usually included in subchronic studies in order to examine the possibility of sex-dependent toxicity. In some cases, it may be possible to use fewer than ten animals per sex per group; however, then one accepts the risk of having too few animals within a given dose group for adequate statistical evaluations should the mortality rate be greater than 10% in the group.

All animals, regardless of species, strain, or sex, must be in good health status prior to the initiation of the test and must have some permanent means of identification (ear tags, tattoos, ear notches, etc.). All possible effects must be taken to ensure that this health status is not compromised during the study, except for any effects due to treatment with the test material. Factors such as animal room humidity, temperature, light/dark cycle, water quality and supply, food quality and supply, and cleanliness are more than important, they are essential and cannot be overstressed.

Prior to any study, all animals should be weighed and randomly selected into weight-homogenized treatment groups. If possible, it is recommended that animals with body weights in excess of two standard deviations of the study population mean should be eliminated from the study before group selection begins.

Test animals may be housed singly or in groups. Whenever possible, however, single housing is always preferred. Prior to the first dermal exposure, animals should be acclimated to the housing conditions to be used in the study for at least 5 d. There are guidelines which limit the number of animals of a given species and size which may be grouped together.[17] Group housing has the potential for unwanted oral exposure by grooming cage mates and through wounds from cage-mate fighting. Male mice are particularly prone to fighting. Fighting may also significantly reduce survival and incidence of disease in a study. Individual caging avoids these problems, but it is more costly, especially when large numbers of animals are involved. If the objective is to more accurately control individual exposure conditions, then housing animals alone is the recommended course of action. Animal housing, care, and environmental conditions should follow the recommendations set forth in the U.S. Department of Health and Human Services *Guide for the Care and Use of Laboratory Animals*.[17]

The nutritional status of the test species is also important, especially if metabolism of the test chemical is required for toxicity to occur. Water consumption may also be important because the hydration status of the animals may also influence the absorption, distribution, and metabolism of the test chemical.

Care should also be taken to provide a controlled temperature and humidity environment in which to conduct the study and, if possible, to control the air flowing over the dose sites. Such factors not only affect such things as the metabolic rate, skin condition, and disease susceptibility of the test species, they also can influence the volatility, stability, and physical state of the test material.

Most laboratory animal species used to conduct subchronic dermal toxicology studies have hair coats which must be removed prior to the dermal application of test materials. The procedure usually includes restraining the animal and closely clipping against the direction of hairgrowth, i.e., posterior to anterior using electric clippers fitted with a standard number 40, or other suitable size clipper blade. Even though the clipping procedure is relatively simple, extreme care must be exercised to avoid cuts, "clipper burns," or skin lacerations and very often some practice is required to become proficient. Many laboratory animals, especially rabbits, have variable hair growth patterns which may make hair removal by clipping more difficult for some animals within an experimental population and less so for others.[18] For subchronic studies using rodents or rabbits, this may result in some animals exhibiting a variable "patchiness" in their hair coat during the course of the study. Experience shows that if this phenomenon is observed more frequently in the treatment groups than controls, the unwary investigator may be inclined to conclude that he has discovered a treatment that affects hair growth. Unfortunately this patchy growth is normal for some laboratory animal species (mice, rats, and rabbits) because, unlike humans, these species exhibit periodic waves of hair growth which move caudally and ventrally from the cervical region usually in 2- to 3-week cycles.[19] If patches of hair growth become a significant problem (e.g., if large areas of the skin are required by the study design and there is no option of using "selected" areas within a more generalized region), depilatory agents can be used to remove the patches, but such agents may alter the penetration characteristics of the stratum corneum. Consideration might also be given to changing the source of the animals since it is commonly observed that animals from one source may exhibit variable hair growth problems to a greater extent than another. Another consideration might be to use rodent strains which have sparse hair coats, often referred to as "hairless" or "nude," which have been successfully used as dermal exposure models.[20,21]

B. STUDY ENVIRONMENT

The percutaneous absorption of a test article may occur rapidly (within a few minutes) or it may take a considerable length of time (days). During the time it takes for a dermally applied chemical to be absorbed into the body, the delivered dose of a chemical is subject to alteration (dilution, evaporation, removal, degradation, etc.) by various factors in the study environment. Some of these factors are more easily controlled than others, however, evaporative or volatilization of the test chemical often poses a difficult problem.

Subchronic studies with highly volatile test materials may be conducted using occluded or covered dose sites to maximize absorption of the dose. Occlusion, however, does alter the normal penetration characteristics of the skin and is not always desireable. Another solution is to house the test animals individually in stainless-steel wire mesh cages placed within a chamber or cage rack designed to induce laminar air flow over the test sites in a manner so as to reduce the likelihood of inadvertent inhalation exposure. In such cases, it is advisable to monitor the flow rate and test material concentration in the breathing zone air in order to demonstrate that inhalation exposure is negligable.[22]

C. DOSE SELECTION

Dose selection is usually based on a number of characteristics of the test chemicals. As a prerequisite to dose selection, it is wise to obtain information on the physical and chemical characteristics of the test material and to chemically establish the identity of the test material as being suitable for use in the study. Information on the physical form, purity, pH (where applicable), stability, and melting and boiling points of the test chemical should also be collected. Data concerning the exposure potential, the acute toxicity, and the results of percutaneous penetration studies are particularly important. Data indicating the amount of chemical absorbed per unit skin area relative to time allow the investigator to estimate the percent application dose absorbed. It is usually assumed that any portion of the applied in excess of the absorbed dose at any given time remains on or within the skin or has been volatilized.

As stated earlier, dermal toxicity studies may have local as well as systemic effects. Thus, it may also be important to consider the potential for skin irritation during dose selection. Subchronic studies are usually done to establish appropriate doses for chronic studies, and results from these studies provide data on both local and systemic effects. It is possible that skin irritation occurs at a considerably higher dose than the one established for maximum percutaneous absorption and the one causing systemic effects. Since a local effect is an important toxicological endpoint in dermal studies, dose selection usually includes at least one dose close to the lowest concentration predicted to cause irritation. It is also relevant to select a dose that is estimated to be at the level of human exposure. If a vehicle is to be used, the effect of the vehicle alone (vehicle control) should be included as an independent study group within the study.

In general practice, a 2- to 4-week range finding or pilot study is often performed prior to the initiation of a more comprehensive 13-week subchronic dermal toxicity study. The purpose of the range-finding study is to assist in the selection of doses for the 13-week study and often the same endpoints to be used for the assessment of toxicity are utilized. At least three different dose groups and a control are included in this study, however, more groups may be added if there is a need to evaluate more than one vehicle or application procedure. Alterations in the endpoints are evaluated at the conclusion of the study in order to determine a "no effect" dose is present.

When conducted, the range-finding study may also include the estimation of the rate of dermal penetration if a suitable method of analysis is available. Data concerning the "no effect" dose is considered in conjunction with penetration data as well as any other relevant findings to arrive at a preliminary estimate of a dose at which no effects will be observed in the 13-week study.

The 13-week study has two principle objectives: (1) to estimate a NOAEL for use in subsequent chronic studies and (2) to determine the organs which have greatest susceptibility to the adverse effects of the chemical following subchronic exposure. Thus, selection of the three dose levels to be included in this study must attempt to ensure that these objectives are met. To do this, there are no quantitative or systematic procedures for dose selection and it is here that toxicology becomes more of an art than a science. If there is one thing that practical experience teaches in this situation, it is the more relevant information you have available at the time of dose selection, the more scientific will be the basis for the doses selected. Also, if one of the two dose-selection objectives must be compromised, it should not be the NOAEL estimation. This is because the chronic study can proceed without observed toxicity in the subchronic study, but neither the chronic study nor the risk assessment process is served when the NOAEL cannot be determined. To be conservative, it is advised that the NOAEL from the range-finding study should be the middle dose level in the 13-week subchronic study. The high and low doses can be whole or fractional log increments apart from the middle dose to help increase the likelihood of meeting both objectives of the 13-week study.

The dose groups selected for each subchronic study should include a control group and at least three different dose level groups per sex. When applicable, a vehicle control group should be included. If adverse effects relative to a particular endpoint or a particular chemical class are being monitored in the study, the use of a positive control chemical should be considered.

D. VEHICLE SELECTION

Test chemicals may be applied neat or diluted with a vehicle. When applied neat, the dose volume is adjusted to achieve the desired exposure levels. However, in some studies, it may be important to keep the dose volume constant using a vehicle and to vary the test agent concentration in order to expose a constant area. Some chemicals may be solids in the pure state and it may be desirable to apply them as a solution in a solvent for toxicity studies. The selection of a vehicle should be decided before toxicity studies begin so that the effects on percutaneous absorption can be determined. Water, ethanol, ethyl acetate, dimethylsulfoxide (DMSO), acetone, etc. have been used as vehicles in dermal studies, but such vehicles may alter the penetration characteristics of the skin relative to some test materials.[23]

For a given chemical, there may be several solvents of choice depending on the chemical characteristics of the test agent. In the selection of an appropriate vehicle, one should consider whether there is an enhancement or inhibition of absorption and whether the vehicle increases the toxicity. The rate of skin penetration for a given chemical often varies with different vehicles.[24,25] Thus, if a vehicle is used, a vehicle control group should always be included in the study. Although water is a good solvent for many chemicals, surface tension may cause this vehicle to bead and run off when applied to the skin and, when surface tension is overcome, may promote the absorption of the test agent. Some studies have used solvent mixtures (e.g., ethanol with water, ethanol with ethyl acetate, etc.) to decrease the surface tension and achieve a better exposure. Ethanol has been used as a vehicle, but may cause intoxication if applied in too great an amount (this is especially a problem when ethanol is selected as a vehicle for studies using mice). Ethanol may also alter the toxicity of the test compounds whose toxicity is affected by alcohol dehydrogenase metabolism. Acetone is commonly used because it is a good solvent for many chemicals and spreads easily and evaporates quickly from the skin. But with frequent use, acetone tends to dry the skin at the site of application.

Regardless of the vehicle used, it often may be difficult to conclusively demonstrate that the toxicity of a test chemical is without vehicle effect. Therefore, it is recommended that the use of a vehicle for subchronic dermal toxicity studies with industrial chemicals be limited to cases where absolutely warranted. When a vehicle is required, water, corn oil, ethanol, and carboxymethylcellulose are considered to be some of the best selections.[1]

E. APPLICATION OF TEST MATERIALS

Most guidelines recommend that the exposure site be limited to a maximum of 10% of the body surface area (e.g., about 30 cm^2 in rats and 1 cm^2 in mice). The dose volume and frequency and duration of dosing are a function of the total surface area available for exposure, toxicity and solubility of chemical, chemical characteristics, and effects of the vehicle. In some studies, sequential treating of the specified area throughout the period of exposure and with fixed volumes (e.g., 30 and 10 μl) to maintain a constant concentration/exposed area relationship may be required. When exposure is to cover the entire 10% of the body surface area, doses of 300 μl for rats and 100 μl for mice are considered the maximum dose volumes. Experience has shown that sometimes dose volumes may be excessive. For example, with some vehicles, running of sample will occur with one vehicle and not with another. To achieve a given dose with chemicals having a low solubility in the vehicle, a volume larger than recommended may be required. In these cases, it may be necessary to divide the dose and to apply several sequential, smaller doses and to allow time for drying in between applications.

Doses can be applied to the skin in a variety of ways. Some older studies used an artist's brush to "paint" the chemical onto the site of application. Hence, the term "skinpaint" was used to indicate the method of application. Although a brush is rarely used in contemporary studies, "skinpaint" still commonly appears in the literature to indicate that the chemical was applied to the skin. More recent studies use applicators that will better control the amount and distribution of the dose. If the dose is a free-flowing liquid, it can be applied conveniently with a micropipette or pipettor with a disposable tip. More viscous chemicals may require the use of a positive displacement pipette or syringe.

In some studies, the design calls for occluding the site of application after exposure in order to mimic the exposure of humans (e.g., chemicals in contact with skin under clothing or in dose-maximized irritation and sensitization studies).

The lateral and dorsal skin surfaces of the body are most commonly selected as sites for the dermal application of test chemicals to the test species. A common application site used in rodent or rabbit studies is the interscapular region of the back. This site is difficult for the animal to reach and dose application is usually faster and better tolerated. Regardless of the site selected, the dose should always be applied to a fixed standard area because skin thickness is not uniform over all body locations and the percutaneous penetration rate (and, therefore, the absorbed dose) may vary if the application site is inconsistent.

Subchronic studies are usually done to determine local and systemic toxic effects as well as doses for long-term studies. In order to maximize the potential to observe a toxic effect in these studies, more of the skin may be exposed so that larger dose volumes can be applied to increase the total absorption of the dose. When dose volume is a factor, skin from the interscapular region posterior to near the base of the tail can be used. As supported by the OECD test guidelines,[2] care should be exercised whenever possible to ensure that the application site does not exceed 10% of the total body surface area. When required, total body surface area may be estimated from the body weight using formulas or nomograms available in the literature for the species to be used.[25,26]

Depending on the physical characteristics of the test material, all dermally applied doses should be expressed in terms of test material weight or volume per unit body weight of the test species (e.g., milligrams per kilogram). It is also important to specify, when possible, to provide an estimate of the application rate for the test material per unit skin surface area (milligrams per squared centimeter).

The frequency of dose application can be an important consideration in the design of subchronic dermal toxicity studies. This is because absorption of the test material sufficient to cause systemic toxicity may not occur if doses are administered at low frequency or if the test material is metabolically inactivated very rapidly. In safety studies where there is a need to evaluate the potential for a chemical to cause toxicity under anticipated exposure conditions, the frequency of exposure should attempt to simulate or exceed that of the anticipated conditions. In order to design subchronic studies with optimal dose frequency, studies to estimate the dermal penetration rate and metabolism of the test material should be considered. In most studies, the frequency of application ranges from 5 to 7 d/week depending on chemical-specific characteristics. Application frequency should take into consideration the dose concentrations to be applied and the absorption rate. Thus, for slowly absorbed, nonvolatile compounds, lower application frequency (i.e., 1 d/week) may be warranted, whereas for more volatile compounds, greater frequency may be used (i.e., several applications per day).

The duration (e.g., length of time for skin exposure) is also related to the chemical characteristics and sometimes to the manner in which humans are exposed to the chemical. Thus, the chemical may remain in contact with the skin until the next subsequent exposure or it may be physically and/or chemically removed after every exposure to simulate human use (e.g., cosmetics). The OECD guidelines[2] recommend covering the application site with a porous gauze dressing and nonirritating tape to hold the dose in contact with the skin between applications. Such measures may be necessary to ensure that the dose site is not disturbed by the test animal between applications. There are also commercial restraint devices available for this purpose which can also be used, but care should be exercised to ensure that such devices do not completely immobilize the animal.

When required for subchronic dermal dose estimations, percutaneous absorption studies can be performed using *in vitro* or *in vivo* dermal penetration models (see Section I of this volume). These experiments should be conducted with single doses of the test chemical within the range of magnitude anticipated for the subchronic study. When feasible, multiple dose penetration studies with the frequency of administration being that anticipated for the subchronic study should be considered to determine if previous exposures to the chemical have any significant effect on the rate of penetration. If metabolic activation or inactivation of the chemical is a concern, it may be desirable to conduct such studies *in vivo* using the test species selected for use in the subchronic study. Further details concerning the procedures used to conduct dermal penetration and metabolism studies are provided in Chapters 4, 5, 6, 7, and 8 of this volume.

Absorption and metabolism information for the test material may help to estimate the potential for systemic effects and the need to conduct further dispositional studies. The study of chemical-specific metabolism is also useful to define the metabolites and tissue accumulation and to determine if species differences in metabolism will be a factor in the interpretation of the results. Data from these kinds of studies should be obtained before the design of definitive toxicology studies begins because they provide useful information for selecting doses, vehicles, frequency of exposure, etc. The results from percutaneous absorption/disposition studies may, in fact, determine whether a selected test chemical would be tested for systemic toxicity by the dermal route.

F. OBSERVATIONS

1. Clinical

In-life observations are important in skin exposure studies to obtain an accurate measure of irritation. Scheduled observations of clinical signs are routinely made and recorded for all animals in all study groups on a daily basis for subchronic studies. These data are often captured electronically and may include, among other things, body weight and food consumption measurements and overt signs of toxicity. Body weights and food consumption should be recorded weekly. Clinical examinations should include changes in skin, haircoat, eyes, mucous membranes, respiration, posture, etc. Interim observations are usually made during the time of chemical application and unusual observations are recorded as they occur.

The skin is a target site and close observation is required to help determine effects on skin that may decrease with continued exposure and, therefore, not be detected during the microscopic evaluations. In some cases, it may be useful to create a map of the skin exposure area to track the first appearance, development, and locations of any exposure-related lesions that may occur.

2. Ophthalmology

Ophthalmological examinations are recommended prior to exposure and at the end of the study. At minimum, the high-dose and control groups should be examined and other doses if necessary based on the high-dose findings. Harroff[27] provides an excellent description as to how to perform this type of examination.

3. Hematology

Most guidelines recommend only terminal evaluation of this clinical aspect because hematologic changes due to maturation and growth of the animal during the study may occur. If desired, however, comparison of individual animals with their prestudy values may provide a means of statistically evaluating individual group effects. Further, in studies where there is prior suspicion that the toxicity may involve a hematological effect, monthly or midpoint study samples might be considered.

At a minimum, hematologic parameters should include hematocrit, hemoglobin concentration, erythrocyte count, total and differential leukocyte counts, and a measure of clotting potential (clotting time, prothrombin time, platelet count, etc.).

4. Clinical Chemistry

The frequency of sample collection and analysis is the same as for hematology. Tests to evaluate electrolyte balance, liver function, kidney function, and carbohydrate metabolism are considered most important. These tests are generally conducted with serum from fasted animals and include calcium, sodium, phosphorus, chloride, potassium, glucose, alanine aminotransferase, aspartate aminotransferase, gamma glutamyl transpeptidase (in appropriate species), albumin, urea nitrogen, total protein, total bilirubin, and creatinine. If lactate dehydrogenase or creatine kinase are determined, it is important to state whether or not isoenzyme analyses are performed. Special measurements such as sorbitol dehydrogenase, glutathione, acetylcholinesterase, etc. may be considered for some exposures.

5. Urinalysis

Although usually only recommended for studies where renal toxicity is suspected, urinalysis can often provide almost as much information as serum chemistry if careful attention is given to the collection of samples. Since metabolism cages are required for sample collection, it often is only possible to collect samples from a representative fraction of the animals on study (five animals per dose group, high-dose and control animals, etc.). Sample analysis can be performed qualitatively with commercially available multitest sticks or quantitatively using analysis procedures similar to those used for serum determinations. Endpoints examined include glucose, protein, albumin, lactate dehydrogenase, gamma glutamyl transpeptidase, *n*-cetylglucosaminidase, and a microscopic examination of urine sediment.

6. Gross Pathology

A complete necropsy is usually performed on exposure groups and controls. Organ weights for organs such as liver, gonads, kidneys, and adrenals are recommended. Brain and heart weights are also recommended for inclusion in rodent studies to be submitted to Japanese regulatory agencies. This includes an external examination of the animal body surfaces and orifices and of the major organs and tissues. Because the method of chemical application is dermal, the effects on the skin should have been recorded throughout exposure and a final map charted at necropsy should be made to provide a way to correlate the in-life observations with the microscopic ones. Tumors are identified and selected for histopathology and the apparent state of development since first appearance in the chemical record is described in a pathology narrative. Also at the time of necropsy, color slides of selected animals or skin sites to show details of representative tumors or lesions may provide graphic examples of the verbal description. In this way, the history of skin lesions examined microscopically are traceable from first appearance during the in-life portion of the study.

Gross observations that are frequently recorded for skin at the site of application are often categorized as those lesions that can be tolerated (nonlife-threatening) and those that are considered toxic (potentially life-threatening). Erythema, scaling, subcutaneous edema, alopecia, and thickening are examples of signs of chemical effects that are usually not considered life-threatening and ulcers, fissures, exudate, and crust-formation and necrosis are examples of chemical effects that are regarded as life-threatening.

7. Histopathology

Slides of selected organs and tissues are prepared in the same way as for other routes of exposure. Tissue samples commonly taken in subchronic dermal toxicity studies usually include, at a minimum, gross lesions observed at necropsy, heart, kidney, liver, lungs, skin, and target organs (if known or suspected). Reproductive organs (i.e., testicles) and blood-forming organs (i.e., spleen, bone marrow) should be strongly considered when evaluating the subchronic dermatotoxicity of solvents. Because it is not possible to completely predict the extent of the toxic effects of any chemical, most pathologists advise the taking of a broad spectrum of tissue samples at necropsy to be stored and prepared for examination if toxicity is suspected.

Skin samples are taken from exposed and nonexposed sites. Thus, it is important to prepare skin samples to be able to identify the orientation and site on the animal from which the sample was taken. Skin should be cut from the animal so that the orientation is well-defined (anterior to posterior). In this respect it is very useful to cut the anterior (cranial) border with an arrow shape. Excised skin may be laid on a piece of labeled index card and placed in fixative; this will keep the skin flat. Tumors submitted for histopathology should include, at a minimum, those identified at the gross observation. Also, samples of normal-appearing skin should be selected from the site of application, if possible, and from a designated site away from the site of application. These samples permit comparisons to be made between treated skin with gross lesions, treated skin without gross lesions, and untreated skin. Care should be taken when removing selected samples from the excised skin for histology to retain the remainder of the excised skin intact. The skin tissue is usually trimmed from anterior to posterior (cranial to caudal) and approximately 1 cm in length for slides. If tumors are present and removed for examination, it is important to include nontumorous adjacent skin in the section. The nontumor section is important for assessing nonneoplastic changes, e.g., dermatitis, hyperplasia, etc., compared with control skin. It is important that nonneoplastic diagnoses be made on this nontumor area. Skin away from the site of application will be important to assess any systemic effects on skin.

Examples of common histopathologic findings are inflammation, mild spongiosis, degeneration and necrosis, epidermal hyperplasia, hyperkeratosis, parakeratosis, dyskeratosis, hyperplasia, dermal edema, fibrosis, atrophy, and hyperplasia of adnexa as signs of tolerated exposure concentrations. When doses are in excess microulcers, marked spongiosis, degeneration and mild to moderate necrosis, severe inflammation, and severe edema are conditions that destroy the integrity of the skin.

Histopathologic examinations are usually initiated with an examination of the tissues taken from the high-dose and control animals. If toxicologically significant lesions are observed in the tissues from the high-dose animals, further examinations of tissues taken from the medium- and low-dose animals are then initiated. Although not required, this sequential histopathologic examination procedure can help to increase the rate at which such studies are evaluated and can help to lower the cost of pathology for such studies.

G. DATA SUMMARY AND PRESENTATION

Data for dermal studies, in addition to observations made at necropsy and microscopically on individual organs, usually include all significant clinical observations that have been made during the study. In contrast to other routes of exposure, dermal studies allow for a continuous description of the effects of the test article on a major target organ — skin — to be made from the first day of exposure.

No matter how well the study is conducted, poor presentation of the data can obfuscate the findings. Due to the technical complexity and amount of data produced in subchronic dermal toxicity studies, it is relatively easy to overlook inadverntently the evaluation of some portions of the study data in favor of others. Because observations made in modern scientific studies are dependent upon formal evaluation of the study data, this can result in a failure to observe potentially important findings. Thus, it is recommended that a plan be established ahead of time to ensure that all portions of the study data are evaluated in a systematic fashion. This evaluation should include a dose-related examination of the findings from each parameter included in the study and collective evaluations of all clearly related parameters (i.e., clinical observations, gross pathology, histopathology, clinical chemistry, urinalysis, etc.). When appropriate, pathology findings should be evaluated in context with other observations and related to effects concerning specific organ systems.

Once made, the observations from the study should then be clearly presented and summarized in a written report and made available to the scientific community for review and comment. This report should include species/strain/sex(s) used, survival rate evaluation, a summary of clinical signs and the time of observation of abnormal signs, food consumption and body weight data, hematologic findings, clinical chemical data, necropsy findings, and histopathological findings (including dose-related incidence of significant lesions).

H. LIMIT TESTS

In some cases, where there is some indication that the toxicity of the chemical is very low, a limit test may be considered. Limit tests may be of 2 to 13 weeks in duration, depending on the needs of the investigator. The limit test involves the administration of very high doses (e.g., 1000 mg/kg) of the test chemical to a group of experimental animals. The study is conducted as discussed earlier for subchronic dermal toxicity studies, including the clinical endpoints measured, gross pathology, hematology, histopathology, etc. Upon completion of this study, the data is evaluated, and if there is a clear indication that at this high dose level no signs of toxicity are evident, then it may be decided that further dermal studies are unnecessary.

III. DISCUSSION

Although general guidelines are available for the conduct of subchronic dermal toxicity studies using rodents, such guidelines should not be taken as the final word on the design of such studies. There is a constant need to refine and improve the procedures used to assess all types of toxicity evaluations, including subchronic dermal studies. As mentioned earlier, the guidelines developed for these studies are intended to serve as basis for ensuring that regulatory submissions and product data comparisons have some degree of standardization, not necessarily to serve as the best means of comprehensively assessing the subchronic toxicity of a given chemical or mixture. In order to do this, a standardized subchronic dermal toxicity study might be considered to be only a beginning.

The principal consideration in the design and conduct of subchronic dermal toxicity studies with industrial chemicals should be to obtain information which may be used to estimate the toxicologic risk to humans (or sometimes other animals) exposed via the dermal route. In regulatory studies, however, sometimes other considerations, such as the need to obtain NOAEL estimates and the need for a clear observation of some type of toxicity within any of the target organs monitored, may tend to take precedence to the extent that other aspects of a study with value in understanding the toxicity of a chemical. Thus, regardless of the intended use of the findings, such studies must be carefully designed, conducted, evaluated, and reported in order to provide information on the target organs, cumulative nature of the toxicity, metabolic tolerance, etc. of a chemical to yield results which are of use in the design of further studies and for risk assessment.

In assessing the results of any subchronic dermal toxicity study, it is very important to consider both the strengths and limitations of the design. There is no "ideal" study design that is capable of complete, or even near complete, evaluation of the subchronic toxicity for all chemicals and mixtures, and the suitability of any particular design must be left to the scientific judgment of the toxicologist. Still, there are many areas of controversy present in the design of such studies (i.e., study duration, type of species, number of animals per dose group, etc.).

It is important that all toxicologists interested in the subchronic dermal effects of industrial chemicals appreciate the value and significance of this type of study. The potential for these studies to obtain important information on immunotoxicity, adverse effects on physiologic and metabolic processes, mutagenic events, gametogenesis, nutrient distribution and utilization, and the initiation or promotion of neoplastic processes is there to be evaluated in such studies. There just needs to be means of measurement. These studies can also serve to eliminate some of the need for long-term carcinogenicity studies by development of alternative designs based on subchronic exposure, leaving chronic studies just for chemicals with definite chronic exposure potential. For example, some recommendations include provisions for the inclusion of "satellite groups" in the study design.[2] This group consists of additional animals in the highest dose group which are observed for a specified period at the end of dosing for residual adverse effects. Of course a study can also be designed to include satellite groups at all dose levels and the control.

In addition to the inherent technical problems conducting subchronic dermal studies mentioned earlier, there are many factors which must be taken into account for the successful conduct of these studies. For example, it is often required that a chemical be evaluated for its subchronic dermal toxicity when relatively little is known concerning its acute toxicity or dermal penetrative characteristics. Also, many chemicals to be evaluated actually occur most often as a component of a more complex preparation for which there is a potential dermal exposure, e.g., paints, solvents, lubricants, fuels, etc. Thus, even though many actual exposures are to chemical mixtures or formulations rather than to pure chemicals, most toxicologic evaluations are performed using chemicals in the neat or pure form. It is only rarely that chemicals are evaluated to determine whether or not formulation factors may affect their percutaneous penetration, metabolism, or dermal toxicity. In fact, for some chemicals the likelihood of dermal contact is greatest for the chemical in its formulated state, not in the neat state. When a chemical is selected for study, the dermal route may be in appropriate because of the chemical characteristics in pure form. These potential problems underscore the importance of obtaining information about the intended application of the study data, the acute toxicity of the chemical, and its percutaneous penetration rate before deciding to conduct subchronic dermal exposure studies with industrial chemicals.

Once the skin has been determined to be a practial route to use, there are many problems which must be solved before the study can be initiated. Most of these difficulties have been discussed and can be overcome with a little effort. Due to the likelihood of dermal exposure to many industrial chemicals and the increasing awareness of the skin as a potential route of exposure, it is anticipated that more subchronic dermal exposure studies will be required in the future.

REFERENCES

1. National Academy of Sciences, Report of the Committee for the Revision of NAS Publication 1138. Principles and Procedures for Evaluating the Toxicity of Household Substances, National Academy of Sciences, Washington, D.C., 1977.
2. The Organization for Economic Cooperation and Development, Guidelines for Testing of Chemicals, 1981.
3. **Rall, D. P., Hogan, M. D., Huff, J. E., Schwetz, B. A., and Tennant, R. W.,** *Ann. Rev. Public Health,* 8, 355, 1987.
4. **Haseman, J. K., Huff, J. E., Zeiger, E., and McConnell, E. E.,** *Environ. Health Perspect.,* 74, 229, 1987.
5. **Huff, J. E., McConnell, E. E., Haseman, J. K., Boorman, G. A., Eustis, S. L., Schwetz, B. A., Rao, G. N., Jameson, C. W., Hart, L. G., and Rall, D. P.,** *Chem. World,* in press.
6. **Hobson, D. W., D'Addario, A. P., Bruner, R. H., and Uddin, D. E.,** *Fundam. Appl. Toxicol.,* 6, 339, 1986.
7. **Lebowitz, H. R., Young, R., Kidwell, J., McGowan, J., Langloss, J., and Brusick, D.,** *Drug Chem. Toxicol.,* 6, 379, 1983.
8. International Agency for Research on Cancer Monographs, Long-term and short-term screening assays for carcinogens: a critical appraisal, Suppl. 2, Lyon, France, 1980.
9. Environmental Protection Agency, Proposed health effects test standards for toxic substance control act test rules and proposed good laboratory practices standards to health effects, *Fed. Reg.,* July 26, 1979.
10. **Mosberg, A. T. and Hayes, A. W.,** Subchronic toxicity testing, in *Principles and Methods of Toxicology,* 2nd ed., Hayes, A. W., Ed., Raven Press, New York, 1989, chap. 7.
11. **Kao, J., Patterson, F. K., and Hall, J.,** Skin penetration and metabolism of topically applied chemicals in six mammalian species, including man: an *in vitro* study with benzo[a]pyrene and testosterone, *Toxicol. Appl. Pharmacol.,* 81, 502, 1985.
12. **Lawrence, C. M., Finnen, M. J., and Shuster, S.,** Effect of coal tar on cutaneous aryl hydrocarbon hydroxylase induction and anthralin irritancy, *Br. J. Dermatol.,* 110, 671, 1984.
13. **Webster, R. C. and Maibach, H. I.,** *J. Toxicol. Environ. Health,* 16, 25, 1985.

14. Environmental Protection Agency, Health Effects Test Guidelines, EPA-560/11-82-002, Office of Pesticides and Toxic Substances, 1982.
15. **Weisburger, J. H. and Weisburger, E. K.,** Tests for chemical carcinogens, in *Methods in Cancer Research,* Vol. 1, Busch, H., Ed., 1967, chap. 7.
16. Environmental Protection Agency, Proposed guidelines for registering pesticides in U.S. hazard evaluation: human and domestic animals, *Fed. Reg.,* August 22, 1978.
17. Department of Health and Human Services, Guide for the Care and Use of Laboratory Animals, NIH Publication No. 86-23, National Institutes of Health, 1985.
18. **Harkness, J. E. and Wagner, J. E.,** *The Biology and Medicine of Rabbits and Rodents,* Lea & Febiger, Philadelphia, 1983, 76.
19. **Chase, H. B.,** Growth of the hair, *Physiol. Rev.,* 34, 113, 1954.
20. **Susten, A. S., Dames, B. L., and Niemeier, R. W.,** *In vivo* percutaneous absorption studies of volatile solvents in hairless mice. I. Description of a skin-depot, *J. Appl. Toxicol.,* 6, 43, 1986.
21. **Mershon, M. M., Mitcheltree, L. W., Petrali, J. P., Braue, E. H., and Wade, J. V.,** Hairless guinea pig bioassay model for vessicant vapor exposures, *Fundam. Appl. Toxicol.,* 15, 1990, in press.
22. **Hardin, B. D., Neimeier, R. W., Smith, R. J., Kuczuk, M. H., Mathinos, P. R., and Weaver, T. F.,** *Drug Chem. Toxicol.,* 5, 277, 1982.
23. **Berner, B., Jeung, R., and Mazzenga, G. C.,** *J. Pharm. Sci.,* 78, 472, 1989.
24. **Scheuplein, R. J. and Blank, I. H.,** *Physiol. Rev.,* 51, 702, 1971.
25. **Lee, M. O.,** *Am. J. Physiol.,* 89, 24, 1929.
26. **Ilahi, M. A., Barnes, B. A., and Burke, J. F.,** *J. Surg. Res.,* 11, 308, 1971.
27. **Harroff, H. H.,** Pathological processes of the eye related to chemical exposure, in *Dermal and Ocular Toxicology: Fundamentals and Methods,* Hobson, D. W., Ed., The Telford Press, New York, chap. 15, in press.

Chapter 15

THE DOSE RESPONSE OF PERCUTANEOUS ABSORPTION

Ronald C. Wester and Howard I. Maibach

TABLE OF CONTENTS

0-8493-7357-3/93/$0.00 + $.50
© 1993 by CRC Press, Inc.

I. INTRODUCTION

In most toxicological specialities, the administered dose is defined precisely. This has not always been so in dermatoxicology. This chapter defines our current, albeit far from perfect, understanding of the relation of applied dose to percutaneous absorption.

The interrelationships of dose response in dermal absorption are defined in terms of accountability, concentration, surface area, frequency of application, and time of exposure. *Accountability* is an accounting of the mass balance for each dose applied to skin. *Concentration* is the amount of applied chemical per unit skin surface area. *Surface area* is usually defined in square centimeters of skin application or exposure. *Frequency* is either intermittent or chronic exposure. "Intermittent" can be one, two, and so on exposures per day. Chronic application is usually repetitive and on a continuing daily basis. *Time of exposure* is the duration of the period during which the skin is in contact with the chemical before washing. Such factors define skin exposure to a chemical and subsequent percutaneous absorption.

The interrelationships of these will first be illustrated with the percutaneous absorption of benzene. Table 1 gives the *in vitro* percutaneous absorption and skin distribution of benzene. The skin absorption of benzene is very low. What is different for benzene (as compared to PCBs and nitroaniline) is that accountability is only 2.67% of applied dose. The majority of the dose evaporated off the skin.[1] However, if one considers the interrelationships of concentration, and frequency and time, some caution emerges. Figure 1 shows that the skin absorption of benzene is dose-related; benzene absorption increases with increased concentration. If the exposure situation is such that benzene remains in contact with skin (clothing?), then benzene will continue to be absorbed through skin. Figure 2 shows benzene absorption with more frequent/chronic exposure. With more exposure, the amounts of benzene in skin and subsequent absorption increase. Obviously, time is a major factor. The increased concentration (Figure 1) and chronic exposure (Figure 2) all increase the time of exposure. The benzene doesn't immediately evaporate from skin (decreased skin exposure time). Surface area would be related to the total amount of skin exposed to benzene. It is the great multiplier for exposure calculations because skin is the largest organ in the body.

II. ACCOUNTABILITY (MASS BALANCE)

Table 2 gives the *in vivo* percutaneous absorption of dinoseb in the rhesus monkey and rat. The absorption in the rat over a dose range of 52 to 644 $\mu g/cm^2$ is approximately 90% for all of the doses.[2] Conversely, absorption in the rhesus monkey for the dinoseb dose range of 44 to 3620 $\mu g/cm^2$ is approximately only 5%.[3] There is an obvious difference because of species (rat and rhesus monkey). The question then becomes one of mass balance to determine dose accountability. (If dinoseb was not absorbed through skin, then what happened to the chemical?) Table 3 shows that at least 80% of the applied doses can be accounted for (rat and rhesus monkey), and that (Table 2) in the rhesus monkey, the dinoseb remained on the skin (skin wash recovery 73.8 ± 6.8) and was not absorbed over the 24-h application period. Figure 3 shows that the dose response for the rat and rhesus monkey are different. Aside from species, differences in formulation and time of exposure are variables which also contribute to the dose response.

III. EFFECTS OF CONCENTRATION

Maibach and Feldmann[4] applied increased concentrations of testosterone, hydrocortisone, and benzoic acid from 4 $\mu g/cm^2$ in three steps to 2000 $\mu g/cm^2$ (4 $\mu g/cm^2$ is approximately equivalent to the amount applied in a 0.25% topical application; 2000 $\mu g/cm^2$ leaves

TABLE 1
In Vitro Percutaneous Absorption and Skin Distribution of Contaminants in Water Solution for 30-Min Exposure

Parameter	Chemical contaminant (percent dose)[a]		
	Benzene	54% PCB	Nitroaniline
Percutaneous absorption (systemic)	0.045 ± 0.037	0.03 ± 0.00	5.2 ± 1.6
Surface bound/stratum corneum	0.036 ± 0.005	6.8 ± 1.0	0.2 ± 0.17
Epidermis and dermis	0.065 ± 0.057	5.5 ± 0.7	0.61 ± 0.25
Total (skin/systemic)	0.15	12.3	6.1
Skin wash/residual	2.51 ± 0.94	71.2 ± 5.5	92.4 ± 0.8
Apparatus wash	0.006 ± 0.005	0.25 ± 0.2	0.47 ± 0.03
Total (accountability)	2.67	83.3	99.0

[a] Percent of applied dose ($n = 4$ for each parameter): benzene, 21.7 μg/ml, 54% PCB, 1.6 μg/ml; nitroaniline, 4.9 μg/ml.

EFFECT OF INCREASED BENZENE DOSE APPLIED TO HUMAN SKIN IN VITRO

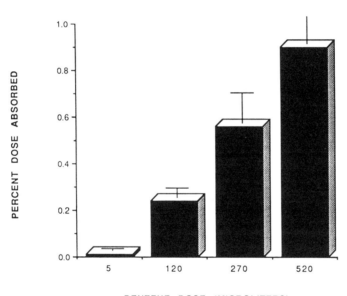

FIGURE 1. The percent dose absorbed increases as the concentration of benzene on skin is increased. The more benzene on skin, the longer (time) that absorption can occur before surface benzene evaporation.

a grossly visible deposit of chemical). Increasing the concentration of the chemical always increased total absorption. These data suggest that as much as gram amounts of some compounds can be absorbed through normal skin under therapeutic and environmental conditions.

Wester and Maibach[5] further defined the relationship of topical dosing. Increasing concentration of testosterone, hydrocortisone, and benzoic acid decreased the efficiency of percutaneous absorption (percent dose absorbed) in both the rhesus monkey and man (Figure 4), but the total amount of material absorbed through the skin always increased with increased concentration. Scheuplein and Ross[6] also showed *in vitro* that the mass of material absorbed across skin increased when the applied dose was increased. The same relationship between dose applied and dose absorbed is also seen with the pesticides parathion and lindane in Figure 5.

BENZENE IN VITRO ABSORPTION SINGLE VS 3X DOSES IN 30 MINUTES

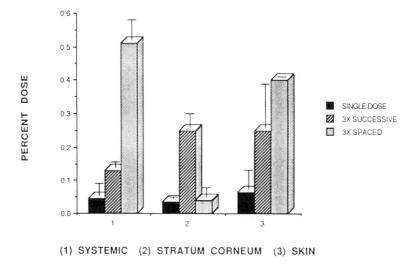

(1) SYSTEMIC   (2) STRATUM CORNEUM   (3) SKIN

FIGURE 2. Multiple dose benzene keeps the concentration of benzene on skin for a longer time period, resulting in more benzene skin absorption.

TABLE 2
In Vivo Percutaneous Absorption of Dinoseb in Rhesus Monkey and Rat[a]

Applied dose (μg/cm^2)	Skin penetration (%)	Dose accountability (%)
Rat		
51.5	86.4 ± 1.1	87.9 ± 1.8
128.8	90.5 ± 1.1	91.5 ± 0.6
643.5	93.2 ± 0.6	90.4 ± 0.7
Rhesus monkey		
43.6	5.4 ± 2.9	86.0 ± 4.0
200.0	7.2 ± 6.4	81.2 ± 18.1
3620.0	4.9 ± 3.4	80.3 ± 5.2

[a] Rat = acetone vehicle; 72-h application. Monkey-Premerge-3 vehicle; 24-h application.

Wedig and co-workers[7] compared the percutaneous penetration of different anatomical sites. A single dose of a ^{14}C-labeled magnesium sulfate adduct of dipyrithione at concentrations of 4, 12, or 40 μg/cm^2 per site was applied to an 8-h contact time to the forearm, forehead, and scalp of human volunteers. The results again indicated that as the concentration increased, more was absorbed. Skin permeability for equivalent doses on different sites assumed the following order: forehead was equal to scalp, which was greater than forearm. The total amounts absorbed increased even when the percentage of dose excreted at two doses remained approximately the same. On the forehead, proportionately more penetrated from the 40 μg/cm^2 than from the 4 and 12 μg/cm^2. Thus, as the concentration of applied dose increased, the total amount of chemical penetrating the skin (and thus becoming sys-

TABLE 3
Percutaneous Absorption and Accountability of Dinoseb
In Vivo Study in the Rhesus Monkey

Disposition parameter	Applied dose ($\mu g/cm^2$)		
	43.6	200.0	3620.0
	Applied dose accountability (%)		
Urine	3.3 ± 1.8	4.4 ± 23.9	3.0 ± 2.1
Feces	0.8 ± 0.5	1.0 ± 0.6	3.0 ± 1.7
Contaminated solids	0.03 ± 0.02	0.02 ± 0.02	0.07 ± 0.08
Pan wash	0.04 ± 0.03	0.8 ± 1.1	0.4 ± 0.3
Skin wash	81.1 ± 4.0	75.0 ± 22.9	73.8 ± 6.8
Total accountability	86.0 ± 4.0	81.2 ± 18.1	80.3 ± 5.2

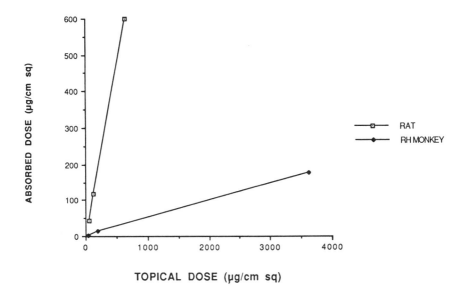

FIGURE 3. Dinoseb skin absorption dose response is dependent on species (as well as formulation used and time of skin exposure).

temically available) also increased for all the anatomical sites studied. Therefore, for exposure of many parts of the body, absorption can take place from all of the sites. As the concentration of applied chemical and the total body exposure increase, the subsequent systemic availability will also increase.

Although the penetration at these doses varied between anatomical sites, the percentage of the dose penetrating was similar at the three doses on the forearm and the forehead. However, on the occiput of the scalp, there was an increasing percentage of penetration with increasing dosage. In other words, at the highest dose, the efficiency of penetration was the greatest.

This effect of concentration on percutaneous absorption also extends to the penetration of corticoids as measured by the vasoconstrictor assay. In this type of assay, Maibach and Stoughton[8] showed that, in general, there is a dose-response relationship with increasing efficacy closely following increased dose. A several-fold difference in dose can override

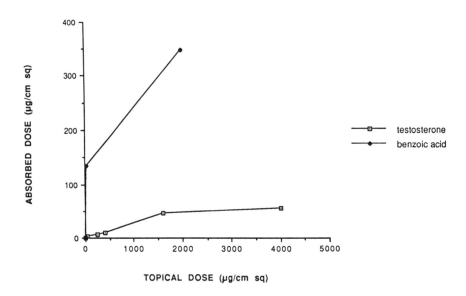

FIGURE 4. Skin absorption dose response in rhesus monkey is compound-dependent.

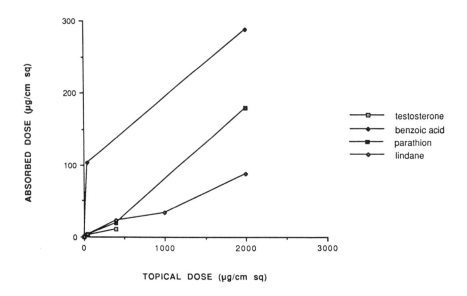

FIGURE 5. Skin absorption dose response in man is compound-dependent.

differences in potency between the halogenated analogues. If this applies to corticoids, it could also apply for other chemicals.

Reifenrath et al.[9] determined the percutaneous penetration of mosquito repellents in hairless dogs. As the topical dose increased in concentration, the penetration in terms of percentage of applied dose was about the same (Table 4). However, the mean total amount of material absorbed increased dramatically. An application of 4 μg/cm² of *N,N*-diethyl-*m*-toluamide gave a 12.8% absorption resulting in a total absorption of 0.5 μg/cm². An increase

TABLE 4
Percutaneous Penetration and Total Absorption of Repellents in Relation to Dose of Chemical Applied to Hairless Dog

Compound	Topical dose ($\mu g/cm^2$)	Penetration (% of applied dose)	Mean total absorbed ($\mu g/cm^2$)
Ethyl hexanediol	4	8.8	0.35
	320	10.3	3.0
N,N-Diethyl-*m*-toluamide	4	12.8	0.51
	320	9.4	30.1
Sulfonamide[a]	100	9.1	9.1
	320	7.5	24.0
	1000	5.4	54.0

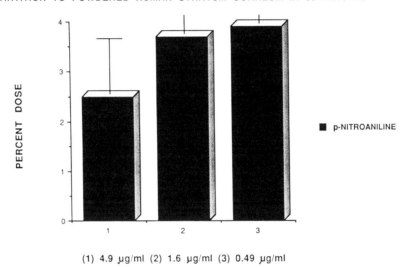

PARTITION TO POWDERED HUMAN STRATUM CORNEUM IN 30 MINUTES

■ p-NITROANILINE

(1) 4.9 μg/ml (2) 1.6 μg/ml (3) 0.49 μg/ml

FIGURE 6. The partitioning dose response of nitroaniline from water to powdered human stratum corneum is linear and dose-independent.

in the dose to 320 $\mu g/cm^2$ decreased the percent absorbed to 9.4; however, the total amount of material absorbed was now up to 30.1 $\mu g/cm^2$, an increase of 60 times!

The dose response of *p*-nitroaniline with skin is shown in Figure 6 and 7. In Figure 6, *p*-nitroaniline partitions from water into powdered human stratum corneum. In Figure 7, the concentrations of *p*-nitroaniline in human skin and subsequently absorbed through the skin are shown. The data show that the percent doses are the same for the tenfold dose range. If the percent doses are the same, then the dose response is linear.[1]

A. CONCENTRATION AND NEWBORNS

Wester and co-workers[10] compared the percutaneous absorption in newborn vs. adult rhesus monkeys. The total amount absorbed per square centimeter of skin again increased with increased applied dose and was further increased when the site of application was occluded. In the newborn, the question of concentration may have special significance because surface area/body mass ratio is greater than in the adult. Therefore, the systemic availability per kilogram of body weight can be increased by as much as threefold.

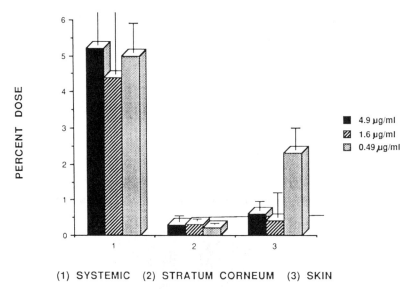

FIGURE 7.   The absorption dose response of nitroaniline from water into skin is linear and dose-independent.

B. CONCENTRATION AND WATER TEMPERATURE

Cummings[11] determined the effect of temperature on rate of penetration on *n*-octylamine through human skin. Increasing the temperature increased the rate of penetration as evidenced by octylamine-induced wheal formation and erythema. The increase in cutaneous blood flow mainly involved areas of the wheal. The increase in cutaneous blood flow mainly involved areas of the epidermal factors. Therefore, increased temperature along with increased concentration will increase the percutaneous absorption.

C. CONCENTRATION AND TIME OF EXPOSURE

Duration of occlusion enhances percutaneous absorption. The significance of time in occlusion was shown by Feldmann and Maibach (1974), who concluded that the longer clothing occludes a pesticide, the greater the contamination potential becomes. The dose response with time is shown in Figure 8.

D. CONCENTRATION, DURATION OF CONTACT, AND MULTIPLE-DOSE APPLICATION

Black and Howes[12] studied the skin penetration of chemically related detergents (anionic surfactants) through rat skin and determined the absorption for multiple variables, mainly concentration of applied dose, duration of contact, and the effect of multiple-dose applications (Table 5).

E. CONCENTRATION AND SURFACE AREA

Sved and co-workers[13] determined the role of surface area on percutaneous absorption of nitroglycerin. As the surface area of applied dose increased, the total amount of material absorbed and systemic availability of nitroglycerin increased. This was confirmed by the percutaneous absorption studies of Noonan and Wester,[14,15] but there was no linear relationship between the size of the surface area and increase in absorption; however, the same information held. The surface area of applied dose determined systemic availability of the chemical.

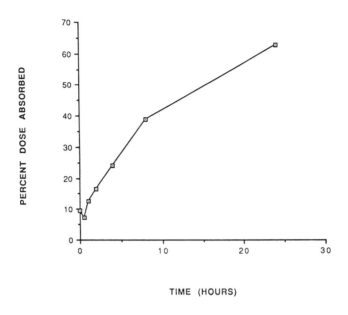

FIGURE 8.   Malathion skin absorption in man increases with time.

TABLE 5
Concentration, Duration of Contact, and
Multiple Application as Variables in
Penetration of Anionic Surfactant Through
Rat Skin

Variable	Penetration ($\mu g/cm^2$)
Concentration (% w/v)	
0.2	0.02
0.5	0.11
1.0	0.23
2.0	0.84
Duration of Contact (min)	
1	0.25
5	0.47
10	0.69
20	0.97
Multiple application ($\times$ 5 min)	
1	0.14
2	0.25
4	0.36

From Black, J. G. and Howes, D., *J. Soc. Cosmet. Chem.,*
30, 157, 1979. With permission.

IV. EFFECT OF APPLICATION FREQUENCY

Wester and co-workers[14] studied the effect of application frequency on the percutaneous absorption of hydrocortisone. Material applied once or three times a day showed a statistical difference ($p < 0.05$) in the percutaneous absorption. One application each 24 h of exposure gave a higher absorption than material applied at a lower concentration, but more frequently,

SKIN ABSORPTION SINGLE (DAY 1) AND MULTIPLE (DAY 8) DOSE IN MAN

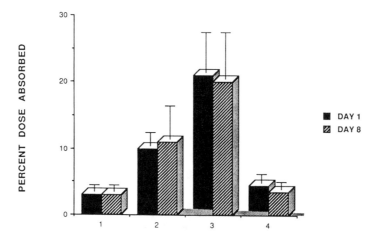

(1)HYDROCORTSONE (2)ESTRADIOL (3) TESTOSTERONE (4)MALATHION

FIGURE 9. Single-dose absorption in man will predict multiple-dose absorption (provided no skin irritation occurs to change skin absorption) properties barrier.

namely, three times a day. This study also showed that washing (effect of hydration) enhanced the percutaneous absorption of hydrocortisone. This relationship between frequency of application and percutaneous absorption is also seen with testosterone.[15]

The aforenoted studies used intermittent application per single day of application. Another consideration is extended vs. short-term administration and the subsequent effect on percutaneous absorption. Wester et al.[16] examined the percutaneous absorption of hydrocortisone with long-term administration. The work suggests that extended exposure had some effect on the permeability characteristics of skin and markedly increased percutaneous absorption.

With malathion, which apparently has no pharmacological effect on skin, the dermal absorption from day 1 was equivalent to day-8 application.[17] Therefore, for malathion, the single-dose application data are relevant for predicting the toxic potential for longer-term exposure. More human data from Bucks et al.[18] further suggest that at least in man, the single dose absorption can be used to predict absorption during chronic exposure (Figure 9).

Two important points are emerging from chronic exposure absorption studies. The first is that absorption data with multiple dosing is different for animals and man. This seems to be related to skin washing. Animal skin washed during chronic exposure react different than human skin washed during chronic exposure.

The second point is that these absorption studies have assumed no skin irritation. Skin irritation during chronic exposure will change skin barrier properties and probably result in higher percutaneous absorption.

V. APPLICATION FREQUENCY AND TOXICITY

There is a correlation between frequency of application, percutaneous absorption, and toxicity of applied chemical. Wilson and Holland[18] determined the effect of application frequency in epidermal carcinogenic assays. Application of a single large dose of a highly complex mixture of petroleum or synthetic fuels to a skin site increased the carcinogenic potential of the chemical compared with smaller or more frequent applications. This carcinogenic toxicity correlated well with the results of Wester et al.,[14,16] in which a single applied dose increased the percutaneous absorption of the material compared with smaller or intermittent applications.

VI. DISCUSSION

Many variables affect percutaneous absorption and subsequent dermal toxicity. Increased concentration of an applied chemical on skin increases the body burden, as does increasing the surface area and the time of exposure. The opposite also holds true, namely dilution of a chemical will decrease the effects of the applied concentration, provided other factors do not change (such as diluting the chemical, but spreading the same total dose over a larger surface area). The body burden is also dependent on the frequency of daily application and on possible effects resulting from long-term topical exposure. Dose accountability (mass balance) completes a dose-response study.

The current data provide a skeleton of knowledge to use in the design and interpretation of toxicological and pharmacological studies, to increase their relevance to the most typical exposures for man. In essence, we have just begun to define the complexity of the interrelationships between percutaneous absorption and dermatoxicology.[19,20] Until an appropriate theoretical basis that has been experimentally verified becomes available, quantitating the various variables listed herein will greatly improve the usefulness of biologically oriented protocols that use the skin as a route of entry.

REFERENCES

1. **Wester, R. C., Mobayen, M., and Maibach, H. I.,** *In vivo* and *in vitro* absorption and binding to powdered stratum corneum as methods to evaluate skin absorption of environmental chemical contaminants from ground and surface water, *J. Toxicol. Environ. Health,* 21, 367, 1987.
2. **Shah, P. V., Fisher, H. L., Sumler, M. R., Monroe, R. J., Chernoff, N., and Hall, L. L.,** A comparison of the penetration of fourteen pesticides through the skin of young and adult rats, *J. Toxicol. Environ. Health,* 21, 353, 1987.
3. **Wester, R. C., McMaster, J., Bucks, D. A. W., Bellet, E. M., and Maibach, H. I.,** Percutaneous absorption in rhesus monkey and estimation of human chemical exposure, in *Biological Monitoring for Pesticide Exposure,* Series 382, Wang, R. G. M., Franklin, C. A., Honeycutt, R. C., and Reinert, J. C., Eds. ACS Publishers, Washington, D.C., 1989, chap 12.
4. **Feldmann, R. J. and Maibach, H. I.,** Systemic absorption of pesticides through the skin of man, in *Occupational Exposure to Pesticides: Report to the Federal Working Group on Pest Management from the Task Group on Occupation Exposure to Pesticides,* Appendix B, 1969, 120.
5. **Wester, R. C. and Maibach, H. I.,** Relationship of topical dose and percutaneous absorption on rhesus monkey and man, *J. Invest. Dermatol.,* 67, 518, 1976.
6. **Scheuplien, R. J. and Ross, L. W.,** Mechanism of percutaneous absorption. V. Percutaneous absorption of solvent-deposited solids, *J. Invest. Dermatol.,* 62, 353, 1974.
7. **Wedig, J. H., Feldmann, R. J., and Maibach, H. I.,** Percutaneous penetration of the magnesium sulfate adduct of dipyrithione in man, *Toxicol. Appl. Pharmacol.,* 41, 1, 1977.
8. **Maibach, H. I. and Stoughton, R. B.,** Topical corticosteroids, *Med. Clin. N. Am.,* 57, 1253, 1973.
9. **Reifenrath, W. G., Robinson, P. B., Bolton, V. D., and Aliff, R. E.,** Percutaneous penetration of mosquito repellents in the hairless dog: effect of dose on percentage penetration, *Food Cosmet. Toxicol.,* 19, 195, 1981.
10. **Wester, R. C., Noonan, P. K., Cole, M. P., and Maibach, H. I.,** Percutaneous absorption of testosterone in the newborn rhesus monkey: comparison to the adult, *Pediatr. Res.,* 11, 737, 1977a.
11. **Cummings, E. G.,** Temperature and concentration effects on penetration of *N*-octylamine through human skin *in situ, J. Invest. Dermatol.,* 53, 64, 1969.

12. **Black, J. G. and Howes, D.,** Skin penetration of chemically related detergents, *J. Soc. Cosmet. Chem.,* 30, 157, 1979.
13. **Sved, S., McLean, W. M., and McGilveray, I. J.,** Influence of the method of application on pharmacokinetics of nitroglycerin from ointment in humans, *J. Pharm. Sci.,* 70, 1368, 1981.
14. **Wester, R. C., Noonan, P. K., and Maibach, H. I.,** Frequency of application on percutaneous absorption of hydrocortisone, *Arch. Dermatol. Res.,* 113, 620, 1977b.
15. **Wester, R. C., Noonan, P. K., and Maibach, H. I.,** Variations in percutaneous absorption of testosterone in the rhesus monkey due to anatomic site of application and frequency of application, *Arch. Dermatol. Res.,* 267, 299, 1980a.
16. **Wester, R. C., Noonan, P. K., and Maibach, H. I.,** Percutaneous absorption of hydrocortisone increases with long-term administration, *Arch. Dermatol. Res.,* 116, 186, 1980b.
17. **Wester, R. C., Maibach, H. I., Bucks, D. A. W., and Guy, R. H.,** Malathion percutaneous absorption following repeated administration to man, *Toxicol. Appl. Pharmacol.,* 68, 116, 1983.
18. **Bucks, D. A. W., Maibach, H. I., and Guy, R. H.,** *In vivo* percutaneous absorption: effect of repeated application versus single dose, in *Percutaneous Absorption,* Bronaugh, R. and Maibach, H., Eds., Marcel Dekker, New York, 1989, chap. 36.
19. **Wilson, J. S. and Holland, L. M.,** The effect of application frequency on epidermal carcinogenesis assays, *Toxicology,* 24, 45, 1982.
19a. **Wester, R. C. and Maibach, H. I.,** *In vivo* percutaneous absorption, in *Dermatotoxicology,* 2nd ed., Marzulli, F. and Maibach, H. I., Hemisphere, Washington, D.C., 1982, 131.
20. **Wester, R. C. and Maibach, H. I.,** Cutaneous pharmacokinetics: 10 steps to percutaneous absorption, *Drug Metab. Rev.,* 14, 169, 1983.

Chapter 16

REPRODUCTIVE AND DEVELOPMENTAL TOXICITY STUDIES BY CUTANEOUS ADMINISTRATION

Rochelle W. Tyl, Raymond G. York, and James L. Schardein

TABLE OF CONTENTS

0-8493-7357-3/93/$0.00 + $.50
© 1993 by CRC Press, Inc.

I. INTRODUCTION

Some chemicals and drugs may be absorbed through the skin and produce a toxic effect. In the pregnant animal, once the agent is in the maternal circulation, an adverse effect on the conceptus is possible. Additionally, the ramifications of dermal absorption of a toxicant on the reproductive capability of the individual are also of concern.

Reproductive and developmental toxicity safety evaluation of agents has become relatively standardized over the past 25 years. The procedures set forth by the 1966 U.S. Food and Drug Agency Guidelines for Reproductive Studies[1] have essentially been followed by the U.S. Environmental Protection Agency,[2] Japan,[3] Canada,[4] and Great Britain.[5] In these guidelines, experimental animal studies are designed to produce data that can be extrapolated to predict human risk. The route of exposure for animal testing is usually required to be the expected route of human exposure. If exposure is anticipated to be by multiple routes, then the oral route is preferred for testing. However, for some agents, the dermal route is expected to be the sole or major route of human exposure, and in this case, conducting the toxicity evaluation by the dermal route would be the most relevant.

A multitude of factors must be considered in developing reproductive and developmental toxicity test procedures by dermal exposure. Many of these factors relate to the general issues of conducting toxicity studies using cutaneous administration, e.g., skin irritation affecting compound administration and absorption, the rate and extent of dermal penetration (absorption and systemic exposure) vs. absorption and exposure by other routes, and the role of first pass cutaneous biotransformation. Other factors are unique to the reproductive system, involving processes which occur during mating, pregnancy, and development of the organism during gestation and the peri- and postnatal period. Experimental research to determine the effects of topically applied compounds is imperative for the protection of public health.

II. TESTING PROCEDURES

A. GENERAL CONSIDERATIONS

Increasing interest in and concern for the cutaneous route of exposure, either intentionally or inadvertently, cross all categories of agents from pharmaceuticals via topical applications or transdermal delivery systems, to pesticides and industrial chemicals via skin contamination on workers or end-use consumers. This concern for evaluating risks from cutaneous administration is reflected in recent considerations by governmental agencies to accept and interpret studies employing this route in the context of existing testing guidelines for reproduction and developmental toxicity evaluations, i.e., U.S. FDA,[1] U.S. EPA TSCA (Toxic Substances Control Act),[6] U.S. EPA FIFRA (Federal Insecticide, Fungicide and Rodenticide Act),[2] OECD (Organization for Economic Cooperation and Development),[7] U.S. EPA Guidelines for the Health Assessment of Suspect Developmental Toxicants,[8] and the most recently proposed amendments to the U.S. EPA Guidelines of Suspect Developmental Toxicants[9] and Reproductive Toxicants.[10,11]

In recognition of this interest and concern, the Office of Research and Development (ORD) and the Office of Pesticides and Toxic Substances (OTS) of the U.S. EPA convened a workshop on the Acceptability and Interpretation of Dermal Developmental Toxicity Studies.[12,13] Current scientific literature was evaluated and research areas identified to support risk assessment for developmental toxicity studies by this route. Three major issues identified were maternal toxicity, pharmacokinetics, and study designs, all of which impact on the conduct, conclusions, and usefulness of studies to be performed by the dermal route.

1. Maternal Toxicity

The current testing guidelines for developmental toxicity evaluations (cited above) indicate gavage administration as the preferred route and require that the top dose result in maternal toxicity defined as significant reduction in weight gain, clinical signs of toxicity, and up to 10% mortality; i.e., systemic toxicity.[14] Skin irritation at the application site, a local reaction to some dermally applied test materials, is viewed as inappropriate as an indication of maternal toxicity.

The requirement for indications of systemic toxicity was confirmed by the Workshop. Maternal systemic toxicity may be more difficult to obtain in dermal studies because of technical limitations: limited absorption (rate and extent) of the test material, possibility of skin irritation, and limitations on useable surface area and therefore the amount of test material (and volume of material and/or vehicle) which can be applied.

a. Skin Irritation Not a Valid Sign of Toxicity

The presence and extent of skin irritation was viewed as useful only to determine whether the dermal route of exposure is feasible or appropriate. The consensus was that the top dose should produce at most moderate erythema and/or moderate edema (area raised approximately 1 mm); marked irritation and/or necrosis was considered unnecessary and excessive.

The presence of marked irritation per se at the dosing site may result in indications of systemic toxicity, e.g., reductions in body weight gain, and/or reductions in food consumption, which would be difficult to distinguish from systemic toxicity due to systemic exposure. Damage to the skin could also affect absorption, although the consequences of a compromised epidermal barrier, increased or decreased absorption, appear to be chemical specific.[15-18] Adaptation to the presence of an irritating agent on the skin, i.e., so-called "skin hardening," may also result in increased or decreased skin penetration.[19] The presence or absence of skin irritation at the dosing site was deemed inappropriate for establishing a "no observable adverse effect level" (NOAEL). Given the real concern for dealing with

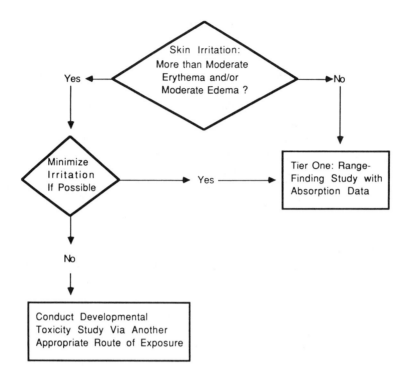

FIGURE 1. Decision matrix for dermal developmental toxicity studies based on properties of skin irritation. (From Kimmel, C. A. and Francis, E. Z., *Fundam. Appl. Toxicol.*, 14, 386, 1990. With permission.)

localized skin irritation, a decision matrix was developed by the workshop (Figure 1). If the chemical, when administered "neat" (undiluted), does not result in erythema and/or edema beyond the recommended upper limit ("moderate," Draize score 3),[20] a range-finding study can be performed with absorption data (see below). If the undiluted test chemical results in unacceptable irritation, i.e., Draize score above 3, attempts should be made to minimize irritation, e.g., dilution with an appropriate solvent, reduction of dosing volume, increased application surface area, rotation of application site. If irritation can be reduced, then the range-finding study can be conducted using dermal application. If irritation cannot be reduced, then the dermal route is not appropriate and another route should be chosen, e.g., subcutaneous or intramuscular injection. In the last case, if dermal exposure is a (or the) major route of potential or actual human exposure, then comparative pharmacokinetic studies using alternative routes can provide a basis for route-to-route extrapolation in risk assessment. If skin irritation was the limiting factor in experimental animals, it is reasonable to assume that it may be in humans as well, and the individuals would remove themselves from the exposure. One caveat in this instance is that human absorption may differ from test animal absorption.

Limitations to how much test chemical can be administered relative to the production of maternal toxicity are also encountered in studies by other routes of exposure, e.g., vapor pressure, solubility or palatability. In reality, the problem of limitations is not unique to studies employing the dermal route of administration and can be resolved by modification of criteria for maternal toxicity in studies which must conform to current testing guidelines.

One of the major potential confounders in the developmental toxicity study is that any developmental toxicity observed may be due to the systemic maternal toxicity per se, which is required at the top dose by current governmental testing guidelines. That is, the adverse embryo/fetal findings may be due, at least in large part, to the compromised physiological status of the maternal animal and not due to any direct action of the test agent, if the developmental and maternal effects are observed at the same dose(s). A recent discussion[8] suggests that developmental toxicity produced at maternally toxic doses should not be ignored if the developmental effects occur as fairly specific types or syndromes and/or if human exposures may occur at toxic levels (e.g., smoking, alcohol, solvents, etc.).

b. *"Stress" Factors*

The physiological bases for maternal toxicity are expected to vary widely in an agent-specific manner; however, one underlying factor that is consistent for most induced toxicity is the presence of generalized stress in the affected animals. Maternal stress and its effects on developmental outcome are directly relevant to studies using dermal administration (because of restraint stress from the dosing procedures per se) and have engendered much interest and research initiatives.

In rats, gestational restraint stress has been associated with increased fetal loss,[21] increased perinatal mortality and increased sexual capacity in male offspring,[22,23] feminized sexual behavior in males,[24] reduced fertility and fecundity in females,[25] delayed growth and development in offspring,[26] and reduced size of the sexually dimorphic nucleus of the preoptic area of the brain in male but not female offspring.[27] In mice, reported effects on reproductive and developmental events of early pregnancy resultant from restraint are more pronounced. In addition to fetal malformation,[28,29] increased embryofetal lethality[30] and increased incidence of supernumerary ribs[30,31] have been reported. Further, reduced pregnancy rate and litter size, increased number of abnormal corpora lutea, decreased serum progesterone levels, fewer implantation sites,[32] and elevated plasma corticosterone[33] have been reported following restraint in this species. In mice, cleft palate produced by restraint or other stress-provoking procedures appears to be mediated through elevated endogenous corticosteroid levels in the dam.[34,35]

Stress has been implicated in the production of human adverse developmental outcomes in general[36] and for cleft palate in children.[37]

It is clear that the induction of stress confounds treatment-caused maternal toxicity and is especially critical in the stress-provoking procedures of restraint involved in studies by the dermal route of administration. It is also clear that the brain via noradrenergic pathways, the hypothalamus, the anterior and posterior pituitary, and the adrenal gland are involved in the "fight-or-flight" response to stress and can affect pregnancy (both in the pre- and postimplantation phases) and postnatal growth and development of the offspring in mammalian species. In fact, one suggestion is that embryonic stress (from whatever situation in the dam) may be a common pathway by which diverse environmental agents result in developmental abnormalities, with the observed specificity of the response dependent on the gestational time of insult.[38]

2. Pharmacokinetics

Penetration or absorption of the test material across the intact epidermal barrier (stratum corneum) is the first and most critical step in systemic exposure via the dermal route. The dermal route shares with other parenteral routes, e.g., inhalation and injection (subcutaneous, intramuscular, intravenous but not intraperitoneal), the lack of first-pass access to the liver, major site of biotransformation in the body. Whereas exposure by non-parenteral routes (e.g., gavage, dosed water or dosed feed) results in absorption of the test material and transport of the parent material and/or metabolite(s) directly to the liver via the hepatic portal system, parenteral routes involve initial venous entry dependent on the route and direct access to the heart and systemic circulation. Once the material is in the systemic circulation, distribution, deposition, transformation, relocation, and excretion are essentially the same as by any other route of administration.

The rate (amount per unit time) and the total amount absorbed via the skin depend on the physicochemical characteristics of the test material. It is generally viewed that absorption through the skin is relatively slow and limited (although the rate and total amount of absorption is clearly chemical specific). Pharmacokinetic data are absolutely essential to determine the rate and total amount of material absorbed, the nature of the material absorbed (including metabolite identification), and the subsequent distribution, biotransformation, and excretion. Important information necessary to interpret study results includes the presence and extent of accumulation with repeated exposure, and the movement from the central compartment to other locations, and, in a pregnant animal, evaluation of transport across the placenta and into the conceptus. Pharmacokinetic data in a non-pregnant animal may not reflect the situation in a pregnant animal with profound physiological changes in many systems, including the cardiovascular system, due to the pregnancy itself.[39] Pharmacokinetic data in a pregnant animal exposed by another route of administration may be useful to select sampling times, duration of sampling, target organs, and metabolites of interest. The suggested pharmacokinetic data in pregnant animals in dermal studies (*vide infra*) would help resolve uncertain results. For example, if a developmental toxicity study is performed via the dermal route in a test animal and there is no demonstrable developmental toxicity, is this finding due to lack of systemic absorption by the dam, transformation of the parent material into innocuous metabolites, or lack of placental transport? If it is not known how the human handles the material as well as how the human compares to the test animal with regard to pharmacokinetic handling of the test material, then the study provides little useful information; in fact, there is the real risk that we think we know more than we do. In the case of a chemical applied dermally which does result in developmental toxicity, the problem is still the same: how are the results to be interpreted? In recognition of the need for pharmacokinetic data, especially in dermal studies where absorption may be questionable, a two-tiered approach (Figure 2) to developmental toxicity testing via the dermal route has been developed.[13]

Tier One was designed to determine if there was absorption and if there was accumulation of the test material during a dose range-finding developmental toxicity study. An additional group of animals (along with the usual dose groups and control group) would be dosed throughout organogenesis. Periodically during the dosing period, beginning at least 6 h after dermal application, blood samples would be repeatedly taken, at least at the high dose. If necessary, the dose on day of sampling would be radiolabeled (although this would mean that only radioactivity would be measured, not identity of the material). Presence of the material or radiolabel in the blood samples would be evidence of absorption. Comparison of data for the first and subsequent days of dosing as well as the slope of the blood clearance curve (and calculation of clearance half-life or half-lives) would determine if accumulation had occurred. The rate of accumulation and the determination if a steady state has been attained and, if so, after how many dosing days, are critical to subsequent study design.

One major aspect of study designs for developmental toxicity assessments by the dermal route is the time of attainment of steady-state blood levels in the dam. The concern is that if dosing was to begin on gestational day 6 or 7 (as required in current testing guidelines depending on the species and relevant guidelines), very slow systemic absorption from the site may result in less than maximal blood levels during most of major organogenesis. Therefore, the perceived risk is that the conceptus will not be exposed to maximal levels of the test chemical and/or metabolites during organogenesis and the response observed, little or no developmental toxicity, may not represent the true risk relative to human exposure and pregnancy outcome. If slow attainment of steady-state blood levels is observed, one

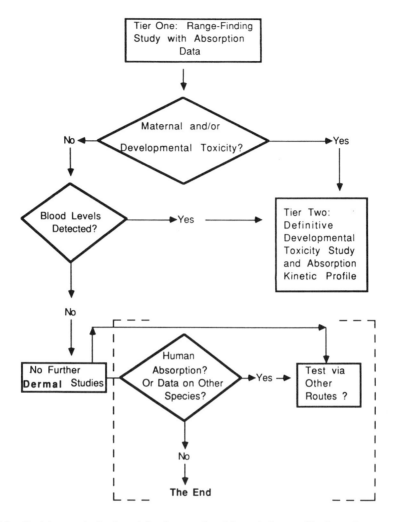

FIGURE 2. Decision matrix for dermal developmental toxicity and pharmacokinetic studies. (From Kimmel, C. A. and Francis, E. Z., *Fundam. Appl. Toxicol.*, 14, 386, 1990. With permission.)

design option would be to begin exposures earlier in the gestational period or even before conception to provide for maximum maternal blood levels during major organogenesis. One risk to this dosing scenario is that if the presence of test agent results in induction of maternal metabolizing enzymes, the implanted conceptus may not be exposed to the most appropriate chemical moiety(ies). For example, if the initial test agent is teratogenic and the metabolites are not, an apparently negative study will result if dosing were initiated early vs. a positive study if dosing were initiated later. If the initial test agent was not teratogenic and the metabolites were, the reverse outcome would result. If the agent and/or metabolites interfered with implantation, few or no conceptuses would be available for evaluation. These latter two scenarios, induction of metabolizing enzymes and possible interference with implantation, are the basis for the current testing guideline requirement for beginning the dosing after implantation is essentially completed.

Since the physiological state of the dam and the structural and functional complexity of the placenta and conceptus change during the pregnancy, the multiple sampling days in the Tier One assessment would provide additional preliminary information on what effect(s) these changes had on absorption and accumulation. Three possible outcomes are envisioned from the Tier One testing. If there was maternal and/or developmental toxicity in the range-finding study, with or without evidence of dermal absorption (although the presence of systemic toxicity is presumptive evidence for absorption), then the definitive developmental toxicity study could and should be performed via the dermal route (Tier Two) with additional pharmacokinetic data collection. If there was evidence of absorption, even if there were no evidence for maternal and/or developmental toxicity, then the definitive study could be performed (Tier Two). If there was no evidence of absorption and no maternal and/or developmental toxicity, then the dermal route of administration was considered inappropriate for subsequent testing. In the last case, subsequent testing would depend on whether there was potential for human exposure via the dermal route. If there were data that human dermal absorption did not occur, then no further dermal studies would be necessary. If there were no human data and there was concern about potential human dermal exposure, a number of options were suggested by the workshop: other species might be employed with exposure by the dermal route; other routes might be employed, subcutaneous or intramuscular to approximate the dermal route of systemic access (to avoid first pass metabolism by the liver); and evaluation of dermal absorption in humans (*in vivo* or *in vitro*).

The Tier Two pharmacokinetic data to be collected from satellite animals include blood sampling at mid-organogenesis after single or repeated dosing at a dose which resulted in maternal and/or developmental toxicity and at a dose which was the NOAEL, with multiple sampling times. The dermal absorption profile of blood-borne chemical would be followed by intravenous dosing to establish comparisons for bioavailability, systemic entry, and clearance. Tissues examined would include, at a minimum, blood, urine, feces, carcass, and the dermal application site. Although not required in the Tier Two scenario, examination of the uterus, placentas, and conceptuses would provide information on the transport of the test material and/or metabolites to the conceptuses. If the conceptuses were large enough, examination of various fetal organs may provide evidence of accumulation and possible target site dosimetry.

3. Selection of Appropriate Test Animal

The selection of an appropriate test animal species for developmental toxicity testing via dermal administration is critical to the usefulness of the data ultimately produced. The usual test animal species for developmental toxicity studies are rats or mice and rabbits since large databases exist and they satisfy the governmental testing requirements. Albino rabbits are the most sensitive species for skin irritation and therefore are useful for assessment of the potential of a test agent for causing skin irritation. Paradoxically, however, by virtue of their sensitivity to skin irritation, they may be inappropriate for dermal developmental toxicity studies since establishment of an upper limit for skin irritation is a major criterion for use of the route in such studies.[13] Because the rabbit is larger than rodents, the site of application could be rotated among several sites to minimize irritation, but the larger size and limits to dosing volume also mean that the dose in terms of milligram per kilogram may be less than for a smaller test species (*vide infra*). The sensitivity of the rabbit to skin irritation may not parallel its sensitivity to developmental toxicants. If other studies are available which employed different routes of administration, the most sensitive species in those studies would be the species of choice for the dermal study. Ideally, a test species should mimic the human in terms of physiology, pharmacology, and handling of the test agent. Pharmacokinetic data on the handling of the test agent by the test animal is minimal at best, and selection of test animal is usually based on other considerations, e.g., cost, availability, and ease of handling.[13] Alternate species may need to be identified and characterized to be used in dermal developmental toxicity studies.

TABLE 1
Approximate Dermal Dosing Parameters In Animals

Species (body weight) (g)	Dimensions of application site	Volume administered (per animal)	Dose
Mouse (30)	1″ × 1″	0.1 ml	3 ml/kg
Rat (300)	2″ × 2″ or 3″ × 3″	0.5–0.6 ml	1.5–1.8 ml/kg
Rabbit (3000)	3″ × 3″ or 4″ × 4″	1–2 ml	0.3–0.6 ml/kg

4. Technical Factors
a. Dosing Considerations

The maximum skin area in experimental animals to which a test chemical is applied is approximately 10% of the total body surface area. This translates into dosing areas, volume, and dose of test agents in the various species as shown in Table 1. The dosing areas are shaved or clipped closely prior to dermal application, and shaving is repeated as needed over the duration of the treatment period.

One concern for limitations on the dosing volume in test animals is whether the suggested volumes would provide an adequate margin of exposure for human risk assessment by governmental agencies. For exposure assessments by OTS,[13] worker exposure, assuming both bare hands heavily contaminated, was estimated at 1300 to 3900 mg of a material (19 to 56 mg/kg for a 70 kg human) per contact per day for intermittent exposure and 6500 to 18,200 mg (93 to 250 mg/kg) for an 8-h workday for routine continuous exposures. Animal testing at 3 g/kg/day (e.g., in rats) would provide a margin of exposure of only approximately 12 (worse case routine continuous exposures) to 54 (worst case intermittent exposure).

b. Occlusion or Non-Occlusion

The decision as to whether to occlude or not occlude the application site has been discussed elsewhere in this volume. There are three major advantages to occlusion:

1. Occlusion reduces the loss of the test material by volatilization; a corollary to this is that the test animal with an occluded site has little or no inadvertent exposure by inhalation of the volatilized material.
2. Occlusion prevents the test animal from gaining access to the application site by grooming. Grooming removes test material from the prepared site, diminishing cutaneous exposure and increasing the risk of contamination of the paws and inadvertent ingestion of the material, which may confound results.
3. Occlusion maximizes cutaneous exposure by retaining the material on the application site; it also may maximize and/or accelerate absorption through the skin by increasing the temperature and humidity at the site.

The only comparison of the reproductive effects of a test material, in this case, ethylene glycol monomethyl ether (EGME) administered dermally to rats with or without occlusion of which we are aware, was performed by Feuston et al.[40] The adverse reproductive effects were more severe and recovery was slower in rats which were occluded than in animals that were not, even at lower doses. Although this was not a developmental toxicity study, it clearly illustrates the utility of occlusion to maximize results; i.e., provide a "worst case scenario" for potential risks of exposure by the dermal route.

There are two major disadvantages inherent in snugly wrapping a pregnant/nursing animal around its circumference in the thoracic and upper abdominal area of the trunk. The wrapping and the restraint (with attendant stress) may impact on the physiological status of the dam as well as on the normal growth and development of the conceptuses.

The EPA workshop on dermal developmental toxicity studies[13] did not endorse occlusion or non-occlusion. The participants did indicate that ingestion or inhalation exposure to a dermally applied chemical should be minimized, and that some form of "containment" would be necessary in the case of a volatile test agent. It was suggested that if restraint were required, the type of restraint should minimize stress as much as possible and the animals should be allowed some period of adjustment to the restraining device prior to the initiation of exposures. No specific technique for restraint was recommended, only the suggestion made that a laboratory select the method deemed appropriate with evaluation of the effect of restraint per se to be determined prior to general use in that laboratory.

Three useful methods of occlusion for pregnant animals are illustrated in Figure 3. Tape occlusion with a collar (Figure 3A) can be used for both rabbits and rats. The "Elizabethan" collar is secured around the animal's neck to prevent it from reaching the dosing site; it should be assured that the collar does not preclude the animal from reaching food and water. The dosing site is occluded by sterile gauze and the gauze is held in place by elastic adhesive tape. The tape, collar, and occlusion are carefully removed to prevent skin injury at the end of each dosing period and replaced at the next dosing time. This technique has been successfully employed in a rabbit dermal study[41,42] with a concurrent vehicle control group (also collared and taped) and an untreated control group which was evaluated similarly to the other groups. The only maternal effects observed in the vehicle control group vs. the untreated control group was a transient reduction in weight gain for the first half of the dosing period (gd 6 to 12) and an increase in weight gain for the post-treatment period. No persistent effects of stress could be determined in the dose and there were no effects of stress on the conceptuses. The collar may also be used, in the absence of occlusion, to prevent access to the dosing site in a non-occluded dermal study design.[43]

A comparison of no restraint, collaring only, or collaring plus wrapping was made on the teratogenicity of dermally applied ethylene thiourea (ETU) or ethylene glycol monoethyl ether (EGEE) in rats.[44] When maternal grooming and presumably ingestion were prevented by collaring or collaring and wrapping, no teratogenicity was observed for dermal ETU exposure; in the absence of wrapping and/or collaring, ETU resulted in developmental toxicity including teratogenicity. The implication is that the teratogenicity of ETU applied dermally without occlusion may be due to ingestional exposure and not to true dermal exposure. EGEE resulted in teratogenicity in the presence of collaring and wrapping. This study included a sham control group (collared and wrapped, no material applied) and a vehicle control group (collared and wrapped with the DMSO vehicle applied) but no untreated control group so the effects of the procedures per se could not be evaluated for all parameters. However, the authors did wrap and/or collar dams on gestational day 5 and began application of the test materials on day 6; a slight reduction in body weight was observed in the dams on day 6 which were collared or collared and wrapped relative to their weights on day 5 (which was not observed in the group which received neither collar nor wrapping).

The second method illustrated (Figure 3B) involves use of a jacket of Lycra® Spandex® (with the elasticity going around the animal). The jacket is cut high in front to prevent effects on the expanding lower thoracic and abdominal areas, with reinforced openings for the forelegs. On the underside of one of the back flaps of the jacket is a Velcro® strip. On the other back flap is a square of plastic sheet (to overlay the dosing site) on the underside and the complementary Velcro® strip on the outside. The animal is placed with its forelegs in the "armholes," the test agent is applied, sterile gauze is placed over the site, if appropriate,

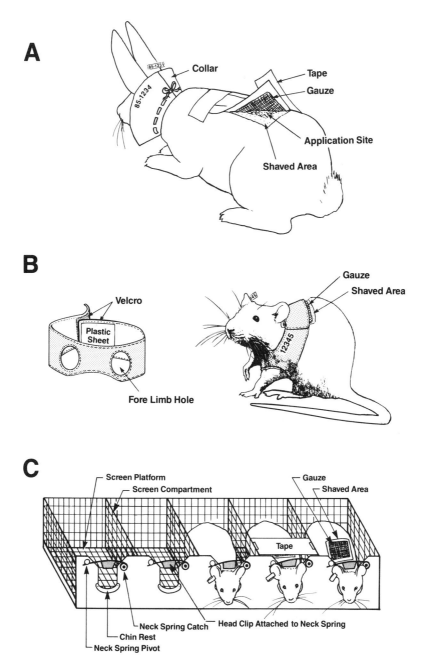

FIGURE 3. Methods of occlusion for pregnant animals. A. Collar and tape (rabbit). (From Tyl, R. W., et al., *J. Toxicol.-Cut. Ocul. Toxicol.*, 5, 263, 1986. With permission.) B. Jacket (rat). C. Restraint (mouse).

and the jacket flaps are brought up over the site and snugly attached by the Velcro® strips. In most cases, a collar is not needed to restrict access to the jacketed site and the animal has unrestricted movement and access to food and water. The jackets may be sized to different rat strains and to rabbits. The jacket was employed in a Fischer 344 rat dermal study[45] with no apparent differences in any maternal, gestational, or fetal measures in the vehicle control group (jacketed) relative to historical control data in that laboratory.

Dermal developmental toxicity studies in mice present rather unique problems in that the mice are too small to wrap, tape, collar, or jacket. A method developed by Tyl et al.[46] is illustrated in Figure 3C. A stainless steel mesh tray was compartmentalized by a stainless steel "fence" between each animal slot. Each compartment has an opening at one end for the animal's head with a chin rest and a neck spring which fits just behind the ears. Once restrained, the mouse is shaved or clipped, the test agent is applied and sterile gauze is placed over the site. Adhesive tape is secured across the dosing site by a metal clip on each "fence" — a single tape strip can be used to cover the sites of all mice in the tray with clips securing each compartment. This procedure was employed in a dermal developmental toxicity study with ethylene glycol in CD-1 mice with a vehicle control and positive gavage control groups.[46] The positive gavage control group exhibited maternal and developmental toxicity including teratogenicity as expected; the mice receiving EG dermally exhibited only an increase in incidence of two skeletal variations; the vehicle control group (restrained) exhibited no untoward maternal or developmental effects.

B. SPECIAL CONSIDERATIONS

The use of the dermal route of exposure in studies which require animals to interact, e.g., reproductive toxicity studies where males and females are cohabited or studies with a postnatal component when the dam is with her litter, creates special concerns. If the animal is being exposed during the time of interaction, one major risk is that the other inhabitant(s) of the cage could groom the application site, thereby removing test material for the dosed animal as well as ingesting the material. The obvious solution is to isolate the dosed animal during the time the material is present on the dosed animal's skin, and return the animal to its home cage and cagemate(s) after exposure is over for the day and the site is cleaned off.

However, even with isolation of the dosed animal to preclude inadvertent exposure to other animals, complications remain. Successful mating (in an FDA Segment I Study or in FDA or EPA multigeneration studies) requires appropriate sensory cues (e.g., smell, tactile, etc.), a cycling female and mounting and intromission, any of which may be compromised by a stressed, shaved animal. Impact of the use of the dermal route of exposure on the lactational period (in an FDA Segment I or III study, in multigeneration studies or in developmental neurotoxicity studies) may include effects on maternal-neonatal interactions including the time spent nursing, the volume and quality of the milk (the pups would be separated from the dam during the time of maternal dermal exposure), as well as other subtler effects from the presence of a shaved dam and possible effects on normal growth and development of the pups, including acquisition of developmental landmarks and/or more complex behaviors (e.g., motor activity, auditory startle, learning, and memory). Again, this concern is not limited to studies involving the dermal route of exposure. Reproductive toxicity studies which employ the inhalation route of exposure also isolate the dam from her litter for 6 to 8 h daily to expose her; results indicate only a slightly more shallow weight gain curve for pups which is shared by the control group (in which the dam is also isolated) in studies requiring removal of the dam.

C. STUDY DESIGN

Additions and/or changes to the current governmental developmental toxicity testing guidelines when the route of administration is by dermal application include:

1. Possible initiation of dosing prior to the time of implantation if there is evidence for relatively slow systemic accumulation in the dam, to ensure the conceptuses are exposed to maximal amounts of the test material and/or metabolites during major organogenesis

2. Examination of the application site twice daily (prior to and subsequent to the daily time of dermal exposure) to evaluate the skin for any evidence of edema and/or erythema (Draize scoring[20])
3. Satellite groups (at an effect level and at no effect level) for pharmacokinetic evaluations including absorption, accumulation, distribution, and excretion (*vide supra*)
4. At necropsy, retention of the area of the cutaneous dosing site for possible subsequent histologic examination
5. Employing two control groups: (a) a vehicle control group with the vehicle or a proven innocuous surrogate for the test material if the test material is dosed "neat" (undiluted), and whatever restraints are employed for the other groups, and (b) an untreated control group which is not shaved, restrained or dosed to ascertain the effects, if any, of the treatment procedures per se. This latter group would not be necessary for every study but would need to be employed in the testing laboratory in the first study or studies for each species by this route.

III. RESULTS OF ANIMAL TESTING

A. DRUGS
1. Developmental Toxicity
a. Corticosteroids

Several corticosteroids have been investigated for potential developmental toxicity following cutaneous exposure.

Cleft palate in rabbit fetuses was reported following administration of 0.625 mg/kg betamethasone benzoate on gestation days 7 to 18.[47] Oral administration of lower doses in this species was embryotoxic, but not teratogenic.[48]

Cutaneous treatment with dexamethasone produced no developmental toxicity in rabbits from application of doses up to 0.1 mg/kg in gestation.[49] Dexamethasone is a known developmental toxicant in animals by intraperitoneal, intramuscular, ocular, intraamniotic, and oral administration.[50] Dosage and/or lack of absorption may be factors in this discrepancy.

Another corticosteroid, diflorasone, induced malformations in rabbits when administered percutaneously at 0.016 mg/kg during pregnancy.[51]

The drug medrysone demonstrated teratogenicity in rats from topical application; no other details were provided.[52]

Triamcinolone acetonide administered topically to rats at 0.03 ml of 0.1% drug on gestation days 6 to 15 produced decreased body weight gain; abortion and death occurred in single animals, but skin irritation was minimal.[53] Litters were characterized by increased embryolethality, decreased fetal weight, and malformed fetuses with cleft palate and omphalocele.

Topical clobetasone 17-butyrate administration to rats at doses of 0.05 to 0.5% on gestation days 7 to 17 resulted in maternal toxicity and fetal skeletal abnormalities, as well as adrenal and thymus pathology.[54]

b. Antibacterials

Alcide allay, when administered topically at doses from 0.32 to 2 g/kg on gestation days 6 to 18 to rabbits, did not induce malformations, but reduced fetal body weight and retarded skeletal ossification.[55]

Alkylbenzenesulfonate sodium applied to the skin at a dose of approximately 500 mg/kg on gestation days 1 to 13 in the mouse induced multiple malformations, including cleft palate and musculoskeletal defects.[56]

Several antibacterial, antifungal agents in shampoos and other commercial products have been examined for toxicity in animals following dermal administration; only one has been developmentally toxic under the experimental conditions utilized. Dipyrithione (omadine disulfide) caused maternal, but not developmental, toxicity in rats and rabbits at doses of 30 and 5 mg/kg/d, respectively, when applied during organogenesis.[57] In the pig, however, dipyrithione induced tail defects when applied dermally on days 8 to 32 of gestation at doses in the range of 10 to 300 mg/kg.[58] Sodium pyrithione (sodium omadine) elicited maternal toxicity without developmental toxicity in the rat at a dermal dose of 7 mg/kg applied on gestation days 6 to 15.[59] Studies with zinc pyrithione in three species did not produce either maternal or developmental toxicity. No adverse effects have been reported in rats when 30 mg/kg was applied during organogenesis,[60] or in rabbits at doses as high as 2.5 g/kg/d.[61] In contrast to the results in the pig from dipyrithione, zinc pyrithione did not induce malformations or developmental toxicity at dermal doses as great as 400 mg/kg/d on gestation days 8 to 32.[62]

Formaldehyde applied topically to hamsters on gestation days 8 to 11 increased resorptions.[63] This chemical has similar toxicity properties in other species by other routes of exposure.[50]

An antibacterial agent and solvent, 2-phenoxyethanol, applied topically at a dose of 600 mg/kg and higher on gestational days 6 to 18 was maternally toxic in rabbits. No developmental toxicity was observed at doses up to 1000 mg/kg.[64]

A standard antibacterial soap solution applied percutaneously to mice, rats, and rabbits in up to 30% concentration during the first 2 weeks of gestation had no potential for toxicity of any kind.[65]

Similarly, 0.5 ml of a 15 or 20% solution of kitchen detergent applied on gestation days 1 to 13 to clipped skin of mice caused no developmental toxicity.[66]

An antibacterial mixture composed of triclocarban and 3-trifluoromethyl-4,4'-dichlorocarbanilide was reported to have no developmental toxicity potential in either rats or rabbits when administered topically (or orally) during organogenesis at dosages at dosages up to 1000 mg/kg.[67]

c. Retinoic Acid/Analogs

Vitamin A acid or retinoic acid (tretinoin) was applied topically to 10% of the body surface area of the rabbit in 0.1% concentration on gestation days 6 to 18; no developmental toxicity was reported.[68] Tretinoin was administered twice on gestational day 12, approximate dose 10 mg/kg in rats; "some retinoid-specific patterns of anomalies" were observed.[69,70] The drug is developmentally toxic and teratogenic in at least five laboratory species by the oral route.[50] Neither all-*trans* retinoic acid or etretinate at low doses were teratogenic in hamsters when administered dermally, although hyperkeratosis was observed. The arotinoid Ro13-6298 elicited dose-dependent terata and embryolethality in the same species by dermal application;[71,72] the differences were attributed to the skin toxicity and attenuated maternal blood levels that limited the quantity of retinoids that reach the embryo.

A vitamin A analog known as RWJ 20257 was teratogenic and also increased developmental variations in the rat when given as 0.1 or 0.25 mg/kg topically during organogenesis.[73]

d. Androgens

Several androgenic hormones have been tested for developmental effects by the topical route. Methyltesterone applied to rats throughout pregnancy increased resorption and disrupted nidation at 10 mg/kg and reduced fetal weights at 1 mg/kg.[74] Doses of 0.05 and 0.1 mg/kg dehydroepiandrosterone applied topically during gestation to rats resulted in functional changes and slight, short-lived circulatory disturbances.[75]

e. Miscellaneous Drugs

A cutaneous study reported in the hamster with methyl salicylate indicated teratogenicity manifested as neural tube defects.[76] The regimen producing the effect was 175 to 525 mg/100 g given by this or the oral route on gestation day 7. The drug was not teratogenic dermally in the rat at doses as high as 6 g/kg.[77]

The widely used vasodilator drug nitroglycerin, administered to rats and rabbits topically at doses up to 4 g/kg during organogenesis, caused no developmental or maternal toxicity.[78,79]

Crufomate, used as a veterinary antihelmintic, has no reported developmentally toxic effects from therapeutic dose levels in cattle,[80] or in rats.[81]

Sesame oil, a chemical used as an additive in food and drug formulations, had no developmental toxicity potential when applied dermally or subcutaneously at dosages in the range of 20 to 1000 mg/kg on gestational days 6 to 18 in the rabbit.[82]

A surfactant used in food and drug formulations, linear alkylbenzene sulfonate (LAS), was tested experimentally for percutaneous toxicity.[65] At concentrations in the range of 0.03 to 3.0%, there was no reported developmental toxicity in mice, rats, or rabbits when LAS was applied over most of gestation. Studies in the rat and mouse were corroborated by other investigators.[83,84] Oral studies in rats and rabbits with LAS had similar negative results,[85] but oral administration of 300 or 600 mg/kg/d on gestation days 6 to 15 in the mouse by the oral route resulted in the production of skeletal anomalies.[85]

Phenethyl alcohol, a chemical with flavoring and antibacterial properties used in drug and food processing, was developmentally toxic in rats by the cutaneous route. At a dosage of 1.4 ml/kg applied on gestation days 6 to 15, terata were observed in all resulting fetuses, along with maternal toxicity.[86]

An experimental antifungal drug known as 710674-S had no developmental toxicity potential in the rabbit when applied dermally during organogenesis at doses up to 20 mg/kg.[87]

The antiviral drug adenine arabinoside (vidarabine) induced the full spectrum of developmental toxicity when applied topically to the pregnant rabbit. Dosages on the order of 200 or 250 mg/kg administered to approximately 5 or 10% of the body surface area induced polymorphic malformations in up to 70% incidence of the resulting fetuses; the fetuses were also reduced in size and viable litters were decreased.[88] The drug was also developmentally toxic by the intramuscular route in several laboratory species.

p-Hydroxyanisole, used in cosmetic formulations, was developmentally toxic in the rat when applied either as a 5% cream or 25% emulsion throughout gestation.[89] Preimplantation mortality was increased, and the fetuses showed subcutaneous hemorrhages.

Nonoxynol 9, a dispersal agent in spermicidal compositions, was given either epicutaneously or orally to rats at 50 to 500 mg/kg.[90] Treatment on gestation days 6 to 15 or 1 to 20 caused maternal toxicity, extra rib formation, and dilated renal pelvices at high doses by either route.

A series of 12 hair dye formulations, designated only by code numbers, was given topically to rats at a standardized dosage of 2 ml/kg on 7 d of gestation; none was developmentally (or maternally) toxic.[91]

2. Reproductive Toxicity

The veterinary antihelmintic drug crufomate is one of only two drugs of which we are aware that have been examined in fertility studies by the cutaneous route. Mice were administered 50 or 100 mg/kg 21, 35, 56, or 70 days prior to mating; no adverse reproductive effects were reported by this regimen.[92]

The other study was conducted on omadine MDS in the rat in the typical U.S. Segment I regimen of treatment of males for 2 months and to females for 2 weeks prior to mating, with treatment continuing in the females through gestation and weaning; no reproductive effects of any kind were observed.[60]

B. CHEMICALS
1. Developmental Toxicity
a. Petrochemicals

Coal-derived complex organic mixtures applied dermally induced cleft palate and increased resorption in mice and rats when administered at dosages of 500 or 1500 mg/kg on gestation days 11 to 15.[93,94] Cleft palate was also elicited in the same two species on an identical regimen with the same dermally applied dosages of solvent-refined heavy distillate coal process hydrocarbons.[95] Process I light oil or process oil and process II light to heavy distillates applied on gestational days 7 to 11 or 12 to 16 produced malformation and reduced fetal weight in rats as well.[96] In contrast, either SRC (first stage) or TSL (second stage) middle distillates were not teratogenic in rats.[97] The former was maternally toxic at doses in the range of 50 to 500 mg/kg on gestational days 6 to 19, while the latter under similar conditions increased fetal mortality, reduced fetal body weight, and retarded ossification. A clarified form of slurry oil, defined as the residual hydrocarbon fraction from fluidized catalytic cracker, when applied topically throughout gestation in rats at maternally toxic doses in the range of 4 to 250 mg/kg, increased resorption and reduced fetal body weight but did not result in terata.[98]

b. Solvents

Two substituted *acetamides* have been studied for effects on development following cutaneous exposure. Dimethylacetamide (DMA) produced ambiguous results in the two species tested, but this was conceivably due to dosage administered. In the rat, 1200 mg/kg applied on gestation days 10 and 11 resulted only in a low incidence of encephaloceles in the fetuses.[99] DMA was also developmentally toxic in this species by oral and intraperitoneal routes of exposure.[100,101] No developmental toxicity was reported in the rabbit following a lower topical dose of 200 mg/kg given on gestation days 8 to 16.[99]

N-Methylacetamide elicited no developmental toxicity when given to rats topically at a dosage of 600 mg/kg on variable gestation days of 9 to 13.[99] This is in contrast to high intraperitoneal dosages, which caused skeletal defects in this species.[100]

Most of the *formamides* tested have had no developmental toxicity potential when tested by the cutaneous route. While methylformamide has been teratogenic in the rat but not the rabbit, this may be related more to dosage than species differences. In the rat, methylformamide applied topically at doses of 200 mg/kg and higher on gestation days 11 and 12 or 12 and 13 caused encephalocele and inguinal hernias.[99] Similar malformations, reduced fetal weight, and diminished viability were also recorded for this species when the chemical was administered orally.[102] As with the acetamides, methylformamide did not elicit developmental toxicity in rabbits at a topical dose of 200 mg/kg.[99]

Dimethylformamide (DMF) at topical doses from 600 to 2400 mg/kg applied to rats during mid-gestation induced increased embryolethality.[99] A dose of 200 mg/kg applied topically to rabbits on gestation days 8 to 16 produced no developmental toxicity.[99] DMF was developmentally toxic in the latter species when given by the oral route, even at approximately 70 mg/kg.[103] Formamide and dibutylformamide were innocuous developmentally when dermally applied to rats at dosages of 600 to 1200 mg/kg on 1 or 2 days in the gestation interval 9 to 13.[99]

With but two exceptions, the *glycol ethers* tested by the cutaneous route have not been developmentally toxic, even when they have been by oral and inhalation exposures. This suggests that the difference may reside in the lack of absorption of these chemicals cutaneously.

Diethylene glycol monobutyl ether (butyl carbitol) was innocuous when applied topically to rabbits on gestation days 7 to 18 at dosages as high as 1 g/kg/day.[104] Similar results were obtained in rats with diethylene glycol monoethyl ether (carbitol) applied on gestation days 7 to 16.[105] Diethylene glycol monomethyl ether (methyl carbitol) elicited some toxicity but only at maternally toxic doses of 750 mg/kg; increased resorption and delayed ossification were observed in rabbit fetuses whose does were dosed on gestation days 6 to 18.[106]

Ethylene glycol has variable effects on mammalian development, depending on the route administered. Notably, the chemical is a developmental toxicant in rats and mice by oral and whole-body inhalational routes of exposure.[107-110] However, when applied cutaneously to mice in concentrations of 12.5 to 100% for 6 h/d on gestation days 6 to 15, the only toxicity elicited was an increased incidence of two variants of skeletal development, in the absence of maternal toxicity.[46] Another study designed to examine this difference between routes of exposure concluded that only oral administration of ethylene glycol is teratogenic, whether from *per os* dosing specifically, or from grooming of fur by the animals following inhalational or cutaneous exposure.[46,109] The difference may be due to the amount of test material available systemically and/or to differences in metabolism from different routes of administration.

Neither the monobutyl ether (butylcellosolve) nor the monoethyl ether acetate forms of ethylene glycol were teratogenic in rats administered high doses on gestational days 7 to 16.[105] In fact, no developmental toxicity occurred with the former, but increased resorption and decreased fetal body weight were seen with the latter. Similar, if not more extensive, developmental toxicity occurred in this species when the chemical was administered by inhalational exposure.[111] The monoethyl ether (ethoxyethanol, EGEE) and monomethyl ether (methylcellosolve, EGME) forms of ethylene glycol were developmentally toxic. The first, EGEE applied dermally at doses of 0.25 or 0.5 ml/animal four times daily on gestational days 7 to 16 to rats, induced cardiovascular malformations, increased the incidence of some skeletal variations, and increased embryonic resorption at doses that were only slightly toxic to the mothers.[112,113] These findings were essentially replicated in an inhalational study of EGEE in both rats and rabbits.[114] EGME induced similar findings. Two separate studies, one in which 0.25 to 1 ml EGME was applied dermally on gestation days 7 to 16, and the other in which the chemical was applied at 500 to 2000 mg/kg on single days (10 to 14) during gestation, resulted in multiple malformations, increased resorption, and depressed fetal weights with only transient maternal weight loss the day after administration.[40,43,115] In addition, EGME was dermally applied undiluted at 150 to 600 mg/kg/day or diluted with water at 37 to 300 mg/kg/d on gestational days 7 to 16 in rats; maternal and developmental toxicity was observed with postnatal loss and visceral and skeletal abnormalities at higher incidence after exposure to EGME in water.[116] EGME was also shown to have predictable developmental toxicity in the rat from dermal administration in a short-term assay.[117] An important consideration from a risk assessment perspective in these studies with EGME is that the developmental toxicity observed in these studies occurred in the presence of only slight or no maternal toxicity, making it hazardous from the developmental perspective.

A number of *miscellaneous solvents* have been investigated for effects on development via cutaneous exposure. Butoxypropanol evidenced no developmental toxicity when given percutaneously to rabbits on gestation days 7 to 18.[118]

Ethyl hexanol caused maternal but no developmental toxicity when given by occlusive dermal application to rats on gestation days 6 to 15 at doses of 1 to 3 ml/kg.[45] The oral route in the same species was developmentally toxic.[119]

N-Methylpyrrolidone induced fetal resorption and skeletal malformations in the rat when applied dermally at doses of 750 mg/kg on gestation days 6 to 15.[120] Maternal toxicity occurred at the same dosage; the developmental toxicity NOEL was 237 mg/kg. The chemical was not developmentally toxic in the same species when given by inhalational exposure as high as 0.36 ml/l.[121]

2-Methoxypropyl-1-acetate was not developmentally toxic in rabbits by cutaneous administration of up to 2000 mg/kg on gestation days 6 to 18.[122] In contrast, the same investigators reported that doses of 500 ppm by inhalation exposure at the same time in gestation elicited congenital malformations, increased incidence of developmental variations, and reduced fetal body weight in association with maternal toxicity.

Tetramethylurea was not developmentally toxic upon cutaneous exposure at various times in pregnancy at doses as high as 6000 mg/kg in rats and 200 mg/kg in rabbits.[99] In contrast, tetrabutylurea was developmentally toxic in the rat at much lower doses. Dermally applied tetrabutylurea on gestation days 6 to 15 caused skin irritation and maternal toxicity at dosages of approximately 170 and 330 mg/kg, and increased resorption and reduced fetal size at approximately 330 mg/kg.[123]

A solvent mixture containing xylene and alkylphenoxy-polyethanol applied topically to rats at doses as high as 9 g/100 g on gestational days 6, 9, or 12 showed no developmental toxicity.[81]

c. Pesticides

A few pesticides have been studied with respect to dermal exposures and effects on developmental toxicity. Coumaphos applied topically to cattle at various stages of gestation at doses of up to 28 g/45 kg had no developmental toxicity.[80]

N,N-Diethyl-*m*-toluamide increased mortality as the only sign of toxicity among rat fetuses whose dams were administered doses of 100 or 1000 mg/kg over gestation.[124] In contrast, only skin irritation occurred when it was applied to rabbits at up to 1000 mg/kg on gestation days 1 to 29.[125]

Imidazolidinethione, a fungicide degradation product of maneb, induced polymorphic malformations and other developmental toxicity when administered topically to rats at only 50 mg/kg on gestation days 10 and 11 or 12 and 13.[99] The developmental toxicity elicited was similar to that observed following oral administration in the rat,[126] whereas the single inhalation exposure study reported in this species did not report developmental toxicity.[127] The chemical is developmentally hazardous, producing developmental effects at a fraction of the dosage eliciting maternal toxicity.

Tetramethyl *O,O'*-thiodi-*p*-phenylene phosphorothioate had no potential for developmental toxicity when administered either dermally or orally to rabbits during days 6 to 18 of gestation at doses of 32 to 164 mg/kg.[128]

The herbicide nitrofen has been investigated in several studies in rats with respect to dermal application and development. It appears that doses in the range of 12 to 18 mg/kg on gestation days 6 to 15 induce eye abnormalities, diaphragmatic hernias, hydronephrosis, and dysmorphic Harderian glands.[129,130] No other classes of developmental toxicity were reported and these results parallel somewhat those reported in mice and rats following oral administration.[131,132]

d. Metals

Chromated copper arsenate is the only metal apparently tested by the cutaneous route. The chemical was not developmentally toxic in the rabbit when applied on gestational days 7 to 20 at the tested dosages.[133]

e. Plastic Chemicals

A number of chemicals used in plastics manufacture have been investigated in animals during development by the cutaneous route. An epoxyresin, bisphenol A diglycidyl ether, induced only skin irritation in rabbits when applied at dosages of 30 to 300 mg/kg on gestational days 6 to 18.[134]

Dimethyl phthalate applied epicutaneously to rats at doses up to 2 ml/day in gestation induced no developmental toxicity.[135] Similarly, a dosage of 5600 mg/kg ethyl phthalate applied to mice in gestation caused no adverse developmental effects, but the high dose utilized did cause maternal toxicity.[136]

A polymeric mixture of phenylmethylcyclosiloxane given dermally at a dosage of 200 mg/kg on gestational days 6 to 18 to rabbits resulted in 9/93 abnormal offspring, an effect termed non-teratogenic by the authors.[82] The same report also considered similar effects produced upon subcutaneous administration in the same species not significant.

Two forms of polydimethylsiloxane used as prosthesis implant materials were not developmentally toxic when administered dermally to rabbits at doses of 200 or 1000 mg/kg of either a 10- or 350-centistroke material.[137] Neither form was developmentally toxic when given subcutaneously at the same dosages.

A cosmetic grade polymer, phenyl trimethacone, was not developmentally toxic in rabbits at doses of 50 or 500 mg/kg during organogenesis; delayed ossification was observed in the rat under identical experimental conditions.[138]

Bis(2-methylaminoethyl)ether (called NIAX®) was toxic to the developing rabbit embryo. One milliliter doses of 1, 5, or 10% of the chemical for 6 h daily on gestation days 6 to 18 caused reversible skin irritation, while maternal renal pathology was observed at 5 and 10%, and other maternal toxicity and reduced fetal weight occurred at 10%.[41,42]

f. Miscellaneous Chemicals

An industrial chemical, hexafluoroacetone, when applied dermally to rats in doses of 1 to 25 mg/kg on gestation days 6 to 16, induced malformations in the offspring.[139] Other classes of developmental toxicity, including fetal resorption and reduced weight, were also reported, as was maternal toxicity. A recent inhalation study in the same species reported malformations plus a wide spectrum of developmental toxicity including increased mortality, reduced fetal weight, and increased skeletal variations, accompanied by maternal toxicity.[140]

Another industrial chemical, a phenylethylphenols (PEP) mixture applied topically in the guinea pig, was reported to induce malformations.[141]

Two other industrial chemicals, tetrahydro-3,5-dimethyl-4H-1,3,5-oxadiazine-4-thione, and 1,1,3,3,-tetramethyl-2-thiourea, were both reported to produce no developmental toxicity when applied topically to rats on 1 or 2 d in gestation.[99] Doses of up to 1500 mg/kg in the former, and up to 500 mg/kg in the latter, were given.

Sodium dodecyl sulfate (SDS) applied dermally to mice on gestation days 6 to 13 reduced fetal body weight at doses of 4.0 and 6.0%.[142]

2. Reproductive Toxicity
a. Miscellaneous Chemicals

Ethylparathion, a pesticide, when applied percutaneously to rats on gestation day 0 decreased the number of implantations but had no adverse effect on development at the dose employed.[143]

An industrial chemical, *p*-butyl-*t*-benzoic acid, given to rats by dermal exposure, caused testicular effects in rats.[144] These effects were reversible, and similar effects were also produced following inhalational exposure.

The industrial solvent ethylene glycol monomethyl ether (methylcellosolve, EGME) was a potent reproductive toxicant in male rats. Animals given the chemical dermally by either occluded or non-occluded technique for 7 days prior to breeding at 625 to 5000 mg/kg/d showed a number of adverse findings including a decline in epididymal weights, an increase in abnormal sperm, and reduced fertility.[145] The reproductive effects were severe and showed slower recovery by the occluded method, even at dosages one-half those applied with no covering.

IV. RESULTS OF HUMAN STUDIES

Testing of drugs in humans by any route for adverse developmental or reproductive outcome cannot be conducted, of course, for ethical reasons. In determining the potential for adverse effects in human populations of exposure to drugs and chemicals, case reports and epidemiological studies provide indications of an association between an agent and adverse reproductive/developmental outcome. However, there is a paucity of information on human exposure by the dermal route. In the following sections, potential effects on reproductive outcome in humans following dermal exposure to drugs and chemicals are speculated if information is available for the agent by other routes of exposure and/or in animals.

A. DRUGS

There are only a few case reports and limited epidemiological studies of percutaneously applied drugs and their association with developmental effects in humans. Only a few therapeutic drugs are currently approved by the FDA for systemic administration by transdermal patch systems; human or animal data of reproductive outcome for these drugs are available primarily for routes other than the dermal route. Agents applied topically for a local therapeutic effect are beyond the scope of this section.

1. Tretinoin

Tretinoin or all-*trans* retinoic acid has been indicated for topical application in the treatment of acne vulgaris for more than 20 years. *In vivo* and *in vitro* investigations have shown that percutaneous absorption of the ethylamide of tretinoin (motretinide) in human skin is limited to 3%.[146] In view of the low systemic absorption, Orfanos et al. concluded that dysmorphogenic effects in humans did not seem to be very likely. Reproduction studies performed in rats and rabbits at dermal doses up to 50 times the human therapeutic dose (assuming 500 mg of gel per day) have revealed no evidence of impaired fertility or developmental toxicity.[147] Embryotoxic effects, however, have been demonstrated after oral administration in numerous animal species.[50] Since vitamin A and its analogs accumulate in the body, increased intake from several routes, such as in those individuals topically applying tretinoin and self-medicating with megadoses of vitamin A, may have the potential for teratogenic response.

2. Triamcinoline

Triamcinoline, a known animal teratogen, is fluorinated glucocorticoid used topically for a variety of dermatologic conditions. Katz et al.[148] reported that a patient who applied 40 mg/kg for dermatitis during weeks 12 to 29 of pregnancy delivered a growth-retarded 700 g baby. Similarly, in monkeys, fetal growth delays of 30 to 40% were produced following intramuscular injection of triamcinolone.[149]

3. Hexachlorophene

The antimicrobial hexachlorophene is an example of a neurotoxic substance that readily penetrates the skin.[150] Exposure to this drug can occur in both industrial and domestic settings. Newborn infants bathed one to three times per day in a 3% hexachlorophene solution died from encephalopathy.[151] Hexachlorophene has been implicated as the causative agent for a "cluster" of malformations in infants of female Swedish hospital employees frequently working with the agent during their pregnancies.[152,153] This association, however, has been questioned by other investigators.[154-156] Currently, it has not been reasonably established whether hexachlorophene has human teratogenic potential.

4. Nitroglycerin

Nitroglycerin is a widely used smooth muscle relaxant and vasodilator with beneficial effects in relief of angina pectoris. It can be applied to the skin in an ointment or by a transdermal patch system. There are no data currently available in the open literature on reproductive/developmental effects of nitroglycerin in humans. Nitroglycerin has not been linked to reproductive dysfunction or developmental toxicity by animal testing (Section III).

5. Scopolamine

Scopolamine is a prophylactic for motion sickness usually applied for therapeutic use by a transdermal patch system. There are no reports of potential adverse reproductive/developmental outcome in humans using scopolamine patches. A teratogenicity study in rabbits and fertility studies in rats, which achieved plasma levels approximately 100 to 500 times greater than those achieved in humans using a transdermal system, showed marginal embryotoxic effects and decreased maternal body weight, respectively.[147] These animal studies suggest current usage of scopolamine in the transdermal system is not likely to produce adverse reproductive/developmental effects.

6. Clonidine

Clonidine is effective by transdermal patch systems for the treatment of hypertension. There are no studies in pregnant women wherein clonidine was topically administered. A negative association between the use of clonidine in early pregnancy and congenital anomalies has been reported.[157] Data from animal teratogenicity and fertility studies by oral administration of clonidine have shown increased post-implantation loss;[147] this endpoint would be difficult to assess in the human population.

7. Estradiol

Estradiol transdermal patch systems can release 17-estradiol through a rate-limiting membrane into systemic circulation.[158] Several reports suggest an association between intrauterine exposure to female sex hormones and congenital anomalies, including congenital heart defects and limb-reduction defects, and is the reason for class labeling restrictions for hormones.

It is reasonable to assume that dermal exposure would not be a great concern for most drugs as they should have an appropriate safety factor. Definitive associations of drugs and adverse developmental or reproductive outcome in the human population are difficult to monitor. As dermal application and the use of transdermal systems for therapeutic drug administration become more widely used, basic research of cutaneous absorption of drugs and better monitoring of the human population will be needed.

B. CHEMICALS

Humans are exposed via the dermal route unintentionally in the home or workplace during the manufacture, handling, and application of industrial and household chemicals, such as solvents, pesticides, and preservatives. Information about adverse reproductive/developmental toxicity is comprised primarily of occupational studies of workers exposed to the chemical by several routes, e.g., dermal and inhalation; it is often difficult to determine which route is the more significant. The following chemicals have been linked to adverse reproductive outcome in humans due primarily to dermal exposure.

1. Dibromochloropropane

This agriculturally important nematocide has been implicated as a cause of human male sterility. Male factory workers in the U.S. and Israel occupationally exposed to dibromo-chloropropane (DBCP) became sterile, showing oligospermia, azoospermia, and germinal aplasia.[159,160] No estimate was made of potential body burden due to dermal exposure and absorption in these studies. DBCP can be absorbed through the skin or by ingestion[161] and low lipid solubility and relatively low vapor pressure of DBCP make skin absorption likely to occur subsequent to skin contamination.[162] Torkelson et al.[161] described male reproductive effects in the rat and rabbit similar to those found in humans.

2. Glycol Ethers

This class of chemicals has been implicated in humans as causing: decreased testicular size without producing semen abnormalities,[163] decreased sperm concentration,[164] and increased oligospermia and azoospermia.[165] Exposure in these studies, however, was both by inhalation and dermal contact. Dermal application of ethylene glycol monoethyl ether induces increased abnormal sperm, reduced sperm count, and reduced fertility in male rats.[145]

3. Vinyl Chloride

Vinyl chloride is a monomer used in the manufacture of polyvinyl chloride (PVC) plastics. Recent data suggest that inhalation and dermal exposure of human males to vinyl chloride in the workplace may be associated with various chromosomal aberrations in lymphocytes and with adverse pregnancy outcome.[166]

4. Inorganic Mercury

Inorganic mercury was widely used in cosmetics and medicated soaps as a bleaching agent and bactericide until 1973 when its use was discontinued due to its toxicity.[167] Mercury is absorbed through the skin and concentrates in the kidney.[168] Skin lightening creams containing inorganic mercury are used in black Africa.[169] Lauwerys et al.[170] reported a case where renal tubular dysfunction associated with cataract and anemia were found in a 3-month-old black child. The mother regularly used a mercury-containing bleach cream and soap containing 1% mercuric iodide during pregnancy and lactation; no mercury-containing products were applied directly to the child's skin. Practically all mercury compounds are teratogenic in animals, producing multiple malformations including eye and kidney malformations.[50]

5. Malathion

Malathion is an organophosphorus insecticide that is not particularly well absorbed through the skin.[171] A case of amyoplasia congenita-like condition is reported from maternal malathion exposure.[172] During the 11th and 12th week after conception, the mother washed her hair repeatedly with large amounts of a hair lotion containing 0.5% malathion for control of lice. The authors speculated that the repeated application and longstanding wetness of the hair on the skin may have resulted in sufficient malathion uptake to result in the severely malformed baby.

6. 2,4,5-Trichlorophenoxyacetic Acid

2,4,5-Trichlorophenoxyacetic acid (2,4,5-T) is a phenoxy herbicide defoliant. Large numbers of civilians and military personnel have been exposed to this chemical both orally and dermally.[173,174] Claims of adverse human reproductive effects have been refuted[175] (see TCDD below). Initial animal studies with 2,4,5-T showed teratogenic effects.[176] However, once the dioxin contamination was removed, the results of teratogenicity testing in animals have been variable.[177,178] Most uses of 2,4,5-T have been halted by the U.S. EPA based on a poorly documented study linking an increased risk of miscarriages to the unavoidable contamination of 2,4,5-T with dioxins.[173]

7. TCDD

The teratogenicity of 2,3,7,8-tetrachlorodibenzo-*p*-dioxin (TCDD) has been the subject of much controversy. Large numbers of humans were dermally exposed to TCDD in the Seveso accident[179] and as a contaminant (approximately 2%) in Agent Orange,[180] the herbicide defoliant used extensively in Vietnam. Initial reports alluded to congenital malformation in both cases[181,182] but subsequent studies have not confirmed any positive relationship.[175,183]

V. FUTURE DIRECTIONS IN DERMAL EXPOSURE

The preceding sections of this chapter have presented the current status of reproductive and developmental studies in animals and humans via dermal exposure. It is clear that the number of drugs and/or chemicals, approximately 80, which have been tested for potential adverse reproductive/developmental effects in animals by this route, is limited, when one considers that almost 4000 drugs and chemicals have been tested and reported for effects on reproduction and development. More limited, still, are the number of occupational studies and case reports available of individuals exposed to drugs and/or to chemicals in domestic or industrial settings by the dermal route.

The dermal route can be a substantial source of intake for both therapeutic agents and chemicals. Some agents may be better absorbed by the dermal route than by other routes, and hence more acutely toxic. Differences in absorption characteristics between routes of exposure may be exploited such as in the administration of certain drugs or may pose an unrecognized threat to individuals working with domestic or industrial chemicals. Too often, little is known about the dermal absorption of agents.

In some workplace situations, individuals may be exposed to a chemical by several routes simultaneously.[184] In these cases it may be difficult to separate the contribution of dermal exposure from that of other routes, such as inhalation, to the individual's total intake of the chemical. For example, percutaneous absorption during the immersion of both hands in xylene for 15 min was approximately equivalent to the pulmonary absorption during exposure to 100 ppm for the same length of time.[185]

The potential for an adverse reproductive/developmental effect of a drug or chemical in contact with human skin is frequently not known, and is often inferred from human reports of the drug by other routes, or from animal data using dermal or other routes of exposure. Case reports or epidemiological studies focusing on the dermal route and animal testing by the same route as human exposure, are the most relevant for discerning whether a reproductive/developmental hazard may exist. Clearly, a future direction should be the increased monitoring and study of human dermal exposure, and animal research and testing of drugs and chemicals by the dermal route. Research should focus on those drugs and chemicals where there is high human dermal exposure potential, such as those agents that are well absorbed or where there is skin contact for prolonged periods of time. Agents that are known or suspected human or animal reproductive/developmental toxicants by other routes of exposure (e.g., oral, inhalation) should receive high priority. Attention should also be directed towards agents where exposure can occur by several routes. Table 2 lists some but not all chemicals which are capable of penetrating the skin causing acute toxicity, have no available human reproductive/developmental toxicity data, and either have data for reproductive/developmental effects in animal studies (by any route) or for which no data are available.

TABLE 2
Some Chemicals with Potential Reproductive and/or
Developmental Effects by the Dermal Route

Chemical	Shepard[186]	Schardein[50]	U.S. EPA[187]
Acrylamide	+	+	+
Aldicarb	ND	ND	ND
Bis(chloroethyl)ether	ND	ND	ND
Carbon disulfide	+	+	+
Carbofuran	−	−	+
Chlordane	−	−	+
Demeton	+	+	+
Dimethyl sulfate	ND	ND	ND
Dinoseb	+	+	+
Disulfoton	ND	ND	+
Endrin	+	+	ND
Fenamiphos	ND	ND	+
Fluridone	ND	ND	+
Furan	ND	ND	ND
Methacrylonitrile	ND	ND	ND
Methamidophos	ND	ND	ND
Methyl bromide	ND	ND	ND
Nickel carbonyl	ND	ND	ND
Nitrobenzene	ND	+	ND
Oxamyl	ND	ND	+
Pentachlorophenol	+	+	+
Phenol	ND	−	+
Pyrene	ND	ND	ND
Selenious acid	ND	ND	ND
Sodium cyanide	ND	+	ND
Tetraethyl lead	ND	−	ND

Note: ND = no data; + = studies showing reproductive/develop-
mental toxicity; − = studies not showing reproductive/devel-
opmental toxicity.

These chemicals may be candidates for further testing as dermal reproductive/developmental toxicants.

Another direction for future research is the influence of route of administration on reproductive/developmental outcome. Understanding these differences may permit route-to-route extrapolation of information, enabling the volume of information on a particular chemical to be utilized. For example, the results of teratogenicity testing conducted via the oral route may be used in lieu of additional testing via the dermal route if potential differences (e.g., pharmacokinetics) are understood.

Conduct of animal studies by cutaneous administration of drugs or chemicals requires consideration of several unique characteristics of reproductive/developmental toxicity testing. Many of these considerations have been discussed in Section II. The need or desire to employ cutaneous administration of agents in reproductive/developmental studies may prompt the development of alternative procedures for restraint, application, or of animal models to bypass these potential confounding influences. Cutaneous application of test agents may affect male-female and mother-offspring interactions which may confound reproductive toxicity studies; the potential for interference with mother-offspring interactions has major implications for developmental neurotoxicologic assessments. During pregnancy, maternal weight, surface area, and plasma volume increase during a short period of time, questioning or precluding the application of a constant amount/volume of the agent (this is true for gavage administration as well). Frequent adjustments may need to be made to maintain a constant dose.

The effect of pregnancy state on metabolism, pharmacokinetics, and pharmacodynamics of cutaneously administered chemicals needs to be addressed. The fact that pregnant rats appear to be less responsive to the induction of hepatic P-450 systems by phenobarbital than are non-pregnant females[188] suggests that non-hepatic (such as cutaneous) P-450 enzymes might also be affected by pregnancy. Metabolism of drugs and chemicals occurring in the skin may be changed during pregnancy exacerbating or ameliorating the fetal insult; new models may need to be developed.

Cutaneous administration of drugs, i.e., transdermal therapy, is a future direction for drug administration and may have some advantages over other routes of administration for certain drugs. Transdermal drug administration can provide continuity of delivery and precise control of drug plasma concentrations by a rate-controlling membrane; better control of plasma concentrations enhances the desired therapeutic effect and may decrease the development of adverse reproductive or teratogenic effects. It offers particular safety advantages for drugs with short half-lives or narrow therapeutic indices. The transdermal route also bypasses the gastrointestinal tract, delivering the drug directly to systemic circulation, eliminating factors that influence drug absorption, such as food content, acidity, and first-pass effect of the liver. Transdermal drug delivery is advantageous when prolonged treatment, or multiday dosage, is needed and also offers additional safety in that therapy can be easily terminated by simply removing the patch from the skin.

The above future applications and directions of reproductive and developmental toxicity studies by cutaneous administration will no doubt lead to more unanswered questions and research.

VI. CONCLUSIONS

The skin is an important portal of entry for numerous drugs and chemicals, both intentionally and accidentally. It can provide an avenue of continuous delivery and control of plasma concentration for therapeutic agents, as well as a portal of entry for the uncontrolled exposure to systemic toxicants found in both domestic and industrial settings. Data on the potential adverse reproductive/developmental effects of agents due to dermal contact in animals and humans are extremely limited. Some drugs and chemicals, with the potential for adverse reproductive outcome when administered orally to humans and which can be absorbed dermally, might also pose a hazard for human fertility and development if administered percutaneously. Additionally, drugs that can be dermally absorbed by humans and have been shown in animal reproductive/developmental studies to produce adverse outcome when administered by dermal or other routes, may also have the potential for an adverse reproductive/developmental outcome were the drug to be administered percutaneously to humans.

Many issues specific to reproductive and developmental toxicity testing by topical application were considered: problems attendant on the requirement for systemic maternal toxicity vs. stress from restraint and localized skin responses; differences in the rate and extent of absorption and systemic exposure; modifications to the standard testing study design; and appropriateness of the typical reproductive/developmental animal models for toxicity assessment. Because of the paucity of human information, animal data become important in assessing the likelihood of reproductive/developmental toxicity of cutaneous absorption of an agent(s) in humans. While there are more animal studies conducted using cutaneous administration of chemicals and drugs, the database is far from comprehensive. Determination of risk in humans becomes dependent on extrapolation of information from other routes of exposure to the dermal route, from animal data to humans and on further research. Future Directions attempts to aid the researcher in prioritizing agents for testing based on available information, focuses on areas of future research for reproductive/developmental testing and the future of drug administration by transdermal systems.

REFERENCES

1. U.S. Food and Drug Administration (FDA), Guidelines for Reproduction Studies for Safety Evaluation of Drugs for Human Use, Washington, D.C., 1966.
2. U.S. Environmental Protection Agency (EPA), Pesticides assessment guidelines, subdivision F, Hazard Evaluation: Human and Domestic Animals (Final Rule), Available from NTIS (PB86-108958), Springfield, VA, 1984.
3. Japanese Guidelines of Toxicity Studies, Notification No. 118 of the Pharmaceutical Affairs Bureau, Ministry of Health and Welfare, 2, Studies of the Effects of Drugs on Reproduction, Yakagyo Jiho Co., Tokyo, 1984.
4. Canada Ministry of Health and Welfare, Health Protection Branch, The testing of chemicals for carcinogenicity, mutagenicity and teratogenicity, The Ministry of Ottawa, 1973.
5. United Kingdom: Committee on Safety of Medicines, Notes for guidance on reproduction studies, Department of Health and Social Security, Great Britain, 1974.
6. U.S. Environmental Protection Agency (EPA), Toxic substances control act test guidelines: Final rule, *Fed. Regis.,* 50, 39412, 1987.
7. Organization for Economic Cooperation and Development (OECD), Guideline for Testing of Chemicals: Teratogenicity, Director of Information, Paris, 1981.
8. U.S. Environmental Protection Agency (EPA), Guidelines for the health assessment of suspect developmental toxicants, *Fed. Regis.,* 51, 34028, 1986.
9. U.S. Environmental Protection Agency (EPA), Proposed amendments to the guidelines for the health assessment of suspect developmental toxicants; request for comments; notice, *Fed. Regis.,* 54, 9386, 1989.
10. U.S. Environmental Protection Agency (EPA), Part II, Proposed guidelines for assessing female reproductive risk; notice, *Fed. Regis.,* 53, 24834, 1988a.
11. U.S. Environmental Protection Agency (EPA), Part III, proposed guidelines for assessing male reproductive risk and request for comments, *Fed. Register,* 53, 24850, 1988b.
12. **Francis, E. Z. and Kimmel, C. A.,** Proceedings of the workshop on the acceptability and interpretation of dermal developmental toxicity studies, *Teratology,* 39, 453, 1989.
13. **Kimmel, C. A. and Francis, E. Z.,** Proceedings of the workshop on the acceptability and interpretation of dermal developmental toxicity studies, *Fundam. Appl. Toxicol.,* 14, 386, 1990.
14. **Kimmel, G. L., Kimmel, C. A., and Francis, E. Z., Eds.,** Evaluation of maternal and developmental toxicity, *Teratogen. Carcinogen. Mutagen.,* 7, 203, 1987.
15. **McCreesh, A. H. and Steinberg, M. K.,** Skin irritation testing in animals, in *Dermatotoxicology,* 3rd ed., Marzulli, F. N. and Maibach, H. I., Eds., Hemisphere, Washington, D.C., 1987, 153–171.
16. **DePass, L. R., Myers, R. C., Weaver, E. V., and Weil, C. S.,** An assessment of the importance of number of dosage levels, number of animals per dosage level, sex and method of LD50 and slope calculation in acute toxicity studies, in *Alternative Methods in Toxicology,* Vol. 2, Mary Ann Liebert, New York, 1984, 139.
17. **Wester, R. C. and Maibach, H. I.,** Cutaneous pharmacokinetics: 10 steps to percutaneous absorption, *Drug Metab. Rev.,* 14, 159, 1983a.
18. **Wester, R. C. and Maibach, H. I.,** *In vivo* percutaneous absorption, in *Dermatotoxicology,* 2nd ed., Marzulli, F. N. and Maibach, H. I., Eds., Hemisphere, Washington, D.C., 1983b, 131–146.
19. **Mathias, C. G. T.,** Clinical and experimental aspects of cutaneous irritation, in *Dermatotoxicology,* 2nd ed., Marzulli, F. N. and Maibach, H. I., Eds., Hemisphere, Washington, D.C., 1983, 167–183.
20. Federal Hazardous Substance Act (FHSA) Standards, Method of testing primary irritant substances, U.S. Code of Federal Regulations, Title 16, Commercial Practices, Chapter II, Part 1500.41, 352, 1980.
21. **Euker, J. S. and Riegle, G. D.,** Effects of stress on pregnancy in the rat, *J. Reprod. Fertil.,* 34, 343, 1973.

22. **Rojo, M., Marin, B., and Menéndez-Patterson, A.,** Effects of low stress during pregnancy on certain parameters of the offspring, *Physiol. Behav.,* 34, 895, 1985.
23. **Rojo-Fernández, M., Marin, B., and Menéndez-Patterson, A.,** Efectos del estrés de inmovilización *in utero* sobre la evolución de la prenez de le rata y varios parámetros de los neonatos, *Rev. Espan. Fisiol.,* 41, 29, 1985.
24. **Ward, I. L.,** Prenatal stress feminizes and demasculinizes the behavior of males, *Science,* 175, 82, 1972.
25. **Herrenkohl, L. R.,** Prenatal stress reduces fertility and fecundity in female offspring, *Science,* 206, 1097, 1979.
26. **Barlow, S. M., Knight, A. F., and Sullivan, F. M.,** Delay in postnatal growth and development of offspring produced by maternal restraint stress during pregnancy in the rat, *Teratology,* 18, 211, 1978.
27. **Anderson, D. K., Rhees, R. E., and Fleming, D. E.,** Effects of prenatal stress on differentiation of the sexually dimorphic nucleus of the preoptic area (SDN-POA) of the rat brain, *Brain Res.,* 332, 113, 1985.
28. **Barlow, S. M., McElhatton, P. R., and Sullivan, F. M.,** The relation between maternal restraint and food deprivation, plasma corticosterone and cleft palate in the offspring of mice, *Teratology,* 12, 97, 1975a.
29. **Beyer, P. E. and Chernoff, N.,** The induction of supernumerary ribs in rodents: role of maternal stress, *Teratogen. Carcinogen. Mutagen.,* 6, 419, 1986.
30. **Chernoff, N., Miller, D. B. Rosen, M. B., and Mattscheck, C. L.,** Developmental effects of maternal stress in the CD-1 mouse induced by restraint on single days during the period of major organogenesis, *Toxicology,* 51, 57, 1988.
31. **Chernoff, N., Kavlock, R. J., Beyer, P. E., and Miller, D.,** The potential relationship of maternal toxicity, general stress and fetal outcome, *Teratogen. Carcinogen. Mutagen.,* 7, 241, 1987.
32. **Wiebold, J. L., Stanfield, P. H., Becker, W. C., and Hillers, J. K.,** The effect of restraint stress in early pregnancy in mice, *J. Reprod. Fertil.,* 78, 185, 1986.
33. **Barlow, S. M., Morrison, P. J., and Sullivan, F. M.,** Effects of acute and chronic stress on plasma corticosterone levels in the pregnant and non-pregnant mouse, *J. Endocrinol.,* 66, 93, 1975b.
34. **Hemm, R. D., Arslanoglou, L., and Pollock, J. J.,** Cleft palate following prenatal food restriction in mice: association with elevated maternal corticosteroids, *Teratology,* 15, 243, 1977.
35. **Barlow, S. M., Knight, A. F., and Sullivan, F. M.,** Diazepam-induced cleft palate in the mouse: the role of endogenous maternal corticosterone, *Teratology,* 21, 149, 1980.
36. **Scialli, A. R.,** Reproductive toxicology review, Is stress a developmental toxin?, *Reprod. Toxicol.,* 1, 163, 1988.
37. **Strean, L. P. and Peer, L. A.,** Stress as an etiologic factor in the development of cleft palate, *Plast. Reconst. Surg.,* 18, 1, 1956.
38. **German, J.,** Embryonic stress hypothesis of teratogenesis, *Am. J. Med.,* 76, 293, 1984.
39. **Warshaw, C. J.,** *Guidelines on Pregnancy and Work,* Government Printing Office, Washington, D.C., 1977.
40. **Feuston, M. H., Kerstetter, S. L., and Wilson, P. D.,** Teratogenicity of 2-methoxyethanol applied dermally to rats, *Teratology,* 39, 451, 1989.
41. **Tyl, R. W., Fisher, L. C., France, K. A., Garman, R. H., and Ballantyne, B.,** Evaluation of the teratogenicity of bis(2-methylaminoethyl) ether after dermal application in New Zealand white rabbits, *J. Toxicol. Cut. Ocul. Toxicol.,* 5, 263, 1986.
42. **Fisher, L. C., Tyl, R. W., France, K. A., Garman, R. H., and Ballantyne, B.,** Teratogenicity evaluation of bis(2-dimethylaminoethyl)ether after dermal application in New Zealand White rabbits, *Toxicology,* 6, 93, 1986.
43. **Feuston, M. H., Kerstetter, S. L., and Wilson, P. D.,** Teratogenicity of 2-Methoxyethanol applied as a single dermal dose to rats, *Fundam. Appl. Toxicol.,* 15, 448, 1990.
44. **Ryan, B. M., Hatoum, N. A., Johnson, W. D., Talsma, D. M., Yernakoff, J. K., and Garvin, P. J.,** Effects of collaring and wrapping on the outcome of dermal teratology studies in rats, *Teratology,* 37, 488, 1988.
45. **Fisher, L. C., Tyl, R. W., and Kubena, M. F.,** Cutaneous developmental toxicity study of 2-ethylhexanol (2-EH) in Fischer 344 rats, *Teratology,* 39, 452, 1989.
46. **Tyl, R. W., Fisher, L. C., Kubena, M. F., Vrbanic, M. A., and Losco, P. E.,** Developmental toxicity evaluation of ethylene glycol (EG) applied cutaneously to CD-1 mice, *Teratology,* 37, 498, 1988a.
47. **Ishimura, K., Honda, Y., Neda, K., Ishikawa, I., Otawa, T., Kawaguchi, Y., Sato, H., and Zenichi, H.,** Teratological studies on betamethasone 17-benzoate (MS-1112). II. Teratogenicity study in rabbits, *Oyo Yakuri,* 10, 685, 1975.
48. **Bertolini, A., Castelli, M., and Genedani, S.,** Pregnancy increases the toxicity of betamethasone and indomethacin, administered singly or together to mice, rats, and rabbits, *Riv. Farmacol. Ter.,* 6, 231, 1975.
49. **Esaki, K., Shikata, Y., and Yanagita, T.,** Effects of dermal administration of dexamethasone 17-valerate in rabbit fetuses, *CIEA Preclin. Rep.,* 7, 245, 1981.

50. **Schardein, J. L.**, *Chemically Induced Birth Defects,* Marcel Dekker, New York, 1985.

51. **Narama, I.**, Reproduction studies of diflorasone diacetate (DDA), 4. Teratogenicity study in rabbits by percutaneous administration, *Oyo Yakuri,* 28, 241, 1984.

52. **Anon.**, Medrysone: a review, *Drugs,* 2, 5, 1971.

53. **Rodwell, D. E., Werchowski, K. M., and Briggs, G. B.**, Triamcinolone acetonide (Aristocort) administered topically to rats as a positive teratogen, *Toxicologist,* 2, 74, 1982.

54. **Aii, S., Kuramoto, M., Shigemi, F., Takeda, K., and Daikoku, S.**, Studies on the safety of topical clobetasone 17-butyrate. 3. Study on the effects of topical clobetasone 17-butyrate continuously applied to pregnant rats in organogenic period, *Shikoku Igaku Zasshi,* 36, 91, 1980.

55. **Abdel-Rahman, M. S., Skowronski, G. A., Gerges, S. E., Turkall, R. M., and von Hagen, S.**, Teratologic studies on alcide allay-gel in rabbits, *J. Appl. Toxicol.,* 7, 161, 1987.

56. **Mikami, Y., Sakai, Y., and Miyamoto, I.**, Anomalies induced by ABS applied to the skin, *Teratology,* 8, 98, 1973.

57. **Johnson, D. E., Schardein, J. L., Mitoma, C., Goldenthal, E. I., Wazeter, F. X., and Wedig, J. H.**, Plasma levels of 2-methylsulfonylpyridine in pregnant rats and rabbits given omadine MDS dermally, *Toxicologist,* 4, 85, 1984.

58. **Wedig, J. H., Kennedy, G. L., Jenkins, D. H., and Keplinger, M. L.**, Teratologic evaluation of the magnesium sulfate adduct of [2,2'-dithio-bis-(pyridine-1-oxide)] in swine and rats, *Toxicol. Appl. Pharmacol.,* 42, 561, 1977.

59. **Rodwell, D. E., Johnson, D. E., and Wedig, J. H.**, Teratogenic evaluation of sodium omadine administered topically to rats, *Toxicologist,* 4, 85, 1984.

60. **Johnson, D. E., Schardein, J. L., Goldenthal, E. I., Wazeter, F. X., and Wedig, J. H.**, Reproductive toxicology of omadine MDS, *Toxicologist,* 2, 75, 1982.

61. **Nolen, G. A., Patrick, L. F., and Dierckman, T. A.**, A percutaneous teratology study of zinc pyrithione in rabbits, *Toxicol. Appl. Pharmacol.,* 31, 430, 1975.

62. **Wedig, J. H., Kennedy, G. L., Jenkins, D. H., Henderson, R., and Keplinger, M. L.**, Teratologic evaluation of dermally applied zinc pyrithione on swine, *Toxicol. Appl. Pharmacol.,* 36, 255, 1976.

63. **Overman, D. O.**, Absence of embryotoxic effects of formaldehyde after percutaneous exposure in hamsters, *Toxicol. Lett.,* 24, 107, 1985.

64. **Scortichini, B. H., Quast, J. F., and Rao, K. S.**, Teratologic evaluation of 2-phenoxyethanol in New Zealand white rabbits following dermal exposure, *Fundam. Appl. Toxicol.,* 8, 272, 1987.

65. **Palmer, A. K., Readshaw, M. A., and Neuff, A. M.**, Assessment of the teratogenic potential of surfactants, Part 3, Dermal application of LAS and soap, *Toxicology,* 4, 171, 1975.

66. **Iimori, M., Inoue, S., and Yano, K.**, Influence of detergent on pregnant mice by application to skin, *Yakagaku,* 22, 807, 1973.

67. **Nolen, G. A. and Dierckman, T. A.**, Reproduction and teratogenic studies of a 2:1 mixture of 3,4,4'-trichlorocarbanilide and 1,1,1-trifluoromethyl-4,4'-dichlorocarbanilide in rats and rabbits, *Toxicol. Appl. Pharmacol.,* 51, 417, 1979.

68. **Zbinden, G.**, Investigations on the toxicity of tretinoin administered systemically to animals, *Acta Derm. Venereol. Suppl. Stockholm,* 74, 36, 1975.

69. **Chahoud, I. and Nau, H.**, Pharmacokinetics and teratogenicity with special drug application techniques: dermal administration, intrauterine and continuous subcutaneous infusion, *Teratology,* 40, 271, 1989.

70. **Chahoud, I., Loefberg, B., Mittmann, B., and Nau, H.**, Teratogenicity and pharmacokinetics of vitamin A acid (tretinoin, all-trans retinoic acid) after dermal application in the rat, *Naunyn-Schmiedeberg's Arch. Pharmacol.,* 339 (Suppl.), R30, 1989.

71. **Sharma, R. P., Willhite, C. C., Barry, D. C., and Allen, P. V.**, Dose-dependent pharmacokinetics and teratogenic activity of topical retinoids, *Toxicologist,* 10, 237, 1990.

72. **Willhite, C. C., Sharma, R. P., Allen, P. V., and Berry, D. L.**, Percutaneous retinoid absorption and embryotoxicity, *J. Invest. Dermatol.,* 95, 523, 1990.

73. **Mitala, J. J., Powers, W. J., Wilson, B. C., Davis, G. J., Higgins, C., Tesh, J. M., and McAnulty, P. A.**, Teratogenic effects of retinoid RWJ 20257, *Toxicologist,* 10, 36, 1990.

74. **Shashkina, L. F. and Remozova, M. I.**, Effect on the fetus of a methyltestosterone preparation applied to the skin of pregnant rats, *Akush. Ginekol. (Moscow),* 49, 49, 1973.

75. **Golubeva, M. I., Shashkina, L. F., Starkov, M. V., and Fedorova, E. A.**, On the condition and development of the progeny of rats undergoing application of androgens to the skin throughout the whole period of their pregnancy, *Gig. Tri. Prof. Zabol.,* 22, 25, 1978.

76. **Overman, D. O. and White, J. A.**, Comparative teratogenic effects of methyl salicylate applied orally or topically to hamsters, *Teratology,* 28, 421, 1983.

77. **Infurna, R., Beyer, B., Twitty, L., Koehler, G., and Daughtrey, W.**, Evaluation of the dermal absorption and teratogenic potential of methyl salicylate in a petroleum based grease, *Teratology,* 41, 566, 1990.

78. **Skutt, V. M., Schardein, J. L., Matsubara, Y., and Ohgo, T.,** Teratology study of nitroglycerin ointment by dermal administration in rats, *Shinyaku to Rinsho,* 34, 2009, 1985.

79. **Miller, L. G., Schardein, J., Matsubara, Y., and Ohgo, T.,** Teratology study of nitroglycerin ointment by dermal administration in rabbits, *Shinyaku to Rinsho,* 34, 2024, 1985.

80. **Bellows, R. A., Rumsey, T. S., Kasson, C. W., Bond, J., Warwick, E. J., and Pahnich, O. F.,** Effects of organic phosphate systemic insecticides on bovine embryonic survival and development, *Am. J. Vet. Res.,* 36, 1133, 1975.

81. **Rumsey, T. S., Cabell, C. A., and Bond, J.,** Effect of an organic phosphorus systemic insecticide on reproductive performance in rats, *Am. J. Vet. Res.,* 30, 2209, 1969.

82. **Palazzolo, R. J., McHard, J. A., Hobbs, E. J., Fancher, O. E., and Calandra, J. C.,** Investigation of the toxicologic properties of a phenylmethylcyclosiloxane, *Toxicol. Appl. Pharmacol.,* 21, 15, 1972.

83. **Masuda, F., Okamoto, K., and Inoue, K.,** Effects of linear alkyl benzenesulfonate spread on mouse skin during pregnancy on fetuses, *J. Food Hyg. Soc. Jpn.,* 14, 580, 1973.

84. **Daly, I. W., Schroeder, R. E., and Killeen, J. C.,** A teratology study of topically applied linear alkyl-benzene sulphonate in rats, *Food Cosmet. Toxicol.,* 18, 55, 1980.

85. **Palmer, A. K., Readshaw, M. A., and Neuff, A. M.,** Assessment of the teratogenic potential of surfactants, Part 1. LAS, AS, CLD, *Toxicology,* 3, 91, 1975.

86. **Ford, R. A., Api, A. M., and Palmer, T. M.,** The effect of phenylethyl alcohol applied dermally to pregnant rats, *Toxicologist,* 7, 175, 1987.

87. **Kobayashi, F., Hara, K., Hasegawa, Y., Takegawa, Y., Yoshida, Y., Yamagata, H., Fukiishi, Y., Ando, M., and Ito, M.,** Reproduction studies of the antimycotic 710674-S in rats and rabbits, *Clin. Rep.,* 18, 4917, 1984.

88. **Schardein, J. L., Hentz, D. L., Petrere, J. A., Fitzgerald, J. E., and Kurtz, S. M.,** The effect of vidarabine on the development of the offspring of rats, rabbits, and monkeys, *Teratology,* 15, 231, 1977.

89. **Akhabadze, A. F., Karoleva, N. B., and Kovanova, E. K.,** Experimental study of substances used in cosmetics containing paragroups, applied to the skin, *Vestn. Dermatol. Venerol.,* 6, 23, 1981.

90. **Meyer, O., Andersen, P. H., Hansen, E. V., and Larsen, J. C.,** Teratogenicity and *in vitro* mutagenicity studies on nonoxynol-9 and -30, *Pharmacol. Toxicol.,* 62, 236, 1988.

91. **Burnett, C., Goldenthal, E. I., Harris, S. B., Wazeter, F. X., Strausburg, I., Kapp, R., and Voelker, R.,** Teratology and percutaneous toxicity studies on hair dyes, *J. Toxicol. Environ. Health,* 1, 1027, 1976.

92. **Khan, M. A.,** Reproduction in mice treated with crufomate: effects of dermal treatments on female mice and on their progeny, *J. Environ. Sci. Health,* (B), 13, 169, 1978.

93. **Mast, T. J., Rommerheim, R. L., and Springer, D. L.,** Coal-derived mixtures: teratogenicity of their dermally applied chemical class fractions in the rat, *Toxicologist,* 7, 4, 1987.

94. **Zangar, R. C., Springer, D. L., Buschbom, R. L., and Mahlum, D. D.,** Comparison of fetotoxic effects of a dermally applied complex organic mixture in rats and mice, *Fundam. Appl. Toxicol.,* 13, 662, 1989.

95. **Mahlum, D. D. and Springer, D. L.,** Teratogenic response of the rat and mouse to a coal liquid after dermal administration, *Toxicologist,* 6, 94, 1986.

96. **Andrew, F. D., Mahlum, D. D., and Petersen, M. R.,** Developmental toxicity of solvent refined coal-related hydrocarbons, *Toxicol. Appl. Pharmacol.,* 48, A27, 1979.

97. **Drozdowicz, B. Z., Schardein, J. L., and Golberg, L.,** Developmental toxicity of dermally-applied SRC-1 coal liquefaction product in the rat, *J. Am. Coll. Toxicol.,* 4, A108, 1985.

98. **Feuston, M. H., Kerstetter, S. L., Singer, E. J., and Mehlman, M. A.,** Maternal and developmental toxicity of dermally applied clarified slurry oil in rats, *Toxicologist,* 7, 174, 1987.

99. **Stula, E. F. and Krauss, W. C.,** Embryotoxicity in rats and rabbits from cutaneous application of amide-type solvents and substituted ureas, *Toxicol. Appl. Pharmacol.,* 41, 35, 1977.

100. **Kreybig, T., Preussmann, R., and Kreybig, I.,** Chemische Konstitution und teratogene Wirkung bei der Ratte. II. N-Alkylharnstoffe, N-alkylsulfonamide, N,N-dialkylacetamide, N-methylthioacetamide, chloracetamide, *Arzneim. Forsch.,* 19, 1073, 1969.

101. **Johannsen, F. R., Levinskas, G. J., and Schardein, J. L.,** Teratogenic response of dimethylacetamide in rats, *Fundam. Appl. Toxicol.,* 9, 550, 1987.

102. **Liu, S. L., Mercieca, M. D., Markham, J. K., Pohland, R. C., and Kenel, M. F.,** Developmental toxicity studies with N-methylformamide (NMF) administered orally to rats and rabbits, *Teratology,* 39, 466, 1989.

103. **Merkle, J. and Zeller, H.,** Studies on acetamides and formamides for embryotoxic and teratogenic activities in the rabbit, *Arzneim. Forsch.,* 30, 1557, 1980.

104. **Nolen, G. A., Gibson, W. B., Benedict, J. H., Briggs, D. W., and Schardein, J. L.,** Fertility and teratogenic studies of diethylene glycol monobutyl ether in rats and rabbits, *Fundam. Appl. Toxicol.,* 5, 1137, 1985.

105. **Hardin, B. D., Goad, P. T., and Burg, J. R.,** Developmental toxicity of four glycol ethers applied cutaneously to rats, *Environ. Health Perspect.,* 57, 69, 1984.
106. **Scortichini, B. H., John-Green, J. A., Quast, J. F., and Rao, K. S.,** Teratologic evaluation of dermally applied diethylene glycol monomethyl ether in rabbits, *Fundam. Appl. Toxicol.,* 7, 68, 1986.
107. **Price, C. J., Kimmel, C. A., Tyl, R. W., and Marr, M. C.,** The developmental toxicity of ethylene glycol in rats and mice, *Toxicol. Appl. Pharmacol.,* 81, 113, 1985.
108. **Price, C. J., Tyl, R. W., Marr, M. C., and Kimmel, C. A.,** Teratologic evaluation of ethylene glycol (EG) in CD rats and CD-1 mice, *Teratology,* 29, 52A, 1984.
109. **Tyl, R. W., Ballantyne, B., Fisher, L. C., Fait, D. L., Savine, T. A., Klonne, D. R., Pritts, I. M., and Dodd, D. E.,** Developmental toxicity of ethylene glycol (EG) aerosol by whole-body exposure in CD rats and CD-1 mice, *Toxicologist,* 9, 270, 1989.
110. **Tyl, R. W., Fisher, L. C., Kubena, M. F., Losco, P. E., and Vrbanic, M. A.,** Determination of a developmental toxicity "no observable effect level" (NOEL) from ethylene glycol (EG) by gavage in CD-1 mice, *Teratology,* 39, 487, 1989b.
111. **Tyl, R. W., Pritts, I. M., France, K. A., Fisher, L. C., and Tyler, T. R.,** Developmental toxicity evaluation of inhaled 2-ethoxyethanol acetate in Fischer 344 rats and New Zealand White rabbits, *Fundam. Appl. Toxicol.,* 10, 20, 1988b.
112. **Hardin, B. D., Niemeier, R. W., Kuczuk, M. H., Mathinos, P. R., and Weaver, T. F.,** Teratogenicity of 2-ethoxyethanol by dermal application, *Teratology,* 25, 46A, 1982a.
113. **Hardin, B. D., Niemeier, R. W., Smith, R. J., Kuczuk, M. H., Mathinos, P. R., and Weaver, T. F.,** Teratogenicity of 2-ethoxyethanol by dermal application, *Drug. Chem. Toxicol.,* 6, 277, 1982b.
114. **Andrew, F. D. and Hardin, B. D.,** Developmental effects after inhalation exposure of gravid rabbits and rats to ethylene glycol monoethyl ether, *Environ. Health Perspect.,* 57, 13, 1984.
115. **Cooper, J. R., Hobson, D. W., and Olson, C. T.,** Teratogenic effects of ethylene glycol monomethyl ether following dermal exposure in rats, *Toxicologist,* 6, 94, 1986.
116. **Cooper, J. C. and Hobson, D. W.,** Enhanced teratogenicity of ethylene glycol monomethyl ether when administered dermally in solution with water, *Toxicologist,* 9, 269, 1989.
117. **Wickramaratne, G. A. S.,** The teratogenic potential and dose-response of dermally administered ethylene glycol monomethyl ether (EGME) estimated in rats with the Chernoff-Kavlock assay, *J. Appl. Toxicol.,* 6, 165, 1986.
118. **Gibson, W. B., Nolen, G. A., and Christian, M. S.,** Determination of the developmental toxicity potential of butoxypropanol in rabbits after topical administration, *Fundam. Appl. Toxicol.,* 13, 359, 1989.
119. **Ritter, E. J., Scott, W. J., Randall, J. L., and Ritter, J. M.,** Teratogenicity of di(2-ethylhexyl)phthalate, 2-ethylhexanol, 2-ethylhexanoic acid, and valproic acid, and potentiation by caffeine, *Teratology,* 35, 41, 1987.
120. **Becci, P. J., Knickerbocker, M. J., Reagan, E. L., Parent, R. A., and Burnette, L. W.,** Teratogenicity study of N-methylpyrrolidone after dermal application to Sprague-Dawley rats, *Fundam. Appl. Toxicol.,* 2, 73, 1982.
121. **Lee, K. P., Chromey, N. C., Culik, R., Barnes, J. R., and Schneider, P. W.,** Toxicity of N-methyl-2-pyrrolidone (NMP): teratogenic, subchronic, and two-year inhalation studies, *Fundam. Appl. Toxicol.,* 9, 222, 1987.
122. **Merkle, J., Klimisch, H.-J., and Jackh, R.,** Prenatal toxicity of 2-methoxy-propylacetate-1 in rats and rabbits, *Fundam. Appl. Toxicol.,* 8, 71, 1987.
123. **Kennedy, G. L., Lu, M.-H., and McAlack, J. W.,** Teratogenic evaluation of 1,1,3,3-tetrabutylurea in the rat following dermal exposure, *Food Chem. Toxicol.,* 25, 173, 1987.
124. **Gleiberman, S. E., Volkova, A. P., Nikolaev, G. M., and Zhukova, E. V.,** Embryonic properties of the repellent diethyltoluamide, *Farmakol. Toksikol.,* 38, 202, 1975.
125. **Angerhofer, R. A. and Weeks, M. H.,** Effect of dermal applications of N,N-diethyl-meta-toluamide (M-DET) on the embryonic development of rabbits, Rep. (AD-A094778) 1981.
126. **Chernoff, N., Kavlock, R. J., Rogers, E. H., Carver, B. D., and Murray, S. J.,** Perinatal toxicity of maneb, ethylene thiourea and ethylenebisisothiocyanate sulfide in rodents, *J. Toxicol. Environ. Health,* 5, 821, 1979.
127. **Dilley, J. V., Chernoff, N., Kay, D., Winslow, N., and Newell, G. W.,** Inhalation teratology studies of five chemicals in rats, *Toxicol. Appl. Pharmacol.,* 41, 96, 1977.
128. **Angerhofer, R. A., Weeks, M. H., and Pope, C. R.,** Prenatal response of rabbits to various Abate formulations, *Pharmacologist,* 20, 201, 1978.
129. **Francis, B. M. and Metcalf, R. L.,** Percutaneous teratogenicity of nitforen, *Teratology,* 25, 41A, 1982.
130. **Costlow, R. D., Hirsekorn, J. M., Stiratelli, R. G., O'Hara, G. P., Black, D. L., Kane, W. W., Burke, S. S., Smith, J. M., and Hayes, A. W.,** The effects on rat pups when nitrofen (4-(2,4-diclorophenoxy)nitrobenzene) was applied dermally to the dam during organogenesis, *Toxicology,* 28, 37, 1983.

131. **Gray, L. E., Kavlock, R. J., Chernoff, N., Ostby, J., and Ferrell, J.,** Postnatal developmental alterations following prenatal exposure to the herbicide 2,4-dichlorophenyl-p-nitrophenyl ether: a dose response evaluation in the mouse, *Toxicol. Appl. Pharmacol.*, 67, 1, 1983.

132. **Costlow, R. D. and Manson, J. M.,** Herbicide-induced hydronephrosis and respiratory distress: effects of *in utero* exposure to nitrofen (2,4-dichloro-4'-nitrophenyl ether), *Toxicol. Appl. Pharmacol.*, A21, 1980.

133. **Hood, R. D., Baxley, M. N., and Harrison, W. P.,** Evaluation of chromated copper arsenate (CCA) for teratogenicity, *Teratology*, 19, 31A, 1979.

134. **Breslin, W. J., Kirk, H. D., and Johnson, K. A.,** Teratogenic evaluation of diglycidyl ether of bisphenol A (DGEBPA) in New Zealand White rabbits following dermal exposure, *Fundam. Appl. Toxicol.*, 10, 736, 1988.

135. **Hansen, E. and Meyer, O.,** No embryotoxic or teratogenic effect of dimethyl phthalate in rats after epicutaneous application, *Pharmacol. Toxicol.*, 64, 237, 1989.

136. **Tanaka, C., Siratori, K., Ikegami, K., and Wakisaka, Y.,** A teratological evaluation following dermal application of diethyl phthalate to pregnant mice, *Oyo Yakuri*, 33, 387, 1987.

137. **Kennedy, G. L., Keplinger, M. L., Calandra, J. C., and Hobbs, E. J.,** Reproductive, teratologic and mutagenic studies with some polydimethylsiloxanes, *J. Toxicol. Environ. Health*, 1, 909, 1976.

138. **FDRL Rep., July 21, 1967,** Teratological tests in rats and rabbits with Dow Corning TX-158E (Phase II)-Dow Corning 556 Cosmetic Grade Fluid Lot No. 7.

139. **Brittelli, M. R., Culik, R., Dashiell, O. L., and Fayerweather, W. E.,** Skin absorption of hexafluoroacetone: teratogenic and lethal effects in the rat fetus, *Toxicol. Appl. Pharmacol.*, 47, 35, 1979.

140. **Mullin, L. S., Valentine, R., and Chromey, N. C.,** hexafluoroacetone developmental toxicity in rats, *Toxicologist*, 10, 41, 1990.

141. **Broitman, A. Y., Danishevskii, S. L., and Robachevskaya, E. G.,** Embryotropic-mutagenic effect of chemical substances, in *Toksikol. Vysokomol. Mater. Khim. Syr'ya Ikh. Sin., Gos. Nauch.-Issled. Inst. Polim. Plast. Mass.*, 1966, 297–317.

142. **Takahashi, A., Ando, H., Kubo, Y., and Hiraga, K.,** Effects of dermal application of sodium dodecyl sulfate (SDS) on pregnant mice and their fetuses, *Tokyo Toritsu. Eisei Kenkyusho Nempo*, 27, 113, 1976.

143. **Noda, K., Numata, H., Hirabayashi, M., and Endo, I.,** Influence of pesticides on embryos. I. On influence of organophosphoric pesticides, *Oyo Yakuri*, 6, 667, 1972.

144. **Lu, C. C., Cagen, S. Z., Darmer, K. I., and Patterson, D. R.,** Testicular effects induced by dermal and inhalation exposure to para-tertiary butyl benzoic acid (ptBBA) in Fischer 344 rats, *J. Am. Coll. Toxicol.*, 6, 233, 1987.

145. **Feuston, M. H., Bodnar, K. R., Kerstetter, S. L., Grink, C. K., Belcak, M. J., and Singer, E. J.,** Reproductive toxicity of 2-methoxyethanol applied dermally to occluded and nonoccluded sites in male rats, *Toxicol. Appl. Pharmacol.*, 100, 145, 1989a.

146. **Orfanos, C. E., Ehlert, R., and Gollnick, H.,** The retinoids. A review of their clinical pharmacology and therapeutic use, *Drugs*, 34, 459, 1987.

147. *Physicians Desk Reference* (PDR), 44th ed., Medical Economics, Oradell, N.J., 1990.

148. **Katz, V. L., Thorp, J. M., and Bowes, W. A.,** Severe symmetric intrauterine growth retardation associated with the topical use of triamcinolone, *Am. J. Obstet. Gynecol.*, 162, 396, 1990.

149. **Hendrickx, A. G., Sawyer, R. H., Terrell, T. G., Osburn, B. I., Hendrickson, R. V., and Steffek, A. J.,** Teratogenic effects of triamcinolone on the skeletal and lymphoid systems in nonhuman primates, *Fed. Proc.*, 34, 1661, 1975.

150. **Marzulli, F. N. and Maibach, H. I.,** *Dermatotoxicology*, 2nd ed., Hemisphere, Washington, D.C., 1983.

151. **Powell, H., Swarner, O., Gluck, L., and Lampert, P.,** Hexachlorophene myelinopathy in premature infants, *J. Pediatr.*, 82, 976, 1973.

152. **Halling, H.,** [Suspected connection between exposure to hexachlorophene and birth of malformed infants], *Lakartidningen*, 74, 542, 1977.

153. **Check, W.,** New study shows hexachlorophene is teratogenic in humans, *J. Am. Med. Assoc.*, 240, 513, 1978.

154. **Kallen, B.,** Hexachlorophene teratogenicity in humans disputed, *J. Am. Med. Assoc.*, 240, 1585, 1978.

155. **Baltzar, B., Ericson, A., and Kallen, B.,** Pregnancy outcome among women in Swedish hospitals, *N. Engl. J. Med.*, 300, 627, 1979.

156. **Janerich, D. T.,** Environmental causes of birth defects: the hexachlorophene issue, *J. Am. Med. Assoc.*, 241, 830, 1979.

157. **LeMoine Parker, M. and Coggins, G.,** The use of clonidine, Catapres, in hypertensive and toxemic syndromes of pregnancy, *Aust. N. Z. J. Med.*, 3, 432, 1973.

158. **Shaw, J. E. and Mitchell, C.,** Dermal drug delivery systems: a review. *J. Toxicol. Cutan. Ocular Toxicol.*, 2, 249, 1983–1984.

159. **Whorton, D., Krauss, R. M., Marshall, S., and Milby, T. H.,** Infertility in male pesticide workers, *Lancet,* 2, 1259, 1977.

160. **Potashnik, G., Ben-Aderet, N., Israeli, R., Yanai-Inbar, I., and Sober, I.,** Suppressive effect of 1,2-dibromo-3-chloropropane on human spermatogenesis, *Fertil. Steril.,* 30, 444, 1978.

161. **Torkelson, T. R., Sadek, S. E., and Rowe, V. K.,** Toxicologic investigation of 1,2-dibromo-3-chloropropane, *Toxicol. Appl. Pharmacol.,* 3, 545, 1961.

162. **Egnatz, D. G., Ott, M. G., Townsend, J. C., Olson, R. D., and Johns, D. B.,** DBCP and testicular effects in chemical workers: an epidemiological survey in Midland, Michigan, *J. Occup. Med.,* 22, 727, 1980.

163. **Cook, R. R., Van Peenen, P. F. D., Bodner, K. M., Dickson, G. S., Kolesar, R. C., Uhlmann, C. S., and Flanagan, K.,** A cross sectional study of ethylene glycol monomethylether process employees, *Arch. Environ. Health,* 37, 346, 1982.

164. National Institute for Occupational Safety and Health (NIOSH), Health hazard evaluation report: Precision Castparts Corporation, HETA 84-415-1688, 1986.

165. **Welch, L. S., Schrader, S. M., Turner, T. W., and Cullen, M. R.,** Effects of exposure to ethylene glycol ethers on shipyard painters. II. Male reproduction, *Am. J. Ind. Med.,* 14, 509, 1988.

166. **Uzych, I.,** Human male exposure to vinyl chloride and possible teratogenic and mutagenic risks: a review, *Hum. Toxicol.,* 7, 517, 1988.

167. **Winter, R.,** *Cancer-Causing Agents. A Preventive Guide,* Crown Publishers, New York, 1979.

168. **Marzulli, F. N. and Brown, D. W. C.,** Potential systemic hazard of topically applied mercurials, *J. Soc. Cosmet. Chem.,* 23, 875, 1972.

169. **Barr, R. D., Woodger, B. A., and Rees, P. H.,** Levels of mercury in urine correlated with the use of skin lightening creams, *Am. J. Clin. Pathol.,* 59, 36, 1973.

170. **Lauwerys, R., Bonnier, C., Evrard, P., Gennart, J., and Bernard, A.,** Prenatal and early postnatal intoxication by inorganic mercury resulting from the maternal use of mercury containing soap, *Hum. Toxicol.,* 6, 253, 1987.

171. **Murphy, S.,** Toxic effects of pesticides, in *Casarett and Doull's Toxicology, The Basic Science of Poisons,* Klassen, C. D., Amdur, M. O., and Doull, J., Eds., Macmillan, New York, 1986.

172. **Lindhout, D. and Hageman, G.,** Amyoplasia congenita-like condition and maternal malathion exposure, *Teratology,* 36, 7, 1987.

173. **U.S. Environmental Protection Agency (EPA),** Epidemiology Studies Division, Six years spontaneous abortion rates in Oregon areas in relation to forest 2,4,5-T spray practices, February 1979.

174. **Stevens, K. M.,** Agent Orange toxicity: a quantitative perspective, *Hum. Toxicol.,* 1, 31, 1981.

175. **Cutting, R. T., Phuoc, T. H., Ballo, J. M., Benenson, M. W., and Evans, C. H.,** Congenital malformations, hydatiform moles and stillbirths in the Republic of Vietnam 1960–1969. GPO, Washington, D.C., 1970.

176. **Courtney, K. D., Gaylor, D. W., Hogan, M. D., Falk, H. L., Bates, R. R., and Mitchell, I.,** Teratogenic evaluation of 2,4,5-T, *Science,* 168, 864, 1970.

177. **Emerson, J. L., Thompson, D. J., Strebing, R. J., Gerbig, C. G., and Robinson, V. B.,** Teratogenic studies on 2,4,5-trichlorophenoxyacetic acid in the rat and rabbit, *Food Cosmet. Toxicol.,* 9, 395, 1971.

178. **Khera, K. S., Huston, B. L., and McKinley, W. P.,** Pre- and postnatal studies on 2,4,5-T, 2,4-D, and derivatives in Wistar rats, *Toxicol. Appl. Pharmacol.,* 19, 369, 1971.

179. **Pocchiari, F., Silano, V., and Zampieri, A.,** Human health effects from accidental release of tetrachlorodibenzo-p-dioxin (TCDD) at Seveso, Italy, *Ann. N.Y. Acad. Sci.,* 320, 311, 1979.

180. **Young, A. L., Calcagni, J. A., Thalker, C. E., and Tremblay, J. W.,** The toxicology, environmental fate, and human risk of herbicide orange and its associated dioxin, U.S. Air Force, OEHL Technical Rep., 78–92, NTIS, 1978.

181. **Commoner, B.,** Seveso: the tradegy lingers on, *Clin. Toxicol.,* 11, 479, 1977.

182. **Whiteside, T.,** *The Pendulum and the Toxic Cloud. The Course of Dioxin Contamination,* Yale University Press, New Haven, CT, 1979.

183. **Reggiani, G.,** Medical problems raised by TCDD contamination in Seveso, Italy, *Arch. Toxicol.,* 40, 161, 1978.

184. **Maddy, K. T., Johnston, L., Smith, C., Schneider, F., and Jackson, T.,** A study of dermal and inhalation exposure of mixer-loaders and applicators to nitrofen in Monterey and Santa Barbara Counties of California, California Dept. of Food and Agriculture Rep. HS-745, 1980.

185. **Engstrom, K., Husman, K., and Riihmaki, V.,** Percutaneous absorption of m-xylene in man, *Int. Arch. Occup. Environ. Health,* 39, 181, 1977.

186. **Shepard, T. H.,** *Catalog of Teratogenic Agents,* 5th ed., The Johns Hopkins University Press, Baltimore, MD, 1986.

187. **U.S. Environmental Protection Agency (EPA),** Integrated Risk Information System (IRIS), Online, Office of Health and Environmental Assessment, Environmental Criteria and Assessment Office, Cincinnati, OH, 1990.

188. **Guenther, T. M. and Mannering, G. T.,** Induction of hepatic monooxygenase systems of pregnant rats with phenobarbital and 3-methylcholanthrene, *Biochem. Pharmacol.,* 26, 577, 1977.

Chapter 17

DERMAL CARCINOGENICITY STUDIES OF PETROLEUM-DERIVED MATERIALS

Richard H. McKee and James J. Freeman

TABLE OF CONTENTS

0-8493-7357-3/93/$0.00 + $.50
© 1993 by CRC Press, Inc.

I. INTRODUCTION

Experimental skin carcinogenesis was first demonstrated in the rabbit ear model in 1915,[1] and, shortly thereafter, in mouse skin.[2] The purpose of these early studies was to develop a method to experimentally confirm putative occupational skin carcinogens. The method proved to be quite valuable; human skin carcinogens which also produced tumors in mouse skin included coal liquids, shale oils, and chimney soot.

Of particular interest to the petroleum industry was the demonstration that unrefined lubricating oils which were carcinogenic in humans also induced mouse skin tumors. In 1924 an unrefined lubricating oil associated with excess cancer among workers in the British textile industry was confirmed as a skin carcinogen by testing in mice.[3] Similar confirmation in mice was obtained in studies of two other petroleum-derived materials believed capable of causing skin cancer in humans, certain cutting fluids[4-7] and some process streams related to the manufacture of paraffin wax.[8,9] These and other similar studies provided evidence that human skin cancer could be reproduced in the mouse skin model.

The materials listed above were complex mixtures of variable composition, and their chemical constituents were largely unknown when the studies were conducted. Fractionation procedures were then developed, and with the mouse skin bioassay it became possible to test the carcinogenic activity of various fractions. Using these procedures, the polycyclic aromatic hydrocarbon (PAH) carcinogens benzo(a)pyrene (BAP) and 7,12-dimethyl-benz(a)anthracene (DMBA) were isolated from coal tar.[10] Subsequent studies also identified PAH as the species responsible for the dermal carcinogenic activity of certain petroleum- and shale-derived liquids. Thus the mouse skin carcinogenesis assay has been a valuable research tool.

These initial biological and chemical data suggested that the dermal carcinogenic potential of petroleum streams could be explained by PAH content. However, not all of the data are consistent with that hypothesis. Several recent studies assessed the carcinogenic potential of petroleum-derived middle distillate fuels which contain very low levels of PAH.[11-13] All of these fuels showed evidence of weak tumorigenic activity characterized by tumor yields of usually less than 20% and tumor latencies often exceeding 2 years. However, the tumor yields were often significantly elevated above control, and the tumors commonly became malignant. The carcinogenic activity appeared to be unrelated to PAH content.[14] Unlike the results of studies of unrefined lubricating oils, the clinical significance of the middle distillate tumor response data is unclear.

The historical evidence indicates that the mouse dermal carcinogenesis assay is a well-validated tool for the confirmation of possible human skin carcinogens and that it has also been useful for certain experimental purposes. However, as the focus of animal testing has shifted from confirmation of human disease to the identification of potential human health hazards, the prospective value of dermal carcinogenicity bioassays has been called into question. Specifically: (1) is the dermal carcinogenesis assay applicable only to those types of materials for which it has been validated; (2) could the tumors induced by chronic administration of middle distillate fuels be an effect without clinical significance; and (3) do all materials which cause tumors in mice necessarily pose a cancer hazard to humans? The objectives of this paper are to summarize the available data on petroleum-derived materials and to describe some research efforts to answer these questions.

II. MATERIALS TO BE DISCUSSED

This paper contrasts two types of petroleum products, lubricating oils and middle distillate fuels. Because these materials may not be well known outside the refinery industry, a short description of the manufacturing processes is provided.

Crude oil is converted into products by distillation and refining. The initial process, atmospheric distillation, fractionates crude oil by boiling point. Typical fractions include gases (methane, ethane, propane, butane; <90°C); naphtha (i.e., gasoline; 90 to 200°C); kerosene (160 to 290°C); light gas oil (215 to 340°C); and residuum.

A. MIDDLE DISTILLATE FUELS

The kerosene and light gas oil fractions are further refined into kerosene, jet fuel, diesel fuel, and residential heating oil, the *middle distllate fuels*. These products distill between approximately 180 to 370°C and contain both saturated and aromatic constituents. However, because of the boiling range, concentrations of carcinogenic polycyclic aromatic hydrocarbons (PAH) are usually very low.

B. LUBRICATING OILS

Lubricating base oils are substances produced by vacuum distillation of atmospheric residuum. These substances are complex mixtures of straight- and branded-chain paraffinic, naphthenic, and aromatic hydrocarbons having carbon numbers of 15 or more and boiling points of approximately 300 to 600°C. The *unrefined lubricating base oils* may contain relatively high levels of PAH. Refining processes such as solvent-extraction and catalytic hydrogen treatment reduce PAH levels, resulting in noncarcinogenic *base stocks*. The lubricating oil base stocks are then combined with additives to produce *lubricants* such as engine oils.

III. COMPARISON OF RESPONSES IN "CONVENTIONAL" EPIDERMAL CARCINOGENESIS BIOASSAYS

A "conventional" epidermal carcinogenesis bioassay is a study in which test material is applied to the backs of mice several times weekly for a period of time approximating the lifetimes of the treated animals (i.e., approximately 18 to 24 months). There are differences in test methods including sex and/or strain of the mice, frequency and volume of application, and application vehicle; any of which might affect the quantitative aspects of the assay.[15] However, these minor differences in protocol are unlikely to produce qualitative differences in study outcome.

Complete carcinogenesis studies as conducted by Exxon utilize the male C_3H mouse, at least 40 animals/group. Test material is applied in doses of 75 μl/week material/week (3×25 μl or 2×37.5 μl); and the animals are treated for at least 24 months. Following sacrifice or spontaneous death, microscopic examination of the grossly identified tumors, sections of skin from the treated area and any other grossly observable lesions is conducted. More extensive microscopic examination is conducted when warranted by the experimental objectives.

A. RESULTS OF STUDIES WITH LUBRICATING BASE OILS

Among the historical references are several reports in which exposure to lubricating oils formulated from unrefined lubricating base oils was associated with the development of human skin cancer. Dermal carcinogenesis studies sponsored by Exxon conclusively demonstrated that the relatively high boiling point material from an atmospheric distillation unit which was used for lubricating oil manufacture was carcinogenic. It was also found that the carcinogenic activity was associated with the aromatic constituents of these distillates, substances which were later shown to be PAH. The paraffinic fraction remaining after removal of the aromatics was inactive in dermal carcinogenesis bioassays.[16]

TABLE 1
Results of Dermal Carcinogenicity Studies of an Unrefined Lubricating Oil and Its Aromatic and Saturated Components[a]

	Unrefined lubricating oil	Aromatic fractions		Paraffinic fractions	
		Phenol	NMP[c]	Phenol	NMP
Fraction of tumor bearing animals[b]	0.58	0.90	0.92	0	0
Median latency (weeks)	69	51	45	—	—

  [a] Data originally published in Reference 17.
  [b] Each study group initially contained 40 C$_3$H mice.
  [c] NMP: *n*-methylpyrrolidone.

TABLE 2
Results of Dermal Carcinogenicity Studies of Unrefined and Severely Hydrotreated Lubricating Oils[a]

Sample	Median survival (weeks)	Total tumor-bearing animals	Median time to tumor (weeks)
Unrefined light naphthenic base oil	70[b]	20/40 (20C, 0P)[c,d]	77 (72–84)[e]
Hydrotreated light naphthenic base stock	80	0/40	—
Hydrotreated light naphthenic base stock	79	0/40	—
Unrefined heavy naphthenic base oil	73[b]	24/40 (21C, 3P)[c]	74 (70–80)[e]
Hydrotreated heavy naphthenic base stock	87	0/40	—
Hydrotreated heavy naphthenic base stock	84	0/40	—
Unrefined heavy naphthenic base oil	86	2/40 (1C, 1P)	136 (110–195)
Hydrotreated heavy naphthenic base oil	80	0/40	—
Hydrotreated heavy naphthenic base oil	76	0/40	—
BAP (0.1% in toluene)	70[b]	22/40 (21C, 1P)[c]	73 (69–78)[e]
White mineral oil (negative control)	87	0/40	—
Toluene	82	0/40	—

  [a] Data taken from Reference 18.
  [b] Survival significantly reduced as compared to control.
  [c] Tumor yield significantly elevated with respect to control.
  [d] Tumor response given as the most advanced tumor type in the treatment area. The tumor distribution is given in parentheses: C = carcinoma, P = papilloma.
  [e] Median latency significantly different from control.

Similar programs continue to the present day. One recent study (Table 1), compared the carcinogenic activity of the aromatic (i.e., solvent-extractable) and paraffinic fractions of crude petroleum oil. The object was to compare two extraction solvents, *n*-methyl pyrrolidone and phenol. The starting material, an unrefined lubricating oil, was carcinogenic, producing tumors in approximately half of the treated mice with a median latency of 69 weeks. This activity was concentrated in the aromatic-extractable fractions; the remaining paraffinic material which was used to formulate lubricants was not carcinogenic.[17] The experiment also showed that the extraction efficiencies of NMP and phenol were equivalent.

In a second study (Table 2) the PAH constituents in unrefined lubricating base oils were removed by hydrotreatment, a process by which the feedstocks were treated with hydrogen under conditions of high temperature and pressure. This process, which saturated the aromatic

TABLE 3
Results of Complete Carcinogenesis Studies
of a Lightly Refined Paraffinic Oil

Year of study initiation	Tumor response[a]	Median latency (weeks)[b]
1975	3/40	133
1979[c]	9/50	115
1985	4/50	105

[a] Tumor response is given as the number of tumor-bearing animals over the number of animals initially on test.
[b] The median latency was estimated by the Weibull method.
[c] Data on this sample previously published in Reference 11.

rings, produced results similar to those obtained by solvent extraction. In carcinogenesis studies the unrefined lubricant base oils produced tumors, but the hydrotreated products did not.[18]

These and many other studies clearly demonstrated that unrefined lubricant base oils contained PAH and could produce skin cancer. That knowledge led to the development of industrial processes which reduced PAH levels to yield non-carcinogenic base stocks as well as to recommendations for industrial hygiene practices which minimized dermal contact. That these high quality lubricating base stocks and the products prepared from them are not skin carcinogens has also been amply demonstrated by mouse skin bioassays.[18-24]

B. RESULTS OF STUDIES WITH MIDDLE DISTILLATE FUELS

In a program conducted by the petroleum industry, petroleum crude oil was separated into a series of fractions of differing boiling range;[25] and each fraction was then tested to assess carcinogenic potential.[12] Finally, the carcinogenic activity of each fraction was quantified and compared to that of BAP.[26] One finding was that the fractions boiling between 175 to 285°C and 285 to 400°C produced tumors. This was unexpected because the PAH constituents present in material boiling below approximately 350°C were at levels well below those which might be expected to produce skin cancer. Further, when the aromatic and saturated constituents of these fractions were tested separately, the majority of the activity in the 175 to 285°C fraction was found to be associated with the saturated components.[12]

It was apparent that, for these middle distillate fuels, the carcinogenic response data could not be explained in terms of PAH content specifically or even in terms of aromatic material in general. Rather the activity appeared to be associated with the paraffinic components, species which were previously regarded as noncarcinogenic.

As further data accumulated, it became apparent that as a class the middle distillate fuels consistently produced tumors in mouse skin. These responses were characterized by low tumor frequency and long median latency, but at least a few tumors were found in every experiment.[11-13,27] As one example, between 1975 and 1989 three complete carcinogenesis studies with a lightly refined petroleum oil (LRPO) having an approximate boiling range of 250 to 320°C (Table 3) were conducted. In all three trials this material produced a weak but consistent tumor response. These data confirmed that, although the tumorigenic activity was weak, it was reproducible. Similar results have been reported in every other published study of this type of material.

IV. COMPARISON OF RESPONSES IN INITIATION/PROMOTION ASSAYS

For purposes of this paper, initiation/promotion assays are defined as studies in which mice are treated for a short period of time with a substance which converts some normal skin cells to a pre-neoplastic state (the initiation phase) and then for a substantially longer period of time with a substance which facilitates the progression of these transformed foci into tumors (the promotion phase). Commonly DMBA is used to initiate mouse skin tumors and 12-*O*-tetradecanoyl-phorbol-13-acetate (TPA) is used for promotion. During the initiation phase the animals are treated with either 10 or 50 μg DMBA or with six 25 μl treatments (over a 2-week period) of substances with unknown initiating potential. The promotion phase consists of treatment three times weekly with either 2.5 μg TPA or 25 μl of test agents. These studies are conducted in CD-1 mice, and usually for 1 year (rather than the more typical 26 weeks) because of an attempt to detect weakly acting agents. Somewhat different protocols are used by other investigators, but, as with the complete carcinogenesis studies, it is anticipated that these differences might result in quantitative but not qualitative differences in study outcome.

A. RESULTS OF STUDIES WITH LUBRICATING BASE OILS

An unrefined lubricating base oil was active both as an initiator and as a promoter in a two-stage carcinogenesis study.[27] Thus, this unrefined lubricating base oil was capable of initiating neoplastic transformation and also of facilitating tumor progression. The role of PAH in these processes was confirmed by a second study in which an oil with the PAH fraction removed was inactive in both the initiating and promoting portions of the assay.[27] In other studies pure PAH have been found to have both initiating and promoting properties. Thus, the results of these investigations were consistent with the known toxicological properties of the constituents of the test material.

B. RESULTS OF STUDIES WITH MIDDLE DISTILLATE FUELS

When mice were initiated with LRPO and then promoted with TPA, fewer tumors were seen than in uninitiated mice which were promoted with TPA (Table 4). However, when DMBA-initiated mice were promoted with LRPO, a statistically significant tumor response was produced. Thus, these middle distillates were capable of promotion, i.e., facilitating tumor growth and development but showed no evidence of initiating potential. Similar results were obtained in a study of furnace oil by Gerhart et al.,[27] and in a study of kerosene by Marino et al.[28]

V. COMPARISON OF RESPONSES IN GENETIC TOXICOLOGY ASSAYS

Genetic toxicology assays are often conducted in bacterial or other cellular systems to determine if test materials are mutagenic. Results of mutagenic studies may have mechanistic significance. Chemicals which are carcinogenic and also mutagenic are believed to produce tumors through genotoxic processes. Chemicals which are carcinogenic but not mutagenic are believed to cause tumor development through secondary (i.e., non-genotoxic) mechanisms.

A. RESULTS OF STUDIES WITH LUBRICATING BASE OILS

In early studies essentially all petroleum-derived materials were inactive in unmodified Salmonella tests regardless of their intrinsic carcinogenic potential.[29] As some of these

TABLE 4
Results of an Initiation/Promotion Study of a Lightly
Refined Paraffinic Oil[a]

	Number of tumor-bearing mice	Number of tumors/mouse
Initiation study[b]		
LRPO/TPA	3 (P)	0.1 (3/30)
Acetone/TPA	9 (P)[c]	0.4 (12/30)[d]
DMBA/TPA	30 (29P, 1C)[c]	4.2 (126/30)[d]
Promotion study		
DMBA/LRPO	5 (P)[c]	0.4 (11/30)
DMBA/Acetone	0	0
DMBA/TPA	30 (P)[c]	4.5 (135/30)[d]

[a] Data taken from Reference 14.
[b] The initiation study was terminated at 26 weeks. The promotion study was continued for 52 weeks, thus accounting for the differences in the positive control data.
[c] Results significantly different from unpromoted control group ($p < 0.05$), one-tailed test.

Abbreviations
LRPO: Lightly refined paraffinic oil.
TPA: 12-O-tetradecanoyl-phorbol-13-acetate.
DMBA: 7,12-dimethyl(benz(a)anthracene).
P: Papilloma.
C: Carcinoma.

materials were known to contain PAH which are highly mutagenic, several research programs were initiated to develop modified assays which were more reflective of carcinogenic activity[30-32] or PAH content.[33]

Using these modified assays, it was shown that unrefined or mildly refined lubricating base oils from paraffinic crude oils and the aromatic constituents of these oils were mutagenic in Salmonella (Table 5).[34] Highly refined base stocks and products derived from them were inactive. Similar results were obtained in other *in vitro* assays including mouse lymphoma and morphologic transformation in Syrian hamster embryo cells (Table 5). The results of these studies are consistent with the hypothesis that the carcinogenic materials within this class are acting by a genotoxic mechanism.

B. RESULTS OF STUDIES WITH MIDDLE DISTILLATE FUELS

In contrast to the lubricating oil data, the carcinogenic responses of middle distillate fuels cannot be explained in terms of mutagenic constituents. Blackburn et al.[35] reported that at least some of these products contained no material which was mutagenic in Salmonella but nevertheless produced tumors following chronic dermal application. They speculated that the tumors might have been due to an indirect genotoxic effect or the result of some other process entirely. Similarly, a considerable effort was made to identify constituents of LRPO which were mutagenic in Salmonella; none were found.[14]

More recently several middle distillate fuels were evaluated in mouse micronucleus tests, and all produced negative results (Table 6). The absence of mutagenic potential suggested that the middle distillates were probably not genotoxic carcinogens.

The mutagenesis data mirror the results of the initiation/promotion studies. The unrefined lubricating base oils are almost certainly genotoxic carcinogens, and are capable of initiating tumors in mouse skin (although these materials apparently can also promote skin tumor

TABLE 5
Summarized Results of Mutagenicity Tests of Unrefined and Refined Lubricating Oils[a]

Sample	Carcinogenicity[b]	Salmonella	Mouse lymphoma	SHE transformation assay
Unrefined lubricating base oil	Positive	Positive	Positive	Positive
Hydrotreated aromatic extract	Positive	Positive	Positive	Positive
Solvent-extracted lubricant base stock (90 SUS)	Negative	Negative	Negative	Negative
Solvent-extracted lubricant base stock (150 SUS)	Negative	Negative	Negative	Negative
Hydrotreated lubricant base stock (150 SUS)	Negative	Negative	NT	Negative
Controls				
Catalytically cracked clarified oil	Positive	Positive	Positive	Positive
White oil	Negative	Negative	Negative	NT

[a]   Results previously summarized in Reference 34.
[b]   A positive result in a dermal carcinogenicity assay was defined as a statistically significant increase in the number of tumor-bearing animals. Samples listed as negative did not produce any tumors at all.

Abbreviations:
SHE: Syrian hamster embryo morphologic transformation assay.
SUS: Saybolt Universal Seconds, a measure of viscosity.
NT: Not tested.

development). The middle distillate fuels are not genotoxic and do not initiate mouse skin tumors although they do have promoting activity. Apparently these materials produce tumors through some non-genotoxic process associated with skin tumor promotion.

VI. STUDIES TO ELUCIDATE THE CARCINOGENESIS MECHANISMS OF MIDDLE DISTILLATE FUELS

It has been suggested that the middle ditillate fuels might produce tumors through a process related to irritation and chronic hyperplasia.[14] Certainly the role of irritation in the promotional process has been described,[36-39] and the irritant properties of petroleum middle distillates have been well documented.[40] Therefore, it is at least plausible that skin irritancy could explain middle distillate promotion. However, two problems remain: (1) how does one account for tumor initiation, and (2) why is it that not all irritants are promoters?

A. STUDIES TO ASSESS SPONTANEOUS INITIATION

Studies have shown that cells are present in the skin of SENCAR mice which are resistant to calcium-induced terminal differentiation. These cells have a number of properties in common with transformed cells, among these is the fact that the frequency of these cells in mouse skin is substantially increased if the mice are treated with DMBA.[41] Thus it is believed that these calcium-resistant cells are initiated and that those isolated from the skin of untreated mice reflect the spontaneous initiation frequency.

The frequency of these cells in the skin of C_3H mice, the strain used for the complete carcinogenesis assay, was recently measured.[42] The rationale for this study was that if the middle distillate fuels induced tumor formation by promotion of spontaneously initiated cells, then such cells had to be present in the skin of the C_3H mouse. The cells were present at a

TABLE 6
Results of Micronucleus Tests of Middle Distillate Fuels
and Related Products in Male CD-1 Mice[a]

Test sample	Treatment (h)	Treatment group	Vehicle control
Lightly refined	24	0.6 ± 0.55	0.6 ± 0.89
Paraffinic oil	48	1.6 ± 1.1	1.2 ± 0.84
	72	1.6 ± 1.5	0.6 ± 0.89
Turbo fuel A	24	3.2 ± 2.0	3.2 ± 0.89
	48	2.0 ± 1.7	1.4 ± 0.55
	72	0.8 ± 0.55	1.0 ± 0.71
Light diesel fuel	24	1.6 ± 1.1	3.2 ± 1.6
	48	1.2 ± 0.84	0.6 ± 0.55
	72	1.6 ± 0.55	1.0 ± 1.2

[a] Results shown are from studies of male CD-1 mice given a single *per os* dose of 5 g/kg test sample. Similar results were found at lower doses (1.0, 2.5 g/kg) and in female mice given the same treatment. None of the treated groups were significantly different from control.

frequency of approximately 0.57 ± 0.68 per skin section evaluated. After correcting for difference in treatment areas, the experiment predicted 0.15 tumors per mouse if all calcium-resistant cells were expressed. By comparison, the average tumor frequency in ten published studies[11] was 0.13 tumors per mouse. Thus the *in vitro* data supported the contention that the middle distillate fuels act by promoting spontaneously initiated cells in mouse skin.

B. STUDIES TO INVESTIGATE THE ROLE OF SKIN IRRITATION

If chronic irritation is sufficient to produce tumors, then all irritants should be carcinogens. However, this does not appear to be the case. We have occasionally used toluene as a diluent control, and, although prolonged and repeated contact does produce irritation, tumors are seldom produced in toluene-treated animals.[18] It has also been found that toluene does not promote DMBA-initiated mouse skin.[43]

As one way of studying the role of chronic skin irritation in the carcinogenic process, the effects of diluting a middle distillate either in toluene or in a non-irritating diluent have been studied.[43] The samples diluted in toluene were about as irritating as the undiluted material. Nevertheless, only the undiluated middle distillate produced any tumors; no activity was found with any of the diluted samples. The overall tumor response was weak, and it is not possible to draw a very firm conclusion from this experiment. However, the simplest interpretation is that skin irritation, at least as measured in this study, had no role in skin carcinogenesis.

Although there are some differences between the irritant responses of aromatic hydrocarbons (such as toluene) and those of saturated hydrocarbons, the relationship between specific cutaneous changes and epidermal carcinogenesis is unknown.[40] Argyris[38] has argued that, at least for promotion, the most important factor is the production of a state of sustained hyperplasia, and this may provide another avenue for investigation. To date, however, it has not been possible to confirm the mechanistic importance of chronic irritation in the development of skin cancer.

VII. SUMMARY AND CONCLUSIONS

The preceding sections reviewed the toxicological data related to the dermal carcinogenic potential of two classes of petroleum-derived materials, unrefined lubricating base oils and middle distillate fuels. Although both types of materials have produced skin tumors in mice, these tumors apparently arise by different mechanisms.

The unrefined lubricating base oils are almost certainly genotoxic carcinogens. Carcinogenic initiation is most likely mediated through a mutagenic mechanism and is due to the presence of PAH in these oils. At least qualitatively similar responses have been obtained in mice,[16] and humans[44] as well as other primates.[45]

The middle distillate fuels are almost certainly non-genotoxic carcinogens. They have no demonstrable carcinogenic initiating potential, but may produce tumors in mice if exposures are prolonged. There is no evidence to date that exposure to these materials produces tumors in humans.[46]

Returning to the questions posed in the introduction:

1. Is the dermal carcinogenesis assay applicable only to the types of materials for which it has been validated?

    Perhaps. For those petroleum-derived materials which contain high levels of PAH, the implications of a positive finding in a mouse dermal carcinogenesis bioassay are well understood and widely accepted. For those materials which contain little or no PAH but still produce tumors, the relevance of the data is much less clear.

2. Could the tumors induced by chronic administration of middle distillates be an effect without clinical significance?

    Perhaps. It seems to be true that the middle distillate fuels produce skin tumors by promoting spontaneously initiated cells through a non-genotoxic process which may be related to repeated skin irritation and/or injury. These spontaneously initiated cells may or may not be present in human skin; that remains to be tested. The role of skin irritation in the carcinogenic process has been examined but has not yet been satisfactorily explained.

3. Do all materials which cause tumors in mice necessarily pose a cancer hazard to humans?

    Perhaps not. It has commonly been stated as a matter of public policy that any animal carcinogen should be regarded as if it were a human carcinogen. However, some animal carcinogens may produce tumors only under such extreme experimental conditions that the data have no real world relevance. If it is true that the middle distillate fuel-induced tumors are the consequence of repeated skin irritation, then exposure to these materials may pose little or no skin cancer hazard to humans except under conditions of prolonged and repeated exposure at high levels. However, a better understanding of mechanism is needed before that claim can be substantiated.

REFERENCES

1. **Yamagawa, K. and Ichikawa, D.,** Experimentelle studie uber die pathogenese der epithelialgeschwulste, *Mitt. Med. Fak. Tokyo,* 15, 295, 1915.
2. **Tsutsui, H.,** Uber das kunstlich erzeugte cancroid bei der maus, *Gann,* 12, 17, 1918.
3. **Leitch, A.,** Mule spinners cancer and mineral oils, *Br. Med. J.,* 11, 941, 1924.
4. **Cruickshank, C. and Squire, J.,** Skin cancer in the engineering industry from the use of mineral oil, *Br. J. Ind. Med.,* 7, 1, 1950.

5. **Mastromatteo, E.,** Cutting oils and squamous-cell carcinoma. I. Incidence in a plant with a report of six cases, *Br. J. Ind. Med.,* 12, 240, 1955.

6. **Gillman, J. and Vasselinovitch, S.,** Cutting oils and squamous-cell carcinoma. II. An experimental study of the carcinogenicity of two types of cutting oils, *Br. J. Ind. Med.,* 12, 244, 1955.

7. **Taylor, W. and Dickes, R.,** The carcinogenic properties of cutting oils, *Ind. Med. Surg.,* 24, 309, 1955.

8. **Eckardt, R.,** *Industrial Carcinogens,* Grune and Stratton, New York, 1959.

9. **Eckardt, R.,** Cancer prevention in the petroleum industry, *Int. J. Cancer,* 2, 656, 1967.

10. **Cook, J., Hewitt, C., and Heiger, I.,** The isolation of a cancer-producing hydrocarbon from coal tar. I, II, and III, *J. Chem. Soc. Trans.,* 1, 395, 1933.

11. **Biles, R., McKee, R., Lewis, S., Scala, R., and DePass, L.,** Dermal carcinogenic activity of petroleum-derived middle distillate fuels, *Toxicology,* 53, 301, 1988.

12. **Lewis, S., King, R., Cragg, S., and Hillman, D.,** Skin carcinogenic potential of petroleum fractions. Carcinogenesis of crude oil, distillate fractions, and chemical class subfractions, in *The Toxicology of Petroleum Hydrocarbons,* MacFarland, H., Holdsworth, C., MacGregor, J., Call, R., and Kane, M., Eds., The American Petroleum Institute, Washington, D.C., 1982, 185.

13. **Witschi, H., Smith, L., Frome, E., Pequet-Goad, M., Griest, W., Ho, C.-H., and Guerin, M.,** Skin carcinogenic potential of crude and refined coal liquids and analogous petroleum products, *Fundam. Appl. Toxicol.,* 9, 197, 1987.

14. **McKee, R., Plutnick, R., and Przygoda, R.,** The carcinogenic initiating and promoting properties of a lightly refined paraffinic oil, *Fundam. Appl. Toxicol.,* 12, 748, 1989.

15. **McKee, R., Nicolich, M., Scala, R., and Lewis, S.,** Estimation of epidermal carcinogenic potency, *Fundam. Appl. Toxicol.,* 15, 320, 1990.

16. **Smith, W., Sunderland, D., and Sugiura, K.,** Experimental analysis of the carcinogenic activity of certain petroleum products, *Arch. Ind. Hyg., Occup. Med.,* 4, 229, 1955.

17. **McKee, R., Lewis, S., and Egan, G.,** Experience gained by the petroleum industry in the conduct of dermal carcinogenesis bioassays, in *Skin Carcinogenesis: Mechanisms and Human Relevance,* Slaga, T., Klein-Szanto, A., Boutwell, R., Stevenson, D., Spitzer, H., and D'Motto, B., Eds., Alan R. Liss, New York, 1989, 363.

18. **McKee, R., Daughtrey, W., Freeman, J., Federici, T., Phillips, R., and Plutnick, R.,** The dermal carcinogenic potential of unrefined and hydrotreated lubricating oils, *J. Appl. Toxicol.,* 9, 265, 1989.

19. **Doak, S., Brown, V., Hunt, P., Smith, J., and Roe, F.,** The carcinogenic potential of twelve refined minearl oils following long term topical application, *Br. J. Cancer,* 48, 429, 1983.

20. **Gradiski, D., Vinot, J., Vissu, J., Limasset, J., and LaFontaine, M.,** The carcinogenic effect of a series of petroleum-derived oils on the skin of mice, *Environ. Res.,* 32, 258, 1983.

21. **Halder, C., Warne, T., Little, R., and Garvin, P.,** Carcinogenicity of petroleum lubricating oil distillates: effects of solvent refining, hydroprocessing, and blending, *Am. J. Ind. Med.,* 5, 265, 1984.

22. **Kane, M., LaDov, E., Holdsworth, C., and Weaver, N.,** Toxicological characteristics of refinery streams used to manufacture lubricating oils, *Am. J. Ind. Med.,* 5, 183, 1984.

23. **McKee, R. and Plutnick, R.,** The carcinogenic potential of gasoline and diesel engine oils, *Fundam. Appl. Toxicol.,* 13, 545, 1989.

24. **McKee, R., Scala, R., and Chauzy, C.,** An evaluation of the epidermal carcinogenic potential of cutting fluids, *J. Appl. Toxicol.,* 10, 251, 1990.

25. **King, R.,** Skin carcinogenic potential of petroleum fractions. I. Separation and characterization of fractions for bioassay, in *The Toxicology of Petroleum Hydrocarbons,* MacFarland, H., Holdsworth, C., MacGregor, J., Call, R., and Kane, M., Eds., American Petroleum Institute, Washington, D.C., 1982, 170.

26. **Holland, J. and Frome, E.,** Statistical evaluations in the carcinogenesis bioassay of petroleum hydrocarbons, in *The Toxicology of Petroleum Hydrocarbons,* MacFarland, H., Holdsworth, C., MacGregor, J., Call, R., and Kane, M., Eds., American Petroleum Institute, Washington, D.C., 1982, 196.

27. **Gerhart, J., Hatoum, N., Halder, C., Warne, T., and Schmitt, S.,** Tumor initiation and promotion effects of petroleum streams in mouse skin, *Fundam. Appl. Toxicol.,* 11, 76, 1988.

28. **Marino, D., Finkbone, H., Strother, D., Mast, R., and Furedi-Machacek, E.,** Tumor initiation and promotion activity of straight run and fractionated kerosenes, *Toxicologist,* 194 (abstr.), 1990.

29. **MacGregor, J., Conaway, C., and Cragg, S.,** Lack of correlation of the Salmonella/microsome assay with the carcinogenic activity of complete petroleum hydrocarbon mixtures, in *The Toxicology of Petroleum Hydrocarbons,* MacFarland, H., Holdsworth, C., MacGregor, J., Call, R., and Kane, M., Eds., American Petroleum Institute, Washington, D.C., 1982, 149.

30. **Carver, J., MacGregor, J., and King, R.,** Mutagenicity and chemical characterization of petroleum distillates, *J. Appl. Toxicol.,* 4, 163, 1984.

31. **Blackburn, G., Deitch, R., Schreiner, C., Mehlman, M., and MacKerer, C.,** Estimation of the dermal carcinogenic activity of petroleum fractions using a modified Ames test, *Cell Biol. Toxicol.,* 1, 40, 1984.

32. **Blackburn, G., Deitch, R., Schreiner, C., and MacKerer, C.,** Predicting tumorigenicity of petroleum distillation fractions using a modified Salmonella mutagenicity assay, *Cell Biol. Toxicol.,* 2, 63, 1986.
33. **Roy, T., Johnson, S., Blackburn, G., and MacKerer, C.,** Correlation of mutagenic and dermal carcinogenic activities of mineral oils with polycyclic aromatic compound content, *Fundam. Appl. Toxicol.,* 10, 466, 1989.
34. **McKee, R. and Przygoda, R.,** The genotoxic and carcinogenic potential of engine oils and refined lubricating oils, *Environ. Mutagen.,* 9 (Suppl.), 72 (abstr.), 1987.
35. **Blackburn, G., Deitch, R., Schreiner, C., and MacKerer, C.,** Testing of petroleum middle distillates in a modified Ames test, *Env. Mutag.,* 9 (Suppl. 8), 15 (abstr.), 1987.
36. **Argyris, T.,** An analysis of the epidermal hyperplasia produced by acetic acid, a weak tumor promoter, in the skin of female mice initiated with dimethylbenzanthracene, *J. Invest. Dermatol.,* 80, 430, 1983.
37. **Argyris, T.,** Promotion of epidermal carcinogenesis by repeated damage to mouse skin, *Am. J. Ind. Med.,* 8, 329, 1985.
38. **Argyris, T.,** Regeneration and the mechanism of epidermal tumor promotion, *CRC Crit. Rev. Toxicol.,* 14, 211, 1985.
39. **Argyris, T. and Slaga, T.,** Promotion of carcinomas by repeated abrasion in initiated skin of mice, *Cancer Res.,* 41, 5193, 1981.
40. **Freeman, J., McKee, R., Phillips, R., Plutnick, R., Scala, R., and Ackerman, L.,** A 90-day study of the effects of petroleum middle distillates on the skin of C3H mice, *Toxicol. Ind. Health,* in press.
41. **Yuspa, S. and Morgan, D.,** Mouse skin cells resistant to terminal differentiation associated with initiation of carcinogenesis, *Nature,* 293, 72, 1981.
42. **Przygoda, R. T., Ribeiro, P. L., Freeman, J. J., and McKee, R. H.,** Mechanisms of skin tumorgenicity of petroleums middle distillates. III, *Toxicologist,* in press.
43. **Freeman, J., McKee, R., Federici, T., Phillips, R., and Scala, R.,** Multidose dermal carcinogenicity bioassay of a petroleum middel distillate (PMD), *Toxicologist,* 194 (abstr.), 1990.
44. **International Agency for Research on Cancer,** Polynuclear aromatic hydrocarbons, Part 2, Carbon blacks, mineral oils (lubricant base oils and derived products) and some nitroarenes, in *IARC Monographs on the Evaluation of the Carcinogenic Risk of Chemicals to Humans,* Vol. 33, International Agency for Research on Cancer, Lyon, 1984, 87.
45. **Sugiura, K., Smith, W., and Sunderland, D.,** Experimental production of carcinoma in Rhesus monkey, *Cancer Res.,* 10, 951, 1956.
46. **International Agency for Research on Cancer,** Occupational exposures in petroleum refining; crude oil and major petroleum fuels, in *IRAC Monographs on the Evaluation of Carcinogenic Risks to Humans,* Vol. 45, International Agency for Research on Cancer, Lyon, 1989, 159–201, 219–237.

Chapter 18

COMPARISON OF RESULTS FROM CARCINOGENICITY TESTS OF TWO HALOGENATED COMPOUNDS BY ORAL, DERMAL, AND INHALATION ROUTES

Elizabeth K. Weisburger

TABLE OF CONTENTS

0-8493-7357-3/93/$0.00 + $.50
© 1993 by CRC Press, Inc.

I. INTRODUCTION

Relatively few environmental carcinogens of economic importance have been tested by multiple routes of administration. Often, if a carcinogenic effect is demonstrated for a compound, by any specific route, interest in further testing is minimal unless the economic value of the substance is very high. However, 1,2-dibromoethane (ethylene dibromide or EDB) and the chloro analog, 1,2-dichloroethane (ethylene dichloride or EDC) have both been examined by three routes, oral administration, dermal application, and inhalation, a rather unusual occurrence. Both were active in inducing tumors, depending on the method of administration and dose, but the degree of tumor response differed under the various protocols.

II. 1,2-DIBROMOETHANE (ETHYLENE DIBROMIDE)

1,2-Dibromoethane (EDB) has been used as a lead scavenger in gasoline, as a fumigant in grain storage, to fumigate citrus and tropical fruits, and as a nematocide in soils. Other uses were as an intermediate in the preparation of Vat Blue 16 and Vat Blue 53, as a solvent for resins, waxes, and gums, and in some fire extinguishers. In 1977 production of EDB reached a level of 244 million pounds.[1] However, the sharp decrease in the production of leaded gasoline, coupled with a ban on pesticidal uses of EDB, has led to relatively low production levels.

A. ORAL ADMINISTRATION

Although the acute toxicity of EDB upon oral administration had been investigated by Rowe et al.,[2] the experiment was not extended for a sufficient period to determine any possible carcinogenic action. A study conducted by the National Cancer Institute (NCI) was one of the more definitive in bringing attention to the effects of EDB.

In a series in which 14 halogenated aliphatic compounds were tested in Osborne-Mendel rats and B6C3F1 mice, the compounds were administered by gavage in corn oil 5 days a week. Male rats at the high or low dose level were treated for 34 to 47 weeks and held for 2 to 15 weeks longer, with the low and high time-weighted average dosages being 38 and 41 mg/kg body weight, respectively. Female rats received 38 or 39 mg/kg body weight dosages for 44 to 57 weeks and were held 4 to 17 weeks longer for the low or high dose regimens, respectively.

Squamous cell carcinomas of the stomach were increased significantly; $p < 0.001$ for male and female rats at each dose level. The incidence of these tumors ranged from 58 to 90% in the four experimental groups.[3-5] Other tumors which increased significantly ($p = 0.017$) were hemangiosarcomas, but only in the low dose male rats.

Mice were less susceptible to the toxicity of EDB and thus received higher doses. Mice received time-weighted average dosages of 62 or 107 mg/kg body weight over treatment periods of 53 weeks, followed by observation periods of 24 to 37 weeks. As was the case with the rats, squamous cell carcinomas were noted at incidences of 56 to 94% in the four male and female groups, with p values similar to those for the rats. In addition, alveolar/ bronchiolar adenomas or carcinomas of the lung were increased in low dose female and high dose male mice. Most striking with the mice was the fact that the number of secondary tumors greatly exceeded the number of mice, an indication of the malignant potential of the primary stomach tumors. Although the tumors in the rats also tended to metastasize, the secondary tumors were not quite so numerous as in the mice.

B. DERMAL APPLICATION

As part of another type of test of halogenated aliphatic compounds, EDB was dissolved in acetone and applied to the skins of female Ha:ICR Swiss at levels of 25 or 50 mg per application per mouse.[6] Two types of experiments were done; in the initiation-promotion study tetradecanoylphorbol acetate (TPA) solution was painted repeatedly after the single application of EDB. In the repeated-dose test, EDB solutions were applied three times weekly until the end of the experiment, approximately 400 days. Immediately after application, the animals were kept in a ventilated hood for at least 2 h to avoid inhalation exposure.

The initiation-promotion test did not lead to any significant number of skin papillomas. However, in the repeated-dose study, there were eight skin papillomas ($p < 0.0001$) among the 30 test mice at the 50 mg level, in addition to 26 lung tumors ($p < 0.0005$) and three stomach tumors. At the 25 mg/application level, there were 24 lung tumors ($p < 0.005$) and three stomach tumors also. This experiment confirms the report by Rowe et al.[2] that EDB can be absorbed readily through the skin.

C. INHALATION EXPOSURE

As a consequence of the positive results with EDB in the study by oral administration, it was important to do an inhalation experiment, since this method of exposure was more relevant to workers on farms and orchards or in grain storage facilities. Two studies of this type have been reported. In the first, initiated by the NCI, and continued by the National Toxicology Program (NTP), groups of 50 F344 rats and B6C3F1 mice of each sex were exposed by inhalation to concentrations of 10 or 40 ppm of EDB for 78 to 103 weeks. Exposure was for 6 h/day for 5 days/week. Because of poor survival of the male mice at either dose level, this part of the experiment was terminated at 78 weeks. High dose male and female rats and female mice were continued until 88 to 91 weeks, but low dose male and female rats and female mice lived to the end of the experimental period at 104 weeks.

Examination of the tumors seen in this experiment showed that in both male and female rats there was a dose-related and highly significant ($p < 0.001$) increase in the incidence of nasal cavity tumors, with incidences ranging from 68 to 86%. Hemangiosarcomas in the circulatory system in both sexes, mesotheliomas in multiple organs and tunica vaginalis of males, and alveolar/bronchiolar carcinomas or adenomas in the lungs of females were all associated with exposure to EDB.[1,7]

In male mice, which had had poor survival, alveolar/bronchiolar adenomas and carcinomas were significantly increased at the high dose level with up to 54% of the animals affected, but the low dose males showed relatively little. In the female mice which had survived better, there were various types of tumors, including fibrosarcomas of subcutaneous tissue on the ribs, nasal cavity, hemangiosarcomas of the circulatory system, and adenocarcinomas of the mammary gland, especially at the higher dose level.

The greater response in the female mice was probably due to their lower susceptibility to the toxicity of EDB thus permitting longer survival and development of additional tumors. There also were decided species differences in response. In rats the nasal cavity appeared more vulnerable, but in mice the lung was primarily affected initially. However, in animals which survived longer, these species differences tended to disappear.

The second EDB inhalation experiment was designed to investigate whether additional exposure to disulfiram (DS), an enzyme inhibitor used to deter alcohol abuse, would alter the results.[8] EDB was given at a median level of 20 ppm to groups of 48 Sprague-Dawley rats of each sex with controls, rats fed 0.05% DS, rats exposed to EDB, and rats fed DS and exposed to EDB also, allo part of the study. As in the previously described study with EDB alone, there were increases in hemangiosarcomas (spleen) and mesotheliomas in males exposed to EDB. Mammary tumors were higher in female EDB rats compared to controls, while adrenal tumors were increased in both males and females. However, significant numbers of tumors of the nasal cavity and lung were not observed, in contrast to the study in the F344 rats. The interesting point was that both males and females fed DS and exposed to EDB had different tumor incidence patterns, namely, a high incidence of hepatocellular tumors, adenomas and carcinomas of the kidney, and thyroid tumors. Males also had an increase in lung tumors.

Thus the simultaneous exposure to EDB and an enzyme inhibitor altered the pattern of tumor response to EDB. It might be surmised that inhibiting the metabolism of EDB would allow longer exposure of the unchanged compound to the liver and, likewise, more EDB may be excreted through the kidneys, with resultant tumor responses in those organs.

III. 1,2-DICHLOROETHANE (ETHYLENE DICHLORIDE)

1,2-Dichloroethane (EDC) is widely used in industry, largely as an intermediate in the production of vinyl chloride. It has other applications as a fuel additive, although that use may decrease also with the banning of leaded gasoline, as a solvent, and as part of a fumigant mixture. Production is estimated to range from 3.6 to 4.3 million tons annually with as many as 200,000 workers exposed.[9] An indication of the importance of EDC is the issuance of a Banbury Report on the substance, covering various bioassays, metabolism, and other studies with EDC.[10]

A. ORAL ADMINISTRATION

The NCI study of EDC was conducted in an analogous manner to that for EDB in that male and female Osborne-Mendel rats and B6C3F1 mice were given a corn oil solution of EDC by gavage 5 days a week.[11] It was evident that EDC had a lower toxicity than EDB, for the survival of the animals was better than with EDB. Time-weighted doses (mg/kg) over a treatment period of 78 weeks were 47 and 95 mg for the low and high levels, respectively. There was an additional observation period of 15 to 32 weeks. However, different from the situation with EDB, stomach tumors were significantly increased only in the high dose males. Hemangiosarcomas and subcutaneous fibromas were more numerous in males at both dose levels than in controls. Female rats at both dose levels showed increased mammary tumors, mainly fibroadenomas.

Mice received higher doses of EDC than did the rats; 97 and 195 mg/kg as time-weighted average doses for males and 149 and 299 mg/kg for females. A treatment period of 78 weeks was also followed by an observation period of 12 to 13 weeks. Lung and liver tumors occurred at statistically higher incidence in the high dose males; in females, lung and mammary tumors were higher at both dose levels. Unlike the case of EDB, there were relatively fewer secondary tumors.

B. DERMAL APPLICATION

EDC was tested under the same protocol as EDB, but at higher dose levels of 42 to 126 mg/application/mouse. In the single dose initiation-promotion test, there was no increase in skin papillomas; repeated skin application of EDC did not cause skin tumors. However, at the higher dose level 26 lung tumors ($p < 0.0005$) and 3 stomach tumors were observed. The lower dose led to 17 lung tumors among the 30 test mice. Thus the lung appeared more prone to tumor induction than did the skin. The results also indicated that EDC is absorbed through the skin.[6]

C. INHALATION EXPOSURE

As with EDB, there have been several tests of EDC by inhalation. In the first of these, Sprague-Dawley rats and Swiss mice of both sexes were exposed to four concentrations of

EDC: 5, 10, 50, and 150 ppm for 7 h a day, 5 days a week, for 78 weeks. The animals were observed until spontaneous death. Various control groups were also included in the protocol.

Despite the large numbers of animals, the various control groups, and the thorough histopathological examination of many organs, there was no increase in tumor incidence noted in either the mice or the rats.[12]

In a second inhalation study, male and female Sprague-Dawley rats were exposed to 50 ppm of EDC for 7 h/day, 5 days/week for 24 months. Evaluation of the results showed no increase in tumors despite careful examination of over 40 separate tissues from each animal. However, in animals exposed to both EDC at 50 ppm and 0.05% DS in the diet, there were significant increases in intrahepatic bile duct cholangiomas in both male and female rats.[9] Male rats exposed to the combination also had a higher incidence of subcutaneous fibromas, neoplastic nodules of the liver, and interstitial cell tumors of the testis, while in females this regimen led to an increase in mammary adenocarcinomas.

A comparative metabolic study at the end of the 24-month period demonstrated that blood levels of unchanged EDC were higher in rats on the EDC/DS combination than in those on EDC alone, while the urinary excretion rate was lower. The profile of urinary metabolites was not altered by the EDC/DS treatment, but EDC/DS rats exhaled more unchanged EDC in the breath. This is consistent with the inhibitory action of DS on various metabolic enzymes, leading to a greater excretion of unchanged material.

IV. DISCUSSION

The available studies with EDB and EDC demonstrate appreciable differences in their effects in animals, despite their structural similarity. EDB was active by all routes of administration, oral, dermal, or by inhalation. On the other hand, EDC, although having an effect by the oral and dermal routes, was less potent than EDB. Furthermore, by inhalation EDC alone was not carcinogenic, even at levels up to 15-fold of that allowed by the Threshold Limit Values of the ACGIH.[13]

The data from dermal application of either EDB or EDC indicate, as reported previously, that a toxic dose can be absorbed through the skin.[2] Allowing for a larger dose of EDC due to lower toxicity, these studies certainly indicated that the lung was a target organ and that both compounds induced a similar number of such tumors.[6] It appears that the hazard of skin exposure may be as great as from oral administration of comparable doses of EDB and perhaps greater than oral administration of EDC. However, due to the irritant properties of EDB, worker exposure by this route was probably minimal, because workers are less likely to expose themselves to an unpleasant material.

Despite the differences in toxicity and carcinogenicity, EDB and EDC are metabolized similarly. Either compound is activated by conjugation with glutathione, yielding a hemisulfur mustard-type compound, followed by cyclization to an episulfonium ion which may be the actual alkylating moiety, which forms adducts with cellular nucleic acids.[14-16] In addition both EDB and EDC may undergo oxidative metabolism to yield halo-aldehydes which may also interact with cellular nucleophilic macromolecules.[17] For EDC, at least, the metabolic pattern was similar whether given by the oral or the respiratory route.[18] Further metabolism of the products from glutathione conjugation affords such compounds as hydroxyethylmercapturic acids, thiodiglycolic acid, and its sulfoxide,[18,19] while *in vitro* ethylene is also formed from either EDB or EDC.[20] Both compounds are hepatotoxic, leading to DNA damage.[21,22] Nevertheless, there was no sharp distinction in metabolic pattern between oral and inhalation exposure which would explain the positive response by one route and the lack of response by the second route.[18] Furthermore, by autoradiography, both EDB and EDC bound to the respiratory and upper alimentary tracts of mice and rats, thus reinforcing the results from the inhalation bioassay.[23,24] However, the binding from EDC was lower than that with EDB, also confirming what had been noted in separate animal bioassays.[9,12]

There has been controversy over the application of the results from the animal studies to the estimation of risk in exposed humans, especially for EDB. Although the total number of EDB-exposed workers was relatively small, cancer incidence in these groups was not higher than in control populations.[25,26] Risk estimates by regulatory agencies, using linear non-threshold additive models, indicated a risk from cancer manyfold that actually noted, which was pointed out by others as overpredicting the risk from EDB.[27] One such regulatory group claimed that if the animal inhalation study of EDB were the basis for an estimate, a one-hit model predicted cancer risks that were compatible with epidemiologic data.[28] However, using the gavage data, the one-hit model grossly overestimated the number of cancer deaths. Unfortunately, only two dose levels were tested in the oral study of EDB, and the regulatory agencies extrapolated from these high doses in a non-threshold model, which is contrary to some toxicological principles.[29,30] Furthermore, denial of a threshold as a means to protect the entire public, including unduly sensitive or susceptible individuals, from carcinogens is also contrary to observation. Everyone is exposed to traces of numerous carcinogens, both those that occur naturally and those produced endogenously and synthetically.[31] The use of overly conservative models in order to protect the public from cancer leads to various types of losses. Some are economic through the banning of the use of a valuable material, such as a pesticide, or lack of purchase of products made with or containing traces of a suspect material.[32] Another type of loss, not usually considered, is the loss of public confidence when it is learned how data were extrapolated in an unrealistic fashion. A case such as that of EDB illustrates the need for more realistic models in risk estimation that are scientifically valid.[33,34]

REFERENCES

1. National Toxicology Program, Technical Rep. Ser. No. 210, Carcinogenesis Bioassay of 1,2-Dibromoethane (CAS No. 106-93-4) in F344 Rats and B6C3F₁ Mice (Inhalation Study), U.S. Department of Health and Human Services, 1982.
2. **Rowe, V. K., Spencer, H. C., McCollister, D. D., Hollingsworth, R. L., and Adams, E. M.,** Toxicity of ethylene dibromide determined on experimental animals, *Arch. Ind. Hyg. Occup. Med.,* 6, 158, 1952.
3. **Olson, W. A., Habermann, R. T., Weisburger, E. K., Ward, J. M., and Weisburger, J. H.,** Induction of stomach cancer in rats and mice with halogenated aliphatic fumigants, *J. Natl. Cancer Inst.,* 51, 1993, 1974.
4. **Weisburger, E. K.,** Carcinogenicity studies on halogenated hydrocarbons, *Environ. Health Perspect.,* 21, 7, 1977.
5. National Cancer Institute, Technical Rep. Ser. No. 86, Bioassay of 1,2-Dibromoethane for Possible Carcinogenicity, U.S. Department of Health, Education, and Welfare, 1978.
6. **Van Duuren, B. L., Goldschmidt, B. M., Loewengart, G., Smith, A. C., Melchionne, S., Seidman, I., and Roth, D.,** Carcinogenicity of halogenated olefinic and aliphatic hydrocarbons in mice, *J. Natl. Cancer Inst.,* 63, 1433, 1979.
7. **Stinson, S. F., Reznik, G., and Ward, J. M.,** Characteristics of proliferative lesions in the nasal cavities of mice following chronic inhalation of 1,2-dibromoethane, *Cancer Lett.,* 12, 121, 1981.
8. **Wong, L. C. K., Winston, J. M., Hong, C. B., and Plotnick, H.,** Carcinogenicity and toxicity of 1,2-dibromoethane in the rat, *Toxicol. Appl. Pharmacol.,* 65, 155, 1982.
9. **Cheever, K. L., Cholakis, J. M., El-Hawari, A. M., Kovatch, R. M., and Weisburger, E. K.,** Ethylene dichloride: the influence of disulfiram or ethanol on oncogenicity, metabolism, and DNA covalent binding in rats, *Fundam. Appl. Toxicol.,* 14, 243, 1990.
10. **Ames, B., Infante, P., and Reitz, R., Eds.,** Banbury Rep. 5, Ethylene Dichloride: A Potential Health Risk?, Cold Spring Harbor Laboratory, Cold Spring Harbor, NY, 1980.

11. National Cancer Institute, Technical Rep. Ser. No. 55, Bioassay of 1,2-Dichloroethane for Possible Carcinogenicity, U.S. Department of Health, Education, and Welfare, 1978.

12. **Maltoni, C., Valgimigli, L., and Scarnato, C.,** Long-term carcinogenic bioassays on ethylene dichloride administered by inhalation to rats and mice, in Banbury Rep. 5, Ethylene Dichloride: A Potential Health Risk?, Ames, B., Infante, P., and Reitz, R., Eds., Cold Spring Harbor Laboratory, Cold Spring Harbor, NY, 1980, 3.

13. American Conference of Governmental Industrial Hygienists, TLVs®, Threshold Limit Values and Biological Exposure Indices for 1988–1989, ACGIH, Cincinnati, OH.

14. **Ozawa, N. and Guengerich, F. P.,** Evidence for formation of an S-[2-(N[7]-guanyl)ethyl] glutathione adduct in glutathione-mediated binding of the carcinogen 1,2-dibromoethane to DNA, *Proc. Natl. Acad. Sci. U.S.A.,* 80, 5266, 1983.

15. **Koga, N., Inskeep, P. B., Harris, T. M., and Guengerich, F. P.,** S-[2-(N[7]-Guanyl)ethyl] glutathione, the major DNA adduct formed from 1,2-dibromoethane, *Biochemistry,* 25, 2192, 1986.

16. **Foureman, G. L. and Reed, D. J.,** Formation of S-[2-(N[7]-guanyl)ethyl] adducts by the postulated S-(2-chloroethyl)cysteinyl and S-(2-chloroethyl) glutathionyl conjugates of 1,2-dichloroethane, *Biochemistry,* 26, 2028, 1987.

17. **Hill, D. L., Shih, T.-W., Johnston, T. P., and Struck, R. F.,** Macromolecular binding and metabolism of the carcinogen 1,2-dibromoethane, *Cancer Res.,* 38, 2438, 1978.

18. **Reitz, R. G., Fox, T. R., Ramsey, J. C., Quast, J. F., Langvardt, P. W., and Watanabe, P. G.,** Pharmacokinetics and macromolecular interactions of ethylene dichloride in rats after inhalation or gavage, *Toxicol. Appl. Pharmacol.,* 62, 190, 1982.

19. **van Bladeren, P. J., Hoogeterp, J. J., Breimer, D. D., and van der Glen, A.,** The influence of disulfiram and other inhibitors of oxidative metabolism on the formation of 2-hydroxyethyl-mercapturic acid from 1,2-dibromoethane in the rat, *Biochem. Pharmacol.,* 30, 2983, 1981.

20. **Livesey, J. C. and Anders, M. W.,** In vitro metabolism of 1,2-dihaloethanes to ethylene, *Drug Metab. Dispos.,* 7, 199, 1979.

21. **Kluwe, W. M., McNish, R., and Hook, J. B.,** Acute nephrotoxicities and hepatotoxicities of 1,2-dibromo-3-chloropropane and 1,2-dibromoethane in male and female F344 rats, *Toxicol. Lett.,* 8, 317, 1981.

22. **Banerjee, S.,** DNA damage in rodent liver by 1,2-dichloroethane, a hepatocarcinogen, *Cancer Biochem. Biophys.,* 10, 165, 1988.

23. **Kowalski, B., Brittebo, E. B., and Brandt, I.,** Epithelial binding of 1,2-dibromoethane in the respiratory and upper alimentary tracts of mice and rats, *Cancer Res.,* 45, 2616, 1985.

24. **Brittebo, E. B., Kowalski, B., Ghantous, H., and Brandt, I.,** Epithelial binding of 1,2-dichloroethane in mice, *Toxicology,* 56, 35, 1989.

25. **Ter Haar, G.,** An investigation of possible sterility and health effects from exposure to ethylene dibromide, in Banbury Rep. 5, Ethylene Dichloride: A Potential Health Risk?, Ames, B., Infante, P., and Reitz, R., Eds., Cold Spring Harbor Laboratory, Cold Spring Harbor, NY, 1980, 167.

26. **Ott, M., Scharnweber, H. C., and Langner, R. R.,** Mortality experience of 161 employees exposed to ethylene dibromide in two production units, *Br. J. Ind. Med.,* 37, 163, 1980.

27. **Ramsey, J. C., Park, C. N., Ott, M. G., and Gehring, P. J.,** Carcinogenic risk assessment: ethylene dibromide, *Toxicol. Appl. Pharmacol.,* 47, 411, 1979.

28. **Hertz-Picciotto, I., Gravitz, N., and Neutra, R.,** How do cancer risks predicted from animal bioassays compare with epidemiologic evidence? The case of ethylene dibromide, *Risk Anal.,* 8, 205, 1988.

29. **Falk, H. L.,** Biologic evidence for the existence of thresholds in chemical carcinogenesis, *Environ. Health Perspect.,* 22, 167, 1978.

30. **Gori, G. B.,** Cancer hazards, risks and thresholds: an assessment, *J. Am. Coll. Toxicol.,* 2(3), 219, 1983.

31. **Yesair, D. W.,** The importance of pharmacokinetic principles in characterizing carcinogenic thresholds for naturally occurring and synthetic chemicals, *J. Am. Coll. Toxicol.,* 2(3), 55, 1983.

32. **Johnson, F. R.,** Economic costs of misinforming about risk: the EDB scare and the media, *Risk Anal.,* 8, 261, 1988.

33. **Paustenbach, D. J.,** A survey of health risk assessment, in *The Risk Assessment of Environmental Hazards,* Paustenbach, D. J., Ed., John Wiley & Sons, New York, 1989, 27.

34. **Cox, L. A., Jr. and Ricci, P. F.,** Risk, uncertainty, and causation: quantifying human health risks, in *The Risk Assessment of Environmental Hazards,* Paustenbach, D. J., Ed., John Wiley & Sons, New York, 1989, 125.

Chapter 19

THE OBJECTIVES AND GOALS OF DERMAL CARCINOGENICITY TESTING OF PETROLEUM LIQUIDS

James J. Freeman and Richard H. McKee

TABLE OF CONTENTS

0-8493-7357-3/93/$0.00 + $.50
© 1993 by CRC Press, Inc.

I. INTRODUCTION

The overall goal of conventional dermal carcinogenicity testing, i.e., the chronic skin painting bioassay in mice, is to detect potential human skin carcinogens. This animal model is believed to be a reasonably good predictor of human response because human carcinogens such as coal tar and certain fractions of unrefined petroleum have been demonstrated to elicit carcinogenic responses in mouse skin.[1] Further, the active components of these complex carcinogens, polycyclic aromatic hydrocarbons (PAH), elicit skin cancers in mouse skin.

The development of the mouse bioassay occurred largely in the petroleum industry,[2] where it has been used in several applications. These include product safety testing as well as studies to determine which industrial processes may contribute to, or eliminate, carcinogenic activity. As reviewed by McKee et al.,[2] the mouse bioassay has its limitations. As used by the petroleum industry, it is a qualitative tool and as such may not be appropriate for quantitative risk assessment purposes. There are correlations between human cancer and the mouse bioassay for only a few materials, as mentioned above, and there are current questions regarding its validity in testing of materials which produce chronic dermal irritation. Regardless, for those materials for which the mouse bioassay is validated (e.g., PAH-containing materials), it has proven to be a useful tool for answering the question, "can this material elicit skin tumors?".

II. STUDY DESIGN

Several factors need be considered to assume an adequate study design. These include species/strain selection, sex, group size, animal husbandry considerations, test material application (frequency and dose), study duration, in-life measurements, and histopathology.

The conventional dermal carcinogenesis bioassay, as conducted through 1990, is a 2-year study conducted in mice. Because the objective of the bioassay has been hazard identification, it is often conducted in one sex of mice only and at only one dose level (e.g., undiluted test material). Thus, compared to traditional toxicologic approaches to chronic toxicity/oncogenicity testing, the mouse skin painting study represents an abbreviated protocol.

The C3H mouse is the strain of choice in studies conducted by the petroleum industry. As such, a large database supports its use. As reviewed by McKee et al.,[2] the animal is docile, relatively hardy, readily available, and responds rather predictively to known carcinogenic treatments. Although the C3H mouse is generally regarded as having good survival, it may not be as long-lived as some other strains. Exxon's recent experience with the male C3H/HeJ mouse indicates that, under conditions of individual housing in a modern vivarium, 50% survival occurs at approximately 20 months and that less than 33% of a group of animals should be expected to survive 2 years.[3] In addition to the C3H mouse, other strains have also been used successfully in dermal carcinogenesis bioassays, most notably the CD-1, Swiss, and the SENCAR. The B6C3F$_1$ is now finding use as well in studies conducted by the National Toxicology Program.

Animals should be housed individually in appropriate caging, and the environment (including temperature, humidity, and diurnal cycle) should be maintained and monitored in accordance with the Guide for the Care and Use of Animals[4] and the Animal Welfare Act.[5] The use of sentinel animals to monitor health is also recommended.

The test material is applied to the clipped interscapular region on the dorsum of the back. The animals are clipped periodically, with care to avoid injury, and if possible, at least 24 h before a scheduled test material application. Mice are commonly treated with test material either two or three times per 5-day work week (NTP has used 5 ×/week treatment[6]).

For some materials, more frequent application may lead to severe skin irritation and frank necrosis, and may even decrease the sensitivity of the bioassay.[7] For non-irritating materials, experience suggests that tumor response is more dependent upon total weekly dose rather than frequency of application.[2]

The typical mouse dermal bioassay is conducted using only one dose level. This approach works well for test materials that are relatively non-irritating and non-toxic (e.g., many petroleum hydrocarbons) and can be tested undiluted. By using this test, petroleum distillates have been classified as either carcinogenic or non-carcinogenic.

For some test materials, it may be desirable to use a vehicle. The preferred vehicles are water, acetone, and mineral oil. Highly refined mineral oils are not carcinogenic (see Chapter 17), are practically non-irritating, and are a reasonably good solvent for hydrocarbons. Toluene has also been used but is rather irritating with repeated application[8] and has been reported to inhibit tumor promotion in mice.[9] Mixtures of acetone and cyclohexane have also been used in dermal bioassays, but cyclohexane was recently reported to have activity as a Stage II tumor promoter.[10]

It is recommended that the dose volume be minimized as much as possible. Following application, the test material tends to spread to the interscapular region, and even relatively small dose volumes can spread onto the flanks and into the hair and skin folds in the axillary region. This was demonstrated in our laboratory using petroleum distillates ranging in viscosity from 4.1 cSt to 30 cSt at 40°C (these viscosities approximate that of fuel oil to crankcase oil). Dose volumes of 25 µl rapidly spread down the flanks and into the axillary region. This type of test material spread is of concern because entrapment in the skin folds in the axillary region may result in an increased level of irritation and associated scratching. More importantly, in dermal carcinogenesis bioassays it is sometimes questioned whether skin tumors which occasionally occur outside of the interscapular ''treatment area'' are treatment related. The potential spread of test material suggests that some may be. In addition, large volumes applied to the skin are of concern because of the possibility of ingestion via preening. The standard bioassay at Exxon utilizes a twice weekly dosing with 37.5 µl, which is sufficient to detect relatively weak (i.e., ≤10% of the mice develop tumors) skin carcinogens of petroleum origin (25 µl three times/week is equally effective).

The standard mouse bioassay is conducted for 2 years. Shorter study periods may not be sufficient to detect relatively weak skin carcinogens. Longer periods appear to be unjustified because survival decreases rapidly, as noted above.

The standard group size for the mouse bioassay is 50. However, in our experience 40 animals per group also yields a statistically reliable study.

All animals are examined for viability twice daily, and moribund animals are sacrificed. Examinations for clinical signs of toxicity and cutaneous growths are conducted on a weekly basis. Body weights can be measured on a monthly basis. In dermal carcinogenesis bioassays conducted with materials which cause extensive irritation with repeated application there may be a need to evaluate the irritation during the course of the study. Our laboratory has utilized a simple (albeit arbitrary) system of recording and summarizing the various indices of irritation (Table 1). Interestingly, gross indices of irritation have correlated well with histopathological indices.[11]

Histopathological evaluations are conducted on untreated and all treated skin specimens. Tumor numbers are confirmed and tumor types are identified. In the infrequent occasions in which a presumed squamous cell papilloma is identified during the course of the study but which subsequently regresses or is otherwise lost for histopathological evaluation, it is included in the final tumor tally. Any other relevant changes to the skin, e.g., irritation, are also characterized histopathologically.

TABLE 1
Cutaneous Irritation Scoring System[a]

Degree of Irritation[b]		
Slight ("1")	Moderate ("2")	Extreme ("4")
Slight erythema[c]	Moderate erythema	Extreme erythema
Slight edema	Moderate edema	Extreme edema
Desquamation	Atonia	Eschar
	Blanching	Exfoliation
	Cracking	Fissuring
	Leathery	Necrosis
	Thickening	Open sores

[a] Originally published in Reference 11.

[b] Each sign of cutaneous irritation was classified as being representative of slight, moderate or extreme changes overall, and assigned numeric values of "1", "2", or "4", respectively. Each sign within a column was so graded and the maximum possible numeric irritation score in an individual animal at a given time point was 39. The Irritation Index was calculated as:

$$\text{Irritation index} = \frac{\text{Sum of numeric irritation scores in group}}{\text{number animals in group}}$$

and could also range from 0 to 39.

[c] Erythema and edema were graded as slight, moderate or extreme. All other signs of irritation were recorded as present or not present.

In most cancer studies, internal organs are also evaluated histopathologically. This is particularly important for classes of materials that have not been well investigated for carcinogenic potential. Indeed, examples exist where the dermal route of exposure resulted in visceral tumors.[12-14] However, re-analysis of several series of Exxon skin painting studies conducted between 1977 to 1983 indicates that petroleum-derived materials are unlikely to induce systemic tumors following cutaneous application and thus, complete histopathological evaluation of the internal organs may be considered optional.

In these studies, 131 groups of C3H/HeJ mice (50/group, housed five per cage) were treated $3\times$/week for life with 25 µl of various petroleum liquids. There included home heating oils, lubricating oils, distillate aromatic extracts, and cracked petroleum streams. USP white mineral oil and catalytically cracked clarified oil (CCCO; known to contain as much as several weight percent PAH) were the negative and positive controls, respectively. A total of 16 vehicle/negative control and 13 positive control groups (CCCO concentrations ranging from 1 to 20%) were included in this series of studies. Histopathology was conducted on skin, external and internal masses, brain, heart, lungs, liver, spleen, kidneys, cervical and mesenteric lymph nodes, and any abnormal tissues.

In individual studies, several types of primary tumors other than squamous cell tumors of the skin initially appeared to be elevated when compared to their concurrent controls. Several tumor types were apparently elevated in only a single treatment group: liver hemangiosarcoma, pulmonary adenocarcinoma, mesothelioma, and astrocytoma. For each of these tumor types, there was no dose-response relationship observed when multiple doses were tested, there was no association of the tumor types with skin carcinogenic potential, and/or similar petroleum liquids did not elicit the tumor type. Thus, it was concluded that these off-site tumors were spontaneous in origin. This frequency of spontaneous tumors is not altogether unexpected, e.g., may be considered Type I statistical errors.

TABLE 2
Relationship of CCCO Concentration
and Mouse Survival to Incidence of
Hepatocellular Tumors

Relationship	Correlation[a]
[CCCO] vs. tumor incidence	−0.64
[CCCO] vs. survival	−0.83
Survival vs. tumor incidence	+0.70[b]

[a] Correlation coefficient derived by least squares method.
[b] Includes positive, negative, and vehicle control groups.

Four additional tumor types were apparently elevated in more than one treatment group: spindle cell sarcomas of the skin (three treatment groups), myelogenous leukemia (two groups), lymphoreticular tumors (three groups), and hepatocellular tumors (five groups). However, as above, the presence of these tumors was not consistently associated with the occurrence of squamous tumors of the skin, and the tumors did not occur in a dose-dependent manner. With the exception of myelogenous leukemia, each of these tumor types was frequently observed in negative control mice (although not necessarily in the negative control group run concurrently with the particular treatment group). With the exception of lymphoreticular tumors, there was no reliable literature evidence linking PAH with development of that tumor type. The most frequently observed offsite tumors were hepatocellular adenomas and carcinomas. However, in the negative and positive control groups, age rather than treatment appeared to be the primary cause of tumor development (Table 2). Also, the incidence of hepatocellular tumors in treatment groups was within the ranges observed in the control groups. In summary, and with the knowledge that PAH are the proximate carcinogens in petroleum liquids, it was concluded for each of these tumor types that their occurrence was spontaneous in origin and fortuitous in distribution.

III. INTERPRETATION

A number of statistical analyses can be conducted on the results of the dermal carcinogenesis bioassay.[15] The Fisher Exact Test[16] is used to compare the number of tumor-bearing animals (TBA) in treatment and control groups. The spontaneous rate of skin tumors in control C_3H mice is less than approximately 1%.[2,17] Mean survival and median latency are often determined by either the Weibull distribution function[18] or Kaplan-Meier method.[19]

Survival, latency, and tumor yield may also be suggestive of mechanism. PAH-rich materials elicit tumors in a large proportion of the test animals in a short period of time (weeks to months), probably via genotoxic action. In contrast, a weak tumor response with a long latency may be suggestive of a secondary mechanism (e.g., tumor promotion), particularly if there is also evidence that the test compound is not genotoxic.

For most materials that have been tested via the dermal route, the skin would appear to be the primary target organ. For sites other than skin, experience with petroleum hydrocarbons suggests that the dermal bioassay is an insensitive screening tool (but probably very sensitive for skin carcinogens). An analysis by Tobin et al.[20] of IARC carcinogens appears to support this view more broadly. Finally, the possibility of oral ingestion from grooming could confound possible conclusions regarding risk of cancer development in the viscera following dermal exposure.

IV. SUMMARY

A number of considerations were presented regarding the design of the mouse dermal carcinogenesis bioassay. The protocol specifics for a given study should continue to be driven by the questions that need to be answered experimentally. Thus, for a petroleum hydrocarbon, a 2-year study in male mice, at one dose level (undiluted) and with histopathology limited to the skin, may suffice, whereas expansion of these experimental data points may be necessary for another material for which less is known.

The interpretation of a given bioassay will depend, in part, upon the protocol and certainly upon the study's findings. In any case, the conventional dermal carcinogenicity bioassay would not appear to be suitable for quantitative risk assessment purposes.[15] Rather, for those materials for which it is validated, it provides a qualitative assessment regarding potential skin carcinogenicity.

REFERENCES

1. **Eckhardt, R. E.,** *Industrial Carcinogens,* Grune and Stratton, New York, 1959.
2. **McKee, R. H., Lewis, S. C., and Egan, G. F.,** Experience gained by the petroleum industry in the conduct of dermal carcinogenesis bioassays, in *Skin Carcinogenesis: Mechanisms and Human Relevance,* Slaga, T. J., Klein-Szanto, A. J. P., Boutwell, R. K., Stevenson, D. E., Spitzer, H. L., and D'Motto, B., Eds., Alan R. Liss, New York, 1989, 363.
3. Exxon Biomedical Sciences, Inc., unpublished data, 1988.
4. Guide for the Care and Use of Animals, U.S. Department of Health, Education and Welfare, Public Health Service, National Institutes of Health, NIH publication No. (NIH) 85-23, revised 1985.
5. CFR, Title 9 Animals and Animal Products, [Subchapter A-Animal Welfare Parts 1, 2 and 3]. Animal Welfare Act of 1966 (P.L. 89-544), as amended by the Animal Welfare Act of 1970 (P.L. 91-579), and by the 1976 Amendments to the Animal Welfare Act (P.L. 94-279).
6. National Toxicology Program, Toxicology and Carcinogenesis Studies of Marine Diesel Fuel and JP-5 Navy Fuel in B6C3F$_1$ Mice, NTP TR 310, NIH Publ. No. 86-2566, USDHSS, 1986.
7. **Wilson, J. S. and Holland, L. M.,** The effect of application frequency on epidermal carcinogenesis assays, *Toxicology,* 24, 45, 1982.
8. **Freeman, J. J., McKee, R. H., Federici, T. M., Phillips, R. D., and Scala, R. A.,** Multidose dermal carcinogenicity bioassay of a petroleum middle distillate, *Toxicologist,* 10, 194, 1990.
9. **Weiss, H. S., O'Connell, J. F., Hakaim, A. G., and Jacoby, W. T.,** Inhibitory effect of toluene on tumor promotion in mouse skin, *Proc. Soc. Exp. Biol. Med.,* 181, 199, 1986.
10. **Gupta, K. P. and Mehrotra, N. K.,** Mouse skin ornithine decarboxylase induction and tumor promotion by cyclohexane, *Cancer Lett.,* 51, 227, 1990.
11. **Freeman, J. J., McKee, R. D., Phillips, R. D., Plutnick, R. T., Scala, R. A., and Ackerman, L. J.,** A 90-day toxicity study of the effects of petroleum middle distillates on the skin of C3H mice, *Toxicol. Ind. Health,* 6(3), 475, 1990.
12. **Cioli, V., Barcellona, P. S., and Silvestrini, B.,** Study of the potential carcinogenicity of topical drugs-suggestions for a new experimental approach, *Exp. Mol. Pathol.,* 15, 1, 1971.
13. **Holland, J. M., Gosslee, D. G., and Williams, N. J.,** Epidermal carcinogenicity of bis(2,3-epoxycyclopentyl)ether, 2,2-bis(p-glycidyloxyphenyl)propane, and m-phenylenediamine in male and female C3H and C57BL/6 mice, *Cancer Res.,* 39, 1718, 1979.
14. **Segal, A., Seidman, I., and Melchionne, S.,** Induction of thymic lymphomas and squamous cell carcinomas following topical application of isopropyl methane sulfonate to female Hsd:(ICR)BR mice, *Cancer Res.,* 47, 3402, 1987.
15. **McKee, R., Nicolich, M., Scala, R., and Lewis, S.,** Estimation of epidermal carcinogenic potency, *Fundam. Appl. Toxicol.,* 15, 320, 1990.
16. **Bradley, J.,** *Distribution Free Statistical Tests,* Prentice Hall, Englewood Cliffs, NJ, 1968, 195.
17. **Blackburn, G. R., Deitch, R. A., Schreiner, C. A., and Mackerer, C. R.,** Predicting Carcinogenicity of petroleum distillation fractions using a modified Salmonella Mutagenicity Assay, *Cell. Biol. Toxicol.,* 2, 63, 1986.

18. **Whitmore, A. and Keller, J.,** Quantitative theories of carcinogenesis, *Soc. Ind. Appl. Math. Rev.,* 20, 1, 1978.
19. **Kaplan, E. and Meier, P.,** Nonparametric estimation from incomplete observations, *J. Am. Stat. Assoc.,* 53, 457, 1958.
20. **Tobin, P. S., Kornhauser, A., and Scheuplein, R. J.,** An evaluation of skin painting studies as determinants of tumorigenesis potential following skin contact with carcinogens, *Regul. Toxicol. Pharmacol.,* 2, 22, 1982.

Chapter 20

CHEMICAL CARCINOGENESIS IN SKIN: CAUSATION, MECHANISM, AND ROLE OF ONCOGENES

Rajesh Agarwal and Hasan Mukhtar

TABLE OF CONTENTS

0-8493-7357-3/93/$0.00 + $.50
© 1993 by CRC Press, Inc.

I. INTRODUCTION

The most common cancer occurring in the human population is the skin neoplasia which, among other factors, is believed to be associated with the skin's continuing direct exposure to the environment. Of all the new cancers diagnosed annually in the U.S. and Europe, almost one third have been estimated to originate in the skin. The role of environmental agents in causation of skin cancer is gaining increasing attention due to the alarming rise in skin cancer incidences. While solar radiation is the major cause of growing incidences of skin cancer, it is beyond doubt that significant levels of cancer-causing chemical agents, especially automobile exhaust and tobacco smoke present in the environment, contribute substantially to the skin cancer risk in humans. The risk factors for human skin cancer include environmental pollutants, occupational exposure to coal tar, pitch, creosote, arsenic compounds, radium, and excessive exposure to the sun.

Although due to ethical reasons for direct human experiments, a clear demonstration for the role of chemical carcinogens in eliciting human cancer is difficult to obtain, a mounting volume of experimental evidences support the fact that chemical agents are causative factors for human skin cancer. One of the major classes of environmental chemical carcinogens includes polycyclic aromatic hydrocarbons (PAHs), the products of incomplete combustion of fossil fuels and other organic matter. The potential involvement of PAHs in human skin cancer is strongly suggested by their exceptionally high carcinogenic potency in experimental animals.

II. HISTORY AND GENERAL CONCEPTS OF INITIATION AND PROMOTION

The role of environmental exposure in the causation of human cancer was demonstrated by Sir Percival Pott, who in 1775 reported the relationship between the high incidence of scrotal skin cancer in chimney sweeps exposed to soot.[1] In 1914 Japanese investigators for the first time experimentally demonstrated the induction of cancer by a chemical agent. In this study repeated application of crude coal tar to the ears of rabbits was shown to produce skin neoplasms.[2] Later, Kennaway and Hieger showed that repetitive skin application of 1,2,5,6-dibenzanthracene, a PAH isolated from coal tar, produced skin tumors in experimental animals.[3] In 1933, benzo(a)pyrene (BP), another PAH, was isolated from crude coal tar[4] which led to extensive studies confirming the fact that PAHs are responsible for the development of skin cancers.[5] Humans are constantly exposed to PAHs through polluted air, cigarette smoke, automobile exhaust, and other airborne environmental pollutants. For experimental chemical carcinogenesis, the skin of mice has proven to be a useful model because of its high sensitivity to all kinds of carcinogenic agents, the ease of manipulation, and the availability of different inbred strains with graded susceptibility.

In the past four decades it has become clear that in murine skin and possibly in human skin the induction of cancer is a multistep process requiring both initiating and promoting substances for the development of a malignant lesion. In murine skin, tumors are generally induced by a single topical application of an initiating agent which is then followed by repeated skin applications of the promoting agent. This is known as the two-stage carcinogenesis protocol. The concept of initiation-promotion leading first to the formation of papillomas and then their conversion to the carcinomas is illustrated in Figure 1. Essentially the initiation stage is an irreversible step where genetic changes have been shown to occur in gene(s) controlling differentiation. Whereas the stage of promotion is a reversible step in which it is believed that clonal expansion of the initiated cell population occurs, which in turn leads to the formation of visible lesions. Only small numbers of the benign papillomas

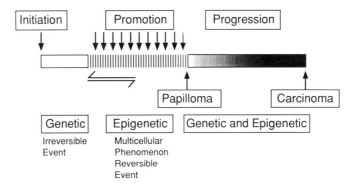

FIGURE 1. Multistage carcinogenesis showing initiation-promotion leading to the induction of papillomas and their subsequent progression to carcinomas.

thus formed by the initiation-promotion protocol progress to malignant tumors, a step which is irreversible and probably requires an additional insult to cellular genome of the papilloma cell. 7,12-Dimethylbenz(a)anthracene (DMBA), a synthetic PAH, is a conventional tumor-initiating agent employed in a majority of experimental studies on murine skin carcinogenesis. The initiation of cutaneous chemical carcinogenesis can be achieved by a single topical application of subthreshold dose of a carcinogen such as DMBA, a dose which does not produce tumors over the life span of mice. Promotion with repeated skin applications of small doses of croton oil or a phorbol ester such as 12,*O*-tetradecanoylphorbol-13-acetate (TPA), after short latency period, leads to the development of benign papillomas. The majority of the benign papillomas thus formed regress, whereas only a few of them progress to carcinomas. A lapse of as long as 1 year between the application of initiator and the beginning of promoter treatment has been shown to produce tumor response similar to that observed when promotion is begun only 1 week after initiation with the carcinogen.[6] This provided evidence that initiation is an irreversible process. In general, tumor promoters by themselves are non-mutagenic and non-carcinogenic agents because their application to normal mouse skin does not produce any tumors. A single topical application of a large dose of a chemical carcinogen on mouse skin can also lead to development of cutaneous tumors. This is often called complete carcinogenesis protocol. The repeated applications of subthreshold doses of a carcinogen can also produce cutaneous tumors and the number of carcinomas developed with this protocol is much higher than that developed in a two-stage tumor protocol in which DMBA is used as the initiating agent and TPA is employed as a promoting agent.[5-7]

III. CHEMICAL AGENTS AS SKIN CARCINOGENS

A wide range of chemicals have been identified as tumor-initiating agents for cutaneous carcinogenesis in murine skin. These include PAHs, nitrosamines, nitrosamides, aromatic amines, and various other alkylating agents.[6,8] The promoting agents include various plant products, PAHs, tobacco products, surface active agents, anthrones, organic peroxides and hydroperoxides, long chain fatty acids and their esters, and phenolic compounds.[5,8] Humans are exposed to most of these agents through environment, occupation, drugs, or through diet. A detailed summary of these agents is provided in reviews by Mukhtar et al.[5] and by Yuspa.[8]

IV. TUMOR INITIATION

In general the initiation of skin cells is accomplished by a single, local, intragastric,[7,9,10] or transplacental[11] application of a chemical carcinogen in a "subthreshold" dose. The key event of chemical carcinogenesis is the covalent binding of the carcinogen to macromolecules especially to DNA.[12] Most carcinogens, especially PAHs which are chemically inert, become metabolically activated primarily by the oxidative reactions.[13] The reactive metabolites which bind to DNA and initiate the carcinogenic response are known as "ultimate" carcinogens, whereas their immediate metabolic precursors are called "proximate" carcinogens.[14] The first step in the oxidative metabolism of skin carcinogens such as PAHs is catalyzed by the cytochrome P-450-dependent monooxygenase system.

All carcinogenic PAHs are known to contain a phenanthrene structure known as "bay region", a prerequisite for a PAH to be metabolically activated to exert its carcinogenic action.[15] It has been confirmed that dihydrodiol epoxides with the epoxy group in this region exhibit the highest biological activity of all isomeric compounds.[15] Extensive studies from several laboratories have identified a specific diol-epoxide derivative of BP, known as $7\beta,8\alpha$-(+)-dihydroxy-9α-10α-epoxy-BP as the "ultimate" carcinogen of the parent compound.[16] A potent skin carcinogen and most frequently used initiator in animal experiments, DMBA is also metabolized quite similarly by the same enzyme reactions as those for BP resulting in the formation of DMBA-3,4-diol as the "proximate" carcinogen and DMBA-3,4-diol-1,2-epoxide as the "ultimate" carcinogen.

The bay region dihydrodiol epoxides react spontaneously with nucleophilic groups of cellular macromolecules forming covalent bonds, a reaction thought to be essential for the neoplastic transformation of the target cell. In view of the irreversibility of tumor formation, at least in the initiation step, the close correlation between mutagenic and carcinogenic efficacy, and the high incidence of cancers in patients with defective DNA repair (e.g., in *xeroderma pigmentosum*), DNA (mainly the hydroxyl and amino groups of the nucleobases) is generally assumed to be the critical target molecule for ultimate carcinogens. In general, the DNA binding of carcinogens possesses a high degree of specificity, for example, the ultimate carcinogenic metabolite of BP has been shown to bind exclusively to the 2-amino group of guanine residues in epidermal DNA, whereas the metabolites of *N*-acetoxy-2-acetylaminofluorene prefers the C-8 position of guanine.[5]

V. DEVELOPMENT OF TUMORS IN INITIATED SKIN

Tumor development in initiated skin is accomplished by wounding and by local application of tumor promoters. In general, tumor promoters induce skin inflammation and epidermal hyperplasia, and are repetitively applied after initiation over a period of several weeks for the development of tumors. Most active tumor-promoting agents include certain diterpene esters such as phorbol ester TPA of plant origin[17] and alkaloids and polyacetate-type compounds such as teleocidin, lyngbyatoxin, and aplysiatoxin, found in marine organisms.[18] In addition, anthralin, phenolic compounds, detergents such as Tween-60, iodoacetic acid, several fatty acid esters, and organic peroxy compounds belong to the group of "weak skin tumor promoters".[19,20]

The most commonly used skin tumor promoter is TPA, the active compound of croton oil, the seed oil of the tropical *Euphorbiacea Croton Tiglium*. Based on the mouse strain, TPA application produces tumors in initiated skin when given at a dose of 0.5 to 20 nmol per local application twice weekly for several weeks. TPA applied twice weekly at a dose of 2 to 3 nmol to DMBA-initiated SENCAR mouse results in the induction of tumors (benign papillomas) after 4 to 6 weeks, and the maximal response is observed after 12 to 18 weeks.

However, extended time interval between the application of tumor promoter leads to a considerable reduction in tumor yield, suggesting that tumor promotion is a quickly reversible step in cutaneous chemical carcinogenesis.[7]

Structure-activity relationships within the phorbol ester series indicate a distinct linkage between promoting potency and chemical structure, and thereby reflect the interaction of phorbol esters with a specific cellular binding site.[21] The two-stage initiation-promotion protocol in mouse skin yields mainly reversibly growing benign papillomas,[22,23] which are known to be of monoclonal origin.[24,25] In the course of the initiation-promotion regimen carcinomas arise directly from the papillomas by the process known as malignant progression.[26] However, the development of carcinoma does not occur in initiated skin without promotion indicating that promotion has to be regarded as a prerequisite for malignant progression. Recently Aldaz et al.[27] provided evidence that the majority of the papillomas produced by initiation-promotion protocol are promoter independent. This study further supported the hypothesis that the large number of papillomas developed by two-stage initiation-promotion regimens in SENCAR mice may have the potential to progress towards malignancy.

VI. MECHANISM OF TUMOR PROMOTION

All skin tumor promoters are strong irritant mitogens, and produce sustained epidermal hyperplasia which has been proposed as a prerequisite for promotion.[28] Several histopathological alterations such as induction of dark basal keratinocytes and dismal infiltration of polymorphonuclear leukocytes, in addition to inflammation and hyperplasia, are believed to be the important events in the process of skin tumor promotion. The most important biochemical response to the brief exposure to TPA is the dramatic induction of epidermal ornithine decarboxylase activity which catalyzes the first step in the biosynthesis of polyamines.[7] Intracellular polyamine levels are regulated in response to cell proliferation, differentiation, and transformation. Inducibility of ornithine decarboxylase and mitogenic activity are considered as important characteristics of tumor promotion. Other biochemical responses to tumor promoters include increased synthesis of DNA, RNA, protein, phospholipids, prostaglandins, increased phosphorylation of histone, increased activity of protein kinase C (PKC), histidine decarboxylase and protease, and decrease in the activities of superoxide dismutase and catalase.[6,8] The importance of several of these individual effects is discussed elsewhere in this chapter.

Tumor promotion in mouse skin can be divided into two stages known as stage I which is often called conversion and stage II, often called as propagation. Stage I is characterized by the induction of dark basal keratinocytes, and the agents capable of producing such effects are known as stage I tumor promoters. Whereas stage II tumor promoters cause biochemical alterations similar to that of TPA, they do not have potential to induce dark basal keratinocytes. The best example of stage II tumor promoter is mezerein which is as potent as TPA in inducing ornithine decarboxylase activity and enhancing DNA synthesis in mouse epidermis. However, it is a weak complete tumor promoter.[6-8] Once stage I tumor promotion is achieved by repeated applications of TPA for a limited time, further applications with mezerein result in comparable tumor yields and incidence of tumors, as are achieved by TPA application. An endogenous second messenger, diacylglycerol (DAG) is another potent stage II tumor promoter and is comparable to mezerein.[29] It mimics in its action with TPA as far as ornithine decarboxylase induction, activation of PKC, DNA synthesis, and arachidonic acid release in mouse epidermis, leukemic cell differentiation, blockage of gap junctional communication, and modulation in intracellular pH is concerned. Additional evidences for the existence of two-stage tumor promotion include the inhibition studies in which fluocinolone acetonide and tosyl phenylalanine chloromethyl ketone have been shown to be potent inhibitors of stage I tumor promotion, whereas retinoic acid is reported to inhibit both stage I and stage II tumor promotion.[5]

Some skin toxins like teleocidin and lyngbyatoxin are other tumor promoters that also bind and activate PKC and induce hyperplasia, inflammation, and epidermal differentiation in mouse skin ultimately leading to the induction of papillomas.[30] Humans are exposed to some of these compounds while swimming in pools. Anthralin[31] and benzoyl peroxides,[20] the clinically important compounds, are also effective tumor promoters in mouse skin. Although the exact mechanism of their action is not well known, they do induce PKC activity, but do not bind to it.[32]

A. MOLECULAR BASIS OF TUMOR PROMOTION

The molecular mechanism leading to tumor promotion by macrocyclic terpenoids such as phorbol esters, mezerein, and teleocidin involves phospholipase C-dependent activation of PKC. It is becoming increasingly clear that PKC is a receptor for these tumor promoters and that tumor promotion by these agents is mediated by their receptor binding. The receptor for these promoters is implicated in the transduction of altered and unscheduled signals of cellular proliferation. PKC is activated by TPA *in vivo* in the presence of Ca^{2+} and phospholipids forming a part of the signal transducing system. Recent evidences have shown that activation of phospholipase C by an external stimuli results in two separate arms of second messengers which include DAG and inositol 1,4,5 triphosphate.[33] Inositol 1,4,5 triphosphate mobilizes Ca^{2+} from internal stores leading to the activation of calmodulin/Ca^{2+}-dependent enzymes which may, in turn, lead to PKC activation via indirect mechanisms. So far, six isozymes of PKC are known, the majority of which are Ca^{2+} dependent.[33] Recently some Ca^{2+} independent PKC are also described, one of which, nPKCη, is predominantly expressed in murine skin.[34] However, the isozyme of PKC involved in skin tumor promotion is not clearly defined. The activation of PKC phosphorylates various growth factor receptors which then receive unscheduled signals leading to the enhanced synthesis of DNA along with the amplified expression of oncogenes. The product of rasHa oncogene (p21) may also play a role in the modulation of activity of phospholipase C.[33] The detailed description of oncogenes is provided separately in this chapter.

The physiological inducers of epidermal differentiation are known to stimulate endogenous phosphatidyl inositol metabolism and to generate DAG which suggest that phosphatidyl inositol-PKC second messenger system may regulate epidermal differentiation and thereby TPA mimics the action of endogenous second messenger system. In response to TPA, all epidermal cells do not differentiate but only immature keratinocytes are stimulated for proliferation;[35] a heterogeneous action which is responsible for the regenerative hyperplasia in TPA-treated murine skin. Thus initiated, papilloma and carcinoma cells behave like immature epidermal cells and proliferate in response to TPA exposure.[35]

In contrast to terpenoid tumor promoters, free radical-generating tumor promoters act through different mechanisms; while they can also activate PKC they do not bind to it.[36] Oxidant tumor promoters are known to activate phospholipase A_2 by oxidizing membrane lipids resulting in the release of free arachidonic acid which leads to the enhanced synthesis of prostaglandins. Some of the metabolites of arachidonic acid such as prostaglandin E_2 and prostaglandin $F_{2\alpha}$ are important for the induction of hyperproliferation and its maintenance in mouse skin.[36] Since hydroperoxides increase lipid peroxidation leading to membrane damage, this may result in an efflux of Ca^{2+} from the intracellular stores in the endoplasmic reticulum and mitochondria. Based on this, it has been suggested that the increase in the levels of prostaglandin E_2 and prostaglandin $F_{2\alpha}$, along with the elevated levels of cytosolic Ca^{2+} by organic hydroperoxides, might be responsible for tumor promotion in mouse skin by these compounds.[36] *t*-Butyl hydroperoxide, superoxide anions, and hydrogen peroxide have been shown to stimulate prostaglandin synthesis in pig aorta epithelial cells and lung fibroblasts and a similar mechanism may also exist in epidermal cells.[36]

B. GAP JUNCTIONAL INTERCELLULAR COMMUNICATION AND TUMOR PROMOTION

Intercellular communication is considered as an important cellular mechanism for regulating the cellular growth and differentiation.[37] This is carried out through the exchange of ions and molecules which act as messenger signals. TPA has been shown to block gap junctional intercellular communication.[33] The precise biochemical mechanism involved in the regulation of gap junction is not clearly understood; however, the down regulation of intercellular communication was found to be associated with an increase in PKC activity,[33,37] suggesting the role of a second messenger system in this process.

The gap junction intercellular communication has been considered to be an important event in the development of lesions during tumor promotion. The initiated cells which are scattered randomly in the population of normal cells remain dormant for long periods of time under the growth-regulating influences of their normal neighbor cells. The blockage of gap junction intercellular communication results in the modulation of repressor effects of normal cells on the adjacent initiated cells leading to their expansion into miniclones.[38] The continuous repeated exposure of tumor-promoting hyperplasiogenic agents results in further expansion of miniclones into tumor mass.

VII. TUMOR PROGRESSION

The majority of malignant carcinomas are known to develop from benign papillomas, and this process is called tumor progression.[39-42] For these reasons papillomas are often called premalignant lesions. However, focal areas of carcinoma *in situ* are often seen without papillomas.[41] It has been shown that by varying the protocol of papilloma induction, subpopulations of papillomas with varying potentia for conversion to malignancy can be accomplished.[42] Thus, carcinoma yield can be increased by increasing the dose of the initiating agent,[43] or by using promoters which differ in their mechanism of action,[42] or by shortening the duration of promoter treatment.[43] It has been suggested that the clinical and biological events of tumor progression represent the result of sequential selection of variant subpopulations within this clone. It has also been hypothesized that such clonal evolution might result from enhanced genetic instability in tumor cells, which may increase the probability of further genetic alterations and their subsequent selection. The gradual increase in malignancy leads to the disappearance or decrease in the organelles and metabolic functions necessary for specialized activities in the cell.[44] Extensive information regarding the alterations in cellular behavior and morphology during this process can be found elsewhere.[44]

The malignant conversion stage of tumor progression occurs spontaneously[45] but may be enhanced by treatment of papilloma-bearing mice with genotoxic agents and free radical generating compounds.[46] It has been shown that after promotion of SENCAR mice with TPA for 10 to 12 weeks which results in large numbers of papillomas, repeated treatments of these mice with certain genotoxic agents substantially increase the rate of conversion of papillomas to carcinomas.[46] This study suggested that those papillomas which possess the highest spontaneous rate of malignant conversion are also sensitive to induction of malignant conversion by treatment with chemicals.

VIII. FREE RADICALS AND CUTANEOUS CARCINOGENESIS

There is substantial evidence to implicate the involvement of free radicals, particularly those derived from molecular oxygen in initiation, promotion, and progression stages of chemical carcinogenesis,[47-49] a brief summary of which is provided here.

A. ROLE IN SKIN TUMOR INITIATION

Since free radicals are known to cause DNA damage which is considered to be an essential obligatory step in tumor initiation,[50] they are thought to play an important role in this process. Though the role of reactive oxygen species (ROS) in tumor initiation is poorly understood, transition metal ions such as iron are involved in the generation of superoxide anion-dependent reactive metabolites and therefore the involvement of ROS in tumor initiation is implicated.[51]

The prototype skin carcinogen BP is metabolized by sequential reactions catalyzed by the cytochrome P-450-dependent monooxygenase system and epoxide hydrolase to BP-7,8 diol,[52,53] which can further be oxidized along two distinct pathways of epoxidation. The stereoselectivity of epoxidation of BP-7,8 diol by peroxyl radicals and by cytochrome P-450 are unique.[52-54] Interestingly, both of these pathways have been shown to occur in mammalian skin.[52-55]

The importance of peroxyl radicals in the initiation of carcinogenesis has been reviewed by Marnett[56] and Eling et al.[57] In these studies it was shown that in the skin of untreated animals where the level of cytochrome P-450 is undetectable or very low, the conversion of BP-7,8 diol to *anti-trans*-BP-7,8 diol-9,10 epoxide (the electrophilic ultimate carcinogenic species of BP) may be carried out preferentially by peroxyl radical dependent pathways. However, in carcinogen treated animals, the P-450-dependent epoxidation pathway predominates.[58]

B. ROLE IN SKIN TUMOR PROMOTION

The involvement of oxygen-centered free radicals in tumor promotion is supported indirectly by several lines of evidence. It is known that skin tumor promoters increase the generation and decrease the degradation of ROS.[47-49,59] The involvement of ROS in skin tumor promotion is also suggested by the observations that certain organic peroxides and free radical-generating systems exert tumor-promoting activities and mimic or enhance certain molecular events related to tumor promotion. For example, many free radical-generating compounds such as benzoyl peroxide, lauryl peroxide, cumene hydroperoxide, 2,2-azobis-(2-amidinopropane), *tert*-butyl hydroperoxide, and *tert*-butyl hydroperoxybenzoate are known tumor promoters in skin.[60,61] Additional evidences for the involvement of ROS in skin tumor promotion come from the studies showing that various antioxidants and free radical scavengers inhibit the biochemical and biological effects of tumor promoters; for example, β-carotene, ascorbic acid, α-tocopherol, butylated hydroxyanisole (BHA), butylated hydroxytoluene (BHT) and other phenolic compounds have been shown to inhibit tumor promotion.[62,63] Similarly, retinoids, though not known to be antioxidants, are well established inhibitors of tumor promotion.[64] The chemical structure of retinoids suggests that they might inhibit promotion by interacting with ROS thereby depleting steady state oxidant concentrations.[65]

The application of tumor promoters to murine skin causes significant depletion of antioxidant enzymes such as superoxide dismutase (SOD), glutathione peroxidase, and catalase,[66] whereas treatment with copper complex of 3,5 diisopropyl salicylic acid, a mimic of SOD, and with dimethylsulfoxide,[68] a potent scavenger of hydroxyl radicals, inhibit some of the biochemical and biological effects of tumor promoters. It is known that depletion of physiological antioxidants such as α-tocopherol, selenium, and glutathione increases tumor induction[59] and treatment with compounds capable of augmenting the cysteine pool such as *N*-acetyl-cysteine provide protection against phorbol ester-mediated tumor promotion in mouse skin.[69] This also suggests the role of ROS in skin tumor promotion. Since tumor promoters are non-mutagenic and do not bind to cellular DNA, their effects on aneuploidy, chromosomal aberration, sister chromatid exchange, DNA strand breakage, and gene amplification have been thought to be mediated through ROS generated in phorbol ester-treated epidermal cells and/or released by activated phagocytic cells infiltrating the skin during inflammation.[59]

C. ROLE IN MALIGNANT CONVERSION

The involvement of ROS in the conversion of benign papillomas to malignant carcinomas is implicated from several lines of evidences. The oxidant tumor promoter, benzoyl peroxide, which is a free-radical generating compound, has been shown to enhance this process.[70,71] Studies have shown that other free-radical generating compounds such as *tert*-butyl hydroperoxybenzoate and 2,2-azobis-(2-amidinopropane) are effective in converting benign papillomas to malignant carcinomas.[72] The relevance of these observations to the role of free radicals generated from these compounds is further supported by recent findings that all-*trans*-retinoic acid affords significant protection against the conversion of benign papillomas to malignant carcinomas in murine skin.[73]

In view of the fact that cutaneous chemical carcinogens and tumor promoters may, directly or indirectly, generate ROS, and that various antioxidants effectively inhibit biochemical and biological events associated with tumor initiation, promotion and/or malignant conversion, it is conceivable that ROS-induced damage in macromolecules may contribute to the transition of epidermal target cells from a preneoplastic state to malignancy.

IX. ONCOGENES IN CUTANEOUS CARCINOGENESIS

The concept of DNA binding as an essential step in chemical carcinogenesis has recently gained considerable support from studies on oncogenes,[74,75] initially identified as the transforming genetic constituents of certain RNA tumor viruses. In general, viral oncogenes closely resemble the genes of eukaryotic cells, known as protooncogenes. At least 50 different protooncogenes are known, which code for proteins required for the reception, transduction, and interpretation of endogenous mitogenic signals such as growth factors. The biochemical route of the "mitotic cascade" governed by protooncogene products may play a key role in embryonic development, tissue regeneration, and wound repair.[74-76] Neoplastic transformations may be related to inappropriate expression of protooncogenes as a result of chromosomal rearrangement, gene amplification, point mutation, or other genetic alterations.[77,78] Of various cellular protooncogenes studied, the Ha-ras gene has been found to be activated in several spontaneous and induced malignant neoplasms.[79] The ras family of protooncogenes was first discovered as oncogenes contained in Harvey and Kirsten rat sarcoma viruses[80] and forms a family of at least three genes[81-83] which are altered in about 10 to 20% of most human tumors.[84] The cHa-, cKi-, and cN-ras genes encode for very similar p21 proteins of 188 to 189 amino acids.[85] The positions at which single amino acid substitution can lead to transforming activity[86] include amino acids 12, 13, 59, 61, and 63. Of these, mutations at three sites have been identified in human and mouse tumors.[87-89] As reported by Barbacid,[90] chemically induced mouse and human skin carcinomas and other epithelial tumors contain activated cHa-ras.

As reviewed by Marks,[91] the transfection of oncogenes from tumor cells into recipient cells leads to neoplastic transformation of the latter, showing the important role of oncogene mutation and expression in tumorigenesis. In human tumors, protooncogenes of the ras family, for example, Ha-ras, N-ras, and K-ras, may be activated frequently by point mutation. Using DNA from 30 basal cell carcinomas and 12 squamous cell carcinomas followed by amplification of genomic DNA using the polymerase chain reaction, van der Schroeff et al.[92] have studied the occurrence of point mutation in these human tumors. Their study demonstrated that in 4 out of 30 basal cell carcinomas, point mutations were detected at codon 12 of the K-ras gene and at codon 61 of the Ha-ras gene. The K-ras mutations involve glycine to cysteine and glycine to asparagine amino acid changes, whereas the mutations at codon 61 of the Ha-ras gene were consistent with substitution of histidine for glutamine. In 1 of 12 squamous cell carcinomas, a point mutation was detected at codon 12 of the K-ras gene involving a glycine to cysteine substitution in the gene product.

A point mutation and an overexpressed Ha-ras protooncogene with transforming potential has been identified in papillomas and carcinomas in mouse skin using the initiation-promotion protocol, indicating that skin carcinogenesis is accompanied by activation of a protoonco-gene.[93,94] This oncogene activation is probably due to a direct attack of the carcinogenic agent on the genome, resulting in changes that lead to papilloma formation. Balmain and colleagues demonstrated that the introduction of activated Ha-ras genes into epidermal cells of mouse skin *in vivo* followed by treatment with TPA can induce benign papillomas, some of which progress to invasive carcinomas.[95] However, as reviewed by Mukhtar et al.,[5] the activation of Ha-ras is not sufficient for oncogenic action unless it is coupled with other important events, as yet not clearly defined.

In keratinocyte cell cultures, ras-containing Harvey murine sarcoma viruses are known to block epidermal cell differentiation,[96] an effect which provides a condition for the selection and clonal expansion of tumor cells during tumor promotion.[97] Moreover, epidermal cells carrying the viral Ha-ras oncogene develop into papillomas when grafted on to athymic nude mice.[98] Although there is a good correlation between Ha-ras activation and papilloma formation, the precise role of Ha-ras in skin carcinogenesis is not known, since the function of the protein coded for by this oncogene is not fully understood. This protein has been identified as a GTP-binding factor similar to the transducing or the regulatory subunit of adenylate cyclase.[91] The activation of ras oncogene however is not sufficient for the development of malignant tumors in skin or other cells and tissues.[74] Instead, the activation of at least one more protooncogene, such as c-myc in addition to c-ras, is required for malignant progression. Whether the enhancement of malignant progression by treating mouse skin papillomas with carcinogenic agents[99] is due to such additional gene activation is still unknown.

Recently it has been suggested that in a fraction of NIH3T3 cells, multiple steps are involved in transformation by the ras oncogene.[100] A similar conclusion based on promotion of Ha-ras-induced transformation of rat embryo cells by TPA has also been shown.[101] Inhibition of ras transformation by protease inhibitors and retinoic acid has been thought to involve mechanisms similar to those operating *in vivo* and in cell transformation systems, especially in the proliferation of NIH3T3 cells during transfection. The ras-induced NIH3T3 cell transformation assay has been suggested to serve as an alternative assay for stage-specific inhibitors or stimulators of carcinogenesis.[100]

In addition, Yuspa and colleagues[102] recently showed that both Ha-ras and fos oncogenes are necessary to produce the malignant phenotype in epidermal cells. Their study clearly links the fos oncogene with malignant conversion, and suggests that since fos acts as a transcriptional regulator of other genes,[102] malignant conversion may be an indirect consequence of the overexpression of the fos-encoded protein leading to a change in the expression of fos-controlled cellular genes. Hashimoto et al.[103] examined the gene expression associated with multistage development of tumors using two- and three-stage models in female CD-1 mice. These investigators have suggested that the regulatory machinery for transcription by PKC-mediated pathway through nuclear oncogenes is altered during the process of tumor promotion and tumor progression, and that genes whose expression is elevated may be associated directly or indirectly with tumor promotion and progression.

The role of Ha-ras oncogene in murine skin tumor initiation has further been established recently by Leder et al.,[104] who developed a transgenic mouse strain that carries the activated v-H-ras oncogene fused to the promoter of the mouse embryonic α-like, ζ-globin gene. Following skin application of TPA, these transgenic mice developed multiple papillomas, some of which progressed on to squamous cell carcinomas and, more frequently, underlying sarcomas. When these mice were treated with retinoic acid, a significant delay, reduction and often complete inhibition in the appearance of TPA-induced papillomas were observed which concluded that this mouse strain may be useful in screening tumor promoters and for assessing antitumor and anti-proliferative agents.

ACKNOWLEDGMENTS

The studies in authors' laboratory were supported by USPHS grants ES-1900, P-30-AR-39750, and by research funds from the Department of Veterans Affairs. We thank Ms. Sandra Evans and Rachel Floyd for preparing the manuscript.

REFERENCES

1. **Pott, P.,** *The Surgical Works of Percival Pott,* Vol. 5, Haws, Clarke and Collins, London, 1775, 60.
2. **Yamagiwa, K. and Ichikawa, K.,** Uber die kunstliche Epithelwucherung, *Gann,* 11, 1914.
3. **Kennaway, E. L. and Hieger, T.,** Carcinogenic substances and their fluorescence spectra, *Br. Med. J.,* 1, 1044, 1930.
4. **Cook, J. W., Hewett, C. R., and Hieger, I.,** The isolation of a cancer-producing hydrocarbon from coal tar, *J. Chem. Soc. Trans.,* I, 395, 1933.
5. **Mukhtar, H., Merk, H. F., and Athar, M.,** Skin chemical carcinogenesis, *Clin. Dermatol.,* 7, 1, 1989.
6. **Slaga, T. J.,** Mechanisms involved in two-stage carcinogenesis in mouse skin, in *Mechanisms of Tumor Promotion,* Slaga, T. J., Ed., CRC Press, Boca Raton, FL, 1984, 1.
7. **Boutwell, R. K.,** Some biological aspects of skin carcinogenesis, *Prog. Exp. Tumor Res.,* 4, 207, 1964.
8. **Yuspa, S. H.,** Cutaneous chemical carcinogenesis, *J. Am. Acad. Dermatol.,* 15, 1031, 1986.
9. **Ritchie, A. C. and Saffiotti, U.,** Orally administered 2-acetyl-aminofluorene as an initiator and as a promoter in epidermal carcinogenesis in the mouse, *Cancer Res.,* 15, 84, 1955.
10. **Goerttler, K., Loehrke, H., Schweizer, J., and Hesse, B.,** Positive two-stage carcinogenesis in female Sprague-Dawley rats using DMBA as initiator and TPA as promoter, *Virchows Arch. A,* 385, 181, 1980.
11. **Goerttler, K. and Loehrke, H.,** Displacental carcinogenesis: tumor localization and tumor incidence in NMRI mice after displacental initiation with DMBA and urethane and postnatal promotion with the phorbol ester TPA in a modified 2-stage Berenblum Mottram experiment, *Virchows Arch. A,* 376, 117, 1977.
12. **Miller, E. C. and Miller, J. A.,** Chemical carcinogenesis: mechanisms and approaches to its control, *J. Natl. Cancer Inst.,* 47, V, 1971.
13. **Levin, W., Wood, A., Chang, R., Ryan, D., Thomas, P., Yagi, H., Thakkar, D., Vyas, K., Boyd, D., Chu, S. Y., Conney, A. H., and Jerina, D. M.,** Oxidative metabolism of polycyclic aromatic hydrocarbons to ultimate carcinogens, *Drug Metab. Rev.,* 13, 555, 1982.
14. **Miller, E. C. and Miller, J. A.,** The metabolism of chemical carcinogens to reactive electrophiles and their possible mechanisms of actions in carcinogenesis, in *Chemical Carcinogens* (ACS Monog. 173), Searle, C. E., Ed., American Chemical Society, Washington, D. C., 1976, 737.
15. **Jerina, D. M., Sayer, J. M., Thakker, D. R., and Yagi, H.,** Carcinogenicity of polycyclic aromatic hydrocarbons: the bay-region theory, in *Carcinogenesis: Fundamental Mechanisms and Environmental Effects,* Pullman, B., Ts'o, P. O. P., and Gelboin, H., Eds., Reidel, Dordrecht, 1980, 1.
16. **Sims, P.,** The metabolic activation of chemical carcinogens, *Br. Med. Bull.,* 36, 11, 1980.

17. **Hecker, E. and Schmidt, R.,** Phorbol esters-the irritants and cocarcinogens of *Croton tiglium* L, *Fortschr. Chem. Org. Naturst.,* 31, 377, 1974.
18. **Fujiki, H., Suganuma, M., Tahira, T., Yoshioka, A., Nahayasu, M., Endo, Y., Shudo, K., Takayama, S., Moore, R. E., and Sugimura, T.,** New classes of tumor promoters, in *Cellular Interactions by Environmental Tumor Promoters,* Fujiki, H., Hecker, E., Moore, R. E. Sugimura, T., and Weinstein, I. B., Eds., Japan Scientific Societies Press, Tokyo/VNU Science, Utrecht, 1984, 37.
19. **Boutwell, R. K.,** The function and mechanisms of promoters of carcinogenesis, *CRC Crit. Rev. Toxicol.,* 2, 419, 1974.
20. **Slaga, T. J., Triplett, L. L., Yotti, L. P., and Trosko, J. E.,** Skin tumor promoting activity of benzoyl peroxide, a widely used free radical generating compound, *Science,* 213, 1023, 1981.
21. **Hecker, E.,** Structure-activity relationships in diterpene esters irritant and cocarcinogenic to mouse skin, in *Mechanisms of Tumor Promotion and Cocarcinogenesis,* Slaga, T. J., Sivak, A., and Boutwell, R. K., Eds., Raven Press, New York, 1978, 11.
22. **Burns, F. J., Vanderlaan, M., Snyder, E., and Albert, R. E.,** Induction and progression kinetics of mouse skin papillomas, in *Mechanisms of Tumor Promotion and Cocarcinogenesis,* Slaga, T. J., Sivak, A., and Boutwell, R. K., Eds., Raven Press, New York, 1978, 91.
23. **Burns, F. J., Albert, R. E., and Altshuler, B.,** Cancer progression in mouse skin, in *Mechanisms of Tumor Promotion,* Vol. II, Slaga, T. J., Ed., CRC Press, Boca Raton, FL, 1984, 18.
24. **Reddy, A. L. and Fialkow, P. J.,** Papillomas induced by initiation-promotion differ from those induced by carcinogen alone, *Nature,* 304, 69, 1983.
25. **Taguchi, R., Yokoyama, M., and Kitamura, Y.,** Intraclonal conversion from papilloma to carcinoma in the skin of PgK-1a/PgK-1b mice treated by a complete carcinogen protocol or by an initiation-promotion regimen, *Cancer Res.,* 44, 3779, 1984.
26. **Marks, F. and Furstenberger, G.,** The conversion stage of skin carcinogenesis, *Carcinogenesis,* 11, 2085, 1990.
27. **Aldaz, C. M., Conti, C. J., Chen, A., Bianchi, A., Walker, S. B., and DiGiovanni, J.,** Promoter independence as a feature of most skin papillomas in SENCAR mice, *Cancer Res.,* 51, 1045, 1991.
28. **Sisskin, E. E., Gray, T., and Barrett, J. C.,** Correlation between sensitivity to tumor promotion and sustained epidermal hyperplasia of mice and rats treated with TPA, *Carcinogenesis,* 3, 403, 1982.
29. **Verma, A. K.,** The protein kinase C activator, L-alpha-dioctanoyl-glycerol: a potent stage II mouse skin tumor promoter, *Cancer Res.,* 48, 1736, 1988.
30. **Arcoleo, J. P. and Weinstein, I. B.,** Activation of protein kinase C by tumor promoting phorbol esters, teleocidin and aplysiatoxin in the absence of added calcium, *Carcinogenesis,* 6, 213, 1985.
31. **DiGiovanni, J., Kruszewski, F. H., Coombs, M. M., Bhatt, T. S., and Pezeshk, A.,** Structure-activity relationships for epidermal ornithine decarboxylase induction and skin tumor promotion by anthrones, *Carcinogenesis,* 9, 1437, 1988.
32. **Donnelly, T. E., Jr., Pelling, J. C., Anderson, C. L., and Dalbey, D.,** Benzoyl peroxide activation of protein kinase C activity in epidermal membranes, *Carcinogenesis,* 8, 1871, 1987.
33. **Nishizuka, Y.,** The family of protein kinase C for signal transduction, *J. Am. Med. Assoc.,* 262, 1826, 1989.
34. **Osada, S.-I., Mizuno, K., Saido, T. C., Akita, Y., Suzuki, K., Kuroki, T., and Ohno, S.,** A phorbol ester receptor/protein kinase, nPKCη, a new member of the protein kinase C family predominantly expressed in lung and skin, *J. Biol. Chem.,* 265, 22434, 1990.
35. **Yuspa, S. H., Ben, T., and Hennings, H.,** The induction of epidermal transglutaminase and terminal differentiation by tumor promoters in cultured epidermal cells, *Carcinogenesis,* 4, 1413, 1983.
36. **Cerutti, P.,** Oxidant tumor promoters, in *Growth Factors, Tumor Promoters and Cancer Genes,* Colburn, N. H., Moses, H. L., and Stanbridge, E. J., Eds., Alan R. Liss, New York, 1988, 239.
37. **Pitts, J. D. and Finbow, M. E.,** The gap junction, *J. Cell Sci.,* 4, 239, 1986.
38. **Kalimi, G. H. and Sirsat, S. M.,** The relevance of gap junctions to stage 1 tumor promotion in mouse epidermis, *Carcinogenesis,* 5, 1671, 1984.
39. **Shubik, P., Baserga, R., and Ritchie, A. C.,** The life and progression of induced skin tumors in mice, *Br. J. Cancer,* 7, 342, 1953.
40. **Burns, F. J., Vanderlaan, M., Snyder, F., and Albert, R. E.,** Induction and progression kinetics of mouse skin papillomas, *Carcinogenesis Compr. Surv.,* 2, 91, 1978.
41. **Ward, J. M., Rehm, S., Devor, D., Hennings, H., and Wenk, M. L.,** Differential carcinogenic effects of intraperitoneal initiation with 7,12-dimethylbenz[a]anthracene or urethane and topical promotion with 12-O-tetradecanoylphorbol-13-acetate in skin and internal tissues of female SENCAR and Balb/C mice, *Environ. Health Perspect.,* 68, 61, 1986.
42. **Hennings, H., Shores, R., Mitchell, P., Spangler, E. F., and Yuspa, S. H.,** Induction of papillomas with a high probability of conversion to malignancy, *Carcinogenesis,* 6, 1607, 1985.

43. **Ewing, M. W., Conti, C. J., Kruszewski, F. H., Slaga, T. J., and DiGiovanni, J.,** Tumor progression in SENCAR mouse skin as a function of initiator dose and promoter dose, duration, and type, *Cancer Res.,* 48, 7048, 1988.

44. **Nowell, P. C.,** Mechanisms of tumor progression, *Cancer Res.,* 46, 2203, 1986.

45. **Burns, F., Albert, R., Altshuler, B., and Morris, E.,** Approach to risk assessment for genotoxic carcinogenesis based on data from the mouse skin initiation-promotion model, *Environ. Health Perspect.,* 50, 309, 1983.

46. **Hennings, H., Shores, R., Balaschak, M., and Yuspa, S. H.,** Sensitivity of subpopulations of mouse skin papillomas to malignant conversion by urethane or 4-nitroquinoline N-oxide, *Cancer Res.,* 50, 653, 1990.

47. **Floyd, R. A.,** Role of oxygen free radicals in carcinogenesis and brain ischemia, *FASEB J.,* 4, 2587, 1990.

48. **Sun, Y.,** Free radicals, antioxidant enzymes, and carcinogenesis, *Free Radical Biol. Med.,* 8, 583, 1990.

49. **Goldstein, B. D. and Witz, G.,** Free radicals and carcinogenesis, *Free Rad. Res. Commun.,* 11, 3, 1990.

50. **Lesko, S. A., Ts'o, P. O., Yang, S. U., and Cheng, R.,** Benzo(a)pyrene radicals and oxygen radical involvement in DNA damage, cellular toxicity and carcinogenesis, in *Free Radicals, Lipid Peroxidation and Cancer,* Mc Brien, D. C. and Slater, T. E., Eds., Academic Press, New York, 1982, 401.

51. **Halliwell, B. and Gutteridge, J. M. C.,** Oxygen toxicity, oxygen radicals, transition metals and disease, *Biochem. J.,* 219, 1, 1984.

52. **Conney, A. H.,** Induction of microsomal enzymes by foreign chemicals and carcinogenesis by polycyclic aromatic hydrocarbons: G. H. A. Clowes memorial lecture, *Cancer Res.,* 42, 4875, 1982.

53. **Mukhtar, M. and Khan, W. A.,** Cutaneous cytochrome P-450, *Drug Metab. Rev.,* 20, 657, 1989.

54. **Marnett, L. J.,** Peroxyl free radicals: potential mediators of tumor initiation and promotion, *Carcinogenesis,* 8, 1365, 1987.

55. **Panthananickal, A., Weller, P., and Marnett, L. J.,** Stereoselectivity of the epoxidation of 7,8-dihydroxy-7,8-dihydrobenzo(a)pyrene by prostaglandin H synthase and cytochrome P-450 determined by the identification of polyguanylic acid adducts, *J. Biol. Chem.,* 258, 4411, 1983.

56. **Marnett, L. J.,** Prostaglandin synthase-mediated metabolism of carcinogens and a potential role for peroxyl radicals as reactive intermediates, *Environ. Health Perspect.,* 88, 5, 1990.

57. **Eling, T. E., Thompson, D. C., Foureman, G. L., Curtis, J. F., and Hughes, M. F.,** Prostaglandin H synthase and xenobiotic oxidation, *Annu. Rev. Pharmacol. Toxicol.,* 30, 1, 1990.

58. **Pruess-Schwartz, D., Nimesheim, A., and Marnett, L. J.,** Peroxyl radical-and cytochrome P-450-dependent metabolic activation of (+)-7,8-dihydroxy-7,8-dihydrobenzo(a)pyrene in mouse skin in vitro and in vivo, *Cancer Res.,* 49, 1732, 1989.

59. **Perchellet, J. and Perchellet, E. M.,** Antioxidants and multistage carcinogenesis in mouse skin, *Free Radical Biol. Med.,* 7, 377, 1989.

60. **Slaga, T. J., Klein-Szanto, A. J. P., Triplett, L. L., and Yotti, P. C.,** Skin tumor promoting activity of benzoyl peroxide, a widely used free radical-generating compound, *Science,* 213, 1023, 1981.

61. **Swauger, J. E., Dolan, P. M., and Kensler, T. W.,** Role of free radicals in tumor promotion and progression by benzoyl peroxide, in *Mutation and the Environment,* Part D, Vol. 340, Mendelsohn, M. L. and Albertini, R. J., Eds., Wiley-Liss, New York, 1990, 143.

62. **Smart, R. C., Huang, M.-T., Han, Z. T., Kaplan, M. C., Focella, A., and Conney, A. H.,** Inhibition of 12-O-tetradecanoylphorbol-13-acetate induction of ornithine decarboxylase activity, DNA synthesis, and tumor promotion in mouse skin by ascorbic acid and ascorbyl palmitate, *Cancer Res.,* 47, 6633, 1987.

63. **Hocman, G.,** Chemoprevention of cancer: phenolic antioxidants (BHA, BHT), *Int. J. Biochem.,* 20, 639, 1988.

64. **Verma, A. K.,** Inhibition of both stage I and stage II mouse skin tumor promotion by retinoic acid and the dependence of inhibition of tumor promotion on the duration of retinoic acid treatment, *Cancer Res.,* 47, 5097, 1987.

65. **Witz, G., Goldstein, B. D., Amoruso, M., Stone, D. S., and Troll, W.,** Retinoid inhibition of superoxide anion radical production by human polymorphonuclear leukocytes stimulated with tumor promoters, *Biochem. Biophys. Res. Commun.,* 97, 883, 1980.

66. **Solanki, V., Rana, R. S., and Slaga, T. J.,** Diminution of mouse epidermal superoxide dismutase and catalase activities by tumor promoters, *Carcinogenesis,* 2, 2555, 1981.

67. **Kensler, T. W., Bush, D. M., and Kozumbo, W. J.,** Inhibition of tumor promotion by a biomimetic superoxide dismutase, *Science,* 221, 75, 1983.

68. **Jacopy, W. T. and Weiss, H. S.,** Inhibition and enhancement of skin tumors in mice by dimethyl sulfoxide depending on method of application, *J. Natl. Cancer Inst.,* 77, 983, 1986.

69. **Perchellet, J.-P., Owen, M. D., Posey, T. D., Orten, D. K., and Schneider, B. A.,** Inhibitory effects of glutathione level-raising agents and D-α-tocopherol on ornithine decarboxylase induction and mouse skin tumor promotion by 12-O-tetra-decanoylphobol-13-acetate, *Carcinogenesis,* 6, 567, 1985.

70. **Reiners, J. J., Jr., Nesnow, S., and Slaga, T. J.,** Murine susceptibility to two-stage carcinogenesis is influenced by the agents used for promotion, *Carcinogenesis,* 5, 301, 1984.

71. **O'Connell, J. F., Klein-Szanto, A. J. P., DiGiovanni, D. M., Fries, J. W., and Slaga, T. J.,** Enhanced malignant progression of mouse skin tumors by the free-radical generator benzoyl peroxide, *Cancer Res.,* 46, 2863, 1986.

72. **Athar, M., Lloyd, J. R., Bickers, D. R., and Mukhtar, H.,** Malignant conversion of UV radiation and chemically induced mouse skin benign tumors by free radical generating compounds, *Carcinogenesis,* 10, 1841, 1989.

73. **Athar, M., Agarwal, R., Wang, Z. Y., Lloyd, J. R., Bickers, D. R., and Mukhtar, H.,** All-trans retinoic acid protects against conversion of chemically induced and ultraviolet B radiation-induced skin papillomas to carcinomas, *Carcinogenesis,* 12, 2325, 1991.

74. **Marks, F. and Furstenberg, G.,** Multistage carcinogenesis in animal skin: the reductionist's approach in cancer research, in *Theories of Carcinogenesis,* Iversen, O. H., Ed., Hemisphere, Washington, D. C., 1987, 179.

75. **Bowden, T. G.,** Oncogene activation during multi-stage carcinogenesis, in *Mutation and the Environment,* Part D, Vol. 340, Mendelsohn, M. L. and Albertini, R. J., Eds., Wiley-Liss, New York, 1990, 1.

76. **Goustin, A. S., Leof, E. B., Shipley, G. D., and Moses, H. L.,** Growth factors and cancer, *Cancer Res.,* 46, 1015, 1986.

77. **Bishop, J. M.,** Retroviruses, *Annu. Rev. Biochem.,* 47, 35, 1978.

78. **Land, H., Parada, L. F., and Weinberg, R. A.,** Cellular oncogenes and multistep carcinogenesis, *Science,* 222, 771, 1983.

79. **Taparowsky, E., Suard, Y., Fasano, O., Shimizu, K., Goldfarb, M., and Wigler, M.,** Activation of the T24 bladder carcinoma transforming gene is linked to a single amino acid change, *Nature,* 300, 762, 1982.

80. **Ellis, R. W., DeFeo, D., Shih, T. Y., Gonda, M. A., Young, H. A., Tsuchida, N., Lowy, D. R., and Scolnick, E. M.,** The P21 src genes of Harvey and Kirsten sarcoma viruses originate from divergent members of a family of normal vertebrate genes, *Nature,* 292, 506, 1981.

81. **Chang, E. H., Gonda, M. A., Ellis, R. W., Scolnick, E. M., and Lowy, D. R.,** Human genome contains four genes homologous to transforming genes of Harvey and Kirsten murine sarcoma viruses, *Proc. Natl. Acad. Sci. U.S.A.,* 79, 4848, 1982.

82. **Hall, A., Marshall, C. J., Spurr, N., and Weiss, R. A.,** The transforming gene in two human sarcoma cell lines is a new member of the ras gene family located on chromosome one, *Nature,* 303, 396, 1983.

83. **Shimizu, K., Goldfarb, M., Perucho, M., and Wigler, M.,** Isolation and preliminary characterization of the transforming gene of a human neuroblastoma cell line, *Proc. Natl. Acad. Sci. U.S.A.,* 80, 383, 1983.

84. **Marshall, C. J.,** Human oncogenes, in *RNA Tumor Viruses,* Weiss, R. A., Teich, N., Varmus, H. E., and Colfin, J. E., Eds., Cold Spring Harbor Laboratory, Cold Spring Harbor, NY, 1985, 488.

85. **Tabin, C. J., Bradley, S. M., Bargmann, C. L., Weinberg, R. A., Papageorge, A. G., Scolnick, E. M., Dhar, R., Lowy, D. R., and Chang, E. H.,** Mechanism of activation of a human oncogene, *Nature,* 300, 143, 1982.

86. **Fasano, O., Aldrich, T., Tamanoi, F., Taparowsky, E., Furth, M., and Wigler, M.,** Analysis of the transforming potential of the human H-ras gene by random mutagenesis, *Proc. Natl. Acad. Sci. U.S.A.,* 81, 4008, 1984.

87. **Balmain, A. and Pragnell, E. B.,** Mouse skin carcinomas induced in vivo by chemical carcinogens have transforming Harvey-ras oncogene, *Nature,* 303, 72, 1983.

88. **Yuasa, Y., Srivastava, S. K., Dunn, C. Y., Rhim, J. S., Reddy, E. P., and Aaronson, S. A.,** Acquisition of transforming properties by alternative point mutations within c-bas/has human proto-oncogene, *Nature,* 303, 775, 1983.

89. **Bos, J. L., Toksoz, D., Marshall, C. J., Vries, M. V., Veeneman, G. H., van der Eb, A. J., van Boom, J. H., Janseen, J. W. G., and Steenvoorden, A. C. M.,** Amino-acid substitutions at codon 13 of the N-ras oncogene in human acute myeloid leukemia, *Nature,* 315, 726, 1985.

90. **Barbacid, M.,** Ras genes, *Annu. Rev. Biochem.,* 56, 779, 1987.

91. **Marks, F.,** Skin cancer (excluding melanomas), in *Pharmacology of the Skin II,* Greaves, M. W. and Shuster, S., Eds., Springer-Verlag, New York, 1989, 165.

92. **Van der Schroeff, J. G., Evers, L. M., Boot, A. J. M., and Bos, J. L.,** Ras oncogene mutations in basal cell carcinomas and squamous cell carcinomas of human skin, *J. Invest. Dermatol.,* 94, 423, 1990.

93. **Quintanilla, M., Brown, K., Ramsden, M., and Balmain, A.,** Carcinogen-specific mutation and amplification of Ha-ras during mouse skin carcinogenesis, *Nature,* 322, 78, 1986.

94. **Balmain, A. and Brown, K.,** Oncogene activation in chemical carcinogenesis, *Adv. Cancer Res.,* 51, 147, 1988.

95. **Brown, K., Quintanilla, M., Ramsden, M., Kerr, I. B., Young, S., and Balmain, A.,** V-ras genes from Harvey and BALB murine sarcoma viruses can act as initiators of two-stage mouse skin carcinogenesis, *Cell,* 46, 447, 1986.

96. **Yuspa, S. H., Kilkenny, A. E., Stanley, J., and Lichti, U.,** Keratinocytes blocked in phorbol ester-responsive early stage of terminal differentiation by sarcoma viruses, *Nature,* 314, 459, 1985.

97. **Yuspa, S. H.,** Molecular and cellular basis for tumor promotion in mouse skin, in *Cellular Interactions by Environmental Tumor Promoters,* Fujiki, H., Hecker, E., Moore, R. E., Sugimura, T., and Weinstein, I. B., Eds., Japan Scientific Societies Press, Tokyo/VNU Science, Utrecht, 1984, 315.

98. **Roop, D. R., Lowy, D. R., Tambourin, P. E., Strickland, J., Harper, J. R., Balaschak, M., Spangler, E. F., and Yuspa, S. H.,** An activated Harvey ras oncogene produces benign tumors on mouse epidermal tissue, *Nature,* 323, 822, 1986.

99. **Hennings, H., Shores, R., Wenk, M. L., Spangler, E. F., Tarone, R., and Yuspa, S. H.,** Malignant conversion of mouse skin tumors is increased by tumor initiators and unaffected by tumor promoters, *Nature,* 304, 67, 1983.

100. **Garte, S. J.,** Oncogene activation in multistage carcinogenesis, *J. Am. Coll. Toxicol.,* 8, 241, 1989.

101. **Hsiao, W.-L. W., Gattoni-Celli, S., and Weinstein, I. B.,** Oncogene-induced transformation of C3HIOT 1/3 cells is enhanced by tumor promoters, *Science,* 226, 552, 1984.

102. **Greenhalgh, D. A., Welty, D. J., Player, A., and Yuspa, S. H.,** Two oncogenes, v-fos and v-ras, cooperate to convert normal keratinocytes to squamous cell carcinoma, *Proc. Natl. Acad. Sci. U.S.A.,* 87, 643, 1990.

103. **Hashimoto, Y., Tajima, O., Hashiba, H., Nose, K., and Kuroki, T.,** Elevated expression of secondary, but not early, responding genes to phorbol ester tumor promoters in papillomas and carcinomas of mouse skin, *Mol. Carcinogen.,* 3, 302, 1990.

104. **Leder, A., Kuo, A., Cardiff, R. D., Sinn, E., and Leder, P.,** v-Ha-ras transgene abrogates the initiation step in mouse skin tumorigenesis: effects of phorbol esters and retinoic acid, *Proc. Natl. Acad. Sci. U.S.A.,* 87, 9178, 1990.

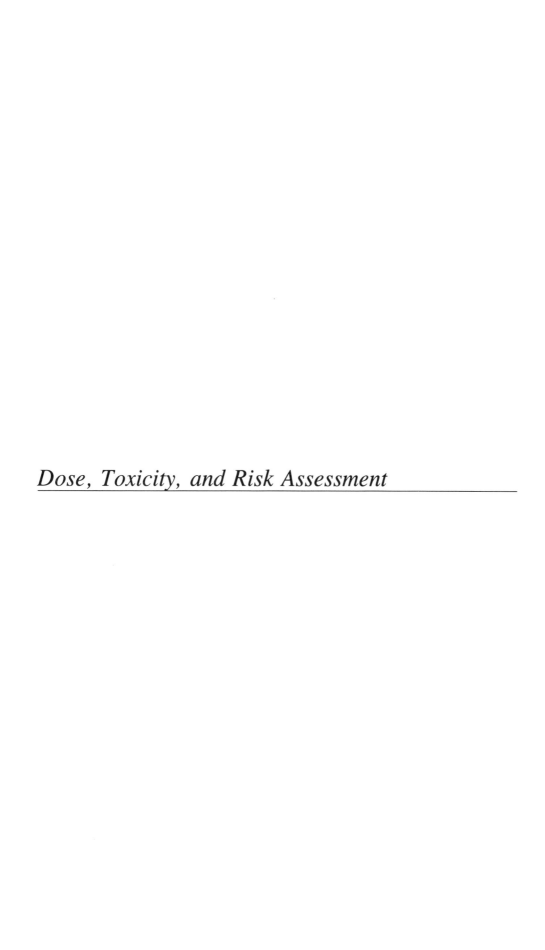

Dose, Toxicity, and Risk Assessment

Chapter 21

INCORPORATING BIOLOGICAL INFORMATION INTO THE ASSESSMENT OF CANCER RISK TO HUMANS UNDER VARIOUS EXPOSURE CONDITIONS AND ISSUES RELATED TO HIGH BACKGROUND TUMOR INCIDENCE RATES*

Chao W. Chen and Kim-Chi Hoang

TABLE OF CONTENTS

---

* The views expressed in this paper are those of the authors and do not necessarily reflect the views or policies of the U.S. Environmental Protection Agency.

I. INTRODUCTION

Ideally, human experience and sound laboratory findings should be used to predict human risk due to exposure to environmental pollutants. However, useful human data are rarely available and one must use animal data to extrapolate human risk to a suspect carcinogen. Extrapolating animal carcinogenic data to human risk is complicated by the high doses used in animal studies and by the lack of information on mechanisms of tumor induction. Since available data usually come from high-dose experiments, tumor incidences in animals exposed at high doses are used to extrapolate risk to animals that might be exposed at low doses. This first step in extrapolation involves constructing the dose-response model. The conventional approach of dose-response modeling is to fit tumor incidence data to a pre-selected mathematical dose-response function which may not reflect the biological basis of carcinogenesis and may not consider the physiological and biochemical properties of the compound under study. Although many different mathematical dose-response models may fit a set of data equally well at the experimental range (usually at high doses), results may differ drastically at the low doses to which humans are exposed. Thus, the selection of a dose-response model constitutes a source of uncertainty. Species to species extrapolation from animal low-dose risk to humans constitutes the second level of uncertainty.

To reduce uncertainties, biological information about the suspect carcinogen and its interaction with the host system should be incorporated into the quantitative risk assessment whenever possible. This biological information includes the mechanism of carcinogenesis, the pharmacokinetics of a compound and its metabolites produced by various routes of exposure, and the interspecies relationships (animal to human) among the first two types of biological information.

In this chapter, approaches to incorporating some biological information into dose-response modeling are demonstrated using perchloroethylene (PERC) as an example. Although data on PERC for biologically based dose-response (BBDR) modeling are much more limited than that available for many other compounds (e.g., diethylnitrosamine), PERC was selected because it induces hepatocellular carcinomas in $B6C3F_1$ mice, and there is controversy regarding the interpretation of increased tumor incidence in mouse liver when a high spontaneous incidence is also present. $B6C3F_1$ mice have been used in National Toxicology Program (NTP) and National Cancer Institute (NCI) carcinogenesis studies since 1971. Many tested compounds are found to be hepatocarcinogens in this strain of mouse. Chemically induced hepatocarcinogenesis may result from genotoxic events through the interaction of reactive metabolites with DNA or from epigenetic events such as increased cellular proliferation due to cytotoxicity. Because of the high spontaneous tumor incidence in male mouse liver and the generally non-positive mutagenic tests for a number of compounds, the relevance of male mouse liver tumors to an inference of human tumor hazard continues to be debated.

Many scientists are concerned that the conventional dose-response model that assumes low-dose linearity (e.g., the linearized multistage model currently used by the U.S. Environmental Protection Agency) unreasonably overestimates risk at low doses when incidences of liver tumors from male $B6C3F_1$ mice are used for extrapolation and when the compound also increases the rate of cell proliferation. From the quantitative viewpoint, the controversy could be minimized if certain relevant biological information could be incorporated into risk extrapolation. For instance, if one assumes that PERC exerts its carcinogenicity through proliferation of normal and precancerous liver cells and if the rates of spontaneous and dose-dependent PERC-induced cell proliferation in animals and humans are known, then there should be no controversy as long as some key biological parameters are properly considered when risk to humans is predicted.

To demonstrate how biological information could be incorporated into quantitative risk assessment, we have summarized several conclusions from Reichert[1] in the next section and have used them as biological bases for dose-response modeling. Reichert's conclusions are used only as illustrations. our goal is to demonstrate how biological information can be used in quantitative risk assessment and how the high background tumor incidence rate can be meaningfully evaluated in dose-response modeling. This approach is also useful for evaluating the effect of using high background tumor incidence rates in animals when extrapolating risk to humans.

II. PRELIMINARY CONSIDERATIONS

Reichert[1] has reached the following conclusions concerning PERC: (These conclusions do not necessarily reflect our views about the compound. They are included here only as an example to demonstrate how the impact of certain assumptions on low-dose extrapolation can be evaluated.)

1. Cytotoxicity seems to be a prerequisite for the carcinogenic action of PERC in the liver. This assumption is supported by the carcinogenic effect in B6C3F$_1$ mice, a strain sensitive to hepatotoxic compounds, and by the lack of a genotoxic effect in short-term test systems. The cytotoxicity of PERC includes cell necrosis that stimulates DNA replication associated with cell division.
2. The results of acute toxicity testing of PERC support the epigenetic mechanism of tumor formation in mice and the assumption of a reactive intermediate in PERC metabolism.
3. The available pharmacokinetic data on PERC indicate that the rate of PERC metabolism is lower for humans than for rats and mice.

From these observations, we can conclude that PERC may exert its carcinogenic effect through an increased rate of cell proliferation, perhaps through cell necrosis, and that human risk due to exposure to PERC may be smaller than that for animals because of the lower human rate of PERC metabolism. There are data on enzyme-altered foci in rat liver (Milman et al.).[2] No comparable data are available for mice. If similar foci data were available for mice, it would be easier to construct a BBDR model. Tumor promotion may exert an effect independent of mutagenesis by expanding the clonal outgrowth of cells that have previously undergone chemically induced or spontaneous genetic damage. Therefore, tumor promotion is an important issue when the liver cancer response from B6C3F$_1$ mice is extrapolated to humans because of the high spontaneous tumor incidence in this strain of mouse and the apparent lack of a similar level of spontaneity in humans.

PERC is absorbed through the lungs, the gastrointestinal tract, and the skin. Irrespective of dose and route of exposure, pulmonary exhalation is the major route for eliminating the parent compound. PERC is biotransformed by microsomal monooxygenase (the cytochrome P-450 system) to reactive metabolites, including the putative carcinogen trichloroacetic acid (TCA) and a short-lived reactive epoxide intermediate that can bind covalently to cellular macromolecules. The hepatotoxic and carcinogenic potentials of PERC are generally thought to reside in its biologically reactive intermediate metabolites rather than in the parent compound itself. Studies showed that metabolic saturation occurs in the range of doses or exposure levels used in the NCI and NTP bioassays.[3,4] Therefore, it is desirable to construct a dose-response model using dosimetry that adjusts for metabolic saturation and that reflects actual body burden.

TABLE 1
Incidence of Hepatocellular Carcinomas in B6C3F₁ Mice in NCI (1977) Gavage Study[a]

Male (0.035 kg)		Female (0.025 kg)	
Administered dose (mg/kg/day)	Tumor incidence	Administered dose (mg/kg/day)	Tumor incidence
0	4/37 (0.11)	0	2/40 (0.05)
536	32/47 (0.68)	386	19/48 (0.40)
1072	27/45 (0.60)	772	19/45 (0.42)

[a] Control groups: a combination of vehicle and untreated animals. Treated groups: animals were exposed to PERC 5 days/week for 78 weeks anD then followed for 12 additional weeks.

TABLE 2
Incidence of Hepatocellular Carcinomas in B6C3F₁ Mice in NTP (1985) Inhalation Study[a]

Male (0.035 kg)		Female (0.025 kg)	
Administered dose (mg/kg/day)	Tumor incidence	Administered dose (mg/kg/day)	Tumor incidence
0	7/49 (0.14)	0	1/46 (0.02)
100	25/47 (0.53)	100	13/42 (0.31)
200	26/50 (0.52)	200	36/47 (0.77)

[a] Control groups: a combination of vehicle and untreated animals. Treated groups: animals were exposed to PERC 6 h/day, 5 days/week for 104 weeks.

III. DOSIMETRY CALCULATIONS FOR BIOASSAY DATA

Hepatocellular carcinomas were induced in both male and female mice when animals were exposed to PERC via gavage or inhalation. The tumor incidences in female mice from gavage[5] and inhalation[6] studies are used to construct a dose-response model for PERC (Tables 1 and 2). The metabolized dose corresponding to each administered dose (via gavage) or exposure concentration (via inhalation) is calculated using a physiologically based pharmacokinetic (PB-PK) model. The tumor incidence data in male mice are not used for dose-response modeling because male mice had a much higher background incidence rate of liver tumors (about 20%) than did female mice (about 4%). As the subsequent discussion will show, high background tumor incidence may lead to a higher risk estimate, even when the background rate is adjusted.

The PB-PK model in Chen and Blancato[7] has been extended to include a skin compartment and is used to calculate the metabolized dose corresponding to an exposure to PERC. The details of the extended PB-PK model are given in the Appendix. The metabolized dose is used as a dosimetry in our dose-response modeling. Under NCI gavage or NTP inhalation exposure patterns, the amount metabolized reaches the steady state during weekdays and is almost totally eliminated during the weekend (nonexposed) period. Tables 3 and 4 give the daily amounts (in milligrams) of total metabolites in female mice exposed to PERC under NCI gavage[5] or NTP inhalation[6] bioassay exposure patterns. The steady state metabolized doses averaged over a week (i.e., the values at the bottom of Tables 3 and 4) are used as dosimetry for BBDR modeling. The use of the dosimetry allows us to combine

TABLE 3
Daily Amount (mg) Metabolized in Female Mice
(0.025 kg) Administered PERC by Gavage
According to the NTP Bioassay Pattern

| | Week 1 | | | | Week 2 | |
| | | Low dose | High dose | | | High dose | Low dose |
Day		Low dose	High dose	Day		High dose	Low dose
1		1.65	2.32	1		1.65	2.32
2		1.66	2.33	2		1.66	2.34
3		1.66	2.34	3		1.66	2.34
4		1.66	2.34	4		1.66	2.34
5		1.66	2.34	5		1.66	2.34
6 (unexposed)		0.04	0.10	6 (unexposed)		0.04	0.10
7 (unexposed)		0.00	0.00	7 (unexposed)		0.00	0.00

Note: Average: low dose, 1.19 mg/day or 47.60 mg/kg/day; high dose, 1.68 mg/day or 67.31 mg/kg/day.

TABLE 4
Daily Amount (mg) Metabolized in Female Mice
Exposed to PERC by Inhalation According to the
NCI Bioassay Pattern

| | Week 1 | | | | Week 2 | |
Day		Low dose	High dose	Day		High dose	Low dose
1		1.75	2.43	1		1.75	2.43
2		1.76	2.44	2		1.76	2.44
3		1.76	2.44	3		1.76	2.44
4		1.76	2.44	4		1.76	2.44
5		1.76	2.44	5		1.76	2.44
6 (unexposed)		0.05	0.11	6 (unexposed)		0.05	0.11
7 (unexposed)		0.00	0.00	7 (unexposed)		0.00	0.00

Note: Average: low dose, 1.26 mg/day or 50.51 mg/kg/day; high dose, 1.76 mg/day or 70.28 mg/kg/day.

tumor response data from gavage and inhalation studies into a single data set in dose-response calculations. The amount metabolized and the corresponding tumor incidence rates in the NTP and NCI bioassays are given in Table 5 for female mice. These data, along with the historical background rate (Figure 1), are used to construct BBDR models for PERC.

IV. BIOLOGICALLY BASED DOSE-RESPONSE (BBDR) MODELS

In the traditional, multistep construct of carcinogenesis, proliferation of initiated (pre-neoplastic) cells (I-cells) is a major factor in tumor formation. Some excellent reviews of the concept of initiation and promotion in hepatocarcinogenesis are found in Pitot and Sirica[8] and Farber and Sarma.[9] There is some evidence that a chemical with only initiation capability may not induce a significant increase of malignant tumors unless the chemically induced I-cells can be proliferated either by the chemical itself (or its reactive metabolites) or by the host condition (e.g., in the infant period).[10] On the other hand, if a chemical has only cell

TABLE 5
Observed and Predicted Incidence Rates of Hepatocellular Carcinomas in Female Mice, Combining Data from Oral and Inhalation Studies of PERC[a]

Age (days)	Metabolized dose (mg/kg/day)	Tumor incidence rates	
		Observed	Predicted
Gavage Study			
672	0	0.05	0.03
672	47.67	0.40	0.24
672	67.20	0.42	0.40
Inhalation Study			
763	0	0.02	0.03
763	50.40	0.31	0.47
763	70.00	0.77	0.73

[a] The mice appear to have been 35 days old when exposure began.
$\chi^2 = \Sigma n_i (\text{Observed} - \text{Predicted})^2 / \text{Predicted} = 8.07$ (d.f. = 5, $p > 0.05$), where n_i is the number of animals in each group.

proliferation potential (i.e., promotion effect), its tumorigenic effect would not be manifested unless spontaneously or chemically induced I-cells already exist. The tumor incidence rate in animals exposed to a chemical having only proliferation potential would depend on the number of spontaneous I-cells in the experimental animals. Animals with more spontaneous I-cells that can be proliferated by the chemical would show more tumor response. Therefore, it is important to consider the number of spontaneous I-cells in mice and their rates of cell proliferation and conversion to malignancy when modeling dose-response for PERC on the basis of liver tumors in B6C3F$_1$ mice.

Several stochastic two-stage models[11-14] that incorporate cell proliferation can be used to construct a dose-response model. The biological basis and data requirements for each of these models are discussed by Chen and Farland.[13] We have selected the Chen-Farland (CF) model to use for the construction of a BBDR model with the intention of stimulating interest and discussions in the scientific community and of generating data for such a modeling. This model is a modified version of the two-event model (MVK model) proposed by Moolgavkar and colleagues[12,14] in which an I-cell can proliferate (spontaneously or via chemical induction) and a tumor is considered to be formed once a malignant cell is born. The CF model allows for the growth of both I-cells and malignant cells, and the rate of cell mitosis is explicitly incorporated by assuming that the time for an I-cell (or a malignant cell) to reach mitosis is a random variable following a certain probability distribution. If the probability distribution is assumed to be exponential and a malignant tumor is considered to be formed once a malignant cell is born, then the CF model is similar to the MVK model, except that mitotic rate is explicitly incorporated in the CF model. The development of the CF model was motivated by the biological concept proposed by Greenfield et al.,[15] which is known as the Cohen-Ellwein model and has been used by the investigators to model bladder tumor incidence data.[16,17] The Cohen-Ellwein model, however, can be used only as a ''semi-simulation'' model because the model does not have a complete mathematical representation of what was intended to be modeled. Because of the lack of a complete mathematical representation, the model cannot be used to estimate parameters; it requires

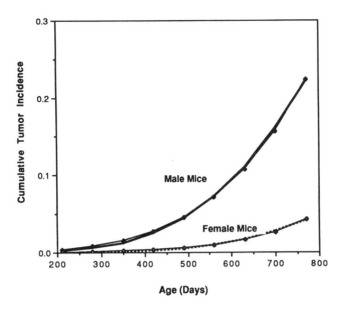

FIGURE 1. Observed vs. predicted cumulative probability of hepatocellular carcinomas in NTP historical control in B6C3F$_1$ mice. (Observed data are from Portier, C., Hedges, J., and Hoel, D., *Cancer Res.*, 46, 4372, 1986.)

all parameters to be known. In this chapter, a special version of the CF model is used to develop a dose-response model. In this version, the time to mitosis is assumed to be exponentially distributed but it allows for the birth and death of malignant cells.

The probability of cancer by time t is given by

$$P(t,d) = 1 - \exp\{\int_{t_0}^{t} \mu_1(s)N(s)[\psi(t - s,d) - 1]ds\} \qquad (1)$$

where

$$\psi(t - s,d) = \frac{y_1(1 - y_2) - y_2(1 - y_1)\exp[mb_1(y_1 - y_2)(t - s)]}{(1 - y_2) - (1 - y_1)\exp[mb_1(y_1 - y_2)(t - s)]}; \qquad (2)$$

$y_1 < y_2$ are two roots of the equation

$$b_1x^2 - [1 - \mu_2(d_m/b_m)]x + d_1 = 0 \qquad (3)$$

N(s) is the number of target cells at age s; μ_1 is rate of initiation per cell per day; m is the mean mitotic rate (i.e., the mean number of cell divisions per cell per day); b$_1$, d$_1$ and μ_2 are the probabilities that, upon division, an I-cell gives birth, dies, or is transformed into a malignancy, b$_1$ + d$_1$ + μ_2 = 1; d$_m$/b$_m$ is the ratio of death to birth for a malignant cell. (Here we consider only a special case in which the mitotic rate for a malignant cell is cancelled out and thus is not needed in the model. (See Reference 13.) Note that this model involves initiation (μ_1), promotion (m, b$_1$, d$_1$), conversion (μ_2), and progression (b$_m$, d$_m$). These terms (initiation, promotion, conversion, and progression) were suggested by Knudson.[18]

The number of target cells for mice at age t is extrapolated from rats (see Reference 19) by assuming that the relative number of liver hepatocytes is the same as the relative liver weights between the two species. This extrapolation is done because the number of hepatocytes in mice over age is not available.

$$N(t) = 1.5 \times 10^7 \exp\left\{\frac{0.118}{0.039} [1 - \exp(-0.039t)]\right\} \qquad (4)$$

Equation 1 can be simplified to a form that does not involve integration if $\mu_1(s) = \mu_1$ and $N(s) = N$ are independent of age s. In addition to the bioassay data (Table 5), the historical background incidence data (Figure 1) for both male and female mice are also used to estimate parameters.

If $\mu_1(s) = \mu_1$ and $N(s) = N$ are assumed to be independent of time, then Equation 1 becomes

$$P(t) = 1 - \exp\Big(\mu_1 N(y_1 - 1)(t - t_0) +$$

$$\frac{\mu_1 N \, \text{Log}\{(y_1 - y_2)/[(1 - y_2) + (y_1 - 1)\exp[mb_I(t - t_0)(y_1 - y_2)]]\}}{mb_I}\Big) \qquad (5)$$

where Log is natural logarithm.

A. ESTIMATION OF PARAMETER VALUES

Ideally, some of the parameters in a BBDR model should be measured directly from laboratory data or estimated indirectly from data other than bioassays. Parameters that can be measured directly from laboratory data include mitotic rates of cells at each stage (normal, initiated, and malignant) of tumor development. The data useful for estimating parameters include number and size of preneoplastic (e.g., enzyme altered) foci and nodules/carcinomas of the liver, obtained by examining animals killed periodically during a study. Unfortunately, none of this information is available for PERC.

One of the issues in the risk assessment of PERC is the high spontaneous incidence of tumors. Since PERC is postulated as non-genotoxic, it is reasonable to assume that the initiation rate, μ_1, is independent of dose. This assumption appears to be supported by the observation that, when μ_1 was assumed to be dose dependent, the PERC-induced mitotic rate was estimated to be negligible. This assumption is unreasonable if the tumor induction by PERC is assumed to occur through cytotoxicity. Since data on mitosis are not available, this information is estimated from tumor incidence data by assuming that the mean mitotic rate, m, for I-cells is related to dose d by $m = 1 - \exp(-m_0 - m_1 \times d)$. This functional form is selected to limit the value of m to be no more than one because it seems unlikely that the mean mitotic rate would exceed one per day *in vivo*. Note that this assumption is not necessary when the actual mitotic rate is measured from laboratory data. Another dose-dependent parameter is $\mu_2 = \mu_{20} + \mu_{21} \times d$.

Parameters are estimated by the least square method using Equation 1 and data in Table 5 and Figure 1. To reduce the number of parameters that must be estimated, b_I and b_m are assumed to be known. The parameter b_m is assumed to be 0.8 (i.e., the ratio of death rate to birth rate for a malignant cell is 0.2/0.8). Other parameters are estimated for a given value of b_I. (Recall that b_I is the conditional probability that, after cell division, an I-cell gives birth, i.e., produces two daughter I-cells). The background incidence of liver carcinomas (Figure 1) is used to estimate background parameters (μ_0, μ_{20}, and m_0. The female data in gavage and inhalation studies are combined (Table 5) and are used to estimate dose-related parameters (μ_{21}, m_1). The duration of exposure and follow-up is taken into account in the modeling. The resultant parameter estimates for mice and humans are given in Table 6 when $b_I = 0.6$. Table 9 displays sets of estimated parameters when different values of b_I are assumed. From this table, we note that a decrease in the b_I value is compensated for by

TABLE 6
Estimated Parameter Values for Mice and Humans

Parameters[a]	Male mice	Female mice	U.S. white men[b]
Background Rates			
μ_1	1.20×10^{-8}	4.08×10^{-9}	2.51×10^{-12}
μ_{20}	2.75×10^{-6}	5.69×10^{-7}	2.45×10^{-5}
m_0	2.53×10^{-2}	3.29×10^{-2}	2.65×10^{-1}
PERC-Induced Rates			
μ_{21}	Not estimated	5.05×10^{-8}	Not estimated
m_1	Not estimated	1.89×10^{-4}	Not estimated

[a] When these parameters are estimated, the death-birth ratio, d_m/b_m, for a malignant cell is fixed at 0.2/0.8; $b_I = 0.60$. The rate unit is per cell per day for animals and per cell per year for humans.

[b] Estimated from the vital statistics on liver cancer mortality in U.S. white men.

an increase in the mitotic rates (m_0 and m_1) and initiation rate (μ_1), and a decrease of the conversion rate (μ_2) for both background and PERC-induced rates. Although all of the models fit data equally well, their biological implications could be different. Therefore, a real BBDR model cannot be constructed unless some of the parameters (e.g., mitotic rate) can be directly measured or estimated from data other than cancer bioassay data. However, even under the uncertainty of parameter determination, the BBDR model provides a useful tool to evaluate the impact of various biological assumptions on risk estimates.

For discussion, we select the set of parameters when $b_I = 0.60$ is assumed. Under this assumption, cell death rate, upon mitosis, is approximately equal to $0.40 \ (= 1 - b_I - \mu_2)$ because μ_2 is numerically small (on the order of 10^{-4} or smaller). However, μ_2 should not be considered negligible. It should be noted that unconditional mean cell birth ($m \times b_I$) and death rates ($m \times d_I$) for I-cells are dose dependent because mitotic rate is does dependent. For instance, using the estimated parameters given in Table 6, we find that the (unconditional) mean birth rate is 1.94×10^{-2} and the mean death rate is 1.31×10^{-2} for control female mice, and that the (unconditional) mean birth rate is 3.02×10^{-2} and the mean death rate is 2.07×10^{-2} for female mice with a 100 mg/kg/day metabolized dose.

The model appears to fit bioassay data (Table 5) and background data (Figure 1) adequately. Figure 2 illustrates the shape of the tumor response for mice over time (age) at 10- and 100-mg metabolized doses/kg/day, assuming that animals were exposed to PERC from the time they were 35 days old. The high-dose group is expected to induce a much higher tumor incidence than the low-dose group. However, even for the high-dose group, most of the tumors are expected to occur late in life, a finding consistent with what is known in the NTP study. For comparison, we have also used a linearized multistage (LMS) model, a procedure used by the U.S. Environmental Protection Agency, to calculate cancer risk at low doses. The cancer risks at low doses predicted by the BBDR and LMS models are given in Table 7. Although the BBDR model (calculated on the basis of female data) tends to predict a smaller risk than the LMS model as dose level decreases, the difference is less than an order of magnitude when the dose is over 0.005 mg/kg/day of metabolized dose (Table 7). As we will show, using animals with higher background tumor incidences will produce higher cancer risks even thought he background incidence rate is adjusted.

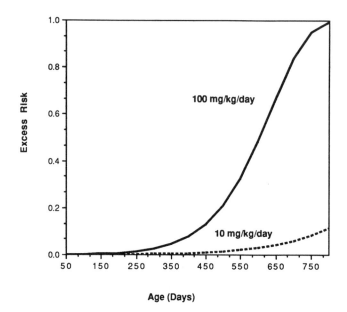

FIGURE 2. Probability of observing tumors over age (day) in female mice exposed to PERC beginning at age of 35 days.

TABLE 7
Comparison of Extra Lifetime Risk Estimates[a] for Mice Exposed to PERC[a]

Amount metabolized (mg/kg/d)	Extra lifetime risk		
	BBDR	95% UB[b]	MLE[b]
1×10^{-3}	7.5×10^{-7}	9.4×10^{-6}	5.4×10^{-6}
5×10^{-3}	5.2×10^{-6}	4.7×10^{-5}	2.7×10^{-5}
1×10^{-2}	1.1×10^{-5}	9.4×10^{-5}	5.4×10^{-5}
5×10^{-2}	5.0×10^{-5}	4.7×10^{-4}	2.7×10^{-4}
1×10^{-1}	3.3×10^{-4}	9.4×10^{-4}	5.4×10^{-4}
5×10^{-1}	1.6×10^{-3}	4.7×10^{-3}	2.7×10^{-3}
1.0	3.5×10^{-3}	9.4×10^{-3}	5.4×10^{-3}

[a] Extra lifetime risk for BBDR = $[p(t,d) - p(t,0)]/[1 - p(t,0)]$, calculated at age $t = 763$ days.

[b] UB: upper bound; MLE: maximum likelihood estimate for LMS (linearized multistage model). Because the LMS model, unlike BBDR, does not take age and duration of exposure into account, the dose values in the gavage study are multiplied by a factor of 78/90 to reflect less than lifetime (90 weeks) exposure. An approximation was made assuming that 90 weeks is the lifespan for mice in the gavage bioassay.

Table 6 suggests that B6C3F$_1$ mice have high spontaneous initiated rates and relatively high background rates (in comparison to the PERC-induced rates) of proliferation and transition from I-cells to malignant cells. If these rates are accepted as accurate, then we would not expect PERC to induce observable liver tumors in a species in which background rates of initiation and promotion are small and when dose is low.

TABLE 8
Comparison of Estimated Extra Risk Due to 0.01 or 1 mg/kg/day of PERC Metabolized Dose When Background Tumor Incidence Rate is Increased or Decreased

Case number	Parameter value[a]	Extra lifetime risk at dose		Background incidence[b]
		0.01 mg/kg/day	1.0 mg/kg/day	
1	Unchanged	8.1×10^{-6}	3.9×10^{-3}	0.038
2	μ_1 increased 6 times	4.7×10^{-5}	2.3×10^{-2}	0.208
3	m_0 increased 0.5 times	6.5×10^{-5}	2.6×10^{-2}	0.235
4	μ_{20} increased 6 times	4.8×10^{-5}	9.2×10^{-3}	0.207
5	μ_1 decreased 13 times	6.3×10^{-7}	3.0×10^{-4}	0.003

[a] Parameter values are increased in order to increase background incidence rate to about 0.20, the background incidence for male mice. In case number 5, μ_1 is decreased 13 times to simulate human background liver cancer rate of 0.003.

[b] Background tumor incidence rates attained due to the change of a parameter value.

B. EFFECT OF HIGH BACKGROUND TUMOR INCIDENCE ON LOW-DOSE EXTRAPOLATION

The extra risks $[p(t,d) - p(t,0)]/[1 - p(t,0)]$ and excess risk $[p(t,d) - p(t,0)]$ due to exposure to a carcinogen are currently used to eliminate the background risk. However, as the following discussion shows, this presumed function of the extra (or excess) risk is not accomplished. Male mice have a background incidence of liver tumors that is about five times higher (about 20%) than that for female mice. The observed high background tumor incidence may be due to high initiation, mitotic, or conversion rates. To evaluate the implication of high background tumor incidence for risk extrapolation, we individually increase the parameter values μ_1, m_0, and μ_{20} derived on the basis of background liver incidence data for female mice so that the background tumor incidence rate is approximately 20%. Table 8 shows the extra cancer risk due to PERC exposure when risk is calculated with one of the three parameters increased. We have also calculated extra risk when μ_1 is decreased 13 times to stimulate the background liver cancer rate (about 0.003) in U.S. white males.

Several conclusions can be drawn from Table 8:

1. The extra risk increases with the increase of background tumor incidence in all the cases (cases 2, 3, and 4). This shows that the conventional approach to adjust for the background rate does not succeed in totally eliminating background effect.
2. The extra risk increases linearly with the value of μ_1 over the dose levels considered. This is generally true when the dose is small. Equation 1 illustrates that the extra (or excess) risk is linearly related to μ_1 over the low probability portion of the curve (equivalently, over the low doses).
3. The extra risk may not increase linearly with the value of μ_2. (See case 4 at a dose of 1 mg/kg/day).
4. A slight increase in the mitotic rate may significantly increase extra risk (case 3).

C. EFFECTS OF CHANGING b_I VALUES

In all of the risk calculations we have presented, the value of b_I is fixed at 0.6. In this section, the effects of changing the b_I value on parameter estimation and on low-dose extrapolation are investigated. Tables 9 and 10 compare the resultant parameter estimates and low-dose extrapolation when $b_I = 0.6$, 0.7, and 0.9. From Table 9, it can be seen that

TABLE 9
Estimated Parameters when b_1 is Fixed at the Given Value

Parameters[a]	b_1 Values		
	0.6	0.7	0.9
μ_1	4.1×10^{-9}	1.8×10^{-10}	7.9×10^{-11}
μ_{20}	5.7×10^{-7}	2.5×10^{-5}	1.2×10^{-4}
μ_{21}	5.1×10^{-8}	1.6×10^{-6}	6.7×10^{-6}
m_0	2.3×10^{-2}	1.6×10^{-2}	8.2×10^{-3}
m_1	1.9×10^{-4}	1.2×10^{-4}	6.5×10^{-5}

[a] When these parameters are estimated, the death-birth ratio for malignant cells is fixed at 0.2/0.8

TABLE 10
Extra Lifetime Risks Calculated with Different Sets of Parameters in Table 9

Metabolized dose (mg/kg/day)	b_1 Values		
	0.6	0.7	0.9
0.001	7.5×10^{-7}	1.0×10^{-6}	1.8×10^{-6}
0.01	8.1×10^{-6}	2.3×10^{-5}	3.9×10^{-5}
0.1	3.6×10^{-4}	3.5×10^{-4}	3.4×10^{-4}
1.0	3.9×10^{-3}	3.6×10^{-3}	3.5×10^{-3}

an increase of b_1 is accompanied by an increase of μ_2 but is offset by a decrease of μ_1 and m (both m_0 and m_1). Table 10 suggests that, as may be expected, the higher the m value, the lower the risk predicted at very low dose levels. However, for 0.1 and 1.0 mg/kg/day metabolized doses, all three models predict almost identical risks. These calculations demonstrate that the change of the b_1 value affects predicted risk value more at lower doses than at higher doses.

D. RISK FROM ANIMALS TO HUMANS

Extrapolating cancer risk from animals to humans poses a great challenge to risk assessors. The main problem arises from the lack of human data. Obviously, the most useful human data are those parameters corresponding to the animal BBDR model. Although the background parameters may be estimated from vital statistics (shown in Table 6), the corresponding PERC-induced rates in humans cannot be obtained. Without human data, we consider two approaches to extrapolating risk from animals to humans. The first approach is an extension of the conventional approach of interspecies extrapolation. The second approach is motivated from the implication of the BBDR model. However, to be valid, these approaches require further support by real data, perhaps through molecular epidemiologic studies. On the basis of the weak carcinogenic potency estimated from animal data, it does not appear possible to obtain positive results from a cancer epidemiologic study on a U.S. population exposed to PERC. These approaches are presented to stimulate discussion on how to obtain data for interspecies extrapolation under the framework of initiation-promotion-conversion-progression carcinogenesis.

1. Approach 1: Interspecies Scaling

This approach is based on the premise that the biological response among species is allometrically related. The most frequently used assumptions are that the same mg/kg of

TABLE 11

Predicted Cancer Risk Due to PERC Exposure via Oral, Inhalation, or Dermal Routes

Medium and exposure condition	Metabolized dose (mg/kg/day)[a]	Biologically based dose-response model		Linearized multistage model	
		Approach I	Approach II	95% upper bound	Maximum likelihood estimate
		Water (1 mg/l)			
Drinking 2 l/day	9.79×10^{-4}	1.3×10^{-5}	1.0×10^{-6}	1.3×10^{-4}	7.4×10^{-5}
		Vapor 24 h/day exposure Inhalation			
1 ppm	1.92×10^{-2}	5.0×10^{-4}	3.9×10^{-5}	1.5×10^{-3}	8.6×10^{-4}
10 ppm	1.89×10^{-1}	1.2×10^{-2}	9.1×10^{-4}	1.4×10^{-2}	8.0×10^{-3}
100 ppm	1.70	1.5×10^{-1}	1.3×10^{-2}	1.3×10^{-1}	7.5×10^{-2}
		8 h/day exposure Inhalation			
1 ppm	5.9×10^{-3}	3.5×10^{-4}	2.7×10^{-5}	4.5×10^{-4}	2.6×10^{-4}
10 ppm	5.9×10^{-2}	3.8×10^{-3}	2.9×10^{-4}	4.5×10^{-3}	2.6×10^{-3}
100 ppm	5.4×10^{-1}	3.7×10^{-2}	2.9×10^{-3}	4.1×10^{-2}	2.4×10^{-2}
		Dermal			
1 ppm	6.16×10^{-6}	1.0×10^{-7}	8.2×10^{-8}	4.7×10^{-7}	2.7×10^{-7}
10 ppm	6.16×10^{-5}	1.2×10^{-6}	9.7×10^{-8}	4.7×10^{-6}	2.7×10^{-6}
100 ppm	6.16×10^{-4}	7.4×10^{-6}	5.7×10^{-7}	4.7×10^{-5}	2.7×10^{-5}

[a] In all calculations, body surface dose equivalence was assumed (i.e., dose in this column is increased 14 times when the BBDR or LMS model is applied).

dose will result in the same response (i.e., the body weight dose equivalent) and that mg/ body surface dose will produce the same response (i.e., body surface dose equivalence). Body surface area is usually expressed in terms of $w^{2/3}$, where w is the body weight. For example, if mice are given a 1 mg/kg/day dose and body weight equivalence is assumed, then the dose is expected to induce the same magnitude of cancer responses in humans as in animals if humans are also given 1 mg/kg/day. However, under the body surface dose equivalence assumption, the equivalent mg/kg/day dose is approximately 14 times smaller (approximately equal to the one third power of the ratio of mouse and human body weights) for humans than for mice to induce the same magnitude of response.

The risk to humans at low doses can be extrapolated from the risk to animals at low doses by using an interspecies scaling factor. For example, the cancer risk due to exposure to PERC at a concentration of 1 ppm in ambient air proceeds as follows. The PB-PK model for humans predicts that 1.92×10^{-2} mg/kg/day of PERC metabolites is produced for a 70 - kg person exposed to PERC at a concentration of 1 ppm in air. Under the body surface dose equivalence assumption, the (extra) risk to humans due to PERC exposure can be calculated from the animal BBDR model at a dose of 2.7×10^{-1} mg/kg/day ($14 \times 1.92 \times 10^{-2}$). Thus, the extra lifetime risk due to 1.92×10^{-2} mg/kg/day of metabolized dose in humans is calculated to be 5.0×10^{-4} based on data for female mice, under the body surface dose equivalence assumption. As shown in Table 11, this approach predicts risks comparable to those derived using the LMS model when exposure is via inhalation. However, this approach predicts lower risk at lower concentration levels or when body burden (metabolized dose) is smaller as in the case of dermal exposure.

2. Approach 2: Animal-Based Model with a Reduced Initiation Rate

When background parameters between mice and humans (white males) are compared (Table 6), the evidence seems to suggest that the lower background liver cancer rate in humans is due to the difference in the initiation rates μ_1 between the two species. Therefore, one method of predicting cancer risk in humans would be to use Approach 1 with μ_1 reduced (and other parameters kept the same) by a factor of 13 to reflect that humans have a background incidence of about 13 times lower. (Recall that Equation 1 implies that the probability of cancer is approximately linearly proportional to the value of μ_1 at the low risk level.) Using this approach, we can calculate the extra risk under various exposure situations (Table 11). As expected, this approach produces lower risk estimates than does Approach 1.

V. SUMMARY AND CONCLUSIONS

Three general observations can be made regarding the extrapolation of risk from animals to humans:

1. Because of the lack of data for some key model parameters, the BBDR model for PERC developed in this chapter can be used only to evaluate the importance of various assumptions on low-dose extrapolation. It is not adequate for low-dose prediction. We have demonstrated the potential utility of the model in risk assessment and what biological information is useful for developing a BBDR model.
2. As demonstrated in Table 8, the risk to humans is likely to be overestimated if it is extrapolated from animals with a high background tumor incidence. Therefore, it would be more appropriate to use bioassay data from animals with lower background rates if the information is available.
3. The most sensitive parameter in the model is the mitotic rate for I-cells. As shown in Table 8, a 50% increase in the background mitotic rate can produce a sevenfold increase of the extra cancer risk. Therefore, we recommend that the information on cell proliferation, including birth and death of I-cells and malignant cells, be obtained in a laboratory setting if a BBDR model is to be used for risk assessment. It should be pointed out, however, that the current practice of measuring the label index alone would not be sufficient for the purpose of modeling.
4. Unless the vapor concentration is very high, the risk due to dermal exposure is negligible. The contribution to total metabolites from dermal exposure is less than 0.1% of that from inhalation.

VI. DISCUSSION

The mathematical model of carcinogenesis that is used to construct the dose-response relationship for PERC is developed on the basis of cellular kinetics/dynamics, under the framework of initiation-promotion-conversion-progression. Although only two critical events (initiation and conversion) are postulated in the model, it does not exclude the possibility that more than two events are actually involved in tumor formation. Initiation (or conversion) may consist of multiple steps. In this case, the rate of initiation (or conversion) could be interpreted as an overall transition rate from target cells to I-cells (or from I-cells to malignant cells); it does not matter how many steps are involved. Recent advances of oncogene research in the laboratory[21-25] have re-enforced the concept of multistep carcinogenesis and have offered a promising possibility of improving the two-event model — either through a better determination of parameters or a better formulation of the model. An obvious advantage of using a model that incorporates explicitly biological information is that scientists from different disciplines can participate in the quantitative risk assessment endeavor, and the impact of certain assumptions on risk estimates can be easily assessed.

Although it is desirable to measure all parameters from laboratory data, it may be preferable to estimate some of the parameters by fitting the mathematical model to bioassay data if the parameters cannot be properly identified or measured in the laboratory. For instance, although it is reasonable to assume the existence of I-cells, the actual measurement of them in the laboratory is difficult if not impossible. This difficulty is further complicated by our lack of precise knowledge about what constitutes the normal target cell population $N(t)$. If one were to assume what constitutes the I-cell and normal target cell population and then use the laboratory measured values that are presumed to be I-cells and normal target cells in the model, the variance could be great if the assumption was wrong. On the other hand, if the initiation rate μ_1 is estimated from tumor incidence data (when other parameters are given), the precise knowledge about normal target cells and I-cells would not be as important because μ_1 and $N(t)$ occur together as $\mu_1 N(t)$ in the mathematical model. Even if the size of the real normal target cell population is only a fraction of the hypothesized normal target cell population, the estimated initiation rate μ_1 will be reduced by the same proportion (from the unknown real initiation rate) if it is statistically estimated; thus the use of a hypothesized normal cell population (which includes real normal target cells as a subpopulation) would have no consequence in final risk prediction.

On the basis of modeling results, we can postulate that PERC induces tumors through an increase of mitotic rate and the rate of conversion from I-cells to malignant cells. In constructing the BBDR model, a PB-PK model is used to calculate the total amount metabolized, and this amount is then used as a dosimetry for dose-response modeling. The use of total metabolites as the dosimetry represents a reasonable approximation of body burden. It does not necessarily represent the "effective" dose, the dose that induces both cell proliferation and transformation from I-cells to malignant cells. A reactive metabolite that induces cell proliferation may differ from the one that induces the production of malignant cells. In this case, one would have a dose-response function, $P(t, d_1, d_2)$, involving two types of doses, d_1 and d_2. Unless the biochemical mechanism of carcinogenic action for PERC is clearly understood, a more specific reactive metabolite(s) should not be given preference as a dosimetry.

We have seen that the BBDR does not totally eliminate the effect of high background tumor incidence unless the background initiation rate (which is much smaller in humans than in mice) is used in the model (Table 8). One of the reasons that the extra (or excess) risk increases with an increase of background rate in the low risk region is that it is approximately linearly proportional to the rate constant μ_1 (see Equation 1). This problem can also be seen intuitively by noting that, when there are many background I-cells in the body, the chemical can transform some of these I-cells (in addition to those that will progress spontaneously) to malignant cells. For this reason, it would be more appropriate to use data from female mice than from males for risk prediction. Another way to see why the extra (or excess) risk increases with background tumor incidence is to note that the extra risk (or excess risk) is derived on the basis of the mid-portion of dose-response curves (the portion above background incidence; e.g., about 20% for male mice and 4% for female mice). In other words, the dose-response model, $P(d)$, for female mice may be viewed as $P(d) = f(d_b + d)$ with $P(0) = f(d_b) = 0.04$, where d_b is considered to be a PERC-equivalent dose that induces 4% of the background tumor incidence. Even when f is nonlinear in the low response region, the curve in the higher response region (e.g., 0.04 or higher) can be adequately approximated by a linear function. Indeed, this was the argument proposed by Crump et al.[26] and many others for using a model with low-dose linearity when the background tumor response is considered "additive". Although the justification for using low-dose linearity is reasonable for situations in which a high background incidence is present, the question remains whether the risk would be overestimated if the background tumor incidence rate is higher in animals than in humans.

In this chapter, we use a PB-PK model that includes a dermal compartment to simulate the total metabolites from PERC exposure via different routes of exposure (dermal, inhalation, and oral). The model is described in the Appendix. There are several publications dealing with PB-PK models for PERC.[7,27-30] Our model, as well as others, involves some uncertainty about the parameters used. For example, Hattis et al.[29] reviewed PB-PK models on PERC from ten reports and found that the metabolic constants V_{max} and K_m are the most variable of all the parameters. These uncertainties would not significantly affect our risk simulation for mice (but may do so for humans) because the body burden (i.e., the amount metabolized) calculated over a broad range of doses appears reasonable when compared to actual observations in the laboratory.[3,4] Since PB-PK modeling is not the main focus of this chapter and space is limited, we chose not to address the issue of uncertainty about PB-PK parameters here.

VII. APPENDIX: CONSTRUCTION OF PB-PK MODELS

A. DESCRIPTION OF THE PB-PK MODEL FOR PERC FROM INHALATION, GAVAGE, AND DERMAL ABSORPTION

The PB-PK model for PERC for inhaled and ingested PERC was described in detail by Chen and Blancato.[7] It is modified in this chapter to take into account dermal absorption as a possible route of exposure with a skin compartment added to the model. The mass balance for this compartment resulted from dermal absorption and systemic distribution. The PB-PK model is shown in Figure 3, with the abbreviations of the variables listed in Table 12. The mass balance for each compartment is mathematically described by a differential equation which quantifies the rate of change of PERC over time in each tissue group.

For the lung, pulmonary uptake and elimination assume a concentration equilibrium between arterial blood and alveolar air. The mass balance of PERC entering and leaving the lung is given by the algebraic equation:

$$Q_a(C_1 - C_a) = Q_t(C_{art} - C_{ven}) \tag{6}$$

Given N = partition coefficient of PERC between arterial blood and alveolar air, N = C_{art}/C_a, the arterial blood concentration can be derived from Equation 6:

$$C_{art} = (Q_aC_{air} + Q_tC_{ven})/(Q_t + Q_a/N) \tag{7}$$

where the venous blood concentration is given by:

$$C_{ven} = (Q_lC_{vl}|P_l + Q_fC_{vf}|P_f + Q_vC_{vr}|P_r + Q_pC_{vp}|P_r + Q_sC_{vs}|P_s)|Q_t \tag{8}$$

For the skin, dermal absorption follows a first order Fick's law diffusion process, governed by the permeability coefficient of PERC, in liquid or vapor state. The rate of

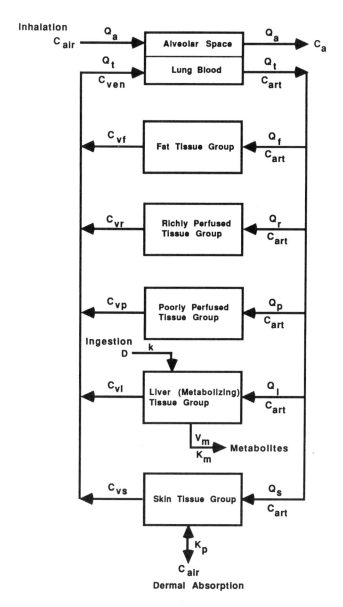

FIGURE 3. Diagram of the physiologically based pharmacokinetic model. Abbreviations are defined in Table 12.

change of PERC concentration in the skin results from the net difference between this dermal absorption process and systemic distribution:

$$V_s \frac{dC_{vs}}{dt} = K_p A \left(C_{air} - \frac{C_{vs}}{P_{S/a}} \right) + Q_s \left(C_{art} - \frac{C_{vs}}{P_s} \right) \tag{9}$$

For the liver, it is assumed that PERC ingested by gavage or by drinking water both follow an exponential elimination process, and all PERC metabolism occurs in the liver, with a rate of metabolism characterized by the Michaelis-Menten type equation. The mass

TABLE 12
Abbreviations Used in Figure 3

Q_a	Alveolar ventilation rate (l/min)
Q_t	Cardiac blood output (l/min)
Q_f	Blood flow rate to fat tissue group (l/min)
Q_r	Blood flow rate to richly perfused tissue group (l/min)
Q_p	Blood flow rate to poorly perfused tissue group (l/min)
Q_l	Blood flow rate to liver tissue group (l/min)
Q_s	Blood flow rate to skin tissue group (l/min)
V_f, V_r, V_p, V_l, V_s	Volumes of tissue groups (liters) corresponding, respectively, to fat, richly perfused, poorly perfused, liver and skin tissue groups
N	Blood/air partition coefficient
$P_{s/a}$	Skin/air partition coefficient
P_f, P_r, P_p, P_l, P_s	Tissue/blood partition coefficient corresponding, respectively, to fat, richly perfused, poorly perfused, liver and skin tissue groups
V_m	Maximum velocity of metabolism (mg/min)
K_m	Michaelis constant (mg/l)
A_m	Amount metabolized (mg)
C_{art}	Concentration in arterial blood (mg/l)
C_{ven}	Concentration in venous blood (mg/l)
$C_{vf}, C_{vr}, C_{vp}, C_{vl}, C_{vs}$	Venous concentrations in tissue groups, corresponding, respectively, to fat, richly perfused, poorly perfused, liver and skin tissue groups
C_{air}	Concentration in inhaled air (mg/l)
C_a	Concentration in alveolar air (mg/l)
K_p	Skin permeability coefficient (cm/h)
A	Surface area of dermal exposure (cm²)
D	Gavage dose (mg)
k	Gut absorption time constant (min⁻¹)

Note: Appropriate unit conversions are provided in the actual program set up for the PB-PK model simulation

balance results from the net difference between ingestion, metabolism, and systemic distribution:

$$V_l \frac{dC_{vl}}{dt} = Dke^{-kt} + Q_l \left(C_{art} - \frac{C_{vl}}{P_l} \right) - \frac{V_m \frac{C_{vl}}{P_l}}{K_m + \frac{C_{vl}}{P_l}} \qquad (10)$$

where D is the administered dose (in milligrams) by gavage, and k is the gut absorption time constant as used by Chen and Blancato.[7]

For the remaining tissue groups, including fat, richly perfused, and poorly perfused, the mass balance equations result simply from the systemic distribution:

Fat:

$$V_f \frac{dC_{vf}}{dt} = Q_f \left(C_{art} - \frac{C_{vf}}{P_f} \right) \qquad (11)$$

Richly perfused:

$$V_r \frac{dC_{vr}}{dt} = Q_r \left(C_{art} - \frac{C_{vr}}{P_r} \right) \qquad (12)$$

TABLE 13
Physiological and Biochemical Parameters Used in the Model

Parameters	Rats	Mice Male	Mice Female	Humans
Body weight (kg)	.35	.035	.025	70
Alveolar ventilation rate (l/min)				
Q_a	.083	.035	.0028	7.5
Blood flow rate (l/min)				
Q_t	.104	.023	.019	6.2
Q_f	.0092	.002	.0017	.31
Q_r	.0434	.0012	.0097	2.76
Q_p	.0074	.0035	.00195	1.26
Q_l	.0389	.0058	.0048	1.55
Q_s	.0052	.00115	.00095	.31
Tissue volume (l)				
V_f	.0315	.0038	.0027	14
V_r	.015	.0021	.0015	3.5
V_p	.220	.0273	.0195	36.4
V_l	.0140	.0017	.0012	1.72
V_s	.035	.00035	.00025	7
Partition coefficient				
N	18.9	16.9		10.3
$P_{s/a}$	—	—		505.4
P_f	108.994	121.893		108.994
P_r	3.179	4.159		3.719
P_p	1.058	1.183		3.72
P_l	3.719	4.159		3.72
P_s	—	—		505.4
Metabolic constant				
V_m (mg/min)	.00586	.0039	.003	.703
K_m (mg/l blood)	2.9378	1.472	1.472	32.043
Absorption coefficient				
K_p (cm/h)	.668	—		.17
k (min^{-1})		11		
Surface area exposed (m^2)				.292

Note: Appropriate unit conversions are provided in the actual program set up for the PB-PK model simulation.

Poorly perfused:

$$V_p \frac{dC_{vp}}{dt} = Q_p \left(C_{art} - \frac{C_{vp}}{P_p} \right)$$ (13)

B. PARAMETERS USED IN THE MODEL

Table 13 presents the physiological and biochemical parameters used in the model. The same assumptions as those used by Chen and Blancato are used here: within a given species, the ventilation and blood flow rates are proportional to the two thirds power of body weight, and the volume of each tissue group is directly proportional to the body weight. Parameters are presented for a 0.35-kg rat, a 0.035-kg male mouse, a 0.025-kg female mouse, and a 70-kg man. The physiological parameters used for this model are similar to those used by Chen and Blancato.

Parameters for the skin compartment are determined using the same approach as McDougal et al.,[30] with the blood flow and volume of the skin compartment delumped from the slowly perfused compartment. The fractions of the total body weight and cardiac output that go to the skin are 10 and 5%, respectively. The skin/air and skin/blood partition coefficients for PERC are evaluated on values reported by Gargas et al.,[31] using the same approach as presented by McDougal et al.:[30] the skin/air partition coefficient is 0.3 times the fat/air partition coefficient plus 0.7 times the muscle/air partition coefficient, the skin/blood partition coefficient is the ratio of the skin/air partition coefficient and the blood/air partition coefficient. The permeability coefficient used in this model was measured by McDougal et al.[30] for PERC vapor in humans; therefore, the risk assessment from dermal exposure is only estimated for humans exposed to PERC in vapor, and comparison of the risk resulting from all three routes of exposures is only provided for humans.

REFERENCES

1. **Reichert, D.,** Biological actions and interactions of tetrachloroethylene, *Mutat. Res.,* 123, 4119, 1983.
2. **Milman, H., Story, D., Riccio, E., Sivak, A., Tu, A., Williams, G., Tong, C., and Tyson, C.,** Rat liver foci and in vitro assays to detect initiating and promoting effects of chlorinated ethanes and ethylenes, *Ann. N.Y. Acad. Sci.,* 534, 521, 1988.
3. **Pegg, D., Zempel, J., Brown, W., and Watanabe, P.,** Deposition of tetrachloro(14C)ethylene following oral and inhalation exposure in rats, *Toxicol. Appl. Pharmacol.,* 51, 465, 1979.
4. **Schumann, A., Quast, T., and Watanabe, P.,** The pharmacokinetics of perchloroethylene in mice and rats as related to oncogenicity, *Toxicol. Appl. Pharmacol.,* 55, 207, 1980.
5. NCI (National Cancer Institute), Bioassay of tetrachloroethylene for possible carcinogenicity, DEHW Publ. No. (NIH) 77–813, Bethesda, MD, 1977.
6. NTP (National Toxicology Program), Toxicology and carcinogenesis of tetrachloroethylene in F344/N rats and B6C3F₁ mice (inhalation study), NTP-TR311, August 1986.
7. **Chen, C. and Blancato, J.,** Role of pharmacokinetic modeling in risk assessment: perchloroethylene as an example, in *Pharmacokinetics in Risk Assessment, Drinking Water and Health,* Vol. 8, 369, National Academy Press, Washington, DC, 1987.
8. **Pitot, H. and Sirica, A.,** The stages of initiation and promotion in hepatocarcinogenesis, *Biochem. Biophys. Acta,* 605, 191, 1980.
9. **Farber, E. and Sarma, D.,** Hepatocarcinogenesis: a dynamics cellular perspective, *Lab. Invest.,* 56, 4, 1987.
10. **Laib, R., Klein, K., and Bolt, H.,** The rat liver foci bioassay. I. Age-dependence of induction by vinyl chloride of the ATPase-deficient foci, *Carcinogenesis,* 6(1), 65, 1985.
11. **Neyman, J. and Scott, E.,** Statistical aspects of the problem of carcinogenesis, 5th Berkeley Symp. on Mathematical Statistics and Probability, University of California Press, Berkeley, 1967, 745.
12. **Moolgavkar, S. and Knudson, A.,** Mutation and cancer: a model for human carcinogenesis, *J. Natl. Cancer Inst.,* 66, 1037, 1981.
13. **Chen, C. and Farland, W.,** Incorporating cell proliferatiol in quantitative cancer risk assessment: approaches, issues, and uncertainties, in *Chemically Induced Cell Proliferation: Implication for Risk Assessment,* Progress in Clinical and Biological Research Series, Butterworth, B., Slaga, T., Farland, W., and McClain, M., Eds., Wiley-Liss, New York, 1991.
14. **Moolgavkar, S. and Venzon, D.,** Two event model for carcinogenesis: incidence curves for childhood and adult tumors, *Math. Biosci.,* 47, 55, 1979.
15. **Greenfield, R., Ellwein, L., and Cohen, S.,** A general probability model of carcinogenesis: analysis of experimental urinary bladder cancer, *Carcinogenesis,* 5(4), 437, 1984.
16. **Cohen, S.,** Multi-stAge carcinogenesis in the urinary bladder, *Food Chem. Toxicol.,* 23, 521, 1985.
17. **Ellwein, L. and Cohen, S.,** A cellular dynamics model of experimental bladder cancer: analysis of the effect of sodium saccharin in rats, *Risk Anal.,* 8(2), 215, 1988.
18. **Knudson, A.,** Biological Factors in Two-Stage Models, presented at the Workshop on Two-Stage Models of Carcinogenesis: Applications and Implications in Risk Assessment, Washington, D. C., November 8, 1990, National Academy of Sciences Committee on Risk Assessment Methodology.

19. **Chen, C. and Moini, A.,** Cancer dose-response models incorporating clonal expansion, in *Scientific Issues in Quantitative Cancer Risk Assessment,* Moolgavkar, S., Ed., Birkhauser, Boston, 1990.
20. **Portier, C., Hedges, J., and Hoel, D.,** Age-specific models of mortality and tumor onset for historical control animals in the National Toxicology Program's carcinogenicity experiments, *Cancer Res.,* 46, 4372, 1986.
21. **Reynold, S., Stowers, S., Patterson, R., Maronpot, R., Aaronson, S., and Anderson, M.,** Activated oncogenes in B6C3F$_1$, *Science,* 237, 1309, 1987.
22. **Klein, G. and Klein, E.,** Oncogene activation and tumor progression, *Carcinogenesis,* 4, 428, 1984.
23. **Klein, G. and Klein, E.,** Conditional tumorigenicity of activated oncogene, *Cancer Res.,* 46, 3211, 1986.
24. **Land, H., Parada, L., and Weinberg, R.,** Tumorigenic conversion of primary embryo fibroblasts requires at least two cooperating oncogenes, *Nature,* 304, 596, 1983.
25. **Weinberg, R.,** Oncogenes, antioncogene, and the molecular bases of multistep carcinogenesis, *Cancer Res.,* 49, 3713.
26. **Crump, K., Guess, H., and Deal, K.,** Confidence interval and test of hypothesis concerning dose response relations inferred from animal carcinogenicity data, *Biometrics,* 33, 437, 1977.
27. U.S. Environmental Protection Agency, Health Assessment Document for Tetrachloroethylene (Perchloroethylene), Final Rep., EPA/600/8-82/005F, 1985.
28. **Bois, F., Zeise, L., and Tozer, T.,** Precision and sensitivity of pharmacokinetic models for cancer risk assessment: tetrachloroethylene in mice, rats and humans, *Toxicol. Appl. Pharmacol.,* 102, 300, 1990.
29. **Hattis, D., White, P., Marmorstein, L., and Koch, P.,** Uncertainties in pharmacokinetic modeling for perchloroethylene. I. Comparison of model structure, parameters, and prediction for low-dose metabolism rates for models by different authors, *Risk Anal.,* 10(3), 440, 1990.
30. **McDougal, J. N., Jepson, G. W., Clewell, H. J., III, MacNaughton, M. G., and Andersen, M. E.,** A physiological pharmacokinetic model for dermal absorption of vapors in the rat, *Toxicol. Appl. Pharmacol.,* 85, 286, 1986.
31. **Gargas, M. L., Burgess, R. J., Voisard, D. E., Cason, G. H., and Andersen, M. E.,** Partition coefficients of low-molecular-weight volatile chemicals in various liquids and tissues, *Toxicol. Appl. Pharmacol.,* 98, 87, 1989.

Chapter 22

PHOTOTOXICITY OF TOPICAL AND SYSTEMIC AGENTS

Francis N. Marzulli and Howard I. Maibach

TABLE OF CONTENTS

0-8493-7357-3/93/$0.00 + $.50
© 1993 by CRC Press, Inc.

I. INTRODUCTION

Phototoxicity is a form of chemically induced non-immunologic skin irritation requiring light (photoirritation). The skin response is likened to an exaggerated sunburn. The involved photoactive chemical may be applied to the skin or reach this target tissue via the blood, following ingestion or parenteral administration. When systemically administered, the chemical may require metabolic conversion to become photoactive, but this is not usually the case. Aftereffects may include erythema, edema, vesiculation, hyperpigmentation, and desquamation. Inflammatory response products that may be released during these processes include histamine and arachidonic acid derivatives such as prostaglandins, leukotrienes, and kinins.

Recent addition of new photoactive therapeutic chemicals to the physicians' armamentarium together with the development of better methods for measuring and delivering solar-simulated radiation have had a stimulating effect on research involving light-related biologic phenomena. This research activity has been accompanied by the development of new test methods and therapies.

Attention of the photobiologist is largely directed at the ultraviolet (UV) area of the electromagnetic spectrum which is subdivided for convenience by the C.I.E.* into UV-A (from 315 to 400 nm), UV-B (from 280 to 315 nm), and UV-C (below 280 nm).** The visible spectrum is important in certain medical conditions such as erythropoietic protophorphyria, porphyria, and light-mediated treatment of solid tumors (Dougherty et al., 1975–1978; Mathews-Roth, 1981).

Tests for phototoxic potential of topically applied chemicals are usually conducted using UV radiation sources within the UV-A range. Some phototoxic substances, however, are activated by wavelengths in the visible spectrum (bikini dermatitis, Hjorth and Moller, 1976); some by UV-B (Jeanmougen et al., 1983) and some (doxycycline) are augmented by UV-B (Bjellerup, 1986).

Accurate measurements of radiation intensity and frequency are important prerequisites for work in phototoxicity. For further details relating to physical measurements of light the reader is directed to other sources. A report by Landry and Anderson (1982) contains useful information in this regard.

Among animals that have proved useful in predicting phototoxic effects are the mouse (hairless or haired), rabbit, swine, guinea pig, squirrel monkey, and hamster, in that approximate order of effectiveness (Marzulli and Maibach, 1970).

Phototoxic events are initiated when a photoactive drug (a chemical that is capable of absorbing UV radiation) enters the skin (via skin penetration or through blood circulation) and becomes excited by appropriate UV radiation that also penetrates skin. In some cases the photoexcited drug transfers its energy to oxygen, exciting it to the singlet oxygen state, which is cytotoxic. In other cases oxygen may not be involved. Chlorpromazine is thought to be activated by a photodynamic process involving molecular oxygen, whereas psoralens do not require molecular oxygen to produce phototoxic effects. Further discussion of photodynamic action is in Spikes and Livingston, 1969; Kearns, 1971; Wennersten, 1977; and Kochevar, 1981.

The clinical identification of phototoxicity largely resides in morphology together with clinical suspicion that is based upon knowledge about phototoxic chemicals. Phototoxicity

---

*   Commision de l'eclairage Pub. No. 17, dated 1970, and Pub. No. 17.4, dated 1987.

** Tests conducted by the National Bureau of Standards at Gainesville, Florida, revealed that radiation was received only down to 286 nm at the earth's surface. (NBS Tech. Note 910-5, Manual on Radiation Measurements, December 1982.)

generally begins with erythema (sometimes even large bullae), increased skin temperature and pruritis. This is followed some time later by hyperpigmentation, which is the chief complaint of patients, as this condition may last for months or years.

The morphology of phototoxic eruptions from skin exposure is not always readily identified by the clinician; the lack of diagnosis in thousands of cases of dimethylparaaminobenzoic acid (sunscreen) dermatitis, until Emmett (1977) brought this to our attention, provides testimony to this issue. A prepared mind is essential; removing the phototoxics is efficient, as clinical identification is often difficult.

With oral agents, photorelated dermatoses present an even greater challenge. Not only is the entity often not suspected on morphologic grounds, but the criteria for defining which are phototoxic vs. photoallergic rests on relatively imprecise grounds.

II. PHOTOTOXIC AGENTS

Naturally occurring plant-derived furocoumarins, including psoralen, 5-methoxypsoralen (bergapten), 8-methoxypsoralen (xanthotoxin), angelicin, and others, constitute the most important class of phototoxic chemicals. Psoralens occur in a wide variety of plants such as parsley, celery, and citrus fruits (Pathak, 1974; Juntilla, 1976).

The Rutaceae (common rue, gas plant, Persian limes, bergamot) and Umbelliferae (fennel, dill, wild carrot, cow parsnip) are prominent among plant families responsible for phytophotodermatitis (Pathak, 1974). Bergapten (5-metho-xypsoralen), psoralen, and xanthotoxin (8-methoxypsoralen) are among the more commonly encountered phototoxic agents. Bergapten is the active component of bergamot oil, a well-known perfume ingredient, whose phototoxic skin effects have been accorded the name berlock dermatitis. Perfume phototoxicity was studied in considerable detail by Marzulli and Maibach (1970). On the basis of these studies, it was suggested that perfumes should contain no more than 0.3% bergamot, which is equivalent to about 0.001% bergapten, to avoid phototoxicity. This study also established that bergapten was the only one of five components isolated from oil of bergamot that was responsible for phototoxic effects of the parent material. Limettin (5,7-dimethoxycoumarin), although more intensely fluorescent than bergapten, did not prove phototoxic. Since publication of this report, the FDA has received few consumer complaints of berlock dermatitis. Bergamot PT continues in some countries where bergapten-free bergamot is not used (Zaynoun et al., 1981). Bergapten has also been implicated as the cause of phototoxicity from skin contact with *Heracleum laciniatum,* a weed that grows in abundance in Norway (Kavli et al., 1983). The giant hogweed *(Heracleum mantegazzianum)* of Denmark is similarly implicated (Knudsen, 1983).

Xanthotoxin, or 8-MOP as it is often called, is the active ingredient of a drug marketed for treatment of vitiligo and psoriasis. It produces photodynamic skin effects both by topical application and by oral ingestion following skin exposure to ultraviolet irradiation. Used orally in crude form in Egypt since ancient times (El Mofty, 1948), impetus for the use of psoralens in this country in the treatment of vitiligo was provided in large part by the work of Lerner et al. (1953). Chronic use in conjunction with exposure to light may enhance prospects for squamous-cell skin cancer, especially in young patients and those whose skin is genetically predisposed (Stern et al., 1979). This potential provides concern and restraint in the now widespread application of psoraln phototherapy (PUVA) for the management of psoriasis (Parrish et al., 1974).

Phototoxicity reactions to psoralen-containing sweet oranges (Volden et al., 1983) and to Ruta graviolens (common rue) have been reported recently (Heskel et al., 1983).

Coal-tar derivatives, another important group of phototoxic agents, produce occupational contact photodermatitis in industrial workers and road workers. Anthraquinone-based disperse blue 35 dye caused such effects in dye process workers. Radiation in the visible spectrum activates the dye (Gardiner et al., 1974). Pyrene, anthracene, and fluoranthene are strongly phototoxic in guinea pigs (Kochevar et al., 1982).

Oral therapeutic use of amiodarone, a cardiac antiarrythmic agent, produced phototoxic effects (Chalmers et al., 1982). Incidence, time course, and recovery from phototoxic effects of amiodarone in humans were studied by Rappersberger et al., 1989.

Quinoline antimalarials appear to be phototoxic, and some of these have been studied *in vitro* and *in vivo* (Moore and Himmens, 1982; Epling and Sibley, 1987; Ljunggren and Wirestrand, 1988). Systemic effects in a 12-year-old boy, who was treated with a combination of chloroquine and sulfadoxine-pyrimethamine, an oral antimalarial prophylaxis, were probably related to the sulfa component.

Tetracyclines — particularly demethylchlortetracycline, but also doxycycline, chlortetracycline, and tetracycline — are phototoxic when orally ingested by humans (Verbov, 1973; Frost et al., 1972; Maibach et al., 1967). Doxycycline was reported more potent than demethylchlortetracycline or limecycline, in one study involving human volunteers (Bjellerup and Ljunggren, 1987).

Cadmium sulfide, used in tattoos for its yellow color, is phototoxic (Bjornberg, 1963).

Thiazide diuretics (hydrochlorothiazide and bendrofluazide) showed phototoxic potential (decrease in minimal erythema dose) in experiments involving cardiovascular patients (hypertension, heart failure, edema) for whom diuretics were prescribed by medical practitioners (Diffey and Langtry, 1989). The incidence of thiazide-induced photosensitivity in clinical practice, however, is rare.

III. NON-STEROIDAL ANTIINFLAMMATORY DRUGS (NSAID)

NSAID have been the subject of extensive recent investigations for phototoxic potential, following reports that benoxaprofen, a suspended British antirheumatic NSAID, has this capability when administered orally or by intradermal injection (Webster et al., 1982; Allen, 1983; Stern, 1983; Anderson et al., 1987). Positive findings have been reported using sheep erythrocytes or human leukocytes *in vitro* (Anderson et al., 1987; Przbilla, 1987).

NSAID that are structurally related to propionic acid have been shown to possess photoxic potential, whereas certain other type NSAID such as tenoxicam and piroxicam were not experimentally phototoxic *in vitro* or *in vivo* (Anderson, 1987; Kaidbey, 1989; Western, 1987). The propionic acid-derived NSAID produces unique immediate wheal and flare in contrast with the much delayed exaggerated sunburn response that typifies psoralen phototoxicity.

Although piroxicam is not phototoxic under experimental test conditions involving human volunteers (Kaidbey, 1989), it has been implicated as a possible photosensitizer (photoallergic or phototoxic) clinically. One explanation for the unexpected photoactivity of piroxicam in skin is that a metabolite of piroxicam is indeed phototoxic when isolated and tested on human mononuclear cells *in vitro* (Western et al., 1987). These positive findings and likely explanation are related to production of singlet oxygen as indicated by emission at 1270 nm when the suspect metabolite is irradiated with UV *in vitro* (Kochevar, 1989; Wester et al., 1987).

Other propionic acid-derived NSAID associated with an immediate phototoxic response are nabumetone, carprofen, naproxen, ketoprofen, and tiaprofenic acid (Kaidbey et al., 1989; Merot et al., 1983; Aloman, 1985 and Differ et al., 1983).

IV. HUMAN TESTING

Because of the universal response, testing of humans with topically applied photoactive chemicals can be done with minimal or no hazard on small test areas in a few subjects. For many years there was confusion on the basic pathophysiology of bergamot dermatitis. Burdick (1966) and Marzulli and Maibach (1970) showed that this dermatitis can be produced in almost all subjects. With sufficient light exposure (from a high output Wood's light) and percutaneous penetration, dermatitis will occur. Obtaining appropriate light sources presents no problem. Penetration is enhanced in several ways: choosing a highly permeable anatomic site such as the scrotum (Feldman and Maibach, 1967), or better still, decreasing the barriers to penetration by removing the stratum corneum by stripping with cellophane tape.

V. CHEMICAL-SKIN CONTACT TIME BEFORE LIGHT EXPOSURE

With many chemicals there is considerable lag time before significant amounts of percutaneous penetration occurs (Feldman and Maibach, 1970). Bioassays on skin must take this delay factor into consideration, e.g., reading the vasoconstrictor corticoid assay at 18 to 24 h after application. One might expect that light exposure could or should be delayed for many hours after application to the skin. This does not appear to be the case, at least with bergamot. Animals exposed within minutes after application will react; degree and frequency of reaction are increased with light exposure 1 to 2 h later. By 4 h, animals are less reactive; at 24 h, light exposure will often produce no response. This time factor must be taken into account for predictive assays; it is of considerable interest in terms of the relationship of pharmacokinetics and site-of-action in skin. The time relationship found optimal for bergamot does not appear to hold for all other phototoxic chemicals. Chemical-skin contact time before light exposure must be altered accordingly.

VI. VEHICLES

Vehicles alter percutaneous penetration. A considerable literature defines chemical and vehicle properties that increase or decrease chemical release and penetration. Much of this work is done with *in vitro* or other model test systems. The experience with vehicles in animal or human *in vitro* test systems is limited. Marzulli and Maibach (1970) showed that reactions to bergamot were greater in the rabbit with 70% than with 95% alcohol. K. John and Y. Gressel (personal communication, 1971) noted that in the guinea pig 8-methoxy-psoralen produced greater reactions in 70% alcohol than in absolute alcohol; in the rabbit the reverse response occurred. Mineral oil gave a similar response to 70% alcohol, but there was a greatly decreased reactivity in castor and olive oil. Kaidbey and Kligman (1974), in studying the phototoxic properties of coal tar, methoxsalen, and chlorpromazine, found the result strongly influenced by vehicle choice. No single base produced optimal effects for the three chemicals; emulsion-type creams were generally poor vehicles.

It is likely that the effect of vehicle on phototoxicity (and penetration) is more complex than generally stated; until all the variables are understood, it is prudent to employ the vehicle intended for human use in the predictive assay. A more realistic evaluation of potential hazard is obtained by also employing an experimental vehicle (such as alcohol) that is likely to release the test compound.

VII. ANIMAL TESTING

By definition, phototoxicity is a form of light-activated irritation that would occur in almost everyone, following exposure to sufficient chemical and light. In practice this rarely occurs. Even bergamot at the highest commercially used concentrations only occasionally produced dermatitis.

The development of an animal model followed that of a similar model in humans (Marzulli and Maibach, 1970). The basic requisites are adequate nonerythrogenic light and percutaneous penetration of the phototoxic agent. Most conveniently, the known phototoxic agents will produce dermatitis at wavelengths that do not ordinarily yield erythema. For this reason, with light sources such as the Wood's light, the light-irradiated negative control site should be free of dermatitis. This convenience makes for an all-or-none effect on reading.

Under appropriate test conditions, several species have been utilized successfully with oil of bergamot (Marzulli and Maibach, 1970). With bergamot the hairless mouse and rabbit appeared somewhat more sensitive than the guinea pig. The pig (swine) was less reactive but "stripping" enhanced the responsiveness; the squirrel monkey appeared quite resistant; and the hamster showed histologic changes that were not apparent on gross inspection of the skin. It is possible that alternate anatomic test sites and different treatment schedules might alter these relative rankings. In practical terms this allows for a more than reasonable choice of test animals for laboratories to choose from. Until additional use experience is obtained, new chemicals that will be used widely in humans might also be examined on human skin.

Forbes et al. (1977) tested 160 fragrance raw materials for phototoxicity using the skin of hairless mutant mice, humans, and swine. Mouse skin was most sensitive; however, humans and swine were more alike in quantitative and qualitative aspects of skin responses. The test methodology was similar to that used by Marzulli and Maibach (1970); however, additional light source and light measurement equipment was used. These results, reported in considerable detail, substantiate the usefulness of the basic test procedure recommended in this report for evaluating phototoxicity in animals. Furthermore, their results support information on the animal models described here.

Stott et al. (1970) used changes in guinea pig ear thickness to establish that chlorpromazine, perchlorperazine, and demethylchlortetracycline are phototoxic by topical application.

Mouse ear swelling, as measured with a caliper, was employed to quantitate phototoxic responses by Sambuco and Forbes, 1984; Cole et al., 1984; and Gerberick and Ryan, 1989. Maurer (1987) reported a preference for back skin rather than ear skin as a measure of phototoxic response when using calipers.

Ljunggren and Moller (1976, 1977a) produced dermatitis on the tail of the mouse by exposing it to UV-A for 5 h after intraperitoneal (i.p.) injection with clorpromazine. Increase in tissue fluid weight was considered a useful quantitative measure of phototoxicity. Griseofulvin, nalidixic acid, imperatorin, kynaurenic acid, and amiodarone were phototoxic in addition to psoralens and tetracyclines, by this technique (Ljunggren and Moller, 1977b).

Saunders et al. (1972) suggested measuring the photoresistance of the mouse ear to evaluate phototoxic potential of candidate agents. By this technique, chlorpromazine (i.p.) produced significant effects when tested on albino or black mice.

Gloxhuber (1970) used hairless mice to study phototoxicity. He found chlorpromazine and tetracycline phototoxic when administered i.p. but not when topically applied. On the other hand, tetrachlorsalicylanilide, acridine, and anthracene were phototoxic by topical application but not by i.p. administration. Differences in metabolic pathways may account for these different findings.

Bay et al. (1970) reported a reproducible technique for studying phototoxicity using white swine.

VIII. PHOTOTOXICITY IN ANIMALS FROM ORALLY ADMINISTERED CHEMICALS

As phototoxicity from orally administered chemicals may involve species-specific differences in absorption and metabolism, a useful animal model for one chemical may not apply to another.

Several animal test systems are in use, hairless mice and albino guinea pigs being animals of choice (Forbes et al., 1970; Kornhauser et al., 1982). As with topical PT, erythema is the toxicological endpoint. Although simple and convenient, this evaluation is subjective and not always comparable in different laboratories (A. Kornhauser, personal communication, 1984).

Kornhauser et al. (1983) prefer the epilated guinea pig. Results are reliable for extrapolating to certain human (clinical) situations. With 8-MOP, serum concentrations marking the onset of PT were within the range of those accompanying the appearance of PT in humans receiving therapeutic doses of 8-MOP (Kornhauser et al., 1982).

The hairless mouse is less effective in detecting weaker PT substances. This is mainly due to difficulties in quantitating erythema and edema. The hairless mouse is well established, however, as a model of choice for photocarcinogenesis studies (Urbach et al., 1974).

Female albino rats have been used with some success in PT tests. The male, whose dorsal skin is thick, patchy, and not uniform, does not lend itself to this type of evaluation (A. Kornhauser, personal communication, 1984). This is the only model for PT where gender of the animals is a limiting condition.

IX. NON-ANIMAL TEST METHODS

There is need for non-animal models for phototoxicity, however, it is important that they be validated prior to extrapolation of results to humans. In some cases, non-animal tests that have limited direct application to humans may be extremely useful in explaining mechanism of action or other aspects of phototoxicity.

A simple method for screening phototoxic agents such as *Candida albicans* (Daniels, 1965) gave discrepant results when used by Mitchell (1971). Tetramethyl thiuram monosulfide appeared to be phototoxic by the *C. albicans* test, but these results could not be confirmed when tested by topical application in mice and humans.

Gibbs (1987) has modified the *C. albicans* phototoxicity test to study action spectra of pyrrolocoumarin derivatives.

Weinberg and Springer (1981) used UV-A irradiation during diffusion of chemicals from a paper disc placed on agar seeded with bakers yeast to measure phototoxic potential of fragrance materials. Resulting light-specific zones of inhibited yeast growth provided test results within 72 h.

Tennebaum et al. (1984) similarly used yeast *(Saccharomyces cerevesiae)* to examine phototoxic potential of fragrance materials.

Salmonella typhimurium has been used to study phototoxicity of chlordiazepoxide (Librium) and its metabolites.

Mammalian cells in culture (human red blood cells, mouse peritoneal macrophages, and Chinese hamster lung cells) were used by Lock and Friend (1983) to study phototoxic effects of representative drugs, dyes, and antiseptics. Their work was intended to investigate the nature and site of phototoxic action within isolated cells — specifically to identify and differentiate substances that damage nucleus, plasma membrane, and cytoplasmic organelles.

X. SPECIAL EFFECTS

Phototoxic reactions sometimes mask or contribute to human photoallergic reaction. Schauder (1985) has proposed a method to distinguish phototoxic from photoallergic reactions with topical chlorpromazine, based on specific test concentrations and UV-A doses.

When the skin temperature of human subjects was experimentally raised during exposure to UV-A, the erythema component of phototoxicity from topical 8-MOP plus UV-A was reduced (Youn et al., 1987). Thus skin temperature may have clinical significance during PUVA treatments for psoriasis, which is reported to increase the skin temperature about 5°C (Ciafone, 1980).

XI. TREATMENT

Acute lesions of human phototoxic dermatitis resolve in days to weeks, whereas recovery from hyperpigmentation is delayed for months or years. Depigmentation therapy is accomplished with 2% hydroquinone. Sunscreens are useful to protect against further hyperpigmentation.

XII. COMMENT

Research was first driven by concerns about topically applied phototoxic substances (bergamot photodermatitis) and attempts to harness the useful aspects of psoralen phototoxicity for therapy of vitiligo and psoriasis. More recently, phototoxic potential of certain nonsteroidal antiinflammatory drugs for oral use has provided impetus for work in this area.

Insights have been gained from bergamot, whose clinical effects long awaited simple resolution by experimental work with animal and human models.

Phototoxicity (photoirritation) is easy to detect and prevent because of the simplicity of predictive animal and human test methods. With increasing use of predictive assays, phototoxicity in humans is expected to become a minor concern to dermatologists. On the other hand, future work with non-animal models may provide further insight regarding mechanism of action of phototoxic chemicals.

REFERENCES

Allen, B., Benoxaprofen and the skin, *Br. J. Dermatol.,* 109, 361–364, 1983.

Aloman, A., Ketoprofen photodermatitis, *Contact Derm.,* 12, 112–113, 1985.

Anderson, R., Eftychis, H., Weiner, A., and Findlay, G., An in vivo and in vitro investigation of the phototoxic potential of tenoxicam, a new non-steroidal anti-inflammatory agent, *Dermatologica,* 175, 229–234, 1987.

Bay, W., Gleiser, C. A., Dukes, T. W., and Brown, R. S., Experimental production and evaluation of drug-induced phototoxicity in Swine, *Toxicol. Appl. Pharmacol.,* 17, 538–547, 1970.

Bjellerup, M. and Ljunggren, B., Photohemolytic potency of tetracyclines, *J. Invest. Dermatol.,* 83, 179–183, 1985.

Bjellerup, M., Medium-wave ultraviolet radiation (UVB) is important in doxycycline phototoxicity, *Acta Derm. Venereol.,* 66, 510–514, 1986.

Bjornberg, A., Reactions to light in yellow tatoos from cadmium sulfide, *Arch. Dermatol.,* 88, 267, 1963.

Burdick, K., Phototoxicity of Shalimar perfume, *Arch. Dermatol.,* 93, 424–425, 1966.

Chalmers, R. J. G., Muston, H. L., Sprinivas, V., and Bennett, D. H., High incidence of amiodarone-induced photosensitivity in Northwest England, *Br. Med. J.,* 285, 341, 1982.

Ciafone, R., Rhodes, A., Audley, M., Freedberg, I., and Abelmann, W., The cardiovascular stress of photochemotherapy (PUVA), *J. Am. Acad. Dermatol.,* 3, 499–505, 1980.

Cole, C., Sambuco, C., Forbes, P., and Davies, R., Response to ultraviolet radiation: ear swelling in hairless mice, *Photodermatology,* 1, 114–118, 1984.

Daniels, F., A simple microbiological method for demonstrating phototoxic compounds, *J. Invest. Dermatol.,* 44, 259–263, 1965.

De Vries, H., Van Henegowen, G., and Wouters, P., Correlations between phototoxicity of some 7-chloro-1, 4-benzodiazepines and their (photo)chemical properties, *Pharm. Weekbl. Sci. Ed.,* 5, 302–307, 1983.

Diffey, B. L., Daymond, T. J., and Fairgreaves, H., Phototoxic reactions to piroxicam, naproxen and tiaprofenic acid, *Br. J. Rheumatol.,* 22, 239–242, 1983.

Diffey, B. L. and Langtry, J., Phototoxic potential of thiazide diuretics in normal subjects, *Arch. Dermatol.,* 125, 1355–1358, 1989.

Dougherty, T., Grindey, G., Field, R., Weishaupt, K., and Boyle, D., Photoradiation therapy. II. Cure of animal tumors with hematoporphyrin and light, *J. Natl. Cancer Inst.,* 55, 115–129, 1975.

Dougherty, T., Kaufman, J., Goldfar, A., Weishaupt, K., Boyle, D., and Mittelman, A., Photoradiation therapy for treatment of malignant tumors, *Cancer Res.,* 38, 2628–2635, 1978.

El Mofty, A. M., A preliminary clinical report on the treatment of leukoderma with Ammi majus, Linn, *J. R. Egypt. Med. Assoc.,* 31, 651, 1948.

Emmett, E. A., Taphorn, B. R., and Kominsky, J. R., Phototoxicity occurring during the manufacture of ultraviolet-cured ink, *Arch. Dermatol.,* 113, 770–775, 1977.

Epling, G. and Sibley, M., Photosensitized lysis of red blood cells by phototoxic antimalarial compounds, *Photochem. Photobiol.,* 46, 39–43, 1987.

Feldman, R. and Maibach, H., Regional variation in percutaneous penetration of hydrocortisone in man, *J. Invest. Dermatol.,* 48, 181–183, 1967.

Feldman, R. and Maibach, H., Absorption of some of organic compounds through the skin of man, *J. Invest. Dermatol.,* 54, 399–404, 1970.

Forbes, P. D., Davies, R. E., and Urbach, F., Phototoxicity and photocarcinogenesis: Comparative effects of anthracene and 8-methoxypsoralen in the skin of mice, *Food Cosmet. Toxicol.,* 14, 243, 1970.

Forbes, P. D., Urbach, F., and Davies, R. E., Phototoxicity testing of fragrance materials, *Food Cosmet. Toxicol.,* 15, 55–60, 1977.

Frost, P., Weinstein, C. D., and Gomex, E. C., Phototoxic potential of minacycline and doxycycline, *Arch. Dermatol.,* 105, 681, 1972.

Gardiner, J. S., Dickson, A., MacLeod, T. M., and Frain-Bell, W., The investigation of photocontact dermatitis in a dye manufacturing process, *Br. J. Dermatol.,* 86, 264–271, 1974.

Gerberick, G. and Ryan, C., Investigations of a predictive mouse ear swelling phototoxicity model, *Food Chem. Toxicol.,* 27, 813–819, 1989.

Gibbs, N., An adaptation of the Candida albicans phototoxicity test to demonstrate photosensitizer action spectra, *Photodermatology,* 4, 312–316, 1987.

Gloxhuber, C., Phototoxicity testing of cosmetics, *J. Soc. Cosmet. Chem.,* 21, 825, 1970.

Heskel, N. S., Amon, R. B., Storrs, F., and White, C. R., Phytophotodermatitis due to Ruta graveolens, *Contact Derm.,* 9, 278–280, 1983.

Hjorth, N. and Moller, H., Phototoxic textile dermatitis (bikini dermatitis), *Arch. Dermatol.,* 112, 1445–1447, 1976.

Jeanmougin, M., Pedreio, J., Bouchet, J., and Civatte, J., Phototoxicity of 5% benezoyl peroxide in man. Evaluation with a new methodology, *Fra Dermatol.,* 167, 19–23, 1983.

Juntilla, O., Allelopathic inhibitors in seeds of Heracleum laciniatum, *Physiol. Plant,* 36, 374–378, 1976.

Kaidbey, K. and Kligman, A., Topical photosensitizers: influence of vehicles on penetration, *Arch. Dermatol.,* 110, 868–870, 1974.

Kaidbey, K. and Mitchell, F., Photosensitizing potential of certain non-steroidal anti-inflammatory agents, *Arch. Dermatol.,* 125, 783–786, 1989.

Kavli, G., Midelfart, G. V. K., Haugsbo, S., and Prytz, J. O., Phototoxicity of Heracleum laciniatum, *Contact Derm.,* 9, 27–32, 1983.

Kearns, D. R., Physical and chemical properties of singlet molecular oxygen, *Chem. Rev.,* 71, 395–427, 1971.

Knudsen, E. A., Seasonal variations in the content of phototoxic compounds in giant hogweed, *Contact Derm.,* 9, 281–284, 1983.

Kochevar, I., Phototoxicity mechanisms: chlorpromazine photosensitized damage to DNA and cell membranes, *J. Invest. Dermatol.,* 76, 59–64, 1981.

Kochevar, I., Phototoxicity of non-steroidal and anti-inflammatory drugs, *Arch. Dermatol.,* 125, 824–826, 1989.

Kochevar, I., Armstrong, R. B., Einbinder, J., Walther, R. R., and Harber, L., Coal tar phototoxicity: active compounds and action spectra, *Photochem. Photobiol.,* 36, 65–69, 1982.

Kornhauser, A., Wamer, W., and Giles, A., Psoralen phototoxicity: correlation with serum and epidermal 9-methoxypsoralen and 5-methoxypsoralen in the guinea pig, *Science,* 217, 733–735, 1982.

Kornhauser, A., Wamer, W., and Giles, A., Jr., Light-induced dermal toxicity: effects on the cellular and molecular level, in *Dermatotoxicology,* Marzulli, F. and Maibach, H., Eds., Hemisphere, Washington, D.C., 1983, 323–355.

Landry, R. and Anderson, F. A., Optical radiation measurements: instrumentation and sources of error, *J. Natl. Cancer Inst.,* 69, 115–161, 1982.

Lerner, A. B., Denton, C. H., and Fitzpatrick, T. B., Clinical and experimental studies with 8-methoxpsoralen in vitiligo, *J. Invest. Dermatol.,* 20, 299–314, 1953.

Ljunggren, B. and Moller, H., Phototoxic reaction to chlorpromazine as studied with the quantitative mouse tail technique, *Acta Derm. Venereol.,* 56, 373–376, 1976.

Ljunggren, B. and Moller, H., Phenothiazine phototoxicity: an experimental study on chlorpromazine and its metabolites, *J. Invest. Dermatol.,* 68, 313–317, 1977a.

Ljunggren, B. and Moller, H., Drug phototoxicity in mice, *Acta Derm. Venereol.,* 1977b.

Ljunggren, B. and Wirestrand, L., Phototoxic properties of quinine and quinidine: two quinoline methanol isomers, *Photodermatology,* 5, 133–138, 1988.

Lock, S. and Friend, J., Interaction of ultraviolet light, chemicals and cultured mammalian cells: photobiological reactions of halogenated antiseptics, drugs and dyes, *Int. J. Cosmet. Sci.,* 5, 39–49, 1983.

Maibach, H. and Marzulli, F., Phototoxicity (photoirritation) from topical agents, in *Animal Models in Dermatology,* Maibach, H., Ed., Churchill Livingstone, Edinburgh, 1975, 84–89.

Maibach, H., Sams, W., and Epstein, J., Screening for drug toxicity by wavelengths greater than 3100 Å, *Arch. Dermatol.,* 95, 12–15, 1967.

Marzulli, F. and Maibach, H., Perfume phototoxicity. *J. Soc. Cosmet. Chem.,* 21, 686–715, 1970.

Mathews-Roth, M. Photosensitization by porphyrins and prevention of photosensitization bt carotenoids, in Photochemical Toxicity. Proc. 7th Symp. Food and Drug Administration Science, U.S. Department of Health and Human Services, Public Health Service, FDA, Rockville, MD, 1981, 279–285.

Maurer, T., Phototoxicity testing — in vivo and in vitro, *Food Chem. Toxicol.,* 25, 407–414, 1987.

Merot, Y., Harms, and Sauvat, J. H., Photosensibilisation au carprofene (Imadyl): un nouvel anti-inflammatoire non-steroidien, *Dermatologica,* 166, 301–307, 1983.

Mitchell, J. C., Psoralen-type phototoxicity of tetramethylthiurammono-sulfide for Candida albicans; Not for man or mouse, *J. Invest. Dermatol.,* 56, 340, 1971.

Moore, D. E. and Himmens, V. J., Photosensitization by antimalarial drugs, *Photochem. Photobiol.,* 36, 71–77, 1982.

Parrish, J. A., Fitzpatrick, T. B., Tannenbaum, L., and Pathak, M. A., Photochemotherapy of psoriasis with oral methoxsalen and longwave ultraviolet light, *N. Engl. J. Med.,* 291, 1207–1211, 1974.

Pathak, M. A., Phytophotodermatitis, in *Sunlight and Man: Normal and Abnormal Photobiologic Responses,* Pathak, M. A., Harber, L., Seiji, M., and Kukita, A., Eds., University of Tokyo Press, Tokyo, 1974.

Przybilla, B., Schwab-Przybilla, V., Ruzicka, T., and Ring, J., Phototoxicity of non-steroidal anti-inflammatory drugs demonstrated in vitro by a photo-basophil-histamine-release test, *Photodermatology,* 4, 73–78, 1987.

Rappersberger, K., Honigsmann, H., Ortel, B., Tanew, A., Konrad, K., and Wolff, K., Photosensitivity and hyperpigmentation in Amiodarone-treated patients: incidence, time course and recovery, *J. Invest. Dermatol.,* 93, 201–209, 1989.

Sambuco, C. and Forbes, P., Quantitative assessment of phototoxicity in the skin of hairless mice. *Food Chem. Toxicol.,* 22, 233–236, 1984.

Saunders, D. R., Miya, T., and Mennear, J. H., Chlotopromazine-ultraviolet interaction on mouse ear, *Toxicol. Appl. Pharmacol.,* 21, 260–264, 1972.

Schauder, S., How to avoid phototoxic reactions in photopatch testing with clorpromazine, *Photodermatology,* 2, 95–100, 1985.

Spikes, J. D. and Livingston, R., The molecular biology of photodynamic action: sensitized photooxidation in biological systems, in *Advances in Radiation Biology,* Vol. 3, Augenstein, L. G., Mason, R., and Zelle, M., Eds., Academic Press, New York, 1969, 29–121.

Stern, R. S., Phototoxic reactions to piroxicam and other nonsteroidal antiinflammatory agents, *N. Engl. J. Med.,* 309, 186–187, 1983.

Stern, R. S., Thibodeau, L. A., Klinerman, R. A., Parrish, J. A., and Fitzpatrick, T. B., Risk of cutaneous carcinoma in patients treated with oral methoxsalen photochemotherapy for psoriasis, *N. Engl. J. Med.,* 300, 809–813, 1979.

Stott, C. W., Stasse, J., Bonomo, R., and Campbell, A. H., Evaluation of the phototoxic potential of topically applied agents using longwave ultraviolet light, *J. Invest. Dermatol.,* 55, 335–338, 1970.

Tenenbaum, S., DiNardo, J., Morris, W., Wolf, B., and Schuetzinger, W., A quantitative in vitro assay for the evaluation of phototoxic potential of topically applied materials, *Cell Biol. Toxicol.,* 1, 1–9, 1984.

Urbach, F., Epstein, J., and Forbes, P. D., Ultraviolet carcinogenesis: experimental, global and genetic aspects, in *Sunlight and Man,* Pathak, M. A., Harber, L. C., Seiji, M., and Kukita, A., Eds., University of Tokyo Press, Tokyo, 1974, 259–283.

Verbov, J., Iatrogenic skin disease, *Br. J. Clin. Pract.,* 27, 310–314, 1973.

Volden, G., Krokan, H., Kavli, G., and Midelfart, K., Phototoxic and contact toxic reactions of the exocarp of sweet oranges: a common cause of cheilitis?, *Contact Derm.,* 9, 201–204, 1983.

Webster, G., Kaidbey, K., and Kligman, A., Phototoxicity from benoxaprofen: in vivo and in vitro studies, *Photochem. Photobiol.,* 36, 59–64, 1983.

Weinberg, E. and Springer, S., The evaluation in vitro of fragrance materials for phototoxic activity, *J. Soc. Cosmet. Chem.,* 32, 303–315, 1981.

Wennersten, G. Photodynamic aspects of some metal complexes, *Acta Derm. Venereol. (Stockholm),* 57, 519–524, 1977.

Western, A., Van Camp, J., Bensasson, R., Land, E., and Kochevar, I., Involvement of singlet oxygen in the phototoxicity mechanism for a metabolite of piroxicam, *Photochem. Photobiol.,* 46, 469–475, 1987.

Youn, S., Maytum, D, and Gange, R., Effect of temperature on the cutaneous phototoxic reaction to 8-methoxypsoralen in human skin, *Photodermatology,* 4, 277–280, 1987.

Zaynoun, S., Aftimos, B., Tenekjian, K., and Kurban, A., Berloque dermatitis — a continuing cosmetic problem, *Contact Derm.,* 7, 111–116, 1981.

ADDITIONAL PERTINENT REFERENCES

Austad, J. and Kavli, G., Phototoxic dermatitis cause celery infected by Sclerotinia sclerotiorum, *Contact Derm.,* 9, 448–451, 1983.

Balato, N., Giordano, C., Montesano, M., and Lembo, G., 8-Methoxypsoralen-induced photo-oncolysis, *Photodermatology,* 1, 202–203, 1984.

Beani, J. C., Gautron, R., Amblarh, P., Bastrenta, F., Harrouch, L., Jardon, P., and Reymond, J. L., Screening for drug photosensitization activity by the variations in oxygen consumption of *Bacillus subtilis, Photodermatology,* 2, 101–106, 1985.

Bjellerup, M. and Ljunggren, B., Studies on photohemolysis with special references to demethyl-chlortetracycline, *Acta Derm. Venereol. (Stockholdm),* 64, 378–383, 1984.

DeVries, H., van Henegowwen, G. M. J. B., and Wouters, P. J. H. H., Correlations between phototoxicity of some 7-chloro-1, 4-benzodiazepines, *Pharm. Weekbl. Sci. Ed.,* 5, 302–307, 1983.

Diette, K. M., Gange, R. W., Stern, R. S., Arndt, K. A., and Parrish, J. A., Coal tar phototoxicity: Kinetics and exposure parameters, *J. Invest. Dermatol.,* 81, 347–350, 1983.

Garge, R., Levins, P., Murray, J., Anderson, R., and Parrish, J., Prolonged skin photosensitization induced by methoxsalen and subphototoxic UVA irradiation, *J. Invest. Dermatol.,* 82, 219–222, 1984.

Golpashin, F., Weiss, B., and Durr, H., Photochemische Modell-Studien an lichtdermatoseninduzierenden Pharmaka: Sulfonamide und Sulfonylharnstoffe, *Arch. Pharm. (Weinheim Ger.),* 317, 906–913, 1984.

Jeanmougin, M., Pedreiro, J., Bouchet, J., and Civatt, K., Phototoxicity of 5% benzoyl peroxide in man. Evaluation with a new methodology (Fren), *Dermatologica,* 167(1), 19–23, 1983.

Kamiole, R., Gigli, I., and Lim, H. W., Participation of mast cells and complement in the immediate phase of hematoporphyrin-induced phototoxicity, *J. Invest. Dermatol.,* 82, 485, 1984.

Kavli, G. and Volden, G., The Candida test for phototoxicity, *Photodermatology,* 1, 204–207, 1984.

Keane, J., Pearson, R., and Malkinson, F., Nalidixic acid-induced photosensitivity in mice: A model for pseudoporphyria, *J. Invest. Dermatol.,* 82, 210–213, 1984.

Knudsen, E. A., The Candida phototoxicity test. The sensitivity of different strains and species of Candida, standardization attempts and analysis of the dose-response curves for 5- and 8-methoxypsoralen, *Photodermatology,* 2, 80–85, 1985.

Kochevar, I., Hoover, K., and Gawienowski, M., Benoxaprofen photosensitization of cell membrane disruption, *J. Invest. Dermatol.,* 82, 214–218, 1984.

Ljunggren, B., Propionic acid-derived non-steroidal antiinflammatory drugs are phototoxic in vitro, *Photodermatology,* 1, 3–9, 1985.

Lock, S. O. and Friend, J. V., Interaction of ultraviolet light, chemicals and cultured mammalian cells: Photobiological reactions of halogenated antiseptics, drugs and dyes, *Int. J. Cosmet. Sci.,* 5, 39–49, 1983.

Marzulli, F. N. and Maibach, H. I., Phototoxicity (photoirritation) of topical and systemic agents, in *Dermatotoxicology,* 2nd ed., Marzulli, F. and Maibach, H., Eds., Hemisphere, Washington, D.C., 1983, 375–389.

Rauterberg, A., Jung, E., Burger, R., and Rauterburg, E., Phototoxic erythema following PUVA treatment: independence of complement, *J. Invest. Dermatol.,* 94, 144–149, 1990.

Sambuco, C. and Forbes, P., Quantitative assessment of phototoxicity in the skin of hairless mice, *Food Chem. Toxicol.,* 22(3), 233–236, 1984.

Tennenbaum, S., DiNardo, J., Morris, W., Wolfe, B. A., and Schnetzinger, R. W., A quantitative in vitro assay for the evaluation of phototoxic potential of topically applied material, *Cell Biol. Toxicol.,* 1, 1–9, 1984.

Chapter 23

TECHNIQUES FOR ASSESSING THE HEALTH RISKS OF DERMAL CONTACT WITH CHEMICALS IN THE ENVIRONMENT

Dennis J. Paustenbach and Hon-Wing Leung

TABLE OF CONTENTS

0-8493-7357-3/93/$0.00 + $.50
© 1993 by CRC Press, Inc.

I. INTRODUCTION

For humans, ingestion, inhalation, and dermal absorption are three typical routes of chemical uptake. Techniques for quantitatively estimating the uptake of environmental contaminants via ingestion and inhalation have been developed and fairly well documented (Kimbrough et al., 1984; Paustenbach et al., 1986; USEPA, 1989a; McKone and Daniels, 1991). In contrast, our knowledge of how well chemicals are absorbed through the skin is less complete (Paustenbach, 1987; McKone, 1990). Although dermal uptake is infrequently the predominant route of exposure at contaminated waste sites, there are certain scenarios, such as farm worker reentry following pesticide application or the occupational handling of neat liquids or solids, where it will be the primary hazard (Poppendorf et al., 1979; Knaak et al., 1989).

Throughout the 1970s, regulatory agencies assumed a value of 10% to estimate the dermal absorption of a liquid chemical agent (e.g., pesticide or solvent) when data were not available (McLaughlin, 1984). In 1983, the Office of Pesticide Program's Scientific Advisory Panel changed its prior position and suggested that if no literature was available on a compound, then a value of 100% should be used to represent the dermal absorption (McLaughlin, 1984). Although it was acknowledged that using this figure was a highly conservative approach and would lead to an overestimation of the actual dermal absorption for nearly all compounds, the rationale was to encourage the regulated community to conduct more research on these chemicals. Such action by agencies is referred to as "science forcing". Although the spirit of such a policy decision is understandable, such a conservative approach to health risk assessment is unnecessary since we have a fairly good understanding of the absorption rate of neat chemicals through the skin and those factors which determine the efficiency of uptake.

Although much is known about the uptake of neat chemicals through the skin, our knowledge of the hazard posed by exposure to chemicals in various environmental media is less clear (Shu et al., 1988a). To understand the dermal uptake of chemicals bound to soil, dust, sludge, sediment, paint, etc., information on the neat chemical is helpful but other factors must be accounted for. The best approach is to conduct specific tests with the contaminated chemical on laboratory animals or using *in vitro* technologies. Since relatively low concentrations of the chemical are typical in the environment and high concentrations are used in laboratory studies, an extrapolation to environmental levels is often necessary. Other factors such as the duration of contact, integrity of the skin, and the chemical properties of the agent must ultimately be considered in the risk assessment.

Although as many as 30 industrial chemicals have been evaluated in tests on volunteers, and perhaps another 30 may have been tested *in vitro* using human skin, nearly all of these

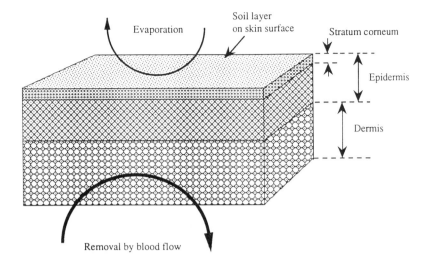

FIGURE 1. The structure of the outer skin illustrating the layers through which a chemical compound passes between the soil-particle layer and the underlying dermis, which is supplied with blood. (From McKone, T. E., *Risk Analysis,* Vol. 10 (No. 3), Plenum Press, New York, 1990.)

studies have involved neat chemicals. In contrast, the dermal uptake of only about six chemicals in soil has been evaluated in rodents (Brewster et al., 1989). Mathematical models are currently being developed to estimate the dermal uptake of chemicals from soil by humans (McKone, 1990). The purpose of this chapter is to discuss the various methods for quantitatively estimating the dermal uptake of chemicals in environmental media by humans. Current approaches and some example calculations are presented.

II. CHARACTERISTICS OF THE SKIN

The skin, or the integumentary system, is one of the largest organs of the body comprising approximately 4% of the normal body weight (Snyder, 1975). The skin consists of two different layers with the thinner outer layer known as the epidermis and the thicker, inner layer known as the dermis (Figure 1). Although skin thickness varies with location in humans, the epidermis is approximately 0.1 mm and the dermis is 2 to 4 mm thick (Rongone, 1983). The outermost layer of the epidermis is called the stratum corneum and it ranges from 10 to 50 μm thick. For purposes of risk assessment, we are mainly concerned about the stratum corneum as it has been shown that it is at least three, and frequently as much as five, orders of magnitude less permeable to most substances than the dermis (Michaels et al., 1975). Also, the permeability of the entire epidermis is indistinguishable from that of the stratum corneum alone. Thus, the skin can be thought of as a three-layer laminate of stratum corneum, the remainder of the epidermis, and dermis with permeation occurring by diffusion through the three layers in series (McLaughlin, 1984).

The stratum corneum is a heterogeneous structure containing about 40% water, 40% protein (primarily keratin), and 15 to 20% lipids (triglycerides, free fatty acids, cholesterol, and phospholipids) (Michaels et al., 1975; Anderson and Cassidy, 1973). The stratum corneum can absorb up to six times its own weight in water, and in its fully hydrated state, its permeability to water and other low molecular weight penetrants increases (Scheuplein, 1967). The lipid components of the stratum corneum may be the chief reason for its uniquely low permeability. For example, when the tissue is extracted with a fat solvent and then rehydrated, the water permeability of the stratum corneum as well as capacity to absorb larger molecules is greatly enhanced (Scheuplein, 1967).

A study by Elias et al. (1981) compared dermal absorption to stratum corneum structure and lipid composition. The extreme sensitivity of this barrier to damage from lipid solvents suggested that the amount of lipid was an important determinant of skin penetration. In addition to lipids, several other stratum corneum structure parameters, including thickness, number of cell layers, and geometric organization, can alter the permeability. Wester and Maibach (1987) also noted the amount of solvent penetrating the skin was related to the area of the skin exposed, the method of application to the surface of the skin, the type of skin exposed, and the duration of exposure.

Elias et al. (1981) correlated the *in vitro* penetration of water and salicylic acid across leg and abdominal stratum corneum with both the lipid composition and the structure of the same samples. One finding from this study was an apparent noncorrelation of penetration of both substances with either stratum corneum thickness or the number of cell layers. Their results suggested that dermal absorption of both substances correlated with the lipid content (by weight) of the sample. Also, their study suggested that relatively small inherent variations in lipid concentration may explain observed variations in permeation across different topographic regions. The nature of skin is that because it is a lipid-containing tissue, it becomes a sink for lipophilic chemicals (Wester and Maibach, 1987). The authors predicted that regions of high permeability such as palms and soles would have low lipid weight percent, whereas those of low permeability such as facial or perineal stratum corneum would have relatively high lipid weight (by percent).

III. TRANSPORT THROUGH THE SKIN

Scheuplein and colleagues wrote one of the landmark papers on transport through the skin and it is the basis for much of their subsequent research. They noted that the "barrier" function of the epidermis resides almost entirely in the stratum corneum which has about 1000 times the diffusive resistance of underlying skin layers for low molecular weight electrolytes. The stratum corneum is formed and continuously replenished by the slow upward migration of cells from the stratum germinativum. The stratum corneum is very thin and very uniform in most regions of the body and ranges from 13 to 15 μm, but on the soles and palms it can be as much as 600 μm in thickness (Snyder, 1975).

The role of skin appendages such as hair follicles and sweat ducts on skin permeability has been studied (McLaughlin, 1984). Although it has frequently been suggested that these "holes" in the skin would facilitate the passage of chemicals, their total area is relatively small in humans, and for most penetrants, absorption through the general skin surface is the preferred route (Dugard, 1983). In short, for humans, hair skin seems to be no more easily penetrated than non-hairy skin. However, in the case of some molecules that penetrate the bulk of the stratum corneum slowly, such as electrolytes and polar molecules with three or more labile polar groups, the route through follicles and ducts may predominate (Scheuplein, 1980; McLaughlin, 1984).

Both Blank (1964) and Scheuplein (1965; 1967) conducted studies to evaluate chemical transport across the outer skin layers. Their work showed that the capacity of these "shunts" to facilitate transfer is limited by the small fraction of skin area (on the order of 0.001) that consists of shafts and pores (Scheuplein and Blank, 1985). However, Kao et al. (1988) noted that for mouse skin, hair appeared to contribute significantly to overall dermal absorption of topical agents. This observation, and others, has shown that caution must be exercised when extrapolating from animal experiments involving the skin to humans.

As noted by McKone (1990), some important conclusions were reached by Scheuplein and Blank (1985) regarding transfer through the skin. These were

1. The stratum corneum is always partially hydrated, with a water content on the order of 15 to 50%.

2. Lipid solubility of the chemical plays a crucial role in determining its permeability through the stratum corneum.

3. The permeability of water through human stratum corneum is on the order of 0.56 to 1.1 mg/m^2/sec at 30°C whether measured *in vivo* from a non-sweating region of the forearm or abdomen, or *in vitro* from excised skin.

4. The stratum corneum is replenished about every 2 weeks in the mature adult.

5. Measurements of skin permeability *in vitro* are not very reliable because degree of hydration, amount of swelling, diameters of sweat ducts, and extent of folding are different in excised skin.

6. The flow of blood to the skin is on the order of 0.05 m^3 blood per m^2 skin/min, which is sufficient to make the mass transfer resistance of perfusion negligible relative to resistance through the stratum corneum (but highly lipid soluble molecules may constitute exceptions to this general rule).

7. The effective diffusion coefficient through the stratum corneum is on the order of 10^{-13} to 10^{-14} m^2/sec for (low) molecular weight compounds and 10^{-15} to 10^{-17} m^2/sec for higher molecular weight compounds.

8. The assumption that human skin has the same permeability as other mammals is not justifiable in general, but pig skin and guinea pig skin in particular provide useful approximations to human skin for some permeability measurements.

9. The stratum corneum has a filament-matrix structure with the intracellular matrix composed of hydrated keratin, which comprises a stable, two-phase system (a water-rich polar region intermingled with a network of nonplar liquid).

It is likely that these nine factors could be mathematically quantitated (modeled) so that the absorption of a chemical could be predicted without using animal or human tests. Several researchers have made significant headway on this issue. Their work should be consulted and hopefully a validated model will be available before the end of the 1990s (McDougal et al., 1986; Paterson and MacKay, 1987; Kissel and McAvoy, 1989; McKone, 1990).

IV. FACTORS AFFECTING ABSORPTION

Absorption is defined as the penetration of substances from the outside into the skin, through the skin, and into the blood stream. Schaefer and Schalla (1980) have broken this process into individual parts: (1) *penetration* is the process of entrance into 1 layer; (2) *permeation* is the migration through 1 or several skin layers; (3) *resorption* is the uptake by the cutaneous microcirculation; and (4) *absorption* is the sum of all these processes (McLaughlin, 1984).

There are a number of parameters which can affect the amount of penetrant that is absorbed. The concentration of applied dose and surface area are the two most important factors, with the greatest potential for absorption occurring when a high concentration of a penetrant is spread over a large surface area of the body (Wester and Noonan, 1980a). The relationship between the concentration of applied dose and the efficiency of absorption is not necessarily a linear one. Several studies have shown that the rate of uptake (efficiency) can decrease with increasing dose for numerous chemicals (Table 1). The site of application was the forearm for both man and rhesus monkey.

The solvent used to deliver a penetrant to the skin (vehicle) will usually have a clear effect on the efficiency of absorption. The ability of a compound to penetrate the skin and exert its effect is dependent on two consecutive physical events. The compound must first

TABLE 1
Relationship Between Dose and Efficiency of Absorption

		Percent of dose absorbed	
Penetrant	Dose (μg/cm^2)	Rhesus monkey	Man
Hydrocortisone	4	2.9	1.9
	40	2.1	0.6
Benzoic acid	4	59.2	42.6
	40	33.6	25.7
	2000	17.4	14.4
Testosterone	4	18.4	13.2
	400	2.4	2.8

From Wester, R. C. and Noonan, P. K., *Int. J. Pharmacol.*, 7, 99–110, 1980a. With permission.

TABLE 2
Ratio of the Absorption of Various Sites Compared to the Forearm

Site	Parathion	Malathion	Hydrocortisone
Forearm	1.0	1.0	1.0
Palm	1.3	0.9	0.8
Foot, ball of	1.6	1.0	—
Abdomen	2.1	1.4	—
Hand, dorsum	2.4	1.8	—
Scalp	4.7	—	3.5
Jaw, angle	3.9	—	13.0
Forehead	4.2	3.4	6.0
Scrotum	11.8	—	42.0

From Maibach, H. I., Feldmann, R. J., Milby, T. H., and Serat, W. F., *Arch. Environ. Health*, 23, 208–211, 1971. With permission.

diffuse out of the vehicle to the skin surface, and then must penetrate the skin en route to the site of action (Ostrenga et al., 1971). If the membrane diffusion constant for the penetrant and the thickness and solvent properties of the membrane are unchanged by the nature of the external vehicle, then the rate of absorption is proportional to the chemical potential of the vehicle. When the chemical potential and penetrant concentration are linearly related, Fick's law of diffusion should hold for the vehicle (Dugard, 1983). Some solvents, such as DMSO, actually dissolve lipids, destroying the barrier function and carrying the penetrant along with it into the body. The efficiency of DMSO in moving chemicals through the skin is why it was once thought that it would revolutionize the delivery of therapeutic drugs. Unfortunately, this approach was abandoned since it caused cataracts in exposed animals.

The site of application of a penetrant can be an important factor in estimating the rate of absorption (Webster and Maibach, 1987). Maibach et al. (1971) measured the absorption of several compounds at various sites on the body. Their results, which compared the forearm to other sites, found that the palm is twice as permeable, the abdomen, and dorsum of the hand, four times as permeable, the scalp, angle of the jaw, postauricular area, and forehead 12 times as permeable, and the scrotum 12 times as permeable (Table 2).

TABLE 3
Penetration Indices for Five Anatomic Sites

Site	Penetration index based on hydrocortisone	Penetration index based on pesticide data
Genitals	40	12
Arms	1	1
Legs	0.5	1
Trunk	2.5	3
Head	5	4

From Guy, R. H. and Maibach, H. I., *J. Appl. Toxicol.,* 4, 26–28, 1984. With permission.

TABLE 4
Body Surface Areas for Five Anatomic Sites

	Adult[a]		Child[b]		Infant[c]	
	Body area (%)	Area (cm²)	Body area (%)	Area (cm²)	Body area (%)	Area (cm²)
Genitals	1	190	1	75	1	19
Arms	18	3420	19	1425	19	365
Legs	36	6840	34	2550	30	576
Trunk	36	6840	33	2475	31	695
Head	36	6840	13	975	19	365
Total		19000		7500		1920

[a] Weight = 70 kg; height = 1.83 m.
[b] Weight = 19 kg; height = 1.10 m.
[c] Weight = 3 kg; height = 0.49 m.

From Guy, R. H. and Maibach, H. I., *J. Appl. Toxicol.,* 4, 26–28, 1984. With permission.

Guy and Maibach (1984) divided the body into five regions for the purpose of calculating correction factors for using the forearm penetration data: genitals, arms, legs, trunk, and head. Table 3 gives "penetration indices" or the ratio of skin penetration for one of the five anatomic sites divided by skin penetration for the forearm. These penetration indices are derived from hydrocortisone skin penetration data and from absorption results using the pesticides malathion and parathion. They also determined the relative proportion of body area and actual skin area for the five anatomic sites (Table 4).

The condition of the skin, such as loss of barrier function of the stratum corneum through disease or damage will also affect absorption. Absorption can be virtually 100% if all barrier function is removed. Damage due to occupations such as bricklaying or covering of the applied dose as with bandaging or putting on clothing after a dermal application will increase absorption. Occlusion changes the hydration and temperature of the skin and also prevents the accidental wiping off or evaporation of an applied dose (Wester and Noonan, 1980a).

The frequency of dermal application is a factor which will often affect the degree of absorption. In one study, the absorption of a single application of a high concentration of penetrant was greater than when the equivalent concentration was applied in equally divided doses (Wester et al., 1977). In another study, the absorption on the eighth day of application of the same penetrant dose was found to be significantly higher than on the first day. The authors suggested that the initial applications of penetrant altered the barrier function of the stratum corneum, resulting in increased absorption for subsequent applications (Wester et al., 1980b). It appears that the effects of repeated application will vary with each chemical and the concentration applied.

The skin contains many of the same enzymes as the liver, thus penetrant metabolism by the skin could affect absorption. The metabolizing potential of skin has been estimated to be about 2% of the liver with most of the enzyme activity localized in the epidermal layer (Pannatier et al., 1978). The slower a penetrant is absorbed through the skin the greater the opportunity for some metabolism to occur. *In vitro* penetrant absorption studies using excised human skin will not show the results of any potential metabolism and, thus, may not reflect the actual *in vivo* absorption for some penetrants.

V. DETERMINING THE RATE OF UPTAKE OF VARIOUS CHEMICALS

Historically, the absorption efficiency or rate of uptake of a chemical through the skin has most often been estimated based on studies of humans. Although these studies were not unusual from 1950 through the 1970s, few have been conducted in recent years. More frequently, estimates of the possible dermal uptake of a chemical are made using animal skin (*in vivo* or *in vitro*) or human skin *(in vitro)*. Recently, some researchers have successfully grafted human skin to rodents. If this is successful, it would dramatically change the way such tests are conducted.

Much of what is known about the types of chemical properties which influence dermal uptake comes from the study of pharmaceutical agents. Perhaps as many as 20 to 30 organic compounds have been studied directly in humans while as many as a hundred have been studied in animals.

Percutaneous absorption in animals *(in vivo)* is usually determined by measuring radio-activity in excreta following the topical application of a labeled compound. The penetrant under study is often labeled with carbon 14 or tritium and, following application, the total amount of radioactivity excreted in urine or urine plus feces is determined. The radioactivity in the excreta will be a mixture of the parent compound and any metabolites. The amount of radioactivity retained in the body or excreted through expiration or sweat can be determined by measuring the amount of radioactivity excreted following an intravenous (i.v.) injection. The fraction determined by i.v. dose is then used to correct the amount of radioactivity following topical administration. This method can be frustrating since, in practice, penetrant concentrations in blood after topical administration are often very low. Other approaches have been used, but only infrequently.

One of the main sources of human *in vivo* data is the work of Feldmann and Maibach (1969 and 1970). Other important human *in vivo* data have been collected by Stewart et al. (1964) and Piotrowski (1977). Feldmann and Maibach studied the percutaneous absorption of some steroids, pesticides, and organic compounds using radiolabeled (^{14}C) doses administered intravenously and topically to human volunteers (Table 5). The intravenous dose was used to correct for the amount of radiolabeled penetrant that was percutaneously absorbed but not recovered in urine samples (i.e., excreted in feces or retained in the body). The topical dose of 4 μg/cm^2 was dissolved in acetone and applied to a 13 cm^2 circular area of the ventral forearm. All urine was collected for 5 days and divided into suitable time periods so that either the absorption rate (%/h) or total absorption (% of dose) could be calculated. The skin sites were not protected and not washed for 24 h. Table 7 is a compilation of the 48 penetrants studied (Fiserora-Bergerova et al., 1990).

Table 6 shows the percutaneous penetration of some steroids in human. For the intravenous dose, 1 μCi (^{14}C) of the steroid was dissolved in 5 to 20 ml of saline (or saline plus ethanol) and injected. The range for total absorption for the compounds in Table 5 is greater

TABLE 5
Percutaneous Absorption of Some Organic Compounds in Humans

Time (h)	Absorption rate (% dose/h)[a]						Total absorption		
	0–12	12–24	24–48	48–72	72–96	96–120	% of dose	SD	No. subjects
Acetylsalicylic acid	0.141	0.438	0.334	0.147	0.760	0.060	21.01	3.31	3
Benzoic acid	3.036	0.340	0.055	0.000	0.000	0.000	42.62	16.45	6
Butter yellow	0.215	0.685	0.289	0.830	0.054	0.022	21.57	4.88	4
Caffeine	0.559	1.384	0.856	0.109	0.032	0.014	47.56	20.99	12
Chloramphenicol	0.007	0.019	0.210	0.022	0.015	0.012	2.04	2.46	6
Colchicine	0.036	0.038	0.033	0.040	0.025	0.004	3.69	2.50	6
Dinitrochlorobenzene	3.450	0.565	0.134	0.045	0.018	0.009	53.14	12.41	4
Diethyltoluamide	0.773	0.331	0.084	0.036	0.016	0.012	16.71	5.10	4
Hexachlorophene	0.029	0.031	0.20	0.028	0.034	0.030	3.10	1.09	7
Hippuric acid	0.005	0.003	0.001	0.001	0.001	0.001	0.21	0.09	7
Malathion	0.313	0.170	0.044	0.017	0.011	0.005	7.84	2.71	7
Methylcholanthrene	0.062	0.329	0.258	0.127	0.064	0.045	16.31	5.16	3
Nicotinic acid	0.000	0.002	0.001	0.001	0.002	0.007	0.34	0.09	3
Nicotinamide	0.019	0.168	0.177	0.088	0.052	0.031	11.08	6.17	7
Nitrobenzene	0.022	0.022	0.013	0.013	0.011	0.006	1.53	0.84	6
Para-aminobenzoic acid	0.159	0.648	0.444	0.196	0.058	0.044	28.37	2.43	13
Phenol	0.254	0.091	0.010	0.601	0.000	0.000	4.40	2.43	3
Potassium thiocyanate	0.051	0.060	0.078	0.097	0.100	0.093	10.15	6.60	6
Salicylic acid	0.116	0.535	0.356	0.156	0.080	0.033	22.78	13.25	17
Thiourea	0.046	0.035	0.010	0.008	0.007	0.007	0.88	0.22	3
Urea	0.008	0.021	0.051	0.073	0.075	0.034	5.99	1.91	4

[a] Corrected for i.v. recovery.

From Feldmann, R. J. and Maibach, H. I., *J. Invest. Dermatol.*, 84,399–404, 1970. With permission.

than 250-fold, much larger than that for the series of steroids in Table 6. However, as with the steroids, the standard deviations are very high for the compounds, indicating a large degree of variability. For example, chloramphenicol had a standard deviation of greater than 100%. The authors pointed out two examples of closely related compounds showing great differences in penetration: benzoic acid was absorbed 200 times more than its glycine conjugate, hippuric acid, nicotinic acid showed minimal penetration while 10% of its amide was absorbed. The authors generally found a good correlation between the maximum penetration rate for each compound and its total absorption.

Table 7 presents similar *in vivo* absorption studies by Feldmann and Maibach (1974) on some pesticides and herbicides. The authors used the same experimental method discussed previously to evaluate five organophosphates, three chlorinated hydrocarbons, two carbamates, and two herbicides. The total excretion from the i.v. dose varied over a wide range within dieldrin, aldrin, and carbaryl at 3 to 7% and malathion, baygon, and 2,4-D being over 80%. For the topical administration, diquat was the only compound with only slight penetration; carbaryl, on the other hand, was nearly completely absorbed after multiplying the topical results by the large correction factor obtained from the i.v. administration.

The authors discussed the rather large standard deviations of $^1/_3$ to $^1/_2$ of the mean value found in their studies. They claimed that the difference was due to actual differences between people rather than experimental error since repeat studies on the same subject gave similar results. Feldman and Maibach (1974) assumed a normal distribution and found that 1 person in 10 will absorb twice the mean value while 1 in 20 will absorb three times this amount. They have also found that experimental subjects differ by a factor of 5 in the amount of percutaneous absorption. It would be interesting to reevaluate these data using a log-normal distribution and more contemporary statistical analysis. They commented that the experimental subjects did not sweat extensively such as a field worker might when exposed to these pesticides. As previously stated, the current EPA policy is to use a value of 100% for the dermal absorption of an environmental contaminant unless a lower value can be supported by scientific studies. For most compounds this will represent a considerable overestimate of the actual dermal absorption rate. As shown in Table 5, hippuric acid, and thiourea have a total dermal absorption of less than 1% of the applied dose after 5 days. On the other hand, compounds like caffeine and dinitrochlorobenzene penetrate to a much higher degree.

A. *IN VIVO* ANIMAL STUDIES

Due to the toxicity of many penetrants, *in vivo* human studies are no longer conducted due to potential health risks from exposure. Further, the legal liabilities involved in such testing usually makes such studies impractical. Consequently, researchers must often use animals to obtain absorption data. Unfortunately, there are a number of difficulties associated with the extrapolation of animal data to humans. Animal species variation, different sites of application, differences between shaved vs. unshaved skin, and difference in skin metabolism (or lack thereof) are some of the factors which need to be considered when using animal studies to predict human absorption via the skin.

Perhaps the most cited study of percutaneous absorption involving rats, rabbits, miniature swine, and man was conducted by Bartek et al. (1972). Table 8 shows the results of their study. Radiolabeled compounds were applied to the shaved skin of the back and protected by a non-occluding device. In general, the penetration through the skin of the pigs and man was similar and much slower than it was through the rat and rabbit skin. For the compounds studied, haloprogin was completely absorbed by the rat and rabbit, while only 11% was absorbed by man. About half of the caffeine applied was absorbed by all the animals plus man. The results of this study showed that absorption in the rabbit and rat would not be predictive of humans, while the miniature swine appears closest to man. This study was

TABLE 6
Percutaneous Absorption of Steroids in Humans

Time (h)	Absorption rate (% dose/h)[a]						Total absorption		
	0–12	12–24	24–48	48–72	72–96	96–120	% of dose	S.D.	No. subjects
Hydrocortisone	0.005	0.023	0.019	0.018	0.016	0.010	1.87	1.59	15
Hydrocortisone acetate	0.020	0.089	0.032	0.024	0.015	0.008	2.65	1.80	6
Cortisone	0.015	0.037	0.039	0.036	0.032	0.024	3.38	1.64	7
Corticosterone	0.013	0.065	0.139	0.070	0.050	0.039	8.78	5.35	6
17-OH Desoxycorticosterone	0.041	0.101	0.084	0.076	0.062	0.055	8.41	4.28	5
Desoxycorticosterone	0.197	0.313	0.143	0.069	0.035	0.020	12.55	8.53	6
17-OH Progesterone	0.042	0.120	0.213	0.211	0.078	0.031	14.76	11.35	7
Progesterone	0.208	0.264	0.135	0.045	0.024	0.011	10.81	5.78	6
Flucinolone acetonide	0.002	0.011	0.012	0.016	0.008	0.005	1.34	1.05	9
Dexamethasone	0.005	0.003	0.004	0.003	0.002	0.002	0.40	0.23	3
Estradiol	0.008	0.056	0.099	0.101	0.107	0.103	10.62	4.86	3
Testosterone	0.147	0.364	0.156	0.66	0.036	0.018	13.24	3.04	17
Testosterone acetate	0.103	0.133	0.048	0.15	0.007	0.004	4.62	2.28	6
Testosterone propionate	0.061	0.096	0.035	0.015	0.009	0.005	3.44	1.03	9
Dehydroepiandrosterone	0.265	0.446	0.249	0.091	0.046	0.028	18.45	7.71	6
Androstenedione	0.183	0.334	0.155	0.076	0.043	0.028	13.47	5.56	11

[a] Corrected for i.v. recovery.

From Feldmann, R. J. and Maibach, H. I., *J. Invest. Dermatol.*, 52, 89–94, 1969. With permission.

TABLE 7
Percutaneous Penetration of Some Pesticides and Herbicides in Humans

Time (h)	Absorption rate (% dose/h)[a]								Total absorption	
	0–4	4–8	8–12	12–24	24–48	48–72	72–96	96–120	% Dose	SD
Azodrin	0.057	0.048	0.092	0.121	0.183	0.147	0.113	0.073	14.7	7.1
Ethion	0.004	0.005	0.026	0.036	0.044	0.041	0.015	0.011	3.3	1.1
Guthion	0.044	0.202	0.294	0.276	0.207	0.125	0.059	0.040	15.9	7.9
Malathion	0.089	0.408	0.396	0.149	0.029	0.008	0.006	0.005	8.2	2.7
Parathion	0.007	0.116	0.243	0.202	0.110	0.062	0.036	0.029	9.7	5.9
Baygon	0.351	0.705	1.204	0.543	0.093	0.023	0.020	0.028	19.6	5.2
Carbaryl	0.005	1.212	3.027	1.944	0.853	0.277	0.154	0.105	73.9	21.0
Aldrin	0.086	0.074	0.078	0.079	0.066	0.053	0.060	0.061	7.8	2.9
Dieldrin	0.143	0.137	0.035	0.093	0.066	0.060	0.043	0.034	7.7	3.2
Lindane	0.064	0.135	0.245	0.113	0.088	0.066	0.051	0.048	9.3	3.7
2,4-D	0.009	0.012	0.020	0.029	0.068	0.082	0.043	0.027	5.8	2.4
Diquat	0.005	0.002	0.005	0.003	0.003	0.003	0.002	0.001	0.3	0.1

[a] Corrected for i.v. recovery.

From Feldmann, R. J. and Maibach, H. I., *Toxicol. Appl. Pharmacol.*, 28, 126–132, 1974. With permission.

TABLE 8
In Vivo Percutaneous Absorption by Rat, Rabbit, Pig, and Human

Penetrant	Percent dose absorbed[a]			
	Rat	Rabbit	Pig	Human[b]
Haloprogin	95.8	113.0	19.7	11.0
N-Acetylcysteine	3.50	1.98	6.00	2.43
Testosterone	47.4	69.6	29.4	13.2[b]
Cortisone	24.7	30.3	4.06	3.38[b]
Caffeine	53.1	69.2	32.4	47.6[b]
Butter yellow	48.2	100.0	41.9	21.6[b]

[a] Corrected for recovery following i.v. administration.
[b] Human data taken from Feldmann and Maibach, 1969 and 1970.

From Bartek, M. J., LaBudde, J. A., and Maibach, H. I., *J. Invest. Dermatol.*, 58, 114–123, 1972. With permission.

important since it was among the first to show that animals were often a poor predictor of the human response.

Bartek and LaBudde (1975) studied the absorption of four pesticides in the rabbit, pig, and squirrel monkey using the same techniques as in their 1972 study. The results were compared to man where the site of application was the forearm (Table 9). They showed that the *in vivo* absorption of pesticides in the rabbit was greater than in man, while absorption in the pig and squirrel monkey was closer to that of humans.

B. *IN VITRO* STUDIES

In the 1980s, *in vitro* studies using human skin began to be used more frequently for estimating dermal absorption (Dugard et al., 1984). In these studies a piece of excised human skin is attached to a diffusion apparatus which often has a top chamber to hold the applied

TABLE 9
In Vivo **Percutaneous Absorption of Several Pesticides**
by Rabbit, Pig, Squirrel, Monkey and Human

Pesticide	Percent dose absorbed[a]			
	Rabbit	Pig	Squirrel, Monkey	Human[b]
DDT	46.3	43.4	1.5	10.4
Lindane	51.2	37.6	16.0	9.3
Parathion	97.5	14.5	30.3	9.7
Malathion	64.6	15.5	19.3	8.2

[a] Corrected for recovery following i.v. administration (except the monkey data).
[b] Human data from Feldmann and Maibach, 1974.

From Bartek, M. J. and LaBudde, J. A., in *Animal Models in Dermatology*, Maibach, H. I., Ed., Churchill Livingstone, New York, 1975. With permission.

dose of the penetrant, an O-Ring holds the skin in place, and a temperature controlled bottom chamber containing saline or other solvents (plus a sampling port to withdraw fractions for analysis). Although human forearm skin is optimal, it is difficult to obtain, so it is common practice to use abdominal skin collected at autopsy.

For most studies, the stratum corneum is heat separated from the epidermis and dermis, then studied by itself. The procedures used to conduct these studies are well described elsewhere (Dugard et al., 1984). It is generally believed that properly conducted *in vitro* tests using human skin, for most classes of chemicals, can be a reasonably good predictor of the absorption rate in humans (Bronaugh et al. 1982). However, due to the fragile nature of the technique, these studies must be carefully interpreted.

VI. DERMAL BIOAVAILABILITY OF CHEMICALS IN VARIOUS MEDIA

Bioavailability is an important parameter when estimating the chemical uptake from complex matrices, especially soil. Bioavailability is the percentage of a chemical which is absorbed by humans (usually as predicted in animal studies) following exposure via inhalation, ingestion, or dermal contact (Paustenbach, 1989). There are a number of parameters which are likely to influence the degree of dermal bioavailability including aging (time following contamination), soil type (e.g., silt, clay, and sand), co-contaminants and the concentration of co-contaminants (e.g., oil and other organics), and the concentration of the chemical contaminant in the media (McConnell et al., 1984; Paustenbach et al., 1986; Shu et al., 1988a, b; Brewster et al., 1989). The typical media of concern for assessing dermal contact to environmental chemicals are fly ash, sediment, house dust, and soil (Van den Berg et al., 1984; Shu et al., 1988a).

The bioavailability of a chemical on soil will be very much affected by its chemical and physical properties. Depending on the chemical of concern and the above-mentioned variables, the bioavailability of a given soil contaminant can vary between 0.05 to 100% (Umbreit et al., 1986; Paustenbach, 1989). Large molecular weight chemicals will often bind to the soil and be less water soluble, while smaller molecules will frequently be water soluble, less tightly bound, and relatively bioavailable (McKone, 1990).

Most dermal bioavailability data for contaminated soil will be obtained in laboratory animals or *in vitro* test systems. This introduces a significant source of uncertainty for predicting the human response. In the interest of conservatism, safety factors have sometimes been applied to dermal bioavailability data obtained in animals to estimate the likely degree of human uptake. However, for purposes of risk assessment, this uncertainty factor is often unnecessary since human skin has generally been shown for a diverse class of chemicals to be about tenfold less permeable than the skin of typical animal species, such as rabbits and rats (Bartek and LuBudde, 1975; Shu et al., 1988a).

A number of studies of the dermal bioavailability of various chemicals in soils have been conducted (Poiger and Schlatter, 1980; McConnell et al., 1984; Umbreit et al., 1986; Shu et al., 1988a; Brewster et al., 1989; Skowronski et al., 1988 and 1989). Other studies have assessed the bioavailability in fly ash since it is often deposited onto or mixed with soil (Van den Berg et al., 1984). These studies showed that differing media and chemicals can yield dramatically different dermal and oral bioavailabilities. One of the problems that has plagued "bioavailability" studies is the lack of a widely held definition of the term or commonly used test procedures. Fries and Paustenbach (1989) have offered a definition which has some merit.

A. A MODEL FOR ESTIMATING DERMAL BIOAVAILABILITY

In a recent paper, McKone (1990) described a generic model for predicting the dermal bioavailability of various industrial chemicals on soil. The approach was based, in part, on fugacity. McKone correctly noted that:

Addressing the mechanism of chemical uptake from soils is a problem of understanding the dynamics of the process. It is rather easy to show using simple chemical thermodynamics or 'level-one fugacity' models.

Most hydrophobic compounds will tend to partition from soil to skin layers and from skin layers to blood to fat tissues. The key question is "at what rate" does the transfer occur? Because total skin has a fat content of about 10% and soil has an organic carbon content of only about 1 to 4%, it can be expected that, without evaporation, a compound placed on the skin in a soil matrix will eventually move from the soil to the underlying fat layers of the skin surface. However, the rate at which this process occurs is crucial for determining whether appreciable uptake occurs during the period of time between deposition on the skin and removal by evaporation, washing, or other processes. It is the mass-transfer coefficients of the soil-skin-layer and the soil-air layer that define the rate at which these competing processes occur (McKone, 1990).

McKone's model is an extension of an approach proposed by Kissel and McAvoy (1989) and Kissel and Robarge (1988). Their model indicated that even as the magnitude of uptake increases, the efficiency of uptake can go down when the thickness of the soil layer (i.e., the magnitude of applied dose) increases. This effect has been measured by Brewster et al. (1989) who observed that dioxins and furans penetrate rat skin more easily at lower doses. For example, they found that 40% of the TCDD placed on skin passes through at high doses while it was as high as 60% at lower doses.

McKone (1990) has developed and presented two models for predicting uptake from soil by the skin. Although preliminary, they have excellent promise. In these, he assumed that the soil loading on the skin was in the range of 0.1 to 100 mg/cm². At the low end, this corresponds to a soil layer of roughly 0.6 μm on the skin surface and at the high end to 600 μm.

For both the continuous and unit-deposition models, the estimate of uptake fraction produced by the two models is particularly sensitive to the values of octanol-water partition coefficient (K_{ow}), Henry's law constant (K_h), and soil deposition (I_s). When K_h is very small (much less than 0.001), the only loss mechanisms for chemicals from the skin soil matrix are mass transfer and wash-off. With the dimensionless K_h much less than 0.001, as illustrated

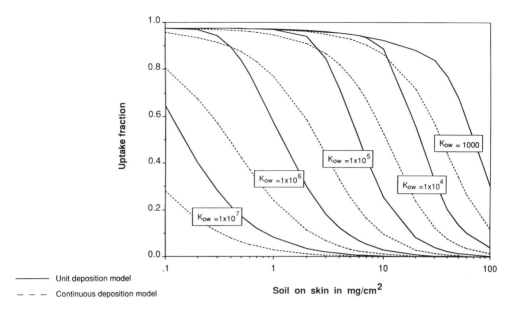

FIGURE 2. Based on the unit and continuous deposition models and the range of octanol-water (K_{ow}) partition coefficients, this graph shows the fractional uptake during 12 h vs. soil deposition per unit area of exposed skin with the dimensionless Henry's law constant $K_h < 0.001$. (From McKone, T. E., *Risk Analysis,* Vol. 10 (No. 3), Plenum Press, New York, 1990.)

in Figure 2, the fraction of uptake across the stratum corneum over a period of 12 h varies significantly with both soil loading and the K_{ow}. Figure 2 reveals that when K_h is below 0.001, the soil uptake fraction increases with decreasing soil loading and with decreasing octanol-water partition coefficient K_{ow}. McKone observed that, for an exposure time of 12 h, the uptake fraction approaches unity for skin loadings up to 10 mg/cm² when K_{ow} is 10^4 or less.

McKone showed how (Figure 3) the estimated uptake fraction varies with soil loading for a range of K_{ow} values when $K_h = 0.001$ (the value for TCDD). When K_h is on the order to 0.001, there is competition between vapor loss and diffusion as removal mechanisms for contaminants bound to soil particles. He noted that the limit on uptake approaches unity only when K_h is 1 or less. However, once again, the soil uptake fraction increases when decreasing soil loading and with decreasing octanol-water partition coefficient K_{ow}. He also showed (Figures 2 and 3) that for K_h less than 100, the unit deposition model gives a higher estimate of uptake fraction than the continuous deposition model. When K_h is above 0.001, the loss of chemical from the soil matrix through vapor mass transfer becomes more important in controlling uptake fraction and the uptake fraction is more sensitive to K_h.

For each of three K_h values (0.001, 0.01, and 0.1), McKone (1990) predicted how the uptake fraction estimated from the unit deposition model varies with soil loading when the octanol-water coefficient is low ($K_{ow} = 10$) (Figure 4). For these same K_h values the uptake fraction varies with soil loading (Figure 5) when the octanol-water coefficient is high ($K_{ow} = 10^7$). By comparing these two figures, it is clear that the uptake fraction decreases strongly with increasing soil loading when K_{ow} is large but is less sensitive to soil loading when K_{ow} is small. As shown in Figure 4, the uptake fraction can actually increase slightly with increasing soil loading when K_{ow} is low and K_h is above 0.01.

McKone's approach is intended for first-order estimates of dermal uptake of chemicals from a soil matrix on the skin. The uptake fraction (bioavailability) from the soil matrix varies with the duration of exposure, soil deposition rate, and chemical properties of the

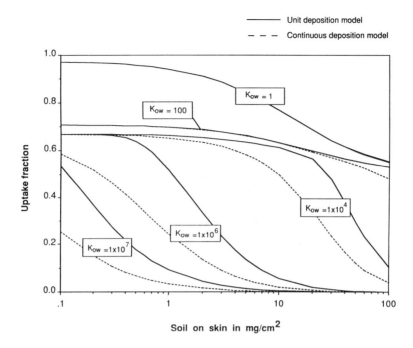

FIGURE 3. Based on the unit and continuous deposition models and the range of octanol-water (K_{ow}) partition coefficients, this graph shows the fractional uptake during 12 h vs. soil deposition per unit area of exposed skin with the dimensionless Henry's law constant $K_h = 0.001$. (From McKone, T. E., *Risk Analysis*, Vol. 10 (No. 3), Plenum Press, New York, 1990.)

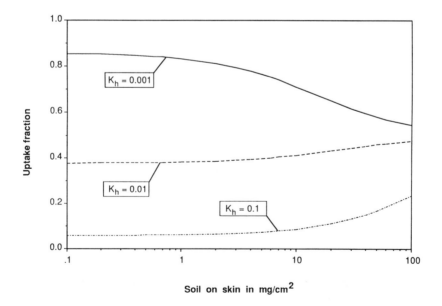

FIGURE 4. Based on the unit and continuous deposition models and the range of Henry's law constant (K_h) values, this graph shows the fractional uptake during 12 h vs. soil deposition per unit area of exposed skin with the octanol-water K_{ow} equal to 10. (From McKone, T. E., *Risk Analysis*, Vol. 10 (No. 3), Plenum Press, New York, 1990.)

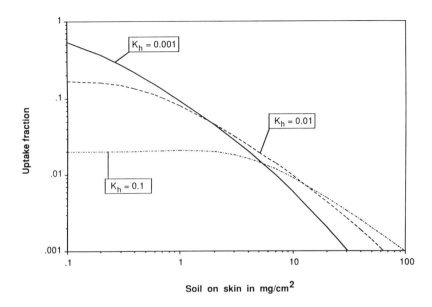

FIGURE 5. Based on the unit and continuous deposition models and the range of Henry's law constant (K_h) values, this graph shows the fractional uptake during 12 h vs. soil deposition per unit area of exposed skin with the octanol-water K_{ow} equal to 10^7. (From McKone, T. E., *Risk Analysis,* Vol. 10 (No. 3), Plenum Press, New York, 1990.)

deposited contaminated media. Results based on this model reveal that the efficiency of contaminant uptake can depend strongly on the amount of soil on the skin surface. When the amount of soil on the skin is less than 1 mg/cm^2, McKone's model (1990) predicts rather large uptake fractions, approaching unity in some cases. For the amount of soil (20 mg/cm^2) used in experiments with rats as reported by Poiger and Schlatter (1980) and Shu et al. (1988a), this model predicts uptake of 0.50% which is comparable to the roughly 1% value they measured for rats. This would suggest that the results of these experiments should not yet be applied to all human soil-exposure settings until the variation of uptake fraction with soil loading is further evaluated.

These various studies and the theoretical analyses suggest that the process by which chemicals pass from soil through human skin is not simple but that the behavior is predictable if the controlling factors are quantitated and accounted for. However, unless the key factors are known, it is not easy to reliably estimate how much lower than unity the uptake fraction will be for a particular organic compound. McKone (1990) suggested that a few generalizations can be made. First, for compounds with a K_{ow} of 10^6 and below, and a K_h below 0.001, it is not unreasonable to assume 100% uptake in 12 h. Second, for compounds with a K_h of 0.01 and above, the uptake fraction is unlikely to ever exceed 40% in 12 h and will be well below this when K_{ow} is above 10. Third, for compounds with a K_h of 0.1 and above, we can expect no more than 3% uptake in 12 h. Finally, he cautioned that none of these conclusions are applicable to metals and inorganic compounds.

VII. FACTORS USED IN EXPOSURE CALCULATIONS

There are a number of other parameters that go into calculating dermal exposure and uptake besides the dermal absorption factor and bioavailability from the matrix. One needs to know the area of exposed skin, the concentration of the penetrant, the duration of the exposure for each event, and the frequency of events expressed as the number of exposures per some unit time (usually per year) (Paustenbach, 1989; Sheehan et al., 1991).

General approaches for calculating the dermal intake of humans have been suggested (McKone and Daniels, 1991). One scenario is that of a thin film of penetrant on the skin. For this finite mass scenario, the exposure is calculated by multiplying (concentration) × (skin surface area) × (frequency) times the thickness of the film layer. The dermal layer is then calculated by multiplying the exposure times the dermal absorption factor expressed as a percent or as %/h times the duration.

Another scenario is where there is an excess amount of penetrant on the skin. In this case, the thickness of the penetrant layer is not calculated; steady-state kinetics are assumed and the dermal intake is calculated by multiplying (skin surface area) × (concentration) × (frequency) × (duration) × (permeability coefficient, Kp).

A. SKIN SURFACE AREA

The surface areas for each region of the body at various ages have been measured (Snyder, 1975). Much of this information was developed to treat persons in burn clinics as well as by persons in the armed forces when designing uniforms.

The "rule of nines" may be used for estimating the surface area of certain regions of the total body (Snyder, 1975):

Head and neck	9%
Upper limbs (each 9%)	18%
Lower limbs (each 18%)	36%
Front of trunk	18%
Back of trunk	18%
Perineum	1%
Outstretched palm and fingers	1%

Highly detailed information on the surface area of numerous regions of the body have been tabulated (Snyder, 1975).

B. SURFACE AREA OF EXPOSED SKIN

An important factor in dermal exposure assessment is the surface area of the human body that is exposed. As noted in the South Coast Air Quality Management District's (SCAQMD) Multi-pathway Health Risk Assessment Input Parameters Guidance Document, "the total exposed skin area is a function of the age and size of the individual, the type of activity in which the individual is engaged, and the prevailing climate conditions at the time of exposure" (Zwiacher and Dennison, 1988). In California, the Department of Health Services (DHS) assumes that the head, neck, lower arms, and feet of children and adults are exposed to soils and dust and that the exposed areas are independent of age. This agency suggests that the likely exposed skin surface of children is 2050 cm^2 and for adults the exposed area is 4656 cm^2. However, the SCAQMD notes that these values "probably overestimate the exposed area of adults and underestimates the exposed area of children" (Zwiacher and Dennison, 1988).

The USEPA has estimated the skin surface area of children and adults (USEPA, 1989a). They suggest using skin surface areas of 2897 cm^2 for children 0 to 2 years and 3400 cm^2 for children 2 to 6 years old (arms, hands, legs, and feet). For adults, they support using a skin surface area of 2940 cm^2 for adults in a residential scenario or a commercial scenario (adult is assumed to wear pants, open-necked, short-sleeved shirt, shoes, with no hat or gloves) (USEPA, 1989a).

Slightly different skin surface areas have been suggested by Hawley (1985). He estimated that for outdoor exposure scenarios young children probably have 2100 cm^2 of exposed skin whereas older children and adults have 1600 cm^2 and 1700 cm^2 of exposed skin, respectively.

C. SOIL LOADING ON THE SKIN

Quantitative estimates of the dermal uptake of chemicals from dusts and soils contain more uncertainty than estimates of uptake through other routes of entry (Paustenbach, 1987). Nonetheless, there are ample data available for making reasonably accurate estimates of the degree of chemical uptake via the skin.

A CDC assessment of dioxin-contaminated soil proposed that dermal exposures probably follow "an age-dependent pattern of deposition similar to soil ingestion" (Kimbrough et al., 1984). Exposure to contaminated soil was assumed to occur primarily through gardening and yard work. In addition, the CDC assumed that dirt could remain on the skin for 24 h. Since 1984, most authors seem to agree that a more likely scenario is that a person would be dermally exposed to soil for 4 to 8 h/day (on days of gardening) with some degree of washing at the end of this period.

Numerous studies have evaluated the amount of soil that is likely to be in contact with skin. Roels et al. (1980) demonstrated that approximately 1.0 mg of soil per cm^2 of skin adheres to a child's hand after playing in and around the home. However, several other studies suggest that soil adheres to skin to a much lesser degree. Lepow et al. (1975), using adhesive tape to sample a defined area of skin, suggested that the amount of soil adhering to human skin was approximately 0.5 mg/cm^2. The California Department of Health Services Toxic Substance Control Division has suggested for purposes of risk assessment that approximately 0.9 mg/cm^2 should be assumed to adhere to the hands of children (average age was 11) (California Department of Health Services (CDHS), 1986).

Sayre et al. (1974) evaluated dust loading and found that lead present in house dust was in sufficient quantities to increase blood lead levels. High blood lead levels were particularly evident in children involving contamination of the hands by repeated contact with floor, walls, and windowsills due to subsequent ingestion by the frequent hand-to-mouth activities of preschool children (age 1 to 6 years old). The authors used paper towels impregnated with liquid (14 × 20 cm) to sample a typical household area where children may play and measured an average of .17 mg of lead per towel specimen. The study yielded a high mean value of .044 mg/100 ml blood lead levels in preschool children exposed to dust loading levels over .03 mg per 280 cm^2 where hand-to-mouth exposure was the most common route of entry for lead contaminated dust.

The bulk of the literature on the degree of soil loading on the skin that can be expected by workers is presented in Tables 10a and b. Soil loading (adherence) appeared to range from approximately 0.33 to 2.8 mg soil/cm^2 (Lepow et al., 1975; Roels et al., 1980; Paustenbach et al., 1986; USEPA, 1989a) for children (Table 11). For adults, lower levels would be anticipated. For the purpose of many risk assessments, a daily loading of 1.8 mg soil/cm^2 on exposed skin for children has been considered amply conservative (Paustenbach, 1989). This value is the measured level of soil on the hands of male youths after playing in a dusty school yard (Roels et al., 1980) and is consistent with factors used in numerous assessments of environmental contaminants (Kimbrough et al., 1984; Hawley, 1985; Paustenbach, 1987; Keenan et al., 1989; Sheehan et al., 1991).

In two often cited assessments of TCDD contaminated soil, it was assumed that dermal exposures would follow "an independent pattern of deposition similar to soil ingestion", as shown in Table 12 (Kimbrough et al., 1984; Paustenbach et al., 1986). In the Kimbrough et al. (1984) assessment (CDC), it was assumed that dirt would remain on the hand for a period long enough to bring about 1% absorption — the percent absorption of TCDD determined in rats exposed for 24 h. A 24-h duration is almost certainly longer than what is likely for humans under normal conditions; therefore, this would represent a value beyond "worst case". Paustenbach et al. (1986) suggested that a more likely scenario (alternate) is that persons would be exposed for 4 to 8 h/day, with some degree of washing at the end of this period.

TABLE 10a
Integrated Dermal Dose to Each Body Location in Field Workers Exposed to Foliar Pesticide Residues (All Values in Milligrams)

Phosalone[b]	Reentry[a]			
	A	B	C	D
Hands	95.0	116.1	84.2	158.0
Forearms	9.8	17.6	12.2	12.3
Upper arms	4.3	14.5	13.4	13.1
Head	7.3	11.6	23.0	24.1
Neck	1.6	2.4	2.4	2.6
Shoulders	0.4	1.7	1.4	0.7
Chest	1.0	1.8	1.1	2.5
Back	0.6	1.7	2.0	3.8
Hips	0.2	0.2	0.6	1.3
Thighs	0.7	1.0	2.3	5.1
Calves	1.4	0.3	0.5	7.9
Feet	0.2	0.1	0.1	0.9
Total without hands	27.3	52.9	58.9	74.3
Overall	122.3	169.0	143.1	232.1

[a]   Reentry: Duration between pesticide application and harvest .A = 22–24 days, B = 13–15 days, C = 7–9 days, D = 3–5 days.

[b]   Phosalone is an organophosphate pesticide with a chemical name of *O,O*-diethyl-*S*-[(6-chloro-2-oxo-benzoxazolin-3-yl)methyl]-phosphorodithioate.

From Popendorf, W. J. et al., *J. Occup. Med.*, 21, 189–194, 1979. With permission.

TABLE 10b
Relative Contributions to Total Dermal Exposure of Body Areas to Pesticides as Studied by the California Department of Food and Agriculture

Chemical	Number of exposures monitored	Average total dermal exposure (%)		
		Hands	Head, face, and neck	Protected area
Parathion	3	23.4	67.4	9.2
Hevinphos	22	48.0	34.6	17.4
TDK	24	49.2	22.7	28.1
DEF/Folex	32	46.8	23.7	29.5
Chlorobenziate	21	27.1	59.4	13.5
Total	102	42.9	33.8	23.3
Job Activity				
Ground Applicator	4	18.1	57.5	24.4
Mixer/Loader	36	50.7	22.0	27.3
Aerial Applicator	18	54.6	27.4	18.0
Ground Applicator	25	30.4	47.9	21.7
Flagger	19	38.7	38.6	22.7
Total	102	42.9	33.8	23.3

From Maddy, K. T., Wang, R. G., and Winger, C. K., Worker Health and Safety Unit, State of California, 1983.

TABLE 11
Summary of Soil Loading Rates on the Palms of Human Test
Subjects

Soil loading rate (mg/cm²)	Type of soil	Age and sex
1.45	Commercial potting soil	Male adult
2.77	Kaolin clay	Male adult
1.48	Playground dirt and dust	Males 10 to 13 years old, rural areas
1.77	Playground dirt and dust	Males 10 to 13 years old, urban areas
0.5	Not available	Children, 4 years old
0.2	Not available	Adults
0.33	House dust	Children, 1–3 years old
0.11	House dust	Adult woman

TABLE 12
Uptake of TCDD from Contaminated Soil in
Residential Site[a] (Case I) and Industrial Site[b]
(Case II)

	Case I TCDD uptake (fg/kg/day)		Case II TCDD uptake (fg/kg/day)	
Route	CDC	Alternate	CDC	Alternate
Oral	603.0	2.8	0	0
Dermal	19.6	2.7	390	78
Inhalation	10.0	5.6	200	57
Total	632.6	11.1	590	135

[a]   10% of land contaminated (10 ppb TCDD).
[b]   20% of land contaminated (100 ppb TCDD).

The EPA cited the work of Roels et al. (1980) in support of their exposure factors (McLaughlin, 1984). Their recommendations are similar to those reported by Day et al. (1975), who observed that 5 to 50 mg of dirt transferred from a child's hand to a sticky sweet. Sayre et al. (1974) found 2000 μg/kg for the lead concentration in rural and urban house dust, respectively. Their data indicated that dust uptake due to mouthing tendencies is about 100 mg/day if all dust on both sides of the hand were ingested or absorbed through the skin. Uptake can also be estimated by multiplying the appropriate absorption rate for a given chemical and the amount of dirt on the skin using table values for skin surface area (Lipsky et al., 1989).

To estimate the annual average dermal uptake of a chemical it is necessary to account for the lack of exposure due to inclement weather conditions (Paustenbach et al., 1986). For example, when assessing exposure of persons in the Midwestern States, one must account for the fact that direct dermal contact with soil can occur only about 180 days/year (due to snow, rain, and frozen soil). About 64 years (age 70 minus age 6) duration is often assumed for adults who garden and do yard work. These values almost certainly overestimate the actual average exposure, since (1) not everyone in the community gardens, (2) many persons wear gloves when working intimately with dirt, (3) gardeners work directly with the soil primarily during only planting and weeding, (4) most people do not garden each day, and (5) the number of days of precipitation during the gardening season further diminishes the frequency of exposure.

Several studies have been published which describe the effect of soil particle size and organic content of soil on the adherence of soil to human skin. For example, Que Hee et al. (1985) using a variety of soils of different particle size suggested that on average, about 0.2 mg/cm^2 of soil adheres to the hands of small adults. Driver et al. (1989) surveyed various soils of different organic content for their ability to adhere to human hands (adult male). They estimated that the average amount of soil retained on human hands was 0.6 mg/cm^2 (unsieved). Thus, the available literature suggests that the amount of soil adhering to the hands of humans is probably between 0.2 and 1.0 mg/cm^2. A value of 0.5 mg/cm^2 appears to imply a conservative estimate for the amount of soil adhering to the human skin.

D. DERMAL UPTAKE FROM LIQUIDS

A number of persons in recent years have described methods for estimating the uptake of liquids through the skin (Scow et al., 1979; Byard, 1989; Paustenbach, 1989). Several examples evaluated the dermal absorption of chemicals by persons when swimming or showering.

Scow et al. (1979) prepared a document for the Office of Water Planning and Standards in which the dermal uptake of heptachlor and chlordane from water by humans was estimated. The water flux (J) through the skin was taken to be between 0.2 to 0.5 mg/cm^2/h while the flux of the solute (penetrant) was estimated by multiplying the water flux by the weight fraction of penetrant in water. At a concentration of 1 ppb, the flux of the penetrant was 0.2×10^{-9} mg/cm^2/h. Using a representative body surface area of 1.8×10^4 cm^2, if one swam for 4 h in any given day through water containing 1 ppb chlordane, these authors estimated that as much as 0.013 to 0.036 µg chlordane might be absorbed.

Beech (1980) estimated the amount of chloroform absorbed over a 3-h period by a 6-year-old boy swimming in water containing 500 µg chloroform/l. The average 6-year-old is assumed to weigh 21.9 kg and have a surface area of 0.88×10^4 cm^2. The permeability coefficient (K_p) for chloroform in aqueous solution was assigned the value of 125×10^{-3} cm/h. The flux through the skin per hour was estimated at 125×10^{-3} cm/h $\times$ 500 µg/l $\times$ 1 l/1000 cm^3 $\times$ 1000 mg/1 µg = 62.5×10^{-6} mg/cm^2/h. The total dermal intake is then 3 h $\times$ 62.5×10^{-6} mg/cm^2 h $\times$ 0.88×10^4 = 1.65 mg chloroform.

Byard (1989) evaluated the dermal and inhalation uptake of xenobiotics due to common daily hygiene activities including bathing and showering. Human exposure to 1,1,1-trichloroethane (TCA), which has been widely used as an industrial degreasing solvent since the mid 1950s in Santa Clara Valley, CA (Silicon Valley), was examined. Slow releases of TCA to soil and groundwater, which served as a domestic water source for local residents, resulted from leakage from underground piping, tanks, and spills from TCA handling. Byard calculated a dermal vapor, dermal water, and inhalation dose of 0.0072 µg/kg/day for a 10 min shower exposure to 50 l of water containing 1 ppb TCA. Using a half-life for volatilization of 10 min, the TCA release would be approximately 24 µg since dermal uptake and volatilization are competing processes, and much of the water does not make direct dermal contact. Therefore, an estimated 1% of TCA in water was available for dermal uptake (Byard, 1989).

In addition, Byard (1989) estimated a dermal dose from a 20 min bath at 0.0475 µg/kg/day by extrapolating data from studies of dermal uptake involving toluene, xylene, ethylbenzene, and styrene. In evaluating the worst case bathroom exposure scenario, which assumes a daily bath and additional hour in the bathroom, he estimated a daily dose of 0.0478 µg/kg/day for water containing 1 ppb TCA. Adding ingestion exposure to this value, the daily dose would be 0.0764 µg/kg/day. Based on the national average weight of a male adult (Abraham et al., 1979), an adult may be exposed to 5 µg of TCA/day. Although a review of the toxicological data in the literature indicates that TCA is relatively nontoxic, it is apparent that the daily uptake due to dermal absorption and inhalation are measurable simply due to routine hygiene activities.

Recently, additional studies have suggested that exposure to volatile organic compounds (VOC) from routes other than direct ingestion may result in a greater exposure than from ingestion. Showering appears to be an event where an individual can be exposed to elevated concentrations of volatile compounds in the air within a confined space and the entire body is dermally exposed to contaminants in the water. Chloroform is the water contaminant normally present due to chlorination and it will generally pose the greatest theoretical risk.

Recent work suggests that the uptake of chloroform from a single, 10 min shower was calculated to be .46 μg/kg/day; .24 μg/kg/day is due to inhalation and .22 μg/kg/day is due to dermal exposure (Travis et al., 1990). If one accepts the EPA cancer potency factor for chloroform, the plausible cancer risk associated with only dermal absorption during a 10 min shower was estimated to be 6.0×10^{-5} or 6 in 100,000. Others have suggested that the cancer risk associated with dermal absorption and inhalation is about 1.22×10^{-4}, while the estimated risk from daily ingestion of tap water ranged from $.130 \times 10^{-4}$ to 1.80×10^{-4} for .15 and 2.0 l, respectively (Jo et al., 1990). These authors predicted that a single daily 10 min shower for a 70-year lifetime could pose an excess cancer risk of 12.2×10^{-5} (Jo et al., 1990).

Calculations of the dermal absorption during showering, based on limited data on the permeability of chloroform through the skin, indicate that dermal exposure could account for between 10 and 25% of the dose due to inhalation exposure (Jo et al., 1990). Comparisons have been made of the concentration of chloroform in exhaled breath after a normal shower with municipal tap water to those after an inhalation-only exposure to chloroform. The resulting concentrations were 6 to 21 μg/kg/day for the shower exposure and 2.4 to 10 μg/kg/day for inhalation-only exposure (Jo et al., 1990). The concentration in exhaled breath in this study after showering was approximately twice as high as that after inhalation-only exposure, indicating that the contribution to the absorbed dose due to dermal absorption was about equivalent to inhalation absorption.

E. DERMAL ABSORPTION STUDIES IN HUMANS

Stewart and Dodd (1964) have studied absorption of chlorinated hydrocarbon solvents (carbon tetrachloride, trichloroethylene, tetrachloroethylene, methylene chloride, and 1,1,1-trichlorethane) through the skin using breath analysis. They found each compound in measurable quantities in the breath following contact with the skin. Wester and Maibach (1987) have conducted similar studies and found measurable levels of organic solvent concentrations in the blood and exhaled air long after skin exposure ended. Other studies examining dermal absorption while bathing found that lipid- and water-soluble chemicals readily bind with the human skin stratum corneum (Wester and Maibach, 1989; Brown et al., 1989) and this will undoubtedly affect the pharmacokinetics of the chemicals in humans.

Piotrowski (1957, 1971, 1973, 1977) studied cutaneous absorption of industrial chemicals and the role of skin in this exposure pathway. He found that absorption of organic compounds may follow surface contamination of the skin and for some compounds it may occur directly from the gaseous phase. Because the transepidermal exposure route for organic compounds in cutaneous absorption seems to be of paramount importance, Piotrowski designed experiments to evaluate dermal absorption of organic compounds and calculate dermal absorption rates of specific chemicals. Some of these rates are listed in Table 13.

Piotrowski also studied the absorption of some chemicals directly through the skin from the gaseous phase; he found this phenomenon to play a significant role for aniline, nitrobenzene, and phenol which otherwise are well absorbed when applied directly (Piotrowski, 1977). Similar to cutaneous absorption, he observed increasing rates of vapor absorption

TABLE 13
Rates of Dermal Uptake of Various Industrial
Chemicals Measured in Human Volunteers

Chemical	Absorption rate	Ref.
Benzene	0.4 mg/cm²/h	Piotrowski, 1977
Toluene	0.5 mg/cm²/h	Piotrowski, 1973
Aniline	0.2–0.7 mg/cm²/h	Piotrowski, 1957
Nitrobenzene	0.2–3 mg/cm²/h	Piotrowski, 1977
Parathion	3.5 mg/cm²/h	Piotrowski, 1977

with increasing ambient temperature and for aniline, with increasing humidity (Piotrowski, 1977). Many chemicals have been studied and evaluated since Piotrowski began evaluating cutaneous absorption. Dermal absorption of organic compounds is now recognized as significantly enhancing the body burden of exposure to chemicals; dermal absorption of vapors is also gaining recognition as a significant exposure pathway as studies continued to evaluate human contact with organic compounds.

F. DERMAL UPTAKE BY PESTICIDE APPLICATORS

One occupation which has been shown to have significant dermal exposure is the pesticide applicator. The other is the worker involved in re-entry to areas where crops have been sprayed. Maddy et al. (1983) has developed exposure assessment techniques designed to quantitate pesticide uptake by workers during the application and re-entry tasks.

The authors conducted dermal monitoring of workers involved in the application of pesticides. Exposures of mixers/loaders, ground applicators, mixer/loader/ground applicators (workers performing all three activities during a single application, aerial application, and flaggers) were determined in a total of 102 individual exposure situations. California regulations require use of clean outer clothing to reduce the potential dermal exposure of workers to pesticides by decreasing the area of bare skin available for contact with chemicals. Rubber or some other type of waterproof gloves were worn by all workers except the aerial applicators and flaggers.

Hand exposure was estimated by rinsing the hands in a predetermined solvent containing either water, soap and water, ethanol, or a combination of the three. Hands were rinsed prior to and upon completion of the application. Exposures to the hands, face, and neck were estimated by placing small patches on the upper collar of the coveralls in the front and back, or, in some cases, placing patches directly on the face. Values were extrapolated to the entire surface area of these body parts. Potential exposure to the skin protected by coveralls was also measured with patches.

Knaak et al. (1989) have assembled an impressive review of the topic of agricultural worker exposure to several groundwater pesticides. This review is worthy of evaluation since it describes retention of soil-like materials.

VIII. CALCULATING DERMAL UPTAKE

The values for body weight, exposure frequency, and fraction of soil from contaminated sources used to estimate dermal uptake are usually the same as those used for soil ingestion (Kimbrough et al., 1984; Paustenbach et al., 1986). The following equation is suggested for estimating the plausible rate of dermal uptake of chemicals on contaminated soil. To estimate the uptake of neat liquids, the higher dermal bioavailability would need to be accounted for.

$$ADD = (C \times SCR \times SA \times FC \times EF \times B)/BW \qquad (1)$$

where

ADD   =   average daily dose (mg/kg/day)
C   =   concentration of contaminant in surface soil (mg/kg)
SCR   =   soil contact rate (soil loading)(mg/cm^2/day)
SA   =   surface area of exposed skin (cm^2)
FC   =   fraction of soil from contaminated source
EF   =   exposure frequency (fraction of year)
B   =   dermal bioavailability factor due to matrix effect (% absorbed)
BW   =   body weight (kg)

A. EXAMPLE CALCULATION 1: ESTIMATING THE UPTAKE OF A CHEMICAL FROM SOIL

What is the plausible daily uptake of toluene by a child who plays in contaminated dirt for 1 h/day for 2 years? The child begins to play at 2 years of age and the concentration of toluene in soil is 0.025 mg/kg. Assume that the contact rate suggested by Roels et al. (1980) is accurate (0.5 mg/cm^2/day), that the legs and arms are not covered, and that the child weighs 16.4 kg.

$$ADD = (C \times SCR \times SA \times FC \times EF \times B)/BW \qquad (1)$$

where

C   =   0.025 mg/kg
SCR   =   0.5 mg/cm^2/day
FC   =   0.51
SA   =   1580 cm^2
EF   =   1 h/day
B   =   0.01
BW   = $\dfrac{(0.025 \text{ mg/kg})(0.5 \text{ mg/cm}^2/\text{day})(1580 \text{ cm}^2)(0.51)(1/24 \text{ day})(0.01)}{16.4 \text{kg}}$

$$ADD = 2.56 \times 10^{-4} \text{ mg/kg/day}$$

The lifetime average daily dose (LADD) would be equal to the total number of days of play in the dirt divided by the number of days in a typical lifetime. The uptake of a chemical from contaminated soil can be calculated. The following example involves this approach:

B. EXAMPLE CALCULATION 2: UPTAKE OF A WATER CONTAMINANT WHILE SWIMMING

Dermal uptake of chemicals through the skin may occur due to the presence of chemicals in water. For example, Jo et al. (1990) have estimated chloroform exposure and the body burden due to showering.

This equation for estimating dermal uptake via contaminated water is very similar to that suggested by McKone and Daniels (1991).

$$Dose_{DW} = \frac{[Water] \times 10^{-3} \times SA \times T \times DAR \times SV \times 10^{-9}}{BW}$$

where

Dose$_{DW}$	=	dose received through dermal contact with water (mg/kg/day)
[Water]	=	contaminant concentration in surface water (μg/l)
10^{-3}	=	conversion factor, mg/μg
SA	=	surface area of exposed skin (cm^2)
T	=	length of exposure (min/day)
DAR	=	dermal absorption rate (μg/cm^2/min)
10^{-9}	=	conversion factor, 10^{-9} μg/kg
SV	=	specific volume of water, 1 l/kg
BW	=	body weight (kg)

Assume that an adult is exposed to ethyl benzene for 1 h/day for 1 week from swimming in a lake on vacation. What is the estimated dermal uptake assuming SA = 4280 cm^2, T = 60 min/day, DAR = 2.7 μg/cm^2/min, SV = 1 l/kg, BW = 70 kg, in [water] = 200 μg/l

$$\text{Dose} = \frac{[200\ \mu g/l] \times 10^{-3}\ mg/\mu g)(4280\ cm^2)(60\ min/day)(2.7\ \mu g/cm^2/min)(1\ l/kg)(10^{-9}kg/\mu g)}{70kg}$$

$$= 1.98 \times 10^{-6}\ mg/Kg$$

The following exposure factors are frequently used to estimate dermal uptake during contact with water:

- A value of 4280 cm^2 is used for the surface area of exposed skin. This value represents the head, forearms, hands, and feet (USEPA 1989b). For future on-site workers, however, a value of 3000 cm^2 is often used, which represents the head, forearms, and hands.
- For the purpose of risk assessment, the length of exposure to surface water is often conservatively assumed to equal 30 to 60 min/day.
- The following chemical-specific, dermal absorption rates are sometimes used:

    Perchloroethylene: 4.0 μg/cm^2/min (Piotrowski, 1977)
    Benzene: 3.2 μg/cm^2/min (Tsuruta, 1982)
    Toluene: 0.78 μg/cm^2/min (Tsuruta, 1982)
    Xylenes: 2.2 μg/cm^2/min (Tsuruta, 1982)
    Ethyl Benzene: 2.7 μg/cm^2/min (Dutkiewicz and Tyras, 1967)

- A value of 70 kg is usually used for the average adult human body weight (USEPA, 1989b).
- A value of 10 kg is used for children 0 to 3 years.

C. EXAMPLE CALCULATION 3: APPROACH SUGGESTED BY CALIFORNIA AGENCIES

As in other approaches, dermal absorption (dose-dermal) is a function of the soil or dust loading of the exposed skin surface, skin surface area exposed, as well as the concentration and bioavailability of the pollutant.

$$\text{Dose-Dermal} = (Cs)(SA)(SL)(ABS)/(ABW \times 10^9)$$

where

Dose-dermal	=	exposure dose through dermal absorption (mg/kg/day)
Cs	=	average concentration of Cr^{6+} in soil (μg/kg)
SA	=	surface area of exposed skin (cm^2)
SL	=	soil loading on skin (mg/cm^2)
ABS	=	fraction absorbed across skin
ABW	=	average body weight (kg)
10^9	=	micrograms to kilogram conversion factor (μg/kg)

In California, the following default values are suggested when chemical specific data are available:

Cs	=	average soil concentration (μg/kg)
SA	=	4656 (cm^2)
SL	=	0.5 (mg/cm^2/d)
ABS	=	0.1
ABW	=	70 kg

Using the above values, what would be the daily dermal dose for an adult if the concentration of Cr^{6+} in soil was 1000 μg/kg.

$$\text{Dose-dermal} = \frac{(1000\ \mu\text{g/kg})(4565\ \text{cm}^2)(0.5\ \text{mg/cm}^2/\text{day})(0.1)}{(70\ \text{kg})(1 \times 10^9\ \mu\text{g/kg})}$$

$$= 3.26 \times 10^{-6}\ \text{mg/kg/day}$$

D. EXAMPLE EQUATION 4: MCKONE AND DANIELS APPROACH

McKone and Daniels (1991) have derived exposure factors used to estimate uptake of contaminants from soil. Dermal-uptake exposure factors F_{sd} suggested by McKone and Daniels (1991) to estimate dermal uptake of chemicals from soil for a child (0 to 15 years) are

Surface area		0.28 m^2 (head and upper extremities exposed to soil for a child)
Body weight	=	27 kg
Skin-soil loading	=	0.03 kg/m^2
Dermal uptake rate	=	1.8% per day

Therefore,

$$F_{sd} = \frac{[0.28\ \text{m}^2] \times 0.018/\text{day} \times 0.03\ \text{kg/m}^2}{27\ \text{kg}}$$

$$= 5.6 \times 10^{-6}(\text{per day})$$

If we were to adopt the McKone and Daniel's exposure factors, what would the lifetime dermal uptake dose be for a child exposed to a soil concentration of hexavalent chromium of 5000 mg/kg?

$$\text{Dose-dermal} = 5000\ \text{mg/kg} \times \frac{15\ \text{years}}{70\ \text{years}} \times 5.6 \times 10^{-6}/\text{day}$$

$$= 0.006\ \text{mg/kg/day}$$

E. EXAMPLE EQUATION 5: UPTAKE OF LIQUIDS

To illustrate the potential importance of skin absorption as a route of entry for the glycol ethers, compared to uptake via inhalation at the TLV, Paustenbach (1988) offered the following example.

Experimental data in animals *(in vivo)* and in humans *(in vitro)* indicate that the glycol ethers pass through the skin much like other solvents. How much 2-methoxyethanol could possibly be absorbed (on an mg/kg basis), if a person has a heavily contaminated glove on one hand for about 30 min:

- Surface area of the hand $400 \ cm^2$ (Snyder, 1975)
- Exposure time   30 min
- Rate of absorption   2.8 $mg/cm^2/h$

Dermal uptake $= (400 \ cm^2)(2.8 \ mg/cm^2/h)(0.5 \ h)(person/70 \ kg) = 8 \ mg/kg/day$

By comparison, how much 2-methoxyethanol will be taken up by a 70-kg employee who is exposed for 8 h/day at the TLV of 5 ppm (16 mg/m^3) assuming an 80% uptake efficiency?

Inhalation uptake $= (16 \ mg/m^3)(8 \ h/workday)(0.8 \ uptake)(person/70 \ kg)$
$= 1.5 \ mg/kg/day$

These calculations indicate that dermal uptake could be as much as 8 mg/kg/day following 30 min of exposure. This dose is about four times greater than the predicted uptake of 2-methoxyethanol when a person works for 8 h in an environment containing 2-methoxyethanol at the TLV concentration. From this example, it is clear that the dermal route of entry can significantly contribute to the total absorbed dose.

IX. PHARMACOKINETICS OF SKIN UPTAKE

Percutaneous absorption is defined as the transport of externally applied chemicals through the cutaneous structures and the extracellular medium to the bloodstream. In reviewing the literature on percutaneous absorption of organic solvents, one of the most interesting phenomena is that levels of solvent often appear in blood and exhaled air long after skin exposure has ended (Wester and Maibach, 1987). This is due to two consecutive processes: (1) a penetration phase, i.e., the passage of a chemical through the superficial skin structures, the stratum corneum and the epidermis, to the extracellular medium; and (2) a resorption phase during which a rapid diffusion occurs from the extracellular fluid to the blood via the cutaneous circulation. The skin behaves as a rate-limiting barrier which only allows penetration of chemicals at a relatively slow rate.

It has been shown that the skin structure that is largely responsible for this barrier function is the stratum corneum (Marzulli and Tregear, 1961). The living cells of the epidermis are relatively more permeable than the stratum corneum and it does not constitute the rate-limiting step under most circumstances.

The simplest way to model the process of skin absorption is to assume that Fick's first law of diffusion at steady-state is applicable:

$$J = dQ/dt = K_p \cdot C = D \cdot k \cdot C/e \qquad (2)$$

where

dQ/dt = rate of chemical absorbed
K_p = permeability coefficient
C = concentration gradient
D = diffusion coefficient in the stratum corneum
e = stratum corneum/vehicle partition coefficient of the chemical
h = thickness of the stratum corneum

The concentration gradient is equal to the difference between the concentration above and below the membrane. Since the concentration below is generally negligible compared to the concentration above, the concentration gradient can be approximated to equal the applied chemical concentration. The above equation summarizes the kinetics of the penetration process through the skin. It must be emphasized that it is an oversimplification and is an approximation for most *in vivo* exposure situations where true steady-state conditions are rarely attained. Nonetheless, this equation includes the most important factors which account for the percutaneous absorption of chemicals.

The two factors which strongly influence the transfer rate are the partition coefficient and the diffusion coefficient of the stratum corneum. Since it is rather difficult to measure skin/vehicle partition coefficients, the parameters D, K, and E are combined together to give a permeability coefficient. From the equation, it can be seen that the absorption intensity is proportional to the concentration of the chemical and to the application area. Thus for the same quantity of chemical, the systemic effects will be greater in proportion when the skin is occluded (Wepierre and Marty, 1979). Table 14 shows the absorption rate of 36 chemicals through the human skin. The cutaneous flux, J, spans over five orders of magnitude, ranging from a low of 0.03 $\mu g/cm^2/h^1$ for sulindac to 1900 $\mu g/cm^2/h$ for salicylic acid.

The diffusion constant represents the rate of migration of a chemical through the stratum corneum. As the stratum corneum has a non-negligible thickness, there is a period of transient diffusion during which the rate of transfer through the skin rises to reach a steady state (Figure 6). The steady state is maintained thereafter indefinitely, provided the system remains constant. The common method of analyzing kinetic profiles such as the one depicted in Figure 6 is to determine the lag time (T_L) by extrapolating the linear portion of the curve to the x-axis. The diffusion coefficient is then given by:

$$D = e^2/6T_L \tag{3}$$

Depending on the type of chemicals, the lag time sometimes can be as long as several hours or even days. From a risk assessment standpoint, if the exposure time is shorter than the lag time, it is unlikely that there will be significant systemic absorption and accumulation (Flynn, 1990).

The partition coefficient in the diffusion equation illustrates the importance of solubility characteristics for a chemical to penetrate the skin (Anderson et al., 1988; Surber et al., 1990). The stratum corneum typically mimics the characteristics of a lipophilic structure (Elias et al., 1981; Raykar et al., 1988). Lipophilic chemicals tend to accumulate in the stratum corneum and a high concentration of the chemical is achieved at the point of contact. Assuming that the chemical is at least slightly soluble in water, penetration at this level will be rather rapid by migration into the intercellular spaces (Scheuplein, 1965).

Examination of the data presented in Table 14 reveals that there is a direct correlation between the *n*-octanol:water partition coefficient, K_{ow} of the chemicals and their skin penetration rate. This clearly illustrates the importance of solubility in determining the percutaneous penetration of chemical substances. With purely lipophilic chemicals, however,

TABLE 14

Physical Properties and Steady States Skin Penetration Rates from Saturated Propylene Glycol Solutions

Compound	MW[a] (daltons)	M.P.[b] (°C)	Volume[c] (Å)³	log K_{ow}[d]	Solubility (g/l)[e]			J[h] μg/cm²/h
					PG[f]	Octanol	IPM[g]	
Acetaminophen	151	170	126	0.47	100	24	0.36	4.6
Benzoic acid	122	122	100	1.95	230	50	46	720
Benzyl alcohol	108	−15	101	1.02	1040	1040	1040	1060
Caffeine	194	238	158	−0.02	9.1	4.6	1.0	1.5
ClonidineHCl	230	305	174	−2.60	120	2.0	0.10	14
Dextromethorphan	271	110	268	4.13	25	170	7.0	10
Dextromethorphan	271	125	268	−0.05	230	14	0.10	9.0
Diazepam	285	130	225	2.80	15	28	12	4.7
Estradiol	272	176	263	2.69	62	28	3.0	11
Ethacrynic acid	303	121	223	3.95	250	210	15	740
5-Fluorouracil	130	283	84	−0.92	15	1.0	0.03	3.2
Furosemide	331	206	227	1.46	13	2.7	0.36	0.21
Griseofulvin	353	220	278	2.18	2.1	1.2	0.42	0.24
HydralazineHCl	161	275	132	−1.53	8.9	1.3	0.45	20
Hydrocortisone	363	218	338	1.61	15	5.8	0.35	0.42
Ibuprofen	206	76	197	3.51	220	250	140	430
Indolyl-3-acetic acid	175	170	142	1.41	160	45	3.4	11
Indomethacin	358	159	270	3.08	8.2	12	2.0	0.25
Isosorbidedinitrate	236	68	166	1.22	3.9	18	20	4.8
Ketoprofen	254	93	210	3.00	250	200	17	12
Methylsalicylate	152	−8	121	2.46	200	1170	1170	1350
Minoxidil	209	248	191	1.24	78	6.3	0.20	0.81
Morphine sulfate	285	254	247	−1.76	6.0	1.1	0.11	0.19
Naproxen	230	155	196	3.18	18	30	4.1	4.8
Nicotinic acid	123	236	100	−0.20	7.4	3.4	0.16	2.2
Nifedipine	346	173	273	2.12	9.0	9.6	2.2	0.07
Pentazocine	285	146	256	2.19	13	74	8.4	0.68
PentazocineHCl	285	254	256	−1.70	100	13	2.0	0.64
Piroxicam	331	198	226	0.05	1.7	4.1	2.3	0.70
PropranololHCl	259	163	235	−0.45	110	7.7	0.12	6.6

	mw[a]	m.p.[b]	volume[c]	log K$_{ow}$[d]	solubility[e]			
Salicylamide	137	142	112	0.89	240	36	11	53
Salicylic acid	138	159	104	2.24	12	160	40	1900
Sulindac	356	180	293	4.53	11	6.7	0.15	0.03
Terbutaline Sulfate	225	247	214	−1.90	64	0.90	0.01	0.04
Testosterone	288	153	293	3.31	8.2	104	9.1	35
Triamcinolone acetate	434	293	379	2.53	8.2	4.9	0.23	0.18

[a] mw = molecular weight in daltons.
[b] m.p. = melting point (°C).
[c] volume − in cubic angstroms.
[d] Log K$_{ow}$ = octanol: water partition coefficient.
[e] solubility = solvents used for solubility in grams/liter.
[f] PG = propylene glycol.
[g] PM = isopropyl myristate.
[h] J = flux in µg/cm^2/h.

From Kasting, G. B., Smith, R. L., and Cooper, E. R., in *Pharmacology of the Skin*, Vol. 1, Shroot, B. and Schaefer, H., Eds., S. Karger, Basel, 1987. With permission.

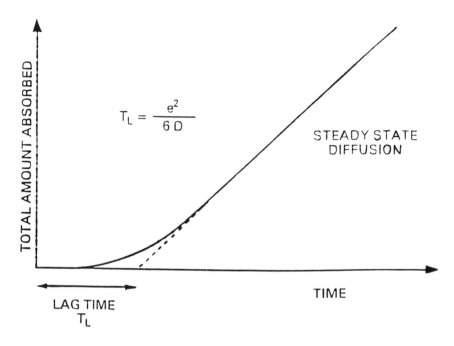

FIGURE 6.   Time course of penetration for drug diffusing through intact human skin. T_L: lag time, e = thickness of the membrane, D = diffusion coefficient. (From Wepierre, J. and Marty, J. P., *Trends in Pharmacological Sciences,* Inaugural issue, Elsevier Biomedical Press, 1979. With permission.)

penetration may not extend readily beyond the stratum corneum. For these chemicals, the stratum corneum behaves as a storage site, and the chemicals can be released over a long period of time. This phenomenon, known as the "reservoir effect" (Vickers, 1963), may explain why sometimes a single exposure to certain chemicals in the skin can lead to prolonged effects (Rougier et al., 1985). Accordingly, stratum corneum is an effective barrier for hydrophilic substances and therefore they have very low skin absorption rates.

X. PREDICTING DERMAL UPTAKE BASED ON PHARMACOKINETICS

To assess the absorption of a chemical via a particular route of exposure, bioavailability studies need to be conducted. There are three indirect methods for estimating the bioavailability of a pure chemical following topical administration.

Area under the curve — If an intravenous dose is assumed to have maximum bioavailability, the extent of absorption following a cutaneous dose can be estimated by comparing the total quantity in the body compartment after giving equivalent doses by both the intravenous and cutaneous routes (Andersen and Keller, 1984). The total dose can be represented as the concentration-time integral carried out to infinite time, i.e., the area under the plasma concentration-time curve (AUC). Bioavailability, or the proportion of administered dose absorbed, can then be calculated as follows:

$$\text{Bioavailability} = \text{AUC (cutaneous)}/\text{AUC (intravenous)} \qquad (4)$$

Since the stratum corneum is a major barrier to percutaneous absorption of chemicals, the plasma levels are usually quite low following cutaneous administration. To improve the sensitivity of detection, it is therefore necessary to use radiolabeled material. Since a radiolabel does not distinguish between the parent chemical and its metabolite(s), the bioavailability determined with this methodology actually represents the composite absorption of the parent chemical and its metabolite(s). Specific chemical assays are necessary to quantify the absolute bioavailability of chemicals which undergo metabolism as they are absorbed through the skin.

Mass Balance Approach — The second method calculates the extent of absorption by monitoring the cumulative amount of the chemical in all routes of excretion (urine, feces, exhaled air, etc.) over infinite time following topical administration. Thus:

$$\text{Bioavailability} = \text{total amount in excreta/applied dose} \qquad (5)$$

In reality, the duration of excreta collection is only for a finite time since collection of excreta over an extremely prolonged period is impractical. For certain chemicals, especially those which have low whole-body clearance, there may be a considerable portion of the absorbed dose remaining in the body at the cessation of the collection period. Thus, the total amount recovered in the excreta does not represent the total dose absorbed. This residual amount and any material excreted in routes not assayed can be corrected for by determining the amount excreted following intravenous administration (Wester and Maibach, 1985).

$$\text{Bioavailability} = \text{Total excreted (cutaneous)/total excreted (intravenous)} \qquad (6)$$

Skin measurement approach — The third method is the most expedient but also the least reliable of the three. It consists of quantifying the amount of chemical recovered from the skin surface after topical administration.

$$\text{Bioavailability} = (\text{Applied dose} - \text{dose remaining on skin})/\text{applied dose} \qquad (7)$$

The limitation of this surface recovery methodology is that it is prone to inaccuracies. Losses of the chemical from the skin may occur as a result of evaporation and total recovery from the skin is difficult to know with high confidence.

XI. ASSESSING THE SKIN SENSITIZATION POTENTIAL OF INDUSTRIAL CHEMICALS

One of the most common occupational diseases is contact dermatitis due to dermal exposure to industrial chemicals (Adams, 1990). Although contact dermatitis is not a life-threatening disease, it can be a persistent condition and debilitating in severe cases. Once sensitized to a chemical, an individual is at risk of dermatitis whenever exposed to the same or an antigenically cross-reactive chemical. Table 15 presents some examples illustrating the wide spectrum of chemicals which are known or suspected to cause allergic contact dermatitis (Emmett, 1986).

Several methods have been developed to identify and to assess the sensitization potential of contact allergens. These include a variation of the patch tests in humans. In animals, the most commonly used procedures are the Draize test, the optimization test, Freund's complete adjuvant test, the maximization test of Magnusson and Kligman, Maguire's modified split adjuvant technique, the Buhler test, and the open epicutaneous test (Schlede et al., 1989). The guinea pig is the most suitable for sensitization testing and is the animal species of

TABLE 15
Selected Important Allergic Contact Sensitizers

Metals
    Nickel and nickel salts
    Chromium salts
    Cobalt salts
    Organomercurials
Plant Sensitizers
    Toxicodendron genus: pentadecylcatechols and other catechols
    Primula obconica: α-methylene-γ-butyrolactone
    Composite family: sesquiterpene lactones
Rubber Additives
    Mercaptobenzthiazole
    Thiuram sulfides
    p-Phenylenediamine and derivatives
    Diphenylguanidine
    Resorcinol monobenzoate
Epoxy oligomer (M.W. 340)
Methyl methacrylate and other acrylic monomers
Pentaerythritol triacrylate and other multifunctional acrylates
Hexamethylenediisocyanate
p-Tertiary butyl phenol
Ethylenediamine, hexamethylenetetramine, and other aliphatic amines
Formaldehyde
Neomycin
Benzocaine
Captan

From Emmett, E. A., in *Casarett and Doull's Toxicology. The Basic Science of Poisons,* 3rd ed., McMillan, New York, 1986. With permission.

choice. Details of these test procedures have been well described (Klecak, 1983) and will not be discussed here.

When assessing the sensitizing potential of chemicals, one must be cognizant of the fact that while there are only a few substances which are obligatory allergens (Paschoud, 1967), the number of facultative allergens are huge and growing. Because the allergenic potentials of facultative allergens are variable, only a small segment of the population is at risk. Whether an effect is manifested in an individual depends on his or her immune responsiveness as well as the condition under which the skin exposure occurs. Thus it is crucial to evaluate the results of sensitization studies with a complete understanding of the experimental conditions under which the test was performed (Robinson et al., 1989).

Many procedures such as the guinea pig maximization test are conducted with extreme or provocative conditions and may not provide a true estimate of the sensitizing potential of a chemical under realistic exposure conditions. Because of these problems, none of the sensitization tests can be applied universally to all chemicals and types of formulation. The proper test selection follows a thorough evaluation of the anticipated use pattern of the chemical and the experimental objective. The use of animal sensitization tests is suggested where they may serve as an indicator of the possible hazard, rather than absolute criteria for safety evaluations of chemicals.

A. CROSS-SENSITIZATION

One facet of evaluating the skin sensitizing potential of chemicals that has been accorded relatively little attention is cross-sensitization. The same is true for some metals. Cross-sensitization refers to the phenomenon where an allergic sensitization engendered by one

TABLE 16

**Dermal Sensitization Response of Nine Alkyleneamines Tested
with the Guinea Pig Maximization Protocol**

Chemical	Concentration[a] used for challenge (%)	Response[b] Percentage	Response[b] Normalized
Ethyleneamine	5	45	9.0
Diethylenetriamine	25	80	3.2
Aminoethylethanolamine	25	40	1.6
Triethylenetetramine	50	74	1.5
Aminoethylpiperazine	25	25	1.0
Pentaethylenehexamine	100	100	1.0
Piperazine	25	5	0.2
Tetraethylenepentamine	50	5	0.1
Hydroxyethylpiperazine	100	10	0.1

[a] Animals were first induced with an intradermal injection (5% of the material in distilled water) followed by a topical treatment 7 days later. The challenge dose was administered topically 2 weeks after the topical induction dose. The concentration of the material used for the topical induction was the highest that would elicit only mild irritation, and the concentration used for the topical challenge was the highest that would not produce irritation.

[b] Percent of animal showing a positive dermal response. Because of different irritation potency, the challenge concentrations used were different among the alkyleneamines. In order to compare the relative sensitizing potencies, the response is normalized with respect to challenge concentrations, i.e., normalized response = response/challenge concentration.

chemical compound extends to one or more other compounds (Bayer, 1954). For example, some persons sensitized to poison ivy can also become concurrently sensitized to poison oak. The chemical causing the primary reaction is referred to as the primary sensitizer or allergen. Secondary allergens are compounds that can cause cross-sensitization and are, in general, chemically related to the primary allergen (Dupuis and Benezra, 1982). Systematic studies of cross-sensitization involving iodinated and methylated compounds were initially performed by Bloch (1911). Subsequent studies by other investigators have resulted in establishing a critical role for chemical function similarities in cross-sensitization.

One study which compared the cross-sensitization potential of a series of alkyleneamines using a modified guinea pig maximization test protocol is illustrative of the approach. The procedures used were as described by Magnusson and Kligman (1969), except with the following modifications. Seven days after the challenge exposure, the cross-challenge treatment was administered. Animals were clipped as before and the test material was administered in a similar manner as in the challenge phase but at a previously untreated site. Smaller patches ($^7/_8$ square in.) were used in order to allow all of the materials to fit on the test site. Materials were applied to saturation (0.03 ml per patch). After 24 h of exposure, the patches were removed and the skin wiped free of any excess test material. Dermal responses were scored 24 and 48 h after removal of patches according to an established scale. Another group of concurrent irritation control animals were also used to differentiate dermal reactions produced by irritation from those produced by sensitization. These control animals were previously untreated (no induction treatment) but were subjected to the same cross-challenge procedure.

Using this maximization protocol, it was found that all the amines tested were capable of inducing allergic skin sensitization. Furthermore, they also cross-sensitized with most of the amines in the group. Tables 16 and 17 show the results of dermal responses in the sensitization and cross-sensitization studies. Since the concentrations used for induction were

TABLE 17

Dermal Cross-Sensitization Response of Nine Alkyleneamines
Tested with the Guinea Pig Maximinization Protocol

Induce with	Challenge with						
	EDA	DETA	AEEA	TETA	AEP	TEPA	PIP
	Normalized Response[a]						
EDA[b]	—	64	72	28	0	0	60
DETA	20	—	14	7	9	6	9
AEAA	12	6	—	3	0	0	1
TETA	10	6	7	—	0	0	3
AEP	4	9	12	4	—	0	5
TEPA	0	0	5	2	3	—	0
PIP	0	0	2	0	0	2	—
PEHA	1	0	1	0	0	1	0
HEP	0	0	1	0	0	1	0

[a] Due to differences in irritation potencies, the concentrations used in the cross-sensitization studies were different among the various alkyleneamines. In order to compare the relative cross-sensitization potencies, the responses have been normalized with respect to the concentrations of the materials used, i.e., normalized response = response/(induction concentration) (challenge concentration).

[b] Abbreviations: EDA, ethylenediamine; DETA, diethylenetriamine; AEEA, aminoethylethanolamine; TETA, triethylenetetramine; AEP, aminoethylpiperzaine; TEPA, tetraethylenepentamine; PIP, piperazine; PEHA, pentaethylenehexamine; HEP, hydroxyethylpiperazine.

defined as the highest concentration which would produce no irritation, they could provide a relative ranking of the irritation potency for these alkyleneamines. Also, as the concentration used in the induction was different for the various amines, the sensitization responses were normalized to provide a qualitative means for comparing the relative potency. The generalizations that emerge from these results are as follows:

1. There appeared to be a direct correlation of the potencies to cause dermal irritation, sensitization, and cross-sensitization in this group of nine chemically related amines.
2. Ethylenediamine (EDA) was the most potent skin sensitizer of the amines tested. Diethylenetriamine (DETA) was next in potency. Triethylenetetramine (TETA), aminoethylethanolamine (AEEA), and aminoethylpiperazine had moderate potency. Tetraethylenepentamine (TEPA), piperazine (PIP), pentaethylenehexamine (PEHA), and hydroxyethylpiperazine (HEP) had low potencies.
3. Potency generally decreased with increasing molecular size, i.e., EDA > DETA > TETA > TEPA > PEHA.
4. Cyclic amines appeared to have lower skin sensitization potencies than olefinic amines.

Although these are not rules set in concrete, they are probably acceptable for purposes of risk assessment.

It must be recognized that risk assessment for skin sensitization potential is not a simple process. Methods which employ intradermal injection in Freund's complete adjuvant, such as the guinea pig maximization test, may be more sensitive than the traditional Buehler test. However, by bypassing the skin barrier at induction, these methods can overstate the sensitization potential of weak sensitizers, as well as understate the potential of very strong sensitizers through tolerance induction.

Skin sensitizers are generally believed to show both dose response and threshold characteristics (Robinson et al., 1989). In order to properly define the risk, one must consider not only the inherent property of the chemical sensitizer, but also the degree and extent of exposure. The key factors in exposure assessment for skin sensitization include chemical dose, concentration (critical for sensitization is the dose/unit surface area, duration, body location, presence of any skin penetration aid or vehicle, primary skin irritation potential, and the extent of occlusion of the exposed skin).

XII. ASSESSMENT OF THE DERMAL ABSORPTION POTENTIAL OF INDUSTRIAL CHEMICALS IN THE WORKPLACE: CRITERION FOR THE SKIN NOTATION ON OCCUPATIONAL EXPOSURE LIMITS

Besides inhalation, skin contact is an important portal of entry for many industrial chemicals used in the workplace. Almost all occupational exposure limits (OELs), such as the Threshold Limit Values (TLVs) of the American Conference of Governmental Industrial Hygienists (ACGIH) and the Permissible Exposure Limits of the United States Occupational Safety and Health Administration (OSHA), are criteria for defining acceptable levels of inhalation exposure. For certain chemicals which are judged to have a significant potential to be absorbed from skin contact, a skin notation is appended to the OEL. The ACGIH (1990) defines the skin notation as follows:

The skin notation refers to the potential contribution to the overall exposure by the cutaneous route including mucous membranes and eye, either by airborne or, more particularly, by direct contact with the substance. This attention-calling designation is intended to suggest appropriate measures for the prevention of cutaneous absorption so that the TLV is not invalidated.

Clearly, this definition is rather loose and lacks a precise criterion for deciding if a chemical warrants a skin notation. The inclusion or exclusion of the skin notation is largely a process of subjective judgement by the deliberating standard-setting committees. Consequently, many inconsistencies exist in the skin notation which have been used by the various bodies engaged in setting the OELs (Grandjean et al., 1988; Scansetti et al., 1988). In order to bring some consistency to the skin notation process, a quantitative criterion for the biological significance of dermal absorption needs to be established. The first such criterion to predict skin notation was proposed by Hansen (1982), who based his criterion on the swelling of psoriasis scales. More recently, Fiserova-Bergerova et al. (1990) proposed another scheme which is toxicologically based and is more consistent with ACGIH's definition. In this approach, the dermal absorption potential of the industrial chemicals, as defined by fluxes predicted from physicochemical properties, are compared with the threshold values, referred to as critical fluxes. Berner and Cooper (1987) proposed the following model for calculating the fluxes.

$$J = (C/15)(0.038 + 0.153 \, P)e^{-0016MW} \tag{8}$$

where

J	=	flux
C	=	saturated aqueous solution of the chemical
P	=	octanol-water partition coefficient
MW	=	molecular weight of the chemical

TABLE 18a
Chemicals with Dermal Absorption Potential

Acetone	Isopropyl alcohol
n-Amyl acetate	Methoxychlor
Atrazine	Methyl acetate
1,3-Butadiene	Methyl *n*-amyl ketone
n-Butyl acetate	Methyl chloroform
p-tert-Butyltoluene	Methyl ethyl ketone
Cumene[a]	Methyl isoamyl ketone
Cyclohexanone[a]	Methyl isobutyl ketone
o-Dichlorobenzene	Methyl propyl ketone
1,1-Dichloroethane	Metribuzin
1,2-Dichloroethylene	Naphthalene
Diethyl ketone	Nitromethane
Diuron	Pentaerythritol
Enflurane	*n*-Propyl acetate
Ethanol	Propylene dichloride
Ethyl acetate	Strychnine
Ethyl benzene	Styrene[a]
Ethyl ether	Toluene
Ethyl formate	1,2,4-Trichlorobenzene
Halothane	Trichloroethylene
n-Heptane	Trimethyl benzene
Hexachloroethane	*o*-Xylene
n-Hexane	*m*-Xylene
Isoamyl alcohol	*p*-Xylene

[a] Skin notation in 1987–1988 TLV-BEI booklet.

From Fiserova-Bergerova, V., Pierce, J. T., and Droz, P. O.,
Am. J. Ind. Med., 17, 617–635, 1990. With permission.

Critical fluxes are determined by comparing the skin uptake rate under specified exposure conditions with the inhalation uptake rate during exposure to the time-weighted average TLV at steady-state conditions.

The specified exposure condition for the critical flux can be calculated assuming exposure of 2% of the body surface area (equivalent to the stretched palms and fingers) to a saturated aqueous solution of the chemical (Fiserova-Bergerova et al., 1990). Two reference values were recommended as criteria for the skin notation:

- *Dermal absorption potential:* If the dermal absorption (flux) of a chemical exceeds the critical flux by 30%, the chemical should be classified as possessing dermal absorption potential. Biological monitoring of these chemicals is recommended.
- *Dermal toxicity potential:* If the flux of a chemical exceeds three times the critical flux, the chemical should be classified as possessing dermal toxicity potential, and should carry a skin notation. Biological monitoring should be instituted.

Based on this scheme, a dermal absorption potential is predicted for 122 chemicals, of which only 43 currently (1991) carry a skin notation with their TLV (Table 18a). A significant dermal toxicity potential is indicated for 77 chemicals, of which only 40 of them currently have skin notations (Table 18b). Generally, a skin notation is lacking for chemicals whose TLVs are based on skin irritation. This is because the threshold for systemic toxicity is usually higher than that for irritation and the critical fluxes for chemical irritants are likely underestimated.

TABLE 18b
Chemicals With Dermal Toxicity Potential

Acetic acid	Ethyl mercaptan
Acetonitrile[a]	Formaldehyde
Acetylsalicylic acid	Formamide[a]
Allylalcohol[a]	Formic acid
2-Aminopyridine	Furfural[a]
Aniline[a]	Glycerin
Benzene	Hexylene glycol
Biphenyl	Hydrazine[a]
n-Butanol[a]	Isobutyl alcohol
sec-Butanol	Isopropylamine
n-Butyl acrylate	Lindane[a]
n-Butylamine[a]	Mesityl oxide
Carbon disulfide[a]	Methanol[a]
Carbon tetrachloride[a]	2-Methoxyethanol[a]
Chloroform	4-Methoxyphenol
Chloropicrine	Methylamine
o-Cresol[a]	*n*-Methyl aniline[a]
m-Cresol[a]	Methyl *n*-butyl ketone
p-Cresol[a]	Methyl chloride
Cyclohexanol[a]	Methyl iodide[a]
Dichloroethyl ether[a]	Morpholine[a]
Dicyclopentadiene	*p*-Nitrotoluene[a]
Dieldrin[a]	Pentachlorophenal[a]
Diethanolamine	*p*-Phenylene diamine[a]
2-Diethylaminoethanol[a]	Phenyl ether
Diethylene triamine[a]	Phenyl mercaptan
Diisopropylamine[a]	Propoxur
N,N-dimethyl aniline[a]	*n*-Propyl alcohol[a]
Dimethyl phthalate	Ronnel
Dioxane[a]	1,1,2,2-Tetrachloroethane[a]
Diphenylamine	Thioglycolic acid[a]
Dimethylacetamine[a]	*o*-Toluidine[a]
Dimethylformamide[a]	*p*-Toluidine[a]
Ethanolamine	Tributyl phosphate
2-Ethoxyethanol[a]	Trichloroacetic acid
Ethylamine	1,1,2-Trichloroethane[a]
Ethyl butyl ketone	Triethylamine
Ethylene glycol	

[a] Skin notation in 1987–1988 TLV-BEI booklet.

From Fiserova-Bergerova, V., Pierce, J. T., and Droz, P. O., *Am. J. Ind. Med.*, 17, 617–635, 1990.

XIII. DISCUSSION

Although an adequate amount of information is known about the uptake of several classes of neat chemicals (as liquids) through human skin more needs to be known about the effects of media on dermal uptake. In the workplace, employees are frequently exposed to liquid chemicals but environmental exposure almost never involves the exposure to neat substances. For example, residents are often exposed to contaminated dust which has been transported through open windows. Children are routinely exposed to soils which have been contaminated by particulate emissions from cars, smelters, foundries, incinerators, or other processes which have been deposited on yards and playgrounds. Adults and children can also be exposed to several organic contaminants in water during showering or swimming.

Progress continues to be made to allow risk assessors to make fairly reasonable estimates of the uptake of chemicals in soil. The development of models which can predict dermal bioavailability and account for media effects would represent a significant step forward. The role of concentration on the rate of dermal uptake is an area that deserves further study. Work conducted thus far suggests that the uptake will depend on the characteristics of the media (% organics, size, etc.) and the properties of the contaminant (lipophilicity, temperature). These parameters need to be quantified and a general model developed. The work of McKone (1990) represents an important step in this direction.

REFERENCES

Adams, R. M., *Occupational Skin Diseases,* 2nd ed., W. B. Saunders, Philadelphia, 1990.

American Conference of Governmental Industrial Hygienists (ACGIH), *Documentation of Threshold Limit Values,* 4th ed., Cincinnati, OH, 1990.

Andersen, M. E. and Keller, W. C., Toxicokinetic principles in relation to percutaneous absorption and cutaneous toxicity, in *Cutaneous Toxicity,* Drill, V. A. and Lazar, P., Eds., Raven Press, New York, 1984, 9–27.

Anderson, S. L. and Cassidy, J, M., Variations in physical dimensions and chemical composition of human stratum corneum, *J. Invest. Dermatol.,* 61, 30–32, 1973.

Anderson, B. D., Higuchi, W. I., and Raykar, P. V., Heterogeneity effects on permeability: partition coefficient relationships in human stratum corneum, *Pharmacol. Res.,* 5, 566–573, 1988.

Bayer, R, L., Cross-sensitization phenomena, in *Modern Trends in Dermatology,* 2nd series, MacKenna, R. M. B., Ed., Butterworths, London, 1954, 232–258.

Bartek, M. J., LaBudde, J. A., and Maibach, H. I., Skin permeability in vivo: comparison in rat, rabbit, pig and man, *J. Intest. Dermatol.,* 58, 114–123, 1972.

Bartek, M. J. and LaBudde, J. A., Percutaneous absorption *in vitro,* in *Animal Models in Dermatology,* Maibach, H. I., Ed., Churchill Livingstone, New York, 1975, 103.

Beech, J. A. Estimated worst-case trihalomethane body burden of a child using a swimming pool, *Med. Hypotheses,* 6, 303–307, 1980.

Berner, B. and Cooper, E R., Models of skin permeability in *Transdermal Delivery of Drugs,* Vol. II, Kydonieu, A. F. and Berner, B., Eds., CRC Press, Boca Raton, FL, 1987, 107–130.

Blank, I. H. Penetration of low-molecular weight alcohols into skin. I. The effect of concentration of alcohol and type of vehicle, *J. Invest. Dermatol.,* 43, 415–420, 1964.

Bloch, B., Experimentelle Studien uber das Wesen der Jodoformidiosynkrasie, *Z. Exp. Pathol. Ther.,* 9P, 509–538, 1911.

Brewster, D. W., Banks, Y. B., Clark, A. M., and Birnbaum, L. S., Comparative dermal absorption of 2,3,7,8-Tetrachlorodibenzo-p-dioxin and three polychlorinated dibenzofurans, *Toxicol. Appl. Pharmacol.,* 97, 156–166, 1989.

Bronaugh, R. L., Stewart, R. F., Congdon, E. R, and Giles, A. L., Methods for *in vitro* percutaneous absorption studies. I. Comparison with *in vivo* results, *Toxicol. Appl. Pharmacol.,* 62, 474–480, 1982.

Byard, J., Hazard assessment of 1,1,1-trichloroethane in groundwater, in *The Risk Assessment of Environmental and Human Health Hazards: A Textbook of Case Studies,* Paustenbach, D. J., Ed., John Wiley & Sons, New York, 1989, 331–344.

California Department of Health Services (CDHS), The development of applied action levels for soil contact: a scenario for the exposure of humans to soil in a residential setting, Toxic Substances Control Division, DHS, Sacramento CA, 1986.

Day, J. P., Hart, M., and Robinson, M. S., Lead in urban street dust, *Nature,* 253, 343–345, 1975.

Driver, J. H., Konz, J. J., and Whitmyre, G. K., Soil adherence to human skin, *Bull. Environ. Contam. Toxicol.,* 17(9), 1831–1850, 1989.

Dugard, P. H., Skin permeability theory in relation to measurements of percutaneous absorption in toxicology, in *Dermatotoxicology,* Marzulli, F. N. and Maibach, H. I., Ed., Hemisphere, Washington, D.C., 1983, 102.

Dugard, P. H., Walker, M., Mawdsleit, S. J., and Scott, R. C., Absorption of some glycol ethers through human skin in vitro, *Environ. Health Perspect.,* 57, 193–198, 1984.

Dupuis, G. and Benezra, C., *Allergic Contact Dermatitis to Simple Chemicals. A Molecular Approach,* Marcel Dekker, New York, 1982.

Elias, P. M., Cooper, E. R., Korc, A., and Brown, B. E., Percutaneous transport in relation to stratum corneum structure and lipid composition, *J. Invest. Dermatol.,* 76, 297–301, 1981.

Emmett, E. A., Toxic responses of the skin, in *Cassarett and Doull's Toxicology, The Basic Science of Poisons,* 3rd ed., Klassen, C. D., Amdur, M. O., and Doull, J., Eds., McMillan, New York, 1986, 412–431.

Feldmann, R. J. and Maibach, H. I., Percutaneous penetration of steroids in man, *J. Invest. Dermatol.,* 52, 89–94, 1969.

Feldmann, R. J. and Maibach, H. I., Absorption of some organic compounds through the skin in man, *J. Invest. Dermatol.,* 54, 399–404, 1970.

Feldmann, R. J. and Maibach, H. I., Percutaneous penetration of some pesticides and herbicides in man, *Toxicol. Appl. Pharmacol.,* 28, 126–132, 1974.

Fiserova-Bergerova, V., Pierce, J. T., and Droz, P. O., Dermal absorption potential of industrial chemicals: criteria for skin notation, *Am. J. Ind. Med.,* 17, 617–635, 1990.

Flynn, G. L., Physicochemical determinants of skin absorption, in *Principles of Route-to-Route Extrapolation for Risk Assessment,* Gerrity, T. R., and Henry, C. J., Eds., Elsevier, 1990, 93–127.

Fries, G. F. and Paustenbach, D. J., Evaluation of potential transmission of 2,3,7,8-tetrachlorodibenzo-p-dioxin contaminated incinerator emissions to humans via food, *J. Toxicol. Environ. Health,* 29, 1–43, 1989.

Grandjean, P., Berlin, A., Gilbert, M., and Penning, W., Preventing percutaneous absorption of industrial chemicals, the "skin" notation, *Am. J. Ind. Med.,* 14, 97–107, 1988.

Guy, R. H. and Maibach, H. I., Correction factors for determining body exposure from forearm percutaneous absorption data, *J. Appl. Toxicol.,* 4, 26–28, 1984.

Hansen, C. M., The Absorption of Liquids into the Skin, Rep. T3-82, Scandinavian Paint Painting Ink Research Institute, Horsholm, Denmark, 1982.

Hawley, J., Assessment of health risk from exposure to contaminated soil, *Risk Anal.,* 5, 289–302, 1985.

Jo, W. K., Weisel, C. P., and Lioy, P. J., Routes of chloroform exposure and body burden from showering with chlorinated tap water, *Risk Anal.,* 10, 575–580, 1990.

Kao, J., Hall, J., and Helman, G., *In vitro* percutaneous absorption in mouse skin: influence of skin appendages, *Toxicol. Appl. Pharmacol.,* 94, 93–103, 1988.

Kasting, G. B., Smith, R. L., and Cooper, E. R., Effect of lipid solubility and molecular size on percutaneous absorption, in *Pharmacology and the Skin,* Vol. 1, Shroot, B. and Schaefer, H., Eds., S. Karger, Basel, 1987, 138–153.

Keenan, R. E., Sauer, M. M., Lawrence, F. H., Rand, E. R., and Crawford, D. W., Examination of potential risks from exposure to dioxin in sludge used to reclaim abandoned strip mines, in *The Risk Assessment of Environmental and Human Health Hazards: A Textbook of Case Studies,* Paustenbach, D. J., Ed., John Wiley & Sons, New York, 1989, 935–998.

Kimbrough, R. D., Falk, H., Stehr, P., and Fries, G., Health implications of 2,3,7,8-tetrachlorodibenzo-p-dioxin (TCDD) contamination of residential soil, *J. Toxicol. Environ. Health,* 14, 47–93, 1984.

Kissel, J. C. and McAvoy, D. R., Reevaluation of the dermal bioavailability of 2,3,7,8-TCDD in soil, *Haz. Waste Haz. Mater.,* 6, 231–248, 1989.

Kissel, J. C. and Robarge, G. M., Assessing the elimination of 2,3,7,8-TCDD from humans with a physiologically-based pharmacokinetic model, *Chemosphere,* 17, 2017–1037, 1988.

Klecak, G., Identification of contact allergens: predictive tests in animals, in *Dermatotoxicology,* 3rd ed., Marzulli, F. N. and Maibach, H. I., Eds., Hemisphere, Washington, D.C., 1983, 193–236.

Knaak, J., Iwati, I., and Maddy, K. T., The worker hazard posed by re-entry into pesticide treated foliage: development of safe re-entry times, with emphasis on chlorthiophos and carbosulfan, in *The Risk Assessment of Environmental and Human Health Hazards: A Textbook of Case Studies,* Paustenbach, D. J., Ed., John Wiley & Sons, New York, 1989, 797.

Lepow, M. L., Bruckman, L., Gillette, M., Markowitz, S., Robino, R., and Kapish, J., Investigations into sources of lead in the environment of urban children, *Environ. Res.,* 10, 415–426, 1975.

Lipsky, D., Assessment of potential health hazards associated with PCDD and PCDF emissions from a municipal waste combustor, in *The Risk Assessment of Environmental Health Hazards,* Paustenbach, D. J., Ed., John Wiley & Sons, New York, 1989, 631–678.

Maddy, K. T., Wang, R. G., and Winger, C. K., Dermal exposure monitoring of mixers, loaders, and applicators of pesticides in California, HS-1069, *Worker Health and Safety Unit,* State of California, 1983, 1–7.

Magnusson, B. and Kligman, A. M., The identification of contact allergens by animal assay: the guinea pig maximization test, *J. Invest. Dermatol.,* 57, 268–276, 1969.

Maibach, H. I., Feldmann, R. J., Milby, T. H., and Serat, W. F, Regional variation in percutaneous penetration in man, *Arch. Environ. Health,* 23, 208–211, 1971.

Marzulli, F. N. and Tregear, R. T., Identification of a barrier layer in the skin, *J. Physiol.,* 157, 52–53, 1961.

McConnell, E. G., Lucier, R., Rumbaugh, P., Albro, D., Harvan, J., Hass, M., and Harris, Dioxin in soil: bioavailability after ingestion by rats and guinea pigs, *Science,* 223, 1077–1079, 1984.

McDougal, J. N., Jepson, G. W., Clewell, H. J., III, MacNaughton, M. G., and Andersen, M. E., A physiological pharmacokinetic model for dermal absorption of vapors in the rat, *Toxicol. Appl. Pharmacol.,* 85, 286–294, 1986.

McKone, J. E., Dermal uptake of organic chemicals from a soil matrix, *Risk Anal.,* 10, 407–419, 1990.

McKone, J. E. and Daniels, J., Estimating human exposure through multiple pathways from air, water, and soil, *Regul. Toxicol. Pharmacol.,* 13, 36–61, 1991.

McLaughlin, T., *Review of Dermal Absorption,* Exposure Assessment Group, U.S. Environmental Protection Agency (EPA), Washington, D.C., 600/8-84/033, 1984.

Michaels, A. S., Chandrasekaran, S. K., and Shaw, J. E., Drug permeation through human skin: theory and *in vitro* experimental measurements, *AIChE,* 2, 985–996, 1975.

Ostrenga, J., Steinmetz, C., and Poulsen, B., Significance of vehicle composition. I. Relationship between topical vehicle composition, skin penetrability, and clinical efficacy, *J. Pharm. Sci.,* 60, 1175–1183, 1971.

Pannatier, A., Jenner, B., Testa, B., and Etter, J. C., The skin as a drug-metabolizing organ, *Drug Metab. Rev.,* 8, 319–343, 1978.

Paschoud, J. M., Externe Kontaktallergene. Alphabetisch geordnete Uebersicht der Fachliteratur 1960 bis 1965, *Hautarzt,* 18, 145–149, 1967.

Paterson, S. and Mackay, D., A steady-state fugacity-based pharmacokinetic model with simultaneous multiple exposure routes, *Environ. Toxicol. Chem.,* 6, 395–408, 1987.

Paustenbach, D. J., Shu, H. P., and Murray, F. J., A critical examination of assumptions used in risk assessment of dioxin contaminated soil, *Regul. Toxicol. Pharmacol.,* 6, 284–307, 1986.

Paustenbach, D. J., Assessing the potential environmental and human health risks of contaminated soil, *Comments Toxicol.,* 1, 185–220, 1987.

Paustenbach, D. J., Assessment of the developmental risks resulting from occupational exposure to select glycol ethers within the semiconductor industry, *J. Toxicol. Environ. Health,* 23, 29–75, 1988.

Paustenbach, D. J., A survey of environmental risk assessment, in *The Risk Assessment of Environmental and Human Health Hazards: A Textbook of Case Studies,* Paustenbach, D. J., Ed., John Wiley & Sons, New York, 1989, 27–124.

Piotrowski, J., Quantitative estimation of aniline absorption through the skin in man, *J. Hyg. Epidemiol. Microbiol. Immunol.,* 1, 1–23, 1957.

Piotrowski, J., Evaluation of exposure to phenol: absorption of phenol vapor in the lungs and through the skin and excretion of phenol in urine, *Br. J. Ind. Med.,* 28, 172–178, 1971.

Piotrowski, J., Exposure Tests for Organic Compounds in Industry, National Institute for Occupational Safety and Health (NIOSH), Publ. 76-111, Cincinnati, OH, 1973.

Piotrowski, J., Exposure Tests for Organic Compounds in Industrial Toxicology, National Institute for Occupational Safety and Health (NIOSH), Publ. 77-144, Cincinnati, OH, 1977.

Poiger, H. and Schlatter, C. H., Influence of solvents and absorbents on dermal and intestinal absorption of TCDD, *Food Cosmet. Toxicol.,* 18, 477–481, 1980.

Popendorf, W. J., Spear, R. C., Leffingwell, J. T., Yager, J., and Kahn, E., Harvester exposure to Zolone (phosalone) residues in peach orchards, *J. Occup. Med.,* 21, 189–194, 1979.

Que Hee, S. S., Peace, B., Scott, C. S., Boyle, J. R., Bornschein, R. L., and Hammond, P. B., Evolution of efficient methods to sample lead sources, such as house dust and hand dust, in the homes of children, *Environ. Res.,* 38, 77–95, 1985.

Raykar, P. V., Fung, M., and Anderson, B. D., The role of protein and lipid domains in the uptake of solutes by human stratum corneum, *Pharm. Res.,* 5, 140–150, 1988.

Robinson, M. K., Stotts, J., Danneman, P. J., Nusair, T. L., and Bay, P. H. S., A risk assessment process for allergic contact sensitization, *Food Chem. Toxicol.,* 27, 479–489, 1989.

Roels, H. A., Buchet, J. P., Lauwerys, R. R., Braux, P., Claeys-Thoreau, F., Lafontaine, A., and Verduyn, G., Exposure to lead by the oral and pulmonary routes of children living in the vicinity of a primary lead smelter, *Environ. Res.,* 22, 81–94, 1980.

Rongone, E. L., Skin structure, function, and biochemistry, in *Dermatotoxicology,* Marzulli, F. N. and Maibach, H. I., Eds., Hemisphere, Washington, D.C., 1983, 2.

Rougier, A. D., Dupuis, D., Lotte, C., and Roguet, R., The measurement of the stratum corneum reservoir. A predictive method for in vivo percutaneous absorption studies: influence of application time, *J. Invest. Dermatol.,* 84, 66–68, 1985.

Sato, A. and Nakajima, T., Differences following skin or inhalation exposure in the absorption and excretion kinetics of trichloroethylene and toluene, *Br. J. Ind. Med.,* 35, 43–49, 1978.

Sato, A. and Nakajima, T., A structure-activity relationship of some chlorinated hydrocarbons, *Arch. Environ. Health,* 34, 69–75, 1979.

Sayre, J. W., Charney, E., Vostal, J., and Pless, B., House and land dust as a potential source of childhood lead exposure, *Am. J. Dis. Child,* 127, 167–170, 1974.

Scansetti, G., Pilatto, G., and Rubino, G. F., Skin notation in the context of workplace exposure standards, *Am. J. Ind. Med.,* 14, 725–732, 1988.

Schaefer, H. and Schalla, W., Kinetics of percutaneous absorption of steroids, in *Percutaneous Absorption of Steroids,* Mauvais-Jarvis, P., Vickers, D. F. H., and Wepierre, J., Eds., Academic Press, New York, 1980, 54.

Scheuplein, R. J., Percutaneous absorption: theoretical aspects, in *Percutaneous Absorption of Steroids,* Mauvais-Jarvis, P. J., Vickers, C. F. H., and Wepierre, J., Eds., Academic Press, New York, 1980, 9.

Scheuplein, R. J., Mechanism of percutaneous absorption. I. Routes of penetration and the influence of solubility, *J. Invest. Dermatol.,* 45, 334, 1965.

Scheuplein, R. J. and Blank, I. H., Permeability of the skin, *Physiol. Rev.,* 51, 702–747, 1971.

Scheuplein, R. J., Mechanism of percutaneous absorption. II. Transient diffusion and the relative importance of various routes of skin penetration, *J. Invest. Dermatol.,* 48, 79–88, 1967.

Schlede, E., Maurer, Th., Potokar, M., Schmidt, W. M., Schulz, K. H., Roll, R., and Kayser, D., A differentiated approach to testing skin sensitization. Proposal for a new test guideline for skin sensitization, *Arch. Toxicol.,* 63, 81–84, 1989.

Scott, K., Wechsler, A. E., Stevens, J., Wood, M., and Callahan, M., Identification and evaluation of waterborne routes of exposure from other than food and drinking water, EPA-440/4-79-016, U. S. Environmental Protection Agency, Washington, D.C., 1979.

Sheehan, P., Meyer, D. M., Sauer, M. M., and Paustenbach, D. J., Assessment of the human health risks posed by exposure to chromium contaminated soils at residential sites, *J. Toxicol. Environ. Health,* in press.

Shu, H. P., Teitelbaum, P., Webb, A. S., Marple, L., Brunck, B., Dei Rossi, D., Murray, F. J., and Paustenbach, D. J., Bioavailability of soil bound TCDD: dermal bioavailability in the rat, *Fundam. Appl. Toxicol.,* 10, 648–654, 1988a.

Shu, H. P., Teitelbaum, P., Webb, A. S., Marple, L., Brunck, B., Dei Rossi, D., Murray, F. J., and Paustenbach, D. J., Bioavailability of soil-bound TCDD: oral bioavailability in the rat, *Fundam. Appl. Toxicol.,* 10, 335–343, 1988b.

Skowronski, G. A., Turkall, R. M., and Abdel, Rahman, M. S., Soil adsorption alters bioavailability of benzene in dermally exposed male rats, *Am. Ind. Hyg. Assoc. J.,* 49, 506–511, 1988.

Skowronski, G. A., Turkall, R. M., and Abdel-Rahman, M. S., Effects of soil on percutaneous absorption of toluene in male rats, *J. Toxicol. Environ. Health,* 26, 373–384, 1989.

Snyder, W. S., *Report of the Task Group on Reference Man, International Commission of Radiological Protection No. 23,* Pergamon Press, New York, 1975.

Stewart, R. D. and Dodd, H. C., Absorption of carbon tetrachloride, trichloroethylene, tetrachloroethylene, methylene chloride, and 1,1,1-trichloroethane through the human skin, *Am. J. Ind. Hyg.,* 25, 439–446, 1964.

Surber, C., Wilhelm, K. P., Maibach, H. I., Hall, L. L., and Guy, R. H., Partitioning of chemicals into human stratum corneum: implications for risk assessment following dermal exposure, *Fundam. Appl. Toxicol.,* 15, 99–107, 1990.

Travis, C. C. and Hester, S. T., Background exposure to chemicals: what is the risk?, *Risk Anal.,* 10, 463–466, 1990.

Tsuruta, H., Percutaneous absorption of organic solvents. III. On the penetration rates of hydrophobic solvents through the excised rat skin, *Ind. Health,* 20, 335–345, 1962.

Umbreit, T. H., Dhun, P., and Gallo, M. A., Bioavailability of Dioxin in soil from a 2, 4, 5-T manufacturing site, *Science,* 232, 497–499, 1986.

U.S. Environmental Protection Agency (USEPA), *Risk Assessment Guidance for Superfund. Vol. I. Human Health Evaluation Manual* (Part A), Interim Final. Office of Emergency and Remedial Response. Washington, D.C., EPA 540/1-89/002, 1989b.

U.S. Environmental Protection Agency (USEPA), Exposure Factors Handbook, *Office of Remedial Response,* EPA/540/1-88/001, Washington, D.C., 1989a.

Van den Berg, M., Olie, K., and Hutzinger, O., Uptake and selective retention in rats of orally administered chlorinated dioxins and dibenzofurans from fly-ash and fly-ash extract, *Chemosphere,* 12, 537–544, 1984.

Vickers, C. F. H., Existence of reservoir in the stratum corneum, *Arch. Dermatol.,* 88, 20–23, 1963.

Wepierre, J. and Marty, J. P., Percutaneous absorption of drugs, *Trends Pharmacol. Sci.,* Inaugural issue, 23–26, 1979.

Wester, R. C., Noonan, P. K., and Maibach, H. I., Frequency of application on percutaneous absorption of hydrocortisone, *Arch. Dermatol.,* 113, 620–622, 1977.

Wester, R. C. and Maibach, H. I., Percutaneous absorption of organic solvents, in *Occupational and Industrial Dermatology,* 2nd ed., Yearbook Medical Publishers, Chicago, 1987, 213–226.

Wester, R. C. and Noonan, P. K., Relevance of animal models for percutaneous absorption, *Int. J. Pharmacol.,* 7, 99–110, 1980a.

Wester, R. C., Noonan, P. K., and Maibach, H. I., Percutaneous absorption of hydrocortisone increases with long-term administration, *Arch. Dermatol.,* 116, 186–188, 1980b.

Wester, R. C. and Maibach, H. I., *In vivo* methods for percutaneous absorption measurements, in *Percutaneous Absorption, Mechanisms, Methodology, Drug Delivery,* Vol. 6, Calnan, C. D. and Maibach, H. I., Eds., Marcel Dekker, New York, 1985, 245–266.

Zwiacher, W. E. and Dennison, W. J., Multi-pathway health risk assessment input parameters guidance document, South Coast Air Quality Management District, 1988.

Chapter 24

INTERSPECIES EXTRAPOLATION of TOXICOLOGICAL DATA

Curtis C. Travis

TABLE OF CONTENTS

I. INTRODUCTION

A fundamental problem in toxicology is the extrapolation of observed experimental results from animal species to humans. Lacking detailed information on interspecies differences, researchers frequently assume that experimental results can be extrapolated between species when administered dosage is standardized as either mg/kg body weight per day (body weight scaling) or mg/m^2 per day (surface area scaling). Several investigators have argued for the efficacy of one or the other of these procedures (Pinkel, 1958; Freireich et al., 1966; Crouch and Wilson, 1978; Hoel, 1979; Crump and Guess, 1980; Hogan and Hoel, 1982; MRI, 1986; FASEB, 1986; and Travis and White, 1988). However, neither of these extrapolation procedures is the most appropriate for all compounds. These procedures do not take into account species- and chemical-specific data which also should be used in risk assessment, when available, to increase the accuracy of the results. In the absence of species- and chemical-specific data, body weight or surface area extrapolations are used with the implicit knowledge that they are only approximately correct.

Andersen (1987) and the National Academy of Sciences (NRC, 1986, 1987) point out that scaling should depend on the kinetic behavior of the particular compound and its mechanism of toxicity. They distinguish three categories of toxic mechanisms depending on whether a parent compound, stable metabolite, or reactive metabolite produces the toxic response. They suggest that if a carcinogenic response is proportional to the area under the curve of the concentration of the toxic moiety in the target tissue, then the appropriate scaling laws for cancer risk assessment interspecies extrapolation are mg/kg$^{0.7}$/day for direct-acting compounds and stable metabolites and mg/kg/y for reactive metabolites.

Travis et al. (1990a) have proposed an alternate definition of dose to target tissue. They proposed that individuals from different species will receive the same dose to target tissue if, and only if, the time profiles in *physiological time* of the concentration of the toxic moiety in the target tissue are the same. Support for this definition is based upon observations that the rate of biological events across species is approximately constant when compared per unit of physiological time (Dedrick, 1973; Boxenbaum, 1986; Mordenti, 1986; Yates and Kugler, 1986; and Lindstedt, 1987). Thus, the pharmacodynamic process should be approximately constant in physiological time across species if dose to target tissue for different species is also constant in physiological time. Based on this definition of dose to target tissue, Travis et al. (1990a) establish that regardless of the mechanism of action of a chemical (direct-acting compound, reactive metabolite, or stable metabolite), the most appropriate interspecies scaling law for toxic effect is mg/kg$^{0.75}$/day.

In this chapter, theoretical and biological data supporting a three fourths power scaling law are reviewed. Before beginning, two aspects of a "universal interspecies scaling metric" are acknowledged. The first aspect is the impossibility to *precisely* predict toxic effects in man based on experimental results in animals. Lack of knowledge concerning interspecies differences precludes such an attempt. The second aspect is that no single interspecies scaling metric will be correct for all chemicals. These two statements, however, do not negate the need for a universal interspecies scaling metric. The general situation in risk assessment is that the assessor has no prior knowledge of a chemical's pharmacokinetics or mechanism of action in either animals or man. In this situation, the most appropriate interspecies scaling metric is that which provides a "Maximum Likelihood Estimate" of toxic effects in man. Such a "universal interspecies scaling metric" will not be exactly correct for all compounds. The attempt is to find a scaling metric which, when considered over the universe of chemicals, will be correct most often.

The best illustration of a maximum likelihood estimate is the reanalysis by Travis and White (1988) of data on acute toxic doses of 27 anticancer agents in mice, rats, dogs, monkeys, and humans. Travis and White found that the best overall interspecies scaling

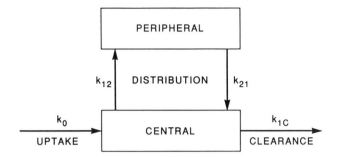

FIGURE 1. Diagram of a data-based model that has been used to describe the pharmacokinetics of styrene. The model has a central blood compartment and a peripheral tissue compartment. Compartment volumes and rate constants are adjusted to yield model-generated curves that fit the empirical data. (Adapted from Young et al., 1978; Ramsey et al., 1980.)

metric for toxicity data was the 0.73 power of body weight with 95% confidence bounds of 0.69 and 0.77. Individual linear regressions performed on each of the 27 compounds found allometric exponents which ranged from 0.53 to 0.96. Thus, while 0.73 is the maximum likelihood estimate of the proper allometric exponent for interspecies scaling of toxicity data, none of the 27 compounds actually scaled with this exponent. However, despite no prior knowledge about a compound, a 0.73 scaling law would be the best estimate of its allometric exponent (Travis, 1990). In this light, the question of interspecies extrapolation of toxicological data is investigated.

The next section begins with a discussion of the use of physiologically based pharmacokinetic (PB-PK) models to investigate the question of an appropriate interspecies scaling metric for an administered dose.

II. PHARMACOKINETIC MODELS

Pharmacokinetics is the study of the absorption, distribution, metabolism, and elimination of chemicals in animals and humans (Gibaldi and Perrier, 1975). A predictive pharmacokinetic model is a set of mathematical equations that can be used to describe the time and course of a parent chemical or metabolite in an animal system. PB-PK models depict the time-dependent distribution and fate of chemicals in living organisms by integrating information on an applied dose (i.e., its magnitude and route of chemical exposure), the physiological composition of the exposed species, and the physicochemical parameters of the chemical of interest to estimate effective dose of the parent compound or its metabolites to a target tissue(s) (Andersen, 1987; Travis, 1987). There are two types of pharmacokinetic models: data-based and physiologically based (NRC, 1986). Physiologically based models will be discussed in the next section.

A. DATA-BASED MODELS
A data-based pharmacokinetic model divides an animal system into a series of compartments which do not represent real, identifiable anatomical regions of the body. In a simply, data-based model, the body is divided into two basic compartments, a central compartment and a peripheral compartment. The central compartment is in equilibrium with arterial and venous blood flows. The peripheral compartment is connected to the central compartment through a series of rate constants that describe the flow of materials in both directions (Clewell and Andersen, 1987). The time course of chemicals in each compartment is represented by a single differential equation (NRC, 1986). Figure 1 shows a simple data-based model that was used to describe the pharmacokinetics of styrene (a volatile organic)

in rats (Young et al., 1978) and humans (Ramsey et al., 1980). A similar model has been used (Reitz et al., 1982) to describe ethylene dichloride kinetics.

The concentration of a parent compound or its metabolite(s) in a given compartment as a function of time is defined by experimental data, not by physiological constraints. Compartment volumes and rate constants are determined by trial and error so that model predictions fit empirical data. Traditionally, these models have been popular because the two equations describing intercompartmental transfer are easy to formulate and solve. Data-based models, however, are only useful for interpolation and limited extrapolation in the *same* species. Because the compartments in these models do not correspond to physiologically distinguishable entities, they do not allow for extrapolation *across* species.

B. PHYSIOLOGICALLY BASED MODELS

Unlike data-based models, predictive PB-PK models rely on actual physiological parameters, such as breathing rates, blood flow rates, and tissue volumes, to describe metabolic processes in physiologically realistic compartments (Andersen et al., 1987; Paustenbach et al., 1988; Ramsey and Anderson, 1984; Reitz et al., 1987; Travis, 1987; Ward et al., 1988, Travis and Hattemer-Frey, 1990). Tissue groups or compartments commonly used in a PB-PK model include (1) organs such as the brain, kidneys, and viscera; (2) muscle; (3) fat; and (4) metabolic organs (principally the liver). Tissue compartments are connected by arterial and venous blood flow, and each compartment is characterized by a unique set of differential equations. Rate constants that describe the flow of materials between compartments and the rate of change in the chemical concentration of each compartment are directly proportional to blood flow, tissue solubility, and organ volumes (NRC, 1986).

PB-PK quantitatively relate exposure concentrations (in air, water, or food) to concentrations of a parent compound or its metabolite in various tissues, thus allowing the relationship between applied dose and effect dose to the target tissue(s) to be characterized. Chemical input can occur either via inhalation or ingestion. Losses can occur via metabolism, exhalation, and excretion (Kissel and Robarge, 1988). Metabolism can be described using a linear, unsaturable component and/or a nonlinear, saturable metabolic term.

One fundamental advance of PB-PK models is that by simply using species-specific physiological, biochemical, and metabolic parameters, the *same* model can be used to describe the kinetics of chemical transport and metabolism in mice, rats, and humans (Clewell and Andersen, 1987). On the other hand, no one model can determine the pharmacokinetics of all chemicals, because the number of and interrelationship between the compartments varies from chemical to chemical.

1. Example of a Physiologically Based Model

A PB-PK model which has been frequently applied to describe the inhalation of volatile organics is shown in Figure 2 (Andersen et al., 1987; Ramsey and Andersen, 1984; Travis et al., 1989; Ward et al., 1988; Travis et al., 1990b). Tissue groups used in this model include organs, muscle, fat, and the liver. The model is described mathematically by a set of differential equations which calculate the rate of change in the amount of chemical in each compartment. If the metabolism of a chemical depends on an enzyme whose supply is limited, then the ability of the liver to metabolize that chemical is also limited, because the pathway along which the chemical is metabolized will become saturated over time as the supply of enzyme is depleted. This type of metabolism is referred to as nonlinear or saturable metabolism and is described by the following equation:

$$\frac{dA_m}{dt} = \frac{V_{max}\, C_{ven}}{(K_m + C_{ven})} \tag{1}$$

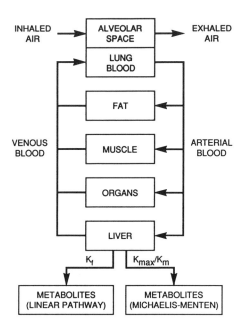

FIGURE 2. Diagram of a physiologically based pharmacokinetic model that has been used to describe the pharmacokinetics of benzene in mice, rats, and humans. The model has tissue groups, breathing rates, blood flow rates, and metabolic parameters. (From Travis, C. C., Quillen, J. L., and Arms, A. D., *Toxicol. Appl. Pharmacol.*, 102, 400–420, 1990b. With permission.)

where A_m is the amount of chemical metabolized in the liver (mg), C_{ven} the concentration of parent compound in mixed venous blood (mg/L blood), V_{max} the maximum metabolic rate constant (mg/h) and K_m is the Michaelis-Menten (nonlinear) rate constant (mg/L blood). K_m is defined as the venous blood concentration at which metabolic activity is half of V_{max}.

When the rate of metabolite formation is nonsaturable and depends only on the concentration of chemical in venous blood (C_{ven}), then a second mathematical representation of metabolism is used:

$$\frac{dA_m}{dt} = K_f \, V \, C_{ven} \qquad (2)$$

where V is the volume (L) of the liver, and K_f is the linear metabolic rate constant (h^{-1}). Linear or nonsaturable metabolism occurs whenever there is an endless supply of the enzyme(s) required to metabolize a given compound. Hence, in general, a PB-PK model can have three kinetic constants that describe metabolism: the linear metabolic constant, K_f, and the two Michaelis-Menten (nonlinear) constants, V_{max} and K_m.

C. SELECTION OF TISSUE COMPARTMENTS

The number of compartments necessary to adequately represent a living organism depends on the objective of the study. In general, however, the number of compartments is kept to as few as possible. A method for combining body tissues into distinct physiological compartments has been described (Fiserova-Bergerova, 1983); tissues in which chemical concentrations increase or decrease at the same rate are treated as a single unit. The biological half-life of a chemical in a tissue (min^{-1}) is determined by dividing the flow rate of the chemical through the tissue by the volume of distribution in the tissue. Flow rate through

TABLE 1
Physiological Parameters for Use in
Pharmacokinetics

	Mouse	Rat	Human
Body weight (kg)	0.024	0.25	70.0
Alveolar ventilation (l/hr)	1.33	7.96	353.5
Total blood flow rate (l/hr)	1.33	5.16	371.6
Blood flow fractions			
Blood flow fraction in liver	0.25	0.25	0.25
Blood flow fraction in fat	0.09	0.09	0.05
Blood flow fraction in organs	0.51	0.51	0.51
Blood flow fraction in muscle tissues	0.15	0.15	0.19
Tissue group volume fractions			
Volume fraction in liver	0.06	0.04	0.04
Volume fraction in fat	0.10	0.07	0.20
Volume fraction in organs	0.05	0.75	0.62
Volume fraction in muscle tissues	0.70	0.75	0.62

the tissue is equal to the blood perfusion rate (L/min) divided by the tissue/blood partition coefficient. The volume of distribution (L) is a hypothetical volume of body fluid required to dissolve the total amount of chemical at the concentration equal to that in blood or plasma (Fiserova-Bergerova, 1983). Table 1 lists tissues, tissue volumes (V), and blood perfusion rates (Q_b) recommended for a 70 kg human. Although the half-life rate constants (Q_b/V) for each tissue group span a wide range, tissues with rate constants that differ by less than a factor of five can be grouped into one compartment (Fiserova-Bergerova, 1983). The tissues in the fat and muscle groups, for example, share this property.

III. PHYSIOLOGICALLY BASED PHARMACOKINETIC MODELS

Physiologically based pharmacokinetic models divide the body into physiologically realistic compartments connected by the arterial and venous blood flow pathways (Gerlowski and Jain, 1983; Ramsey and Andersen, 1984; Andersen et al., 1987; Paustenbach et al., 1988; and Ward et al., 1988). The tissue groups generally include (1) organs such as brain, kidney, and viscera, (2) muscle, (3) fat, and (4) metabolic organs (principally liver). The models use actual physiological parameters such as breathing rates, blood flow rates, blood volumes, and tissue volumes to describe the pharmacokinetic process. These physiological parameters are coupled with chemical specific parameters such as blood/gas partition coefficients, tissue/blood partition coefficients, and metabolic constants to predict the dynamics of a compound's movement through an animal system. An advantage of the physiologically based model is that by simply using the appropriate physiological, biochemical, and metabolic parameters, the same model can be utilized to describe the dynamics of chemical transport and metabolism in any species, including mice, rats, and humans.

The mathematical equations that define the physiologically based pharmacokinetic model used in this paper are presented in Appendix A. The family of pharmacokinetic models characterized by Equations A1 through A15 can be used to describe the transport and metabolism of organics in any mammalian species once the species-specific physiological and metabolic parameters are specified. This model is a variant of the Ramsey and Andersen styrene model (Ramsey and Andersen, 1984), and has been used successfully to characterize the pharmacokinetics of a number of volatile organics (Ramsey and Andersen, 1984; Andersen et al., 1987; Hattis, 1987; Paustenbach et al., 1988; Ward et al., 1988; Travis et al., 1990b) in multiple species. Models with similar structure have also been used to model dioxin (Andersen et al. 1989), PCBs (Lutz et al., 1977), methotrexate (Dedrick et al., 1970), and other nonvolatile organics. It should be understood, however, that the model is not universally valid and may not adequately reflect pharmacokinetic processes related to the toxicity of some compounds.

IV. SCALING PHYSIOLOGICAL AND METABOLIC PARAMETERS

Many of the physiological and metabolic parameters used in pharmacokinetic modeling are directly correlated to the body weight of the particular organism (Adolph, 1949). These physiological parameters generally vary with body weight according to a power function expressed as:

$$y = a \, BW^b \tag{3}$$

where y is a physiological parameter of interest, and a and b are constants (Adolph, 1949; Schmidt-Nielsen, 1970, 1984; and Lindstedt, 1987). If the constant b equals one, the physiological parameter y correlates directly with body weight. If the constant b equals two thirds, the parameter y correlates with surface area. A brief review of the empirical scaling laws for physiological and metabolic parameters are found in Equations A1 through A15.

A. ORGAN VOLUMES
Organ volumes tend to scale across species with the first power of body weight (Schmidt-Nielsen, 1984; NRC, 1986). Examples include total blood volume which scales across species with the 1.02 power of body weight (Stahl, 1967) and the mass of the mammalian heart which scales with the 0.98 power of body weight (Prothero, 1979). The liver is an exception scaling with 0.87 power of body weight (Stahl, 1965). Following the National Academy of Sciences (NRC, 1986), we will assume that the appropriate scaling law for volume of tissue group i is

$$V_i = V_{i\theta}BW^{1.0} \tag{4}$$

where $V_{i\theta}$ is a species-independent allometric constant.

B. CARDIAC OUTPUT
Cardiac output is defined as the volume of blood pumped by each ventricle of the heart per minute. There is considerable evidence that cardiac output is related to metabolic rate (Guyton, 1971) and that metabolic rates across species are related to the three fourths power of body weight (Kleiber, 1961; Holt et al., 1968; White et al., 1968, and Schmidt-Nielsen, 1970). The most commonly assumed scaling law for cardiac output has the form:

$$Q_{b\theta} = Q_{b\theta} \, BW^{0.75} \tag{5}$$

where $Q_{b\theta}$ is a species-independent allometric constant.

Percent of cardiac output distributed to different organs is approximately constant across species (Arms and Travis, 1988). Thus, cardiac output, Q_i, to tissue group i has the form:

$$Q_i = Q_{i\theta} \, BW^{0.75} \tag{6}$$

where $Q_{i\theta}$ is a species-independent allometric constant.

C. ALVEOLAR VENTILATION

Ventilation is a cyclic process of circulation and exchange of gases in the lungs that is basic to respiration. Total ventilation or minute volume is defined as the volume of air exhaled per minute. The fraction of minute volume available for gas exchange in the alveolar compartments is termed the alveolar ventilation rate. Minute volume and, hence, alveolar ventilation have been shown to scale across species with the $^3/_4$ power of body weight (Guyton, 1947; Adolph, 1949; Stahl, 1967). The most commonly assumed scaling law for alveolar ventilation rate has the form:

$$Q_{alv} = Q_{alv\theta} \, BW^{0.75} \qquad (7)$$

where $Q_{alv\theta}$ is a species-independent allometric constant.

D. RENAL CLEARANCE

Clearance is the volume of blood cleared of a substance per unit of time. Renal clearance relates the kidneys' rate of elimination of a given compound to the concentration of the compound in the blood. Adolph (1949) first showed that renal clearance of insulin in four species scaled with body weight to the $^3/_4$ power. Studies by Brody (1945), Edwards (1975), Lindstedt and Calder (1981), Boxenbaum (1982), Schmidt-Nielsen (1984), and Mordenti (1986) support a general scaling law for renal clearance:

$$K_r = K_{r\theta} \, BW^{0.75} \qquad (8)$$

where $K_{r\theta}$ is a species-independent allometric constant.

E. METABOLIC PARAMETERS

Rubner (1883) studied the relationship between the body weight of dogs and their metabolic rates and found that heat production (an indication of metabolic rate) divided by total body surface area (calculated by taking body weight to the two thirds power) remained constant in dogs of various sizes. This finding established body surface area as the basis for predicting metabolic rates. Unfortunately, Rubner's (1883) results were for *intra*specific not *inter*specific scaling. Kleiber (1932, 1947) and Brody (1945) examined several species of animals covering a broad size range (mice to elephants) to ascertain if body weight or surface area ($BW^{0.67}$) was the appropriate interspecific scaling factor for metabolic rates. Results indicated that metabolic rates increased with the three fourths power of body weight (Figure 3). This relationship has been verified by many investigators (Kleiber, 1932, Brody, 1945, Benedict, 1938; McMahon, 1973; Schmidt-Nielsen, 1984; Lindstedt and Calder, 1981; and Lindstedt, 1987). Based on these studies, the $^3/_4$ power of body weight is now well accepted as the correct scaling factor for metabolic constants (Boxenbaum and Ronfeld, 1983; Prothero, 1980; Schmidt-Nielsen, 1984).

Direct empirical data on interspecies scaling of metabolic enzymatic activity are limited. Cytochrome oxidase has been found to scale with the $^3/_4$ power of body weight (Kunkel et al., 1956; Jansky, 1961; Jansky, 1963). The number of mitochondria in mammalian liver scales with the 0.72 power of body weight (Smith, 1956), and mitochondria densities in 13 species of mammals have been shown to closely parallel maximal rates of oxygen consumption (Mathieu et al., 1981). However, information on interspecies scaling of metabolic parameters is inadequate and further studies are needed. Nevertheless, we assume that the appropriate scaling law for metabolic parameters is

$$V_{max} = V_{max\theta} \, BW^{0.75} \qquad (9)$$

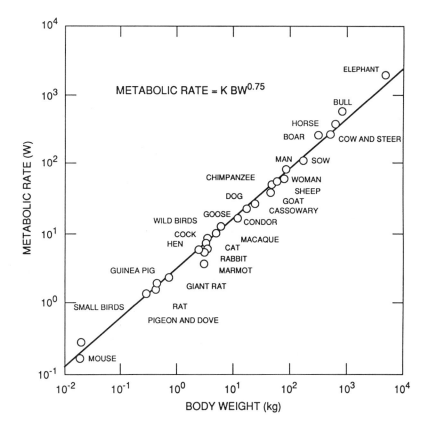

FIGURE 3. Interspecies extrapolation of metabolic rate showing that metabolic rates increase with the three-fourths power of body weight ($BW^{0.75}$) for species covering a broad size range. (From Kleiber, M., *Physiol. Rev.*, 27, 511–541, 1947. With permission.)

The Michaelis-Menten constant, K_m, is generally assumed to be approximately constant across species (NRC, 1986).

F. PARTITION COEFFICIENTS

Partition coefficients are an expression of a chemical's solubility in tissues. The partition coefficient of a given chemical between two media is defined as the ratio of the equilibrium chemical concentration in the first medium to the chemical concentration in the second medium. The most common measurements are blood/air and tissue/air partition coefficients with tissue/blood derived as the ratio of tissue/air to blood/air. Tissue/air partition coefficients tend to be constant across species (NRC, 1986); while blood/air partition coefficients show some species-dependent variability. As a general rule, however, we will assume that partition coefficients are approximately constant across species (NRC, 1986).

G. DISCUSSION OF ALLOMETRIC SCALING

For the purposes of the present study, we have assumed that certain physiological and metabolic processes all scale across species with the three fourths power of body weight. While there is a substantial body of empirical data to suggest that this assumption is at least approximately correct, it is far from universally accepted. Recently, Hayssen and Lacy (1985) have criticized Kleiber's (1961) original work establishing 0.75 as the allometric exponent for basal metabolic rates. They argue that Kleiber's data were insufficient in number, unrepresentative of the class Mammalia and incorrectly analyzed statistically. They analyzed data on 293 mammalian species and found an exponent of 0.7, but note that 22% of the species fell more than 50% above or below the line.

Nevertheless, taken as a whole, existing data (physiological parameters, metabolic parameters, clearance data, half-time data, and data on physiological time) seem to indicate that 0.75 is the single most appropriate allometric scaling parameter describing interspecies scaling of pharmacokinetics.

V. PHYSIOLOGICAL TIME

The presence between species of a biologically variable time scale has been asserted by several authors (Carrell, 1931; Huxley, 1927; Brody, 1945, Hill, 1950; Adolph, 1949; Dedrick, 1973; Boxenbaum, 1982; Mordenti, 1986; and Yates and Kugler, 1986). Hill (1950) first suggested that body size served as the regulating mechanism for an internal biological clock, making the rate of all biological events constant across species when compared per unit physiological time. His conclusions are supported by Adolph (1949), Stahl (1967), Gunther and de la Barra (1966), Calder (1968), Dedrick (1973), Lindstedt and Calder (1981), Boxenbaum (1982 and 1986), Mordenti (1986) and Lindstedt (1987), who have shown that breath duration, heartbeat duration, longevity, pulse time, breathing rates, and blood flow rates are approximately constant across species when expressed in internal time units. These time units have been termed physiological time (t′) and can be defined in terms of chronological time (t) and body weight (BW) as

$$t' = t/BW^{0.25} \qquad (10)$$

Thus, while chronological time is the same for all species, physiological time is different for each species. The value of this concept is that all species have approximately the same physiological and metabolic rates when measured in the physiological time frame (Dedrick, 1973; Boxenbaum, 1986; Mordenti, 1986; Yates and Kugler, 1986; and Lindstedt, 1987).

Dedrick et al. (1970) first observed that plasma concentrations of methotrexate could be standardized across species if tissue concentrations were expressed in terms of physiological time (Figure 4). When the blood concentration of a given chemical in various species are plotted against physiological time, the various lines representing different species converge into a single, species-independent profile. These plots are known as Dedrick plots.

Dedrick plots allow researchers to readily evaluate differences in the pharmacokinetic behavior of chemicals in different species. The convergence of the pharmacokinetic profile for each species into a species-independent line demonstrates the importance of incorporating physiological time into pharmacokinetic analyses and provides a basis for interspecific pharmacokinetic modeling that can be used to predict the behavior of chemical carcinogens in humans (Mordenti, 1986; Travis et al., 1990a).

VI. INTERSPECIES EXTRAPOLATION OF PHARMACOKINETICS

Interspecies extrapolation of toxic effect attempts to find a measure of administered dose (i.e., mg/kg/day or mg/m²/day) which produces the same measure of effect in all species. It is understood that any such extrapolation procedure is only approximately correct and should be used only when species-specific data are unavailable. Historically, it has been assumed that a single extrapolation procedure would work for all chemicals regardless of their mechanism of action. They distinguish three classes of mechanisms, depending on

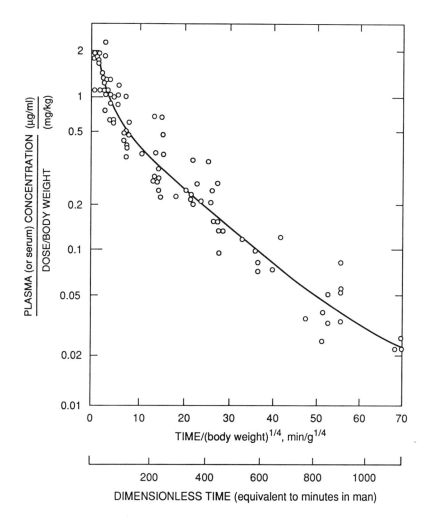

FIGURE 4. Correlation of plasma (or serum) concentration of methotrexate (MTX) in mouse, rat, dog, and human as a foundation of physiological time (Dedrick et al., 1970). (From Dedrick, R. L., Bischoff, K. B., and Zaharko, D. S., *Cancer Chemother. Rep.*, 54 (Part 1), 95–101, 1970. With permission.)

whether the parent compound, stable metabolite, or reactive metabolite produces the toxic response. We will demonstrate that it is not necessary to make such distinctions.

A. MEASURE OF EFFECTIVE DOSE

There can be little disagreement that the most precise measure of dose to target tissue is the time profile of the concentration of the toxic moiety in the target tissue. That is, two individuals within the same species will receive the same dose to target tissue, if and only if the time course of the tissue concentration curve is the same in both individuals. However, this is rarely the case and it is inconvenient to compare tissue concentration curves at all points. Therefore, the area under the tissue concentration curve (AUC) of the toxic moiety is often used as a convenient surrogate. Historical experience indicates that this measure is appropriate when standardizing dosing schedules for the purposes of intraspecies extrapolations of chronic effects. However, its use is questionable for interspecies extrapolation.

We propose that individuals from different species will receive the same dose to target tissue, if and only if the time profiles in physiological time of the concentration of toxic moiety in target tissue are the same. Support for this definition is based upon observations that the rate of biological events across species is approximately constant when compared per unit of physiological time (Dedrick, 1973; Boxenbaum, 1986; Mordenti, 1986; Yates and Kugler, 1986; and Lindstedt, 1987). Thus, pharmacodynamic processes should be approximately constant in physiological time across species if doses to target tissue for different species are constant in physiological time. A convenient surrogate metric would be area under the tissue concentration curve in physiological time of the toxic moiety (AUC_{pt}). To clarify this concept, consider the definition of AUC. Using a change of variable from chronological time to physiological time,

$$AUC = \int_0^\infty C_1(t)dt$$

$$= \int_0^\infty C_1(t')dt'BW^{0.25} \tag{11}$$

Thus, $AUC_{pt} = AUC/BW^{0.25}$. Within a species, no difficulty arises from assuming that toxic effect is proportional to AUC rather than AUC_{pt} because the two measures of dose differ by a constant. For interspecies extrapolation, however, the two different dose measures produce different results. We shall assume that the proper surrogate measure of dose to target tissue is AUC_{pt}. We note that this proposal is in contradiction to current practice (Andersen, 1987; NRC, 1986, 1987; U.S. EPA, 1985, 1986, 1987), where AUC is the standard measure.

Under the assumption that the time profile of the tissue concentration curve in physiological time is the most appropriate metric for dose to target tissue, the question of interspecies extrapolation of pharmacokinetics reduces to "Is it possible to choose a measure of administered dose so that tissue concentration curves are species-independent when measured in physiological time?" We investigate this question using the pharmacokinetic equations expressed in physiological time.

B. DIRECT-ACTING COMPOUNDS

Direct-acting compounds are those which do not require metabolic transformation to be active. As we have stated, we are assuming that the proper metric for dose to target tissue of a direct-acting compound is the tissue concentration curve in physiological time. It is shown in Appendix B that when the administered dose is measured in mg/kg/pt (mg/kg per unit of physiological time) and when metabolic deactivation is assumed, $C_1(t')$ is identical for all species; that is, tissue concentration curves are species-independent when measured in physiological time. In risk assessment applications, it is customary when making interspecies and low dose extrapolations to assume linear pharmacokinetics and to assume a linear relationship between dose and cancer response (Anderson et al., 1983). Under such an assumption, the metric mg/kg/pt can be replaced by $mg/kg^{0.75}/day$ because physiological and chronological time are related by $t' = t/BW^{0.25}$. Thus, for a metabolically deactivated parent compound, $C_1(t')$ will be constant across species if administered dose is measured in $mg/kg^{0.75}/day$. It is understood that this statement will only be true at low dose rates.

As a computational verification of the results of this section, the pharmacokinetic model (A1 to A15), together with the physiological and metabolic scaling laws (1 to 8), were used to predict AUC of the concentration of parent compound in livers of mice, rats, and humans for different intravenously administered doses. Figures 5A and B present the relationship between AUC and $AUC/BW^{0.25}$ of the parent compound, respectively, following an intravenously administered dose (mg/kg). A single pulse dose was used in these calculations, but the results would not be the same for repeated pulse doses in units of mg/kg/pt. In Figure

Parent Compound

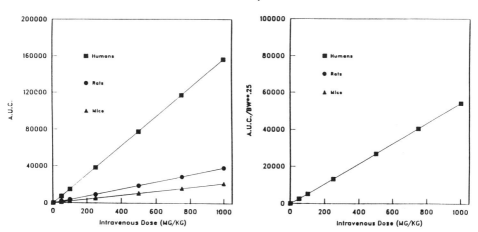

FIGURE 5. Area under the curve (AUC) of the concentration of parent compound in livers of humans, rats and mice for different intravenously administered doses. AUC for the human curve (squares) is larger than that for rats (circles), which is larger than that for mice (triangles). AUC/$BW^{0.25}$ for the human, rat, and mouse curves are identical, demonstrating that when parent compound is the toxic moiety, interspecies scaling should be on the basis of mg/kg per unit of physiological time (mg/kg/pt). At low dose rates, this is equivalent to mg/$kg^{0.75}$/day.

5A, AUC for the human curve (represented by squares) is larger than that for rats (represented by circles), which is larger than that for mice (represented by triangles). Thus, use of AUC as a surrogate dose measure would underpredict toxic effect of a direct-acting compound in humans based on animal data. In Figure 5B, AUC/$BW^{0.25}$ for the human, rat, and mouse curves are identical. Thus, when a metabolically deactivated direct-acting compound is the toxic moiety, interspecies scaling should be on the basis of mg/kg/pt or at low dose rates, mg/$kg^{0.75}$/day.

C. REACTIVE METABOLITE

When the toxic effect is caused by a short-lived by-product of the metabolic process which interacts with the cellular constituents, the toxic moiety of concern is a reactive metabolite. We are assuming that the proper metric for dose to target tissue for a reactive metabolite is the tissue concentration curve in physiological time. It is shown in Appendix B that for metabolic deactivation and administered doses in mg/kg/pt, $C_{rml}(t')$ is identical for all species; that is, tissue concentration curves are species-independent when measured in physiological time. Thus, for metabolically deactivated reactive metabolites, the appropriate interspecies scaling law is mg/kg/pt. For risk assessment purposes (i.e., the low-dose region), this interspecies scaling law is approximately equivalent to mg/$kg^{0.75}$/day.

A surrogate metric is area under the tissue concentration (mg/kg) curve in physiological time (AURMC$_{pt}$). For specificity, we again look at AURMC of liver concentration. Using a change of variable from chronological time to physiological time,

$$AURMC = \int_0^\infty C_{rml}(t)dt$$

$$= \int_0^\infty C_{rml}(t')dt'BW^{=0.25} \qquad (12)$$

Because $C_{nml}(t')$ is identical for all species, AURMC/$BW^{0.25}$ will be constant across species.

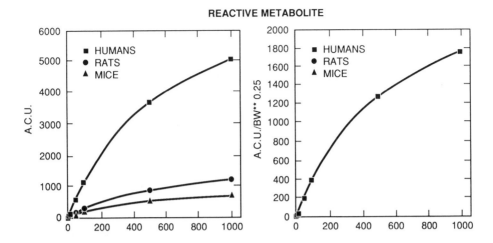

FIGURE 6. Area under the curve (AUC) of the concentration of reactive metabolite in livers of humans, rats and mice for different intravenously administered doses. AUC/BW$^{0.25}$ for the human, rat, and mouse curves are identical, demonstrating that if a reactive metabolite is the toxic moiety, interspecies scaling should be on the basis of mg/kg/pt.

Figures 6 A and B present the AUC and CUC/BW$^{0.25}$ of the concentration of metabolically deactivated reactive metabolite in livers of mice, rats and humans following an intravenously administered dose (mg/kg/pt). These curves were generated using the pharmacokinetic model (A1-A15), together with the physioloigcal and metabolic scaling laws (1 to 8). Figure 6B presents a computational verification of the fact that when a metabolically deactivated reactive metabolite is the toxic moiety, interspecies scaling should be on the basis of body weight per unit of physiological time (mg/kg/pt).

The above results deal with reactive metabolites that are metabolically deactivated. We now consider the case of a spontaneously deactivated reactive metabolite. It is shown in Appendix B that when administered dose is in mg/kg/pt, BW$^{0.25}$ $C_{rml}(t')$ is identical for all species when the reactive metabolite is spontaneously deactivated. Equation 11 shows that in this case, AURMC will be constant across species. However, we are assuming that AURMC$_{pt}$ = AURMC/BW$^{0.25}$ is the proper surrogate metric of dose to target tissue, implying that at low dose rates, mg/kg/day is the proper interspecies scaling factor.

D. STABLE METABOLITE

A stable metabolite is one which does not undergo further metabolite transformation. For specificity, we again look at AUSMC of liver concentration. Again, using a change of variable from chronological time to physiological time,

$$AUSMC = \int_0^\infty C_{sml}(t)dt$$

$$= \int_0^\infty C_{sml}(t')dt'BW^{0.25} \tag{13}$$

Appendix B shows that when dose is administered in units of mg/kg/pt, $C_{sml}(t')$ is identical for all species; that is, tissue concentration curves are species-independent when measured in pysiological time. Thus, AUSMC/BW$^{0.25}$ is constant across species when the administered dose is normalized by body weight per unit of physiological time (i.e., mg/kg/pt).

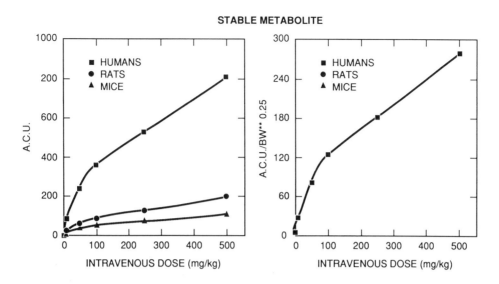

STABLE METABOLITE

FIGURE 7. Area under the curve (AUC) of the concentration of stable metabolite in livers of humans, rats and mice for different intravenously administered doses. AUC/BW$^{0.25}$ for the human, rat, and mouse curves are identical, demonstrating that if a stable metabolite is the toxic moiety, interspecies scaling should be on the basis of mg/kg/pt.

Figures 7A,B present the AUC and AUC/BW$^{0.25}$ of the concentration of stable metabolite in livers of mice, rats and humans following an intravenous administered dose (mg/kg). In Figure 7A, AUC for the human curve (represented by squares) is larger than that for rats (circles), which is, in turn, larger than that for mice (triangles). Thus, traditional body weight scaling (mg/kg/day) would underpredict toxic effect of a stable metabolite in humans based on animal data. In Figure 7B, AUC/BW$^{0.25}$ for the human, rat, and mouse, the curves are identical. Thus, when a stable metabolite is the toxic moiety, interpecies scaling should be on the basis of mg/kg/pt.

VII. BIOLOGICAL DATA SUPPORTING THE THREE FOURTHS SCALING POWER

The most extensive data base on interspecies extrapolation is the Freireich et a. (1966) study of acute toxicity data in mice, rats, dogs, monkeys, and man. Using the LD$_{10}$s in small animals and the maximum tolerated dose (MTD) of 14 anticancer agents in large animals and humans, Freireich et al. (1966) concluded that these measures to toxicity are approximately constant when measured in mg/kg$^{0.67}$/day. A re-analysis of the Freireich et al.(1966) study was conducted by Travis and White (1988) who augmented the Freireich et al. data set with 13 additional chemotherapy agents from a data set by Schein et al. (1979). Travis and White (1988) analyzed the combined data set by multiple linear regression. Using a model with a common slope and individual intercepts, they found that the best overall power of body weight for the total variance (r^2) of the data explained by this model was 0.96. Both of the traditional (surface area and body weight) scaling metrics fall outside the 95% confidence bounds of this estimate. Thus the Travis and White (1988) analysis supports mg/kg$^{0.75}$/day as the appropriate scaling metric for acute toxicity.

Travis and Bowers (1990) analyzed the anesthetic potency data for 11 volatile anesthetics in multiple mammalian species, including man. Anesthetic potency is defined as the minimum (sustained) alveolar concentration required to prevent a muscular response to stimulus for 50% of the subject population. Using a model with a common slope and individual intercepts, they found that the best overall power of body weight for interspecies scaling was -0.008 with 95% confidence bounds of -0.018 and 0.002. The proportion of the total variance (r^2) of the data explained by the model was 0.99. Thus, Travis and Bowers (1990) conclude that the sustained alveolar concentration (mg/m^3) necessary to produce an anesthetized state is approximately independent of body weight across species. Since alveolar ventilation rates scale with the 0.75 power of body weight, this implies that an equal $mg/kg^{0.75}/day$ administered dose rate produces the same anesthetic effect in all species. Thus, the anesthetic potency data provide further empirical support for the use of $mg/kg^{0.75}/day$ as the proper interspecies scaling metric for acute toxic effects.

VIII. CONCLUSIONS

A long-standing problem in toxicology is the extrapolation of observed experimental results between animal species and man. It has historically been assumed that experimental results can be extrapolated between species if administered dose is standardized in one of two metrics: mg/kg body weight/day (body weight scaling) or $mg/m^2/day$ (surface area scaling). More recently, Andersen (1987) and the National Academy of Sciences (NRC, 1986, 1987) point out that scaling should depend on the kinetic behavior of the toxic moiety and its mechanism of toxicity. They suggest that if carcinogenic response is proportional to the area under the curve of the concentration of the toxic moiety in the target tissue, then the appropriate scaling laws for cancer risk assessment interspecies extrapolations are $mg/kg^{0.7}/day$ for direct-acting compounds and stable metabolites, and mg/kg/day for reactive metabolites.

With the advent of biologically based pharmacokinetic models, it has become possible to present a scientifically defensible justification for the choice of an interspecies scaling metric. It has been established that biologically based pharmacokinetic models provide an accurate description of the pharmacokinetics of the parent compound and metabolites in mice, rats and humans (Ramsey and Andersen, 1984; NRC, 1987; Andersen et al., 1987; Paustenbach et al., 1988; Ward et al., 1988; and Travis et al., 1990a). Thus, pharmacokinetic models provide a tool to quantitatively evaluate the scientific bases for interspecies extrapolation of experimental results.

The parameters controlling pharmacokinetics in a given species are physiological (breathing rates, blood flow rates, blood volumes, tissue volumes, etc.), biochemical (partition coefficients), and metabolic, in nature. It is the extrapolation of individual parameters across species that controls interspecies extrapolation of pharmacokinetics. It has long been established that tissue volumes tend to extrapolate across species with the first power of body weight (Schmidt-Nielsen, 1984; NRC, 1987; Lindstedt, 1987), biological time extrapolates with the 1/4 power of body weight (Adolph, 1949; Stahl, 1967; Gunther and Leon de la Barra, 1966; Calder, 1968; Dedrick, 1973; Lindstedt and Calder, 1981; Boxenbaum, 1982, 1986; Mordenti, 1986; and Lindstedt, 1987), and volume rates (i.e., volume divided by time such as clearance rates, cardiac output, alveolar ventilation, etc.) vary with approximately the 3/4 power of body weight (Guyton, 1947; Adolph, 1949; Kleiber, 1961; Stahl, 1967; Lindstedt and Calder, 1981; Boxenbaum, 1982; Schmidt-Nielsen, 1984; Mordenti, 1986; and Lindstedt, 1987). These allometric relationships, together with physiologically based pharmacokinetic models, provide the basis for the derivation of scientifically defensible interspecies scaling laws.

Travis et al. (1990a) have recently argued that if toxic response is a function of the time profile in physiological time of the concentration of the toxic moiety in the target tissue, then regardless of the mechanism of action appropriate interspecies scaling law is mg/kg/pt if the toxic moiety is metabolically deactivated. In the low dose region, this metric is equivalent to mg/kg$^{0.75}$/day. For reactive metabolites which are spontaneously deactivated, the appropriate interspecies scaling law at low dose levels is mg/kg/day.

The above conclusions are based on a series of assumptions outlined in the text of this paper. The most debatable assumption involves the allometric scaling of physiological parameters with the 3/4 power of body weight. It might be argued that if these physiological parameters scaled with the 0.7 power of body weight, then our conclusion would be that interspecies scaling should be on the basis of mg/kg$^{0.7}$/day, a result more consistent with the currently accepted surface area scaling law. The scaling of physiological parameters such as cardiac output and alveolar ventilation rate is only necessary to obtain a closed form analytical proof of our results (Appendix B). If one is willing to accept a computer verification of the result, then it is not necessary to scale the physiological parameters. One can set up three pharmacokinetic models (one each for mice, rats, and human) using actiual physiological parameters. It still must be assumed that (1) partition coefficients are approximately constant across species; (2) V_{max} scales with the 3/4 power of body weight; and (3) K_m is approximately constant across species. We have performed such computer calculations and found that the body weight per unit of physiological time scaling law still holds. Thus our conclusion is, in reality, based only on the above three assumptions plus the assumption concerning the appropriateness of the pharmacokinetic model. The debate (Section IV.F) over the appropriate scaling law for physiological parameters becomes irrelevant.

As we have previously pointed out, information on interspecies scaling of metabolic parameters is inadequate to determine the proper scaling law for V_{max}. The only available data set that sheds light on the appropriateness of the above three assumptions is that concerning clearance of chemicals from the body. Several studies (Edwards, 1975; Bonati et al., 1984; Chung et al., 1985; Boxenbaum, 1982; Mordenti, 1986) have demonstrated that various types of clearance scale across species with body weight to the 0.75 power. Many of the compounds considered are nonvolatile and cleared only through metabolic pathways. The fact that their clearance scales with the 3/4 power of body weight is indirect evidence that this is the proper scaling law for V_{max} and that K_m and partition coefficients are approximately constant across species.

We acknowledge that the present work is not the final word on the very important issue of interspecies scaling laws. Further investigations are needed in the area of pharmacodynamics, in particular, on the question of the appropriateness of AUC_{pt} as the proper measure of dose to target tissue. The present chapter does, however, demonstrate the value of pharmacokinetic models in investigating the underlying biological processes that control interspecies scaling. Similar analyses are needed in the pharmacodynamic area.

IX. APPENDIX A: THE PHARMACOKINETIC MODEL

A. INHALATION MODEL

For the inhalation route, the compound is inspired at concentration C_{inh} with a flow rate equal to the alveolar ventilation rate, Q_{alv}. The model assumes that there is no gas storage in the lungs and that ventilation of the alveoli is continuous, rather than cyclic. The compound in the alveolar air is assumed to equilibrate instantaneously with pulmonary capillary blood so that the compound concentration in lung blood and in alveolar air leaving the lungs maintains a constant ratio specified by the blood/gas partition coefficient, λ_b.

Because the model assumes no chemical storage in the lungs, conservation requires that the quantity of chemical entering the lungs in a time interval, dt, be equal to the quantity leaving the lungs during the same interval of time:

$$Q_{alv}C_{inh}dt + Q_bC_{ven}dt = Q_{alv}C_{alv}dt + Q_bC_{art}dt$$

or   (A1)

$$Q_{alv}(C_{ink} - C_{alv}) = Q_b(C_{art} - C_{ven})$$

where Q_b is the total blood flow rate, the C_{alv}, and C_{ven}, and C_{art} are the alveolar, mixed venous blood, and arterial blood concentrations, respectively. The diffusion equilibrium is assumed to be maintained between air leaving the lungs and blood leaving the lungs:

$$\lambda_b C_{alv} = C_{art}$$ (A2)

In the body tissues, conservation requires that the amount of chemical entering via the arterial blood in an interval of time, dt, be equal to the quantity gained by each tissue group, dA_i, plus the amount leaving in the venous blood:

$$Q_iC_{art}dt = dA_i + Q_iC_{vi}dt$$

or   (A3)

$$\frac{dA_i}{dt} = Q_i(C_{art} - C_{vi})$$

where Q_i is the blood flow rate in tissue group i. The chemial in venous blood leaving the tissue group i is assumed to satisfy the equilibrium equation

$$C_{vi} = \frac{C_i}{\lambda_i} = \frac{A_i}{V_i\lambda_i}$$ (A4)

where V_i is the tissue volume and λ_i is the tissue/blood partition coefficeint. Equation A3 applies to those tissue groups which neither metabolize nor excrete the chemical; in the model, these are the fat, muscle, and organ groups (Ramsey and Andersen, 1984).

In the model, the metabolism of the chemical is assumed to occur mainly in the liver (Ramsey and Andersen, 1984). However, the liver compartment can also represent other tissues with significant metabolic capacity. The metabolic rate, where A_m is the metabolized amount, is thus expressed in terms of the concentration of the chemical in venous blood leaving the liver (C_{vl}) as:

$$\frac{dA_m}{dt} = \frac{V_{max}C_{vl}}{K_m + C_{vl}} + K_fC_{vl}$$ (A5)

Where V_{max} is the Michaelis-Menten metabolic rate, K_m is the Michaelis-Menten constant, and K_f is the linear metabolic rate constant. By combining Equation A3 for the liver compartment with the metabolic removal rate described by Equation A5, the mass balance differential equation for the chemical in the metabolizing tissue group is

$$\frac{dA_l}{dt} = Q_l(C_{art} - C_{vl}) - \frac{dA_m}{dt}$$ (A6)

The concentration of the chemical in the mixed venous blood returning to the lungs is formulated as the sum of the venous contribution from each of the tissue groups

$$C_{ven} = \sum\{(Q_iC_{vi})/Q_b\} \tag{A7}$$

By combining Equations A1 and A2, the equation for arterial blood concentration is

$$C_{art} = \frac{Q_{alv}C_{inh} + Q_bC_{ven}}{Q_b + Q_{alv}/\lambda_b} \tag{A8}$$

Thus, the pharmacokinetics of the parent compound is described by the suystem of equations given by Equations A3 through A8.

1. Stable Metabolite

The blood and tissue concentrations of a stable metabolite (one which circulates in the blood) can be determined through a slight modification of Equations A3 through A8. The relevant equations then become:

$$dA_{smi}/dt = Q_i(C_{smart} - C_{smvi}) \tag{A9}$$

$$dA_{sml}/dt = Q_1(C_{smart} - C_{smv1}) + dA_m/dt \tag{A10}$$

$$C_{smvi} = C_{smi}/\lambda_{smi} \tag{A11}$$

$$\frac{dC_{smven}}{dt} = \sum \frac{dA_{smi}}{dt} Q_i/Q_bV_i\lambda_{smi}) - K_rC_{smven}/V_{ven} \tag{A12}$$

$$C_{smart} = \frac{Q_bC_{smven}}{Q_b + Q_{alv}/\lambda_{smb}} \tag{A13}$$

where K_r is the renal clearance rate constant.

2. Reactive Metabolite

The mass balance equation for an intermediate metabolite which is subsequently metabolized (through i possible pathways) at rates linear with its concentration (NRC, 1986; Andersen, 1987) is given by

$$dA_{rmi}/dt = dA_m/dt - \sum K_iV_1C_{rmi} \tag{A14}$$

where K_i is the rate constant (1/h) for the i-th metabolic pathway of the intermediate.

B. GAVAGE AND INTRAVENOUS MODEL

For the gavage route, absorption of the compound from the gut into the liver is simulated as a first order process. Equation 6 is modified as follows:

$$dA_1/dt = Q_1(C_{art} - C_{vl}) - dA_m/dt + kD_0e^{-kt} \tag{A15}$$

where k is the first order absorption rate constant D_o is the total quantity of compound absorbed.

X. APPENDIX B: PHARMACOKINETICS IN PHYSIOLOGICAL TIME

The mathematical Equations A1 through A15 together with the physiological and metabolic scaling laws (1 to 8) define a physiologically based pharmacokinetic model which is theoretically applicable to any species. The ability of physiologically based models to describe the dynamics of chemical transport and metabolism in multiple species, including man, provides a major tool for investigating the interspecies extrapolation problem (Travis, 1987).

A. PARENT COMPOUND

We first demonstrate that when expressed in physiological time, the pharmacokinetic model A1 through A15 (which, as mentioned before, has species-dependent physiological and metabolic parameters and a species-independent time variable) can be replaced by a pharmacokinetic model with species-independent physiological and metabolic parameters and a species-dependent time variable (Travis et al., 1990a). First make the change of dependent variable

$$A'_i = A_i/BW$$

$$A'_m = A_m/BW$$

so that amounts are expressed in mg/kg body weight. When expressed in physiological time, the balance equation (Equation A3) describing the distribution of the parent compound from the arterial blood to the tissue and mixed venous blood becomes:

$$
\begin{aligned}
\frac{dA'_i(t')}{dt'} &= \frac{1}{BW} \frac{dA_i(t')}{dt} \frac{dt}{dt'} \\
&= BW^{-1.0} Q_i[C_{art}(t') - C_{vi}(t')] LBW^{0.25} \\
&= BW^{-1.0} Q_{i\theta} BW^{0.75} BW^{0.25} [C_{art}(t') - C_{vi}(t')] \\
&= Q_{i\theta} [C_{art}(t') - C_{vi}(t')]
\end{aligned}
\tag{B1}
$$

where $Q_{i\theta}$ is the species-independent allometric blood flow constant in tissue group i. Similar reformulation of the equation for the concentration of the chemical in the metabolizing tissue group (Equation A7) yields:

$$
\frac{dA'_m(t')}{dt} = \frac{V_{max\theta} C_{vl}(t')}{K_m + C_{vl}(t')} + K_{f\theta} C_{vl}(t')
$$

$$
\frac{dA'_l(t')}{dt} = Q_{l\theta}[C_{art}(t') - C_{vl}(t')] - \frac{dA'_m(t')}{dt}
\tag{B2}
$$

where $V_{max\theta}$, $K_{m\theta}$, and $K_{f\theta}$ are allometric metabolic parameters.

Equations for the concentration of compound in the mixed venous blood and arterial blood can also be rewritten in species invariant units:

$$
C_{ven}(t') = \sum C_{vi}(t')Q_{i\theta}/Q_{b\theta}
$$

$$
C_{art}(t') = C_{ven}(t') + [Q_{alv\theta}/Q_{b\theta}][C_{inh}(t') - C_{alv}(t')]
\tag{B3}
$$

where $Q_{i\theta}$, $Q_{alv\theta}$, and $Q_{b\theta}$ are species-independent allometric constants.

B. STABLE METABOLITE

Reformulation of equations describing the stable metabolite yields equations identical to A9 through A13 with A_{smi} replaced by A'_{smi} and t replaced by t'. For example, Equation A9 becomes

$$
\begin{aligned}
\frac{dA_{smi}(t')}{dt'} &= \frac{1}{BW} \frac{dA_{smi}(t')}{dt} \frac{dt}{dt'} \\
&= BW^{-1.0} Q_i[C_{smart}(t') - C_{smvi}(t')]BW^{0.25} \\
&= BW^{-1.0} Q_{i\theta}BW^{0.75} BW^{0.25}[C_{smart}(t') - C_{smvi}(t')] \\
&= Q_{i\theta}[C_{smart}(t') - C_{smvi}(t')]
\end{aligned} \tag{B4}
$$

Equations A10 through A13 can be reformulated in a similar fashion.

C. REACTIVE METABOLITE

Reformulation of Equation A14 yields

$$
\begin{aligned}
\frac{dA_{rmi}(t')}{dt'} &= \frac{1}{BW} \frac{dA_{rmi}(t')}{dt} \frac{dt}{dt'} \\
&= BW^{-1.0}[dA_m(t')/dt - \sum K_i V_i C_{rmi}(t')]BW^{0.25} \\
&= dA_m(t')/dt - \sum K_{i\theta} A'_{rmi}(t') BW^{0.25}
\end{aligned} \tag{B5}
$$

In the case of a metabolically deactivated reactive metabolite, we assume that $K_i = K_{i\theta} BW^{0.25}$ and Equation B5 becomes species-independent. In the case of spontaneous deactivation, $Y(t) = BW^{0.25} A'_{rml}(t/BW^{0.25})$ is the solution to the species-independent equation:

$$
\frac{dY(t)}{dt} = \frac{dA_{m'}(t'(t))}{dt} - \sum K_{i\theta \ Y(t)} \tag{B6}
$$

D. INITIAL CONDITIONS

Chemical input to the model system can occur either via inhalation, gavage, or intravenous injection. We assume that dosing regimens are identical in physiological time. For the gavage or intravenous route, such an assumption is satisfied if the administered dose is the same for all species when measured in mg/kg per unit of physiological time. Because $t' = t/BW^{0.25}$, this metric is equivalent to $mg/kg^{0.75}/day$. For inhalation exposures, assume all species are exposed to the same inhaled concentration, C_{inh}, for the same physiological time interval, T_e'. Then the inhalation exposure

$$
C_{inh}(t') = \begin{cases} C_{inh}, & 0 \leq t' \leq T'_e \\ 0, & T'_e \leq t' \end{cases} \tag{B7}
$$

is identical for all species. This can be shown to produce the same dose in mg/kg per unit of physiological time.

We have thus demonstrated that the pharmacokinetic model characterized by Equations A1 through A15 and 1 to 8 (which has species-dependent physiological and metabolic parameters and a species-independent time variable) can be replaced by a pharmacokinetic model with species-independent physiological and metabolic parameters and a species-dependent time variable. Because the initial conditions for this new set of differential equations are identical in physiological time for all species, the solution to the differential equations will be identical for all species. That is, tissue concentration curves are species-independent when measured in physiological time. This is a mathematical proof of an empirical observation first made by Dedrick et al. (1970).

REFERENCES

Adolph, E. F., Quantitative relations in the physiological constitutions of mammals, *Science,* 109, 579, 1949.

Andersen, M. E., Tissue dosimetry in risk assessment, or what's the problem here anyway? in *Drinking Water and Health, Pharmacokinetics in Risk Assessment,* Vol. 8, National Academy Press, Washington, D.C., 1987, 8–26.

Andersen, M. E., Clewell, H. J., III, Gargas, M. L., Smith, F. A., and Reitz, R. H., Physiologically-based pharmacokinetics and the risk assessment process for methylene chloride, *Toxicol. Appl. Pharmacol.,* 87, 185, 1987.

Andersen, M. E., Leung, H., Ku, R. H., and Paustenbach, D. J., A physiologically-based pharmacokinetic model for 2,3,7,8-tetrachlorodibenzo-p-dioxin in C57BL/6J and DBA/2J mice, *Toxicol. Lett.,* 1989.

Anderson, E. L. and the Carcinogen Assessment Group of the U.S. Environmental Protection Agency, Quantitative approaches in use to assess cancer risk *Risk Anal.,* 3(4), 277–295, 1983.

Arms, A. D. and Travis, C. C., Reference Physiological Parameters in Pharmacokinetic Modeling, EPA 600/6–88/—4, Environmental Protection Agency, available from NTIS No. PB88–196019 (April), Washington, D.C., 1988.

Benedict, F. G., *Vital Energetics: A Study in Comparative Basal Metabolism,* Carnegie Institute of Washington, Washington, D.C., 1938.

Bonati, M., Latinit, A., Tognoni, G., Young, J. F., and Garattini, S., Interspecies comparison of in vivo caffeine pharmacokinetics in man, monkey, rabbit, rat, and mouse, *Drug Metab. Rev.,* 15(7), 1355–1383, 1984.

Boxenbaum, H., Interspecies scaling, allometry, physiological time, and the ground plan of pharmacokinetics, *J. Pharmacol. Biopharmacol.,* 10, 201–227, 1982.

Boxenbaum, H., Time concepts in physics, biology, and pharmacokinetics. *J. Pharm. Sci.,* 75(11), 1053–1062, 1986.

Boxenbaum, H. and Ronfeld, R., Interspecies pharmacokinetic scaling and the Dedrick plots, *Am. J. Physiol.,* 245, R768–R774, 1983.

Brody, S., *Bioenergetics and Growth: With Special Reference to the Efficiency Complex in Domestic Animals,* Reinhold, New York, 1945.

Calder, W. A., Respiration and heart rates of birds at rest, *Condor,* 70, 358–365, 1968.

Carrell, A., Physiological time, *Science,* 74, 618–621, 1931.

Chung, M., Radwanski, E., Loebenberg, D., Lin, C., Oden, E., Symchowicz, S., Gural, R. P., and Miller, G. H. Interspecies pharmacokinetic scaling of SCH 34343, *J. Antimicrob. Chemother.,* 15(c), 227–223, 1985.

Clewell, H. J., III and Andersen, M. E., Dose, species, and route extrapolation using physiologically based pharmacokinetic models, in *Pharmacokinetics and Risk Assessment* National Academy Press, Washington, D.C., 1987, 65–79.

Crouch, E. and Wilson, R., Interspecies comparison of carcinogenic potency, *J. Toxicol. Environ. Health,* 5, 1095–1118, 1978.

Crump, K. S. and Guess, H. A., *Drinking Water and Cancer: Review of Recent Findings and Assessment of Risks,* Science Research Systems Inc., Ruston, LA, CBQ Contract No. EQ10ACO18, 1986.

Dedrick, R. L. Animal scale-up, *J. Pharmacokinet. Biopharmacol.,* 1(5), 435–461, 1973.

Dedrick, R. L., Bischoff, K. B., and Zaharko, D. S., Interspecies correlation of plasma concentration history of methotrexate (NSC–740), *Cancer Chemother. Rep.,* 54(Part 1), 95–101, 1970.

Edwards, N. A., Scaling of renal functions in mammals, *Comp. Biochem. Physiol.,* 52A, 63–66, 1975.

Federation of American Societies for Experimental Biology (FASEB), *Biological Bases for Interspecies Extrapolation of Carcinogenicity Data,* Hill, T., Wands, R., and Leukroth, R. R., Jr., Eds., prepared for the FDA by Life Sciences Research Office of the FASEB, July 1986.

Fiserova-Bergerova, V., Physiological models for pulmonary administration and elimination of inert vapors and gases, in *Modeling of Inhalation Exposure to Vapors: Uptake, Distribution and Elimination,* Fiserova-Bergerova, V., Ed., CRC Press, Boca Raton, FL, 1983, 73–100.

Freireich, E. J., Gehan, E. A., Rall, D. P., Schmidt, L. H., and Skipper, H. E., Quantitative comparison of toxicity of anticancer agents in mouse, rat, hamster, dog, monkey and man, *Cancer Chemother. Rep.,* 50(4), 219–244, 1966.

Gerlowski, L. E. and Jain, R. K., Physiologically based pharmacokinetic modeling: principles and applications, *J. Pharm. Sci.,* 72:1103–1126, 1983.

Gibaldi, M. and Perrier, D., *Pharmacokinetics,* Marcel Dekker, New York, 1975.

Gunther, B. and de la Barra, L., On the space-time continuum in biology, *Acta Physiol. Lat. Am.,* 16, 221–231, 1966.

Guyton, A. C., Measurement of the respiratory volumes of laboratory animals, *Am. J. Physiol.,* 3150, 70–77, 1947.

Guyton, A. C., *Textbook of Medical Physiology,* 4th ed., W. B. Saunders, Philadelphia, 1971.,

Hattis, D., A pharmacokinetic/mechanism-based analysis of the carcinogenic risk of ethylene oxide, Massachusetts Institute of Technology (MIT) Center for Technology, Policy and Industrial Development, CTPID 87–1, August 1987.

Hayssen, V. and Lacy, R. C., Basal metabolic rates in mammals: taxonomic differences in the allometry of BMR and body mass, *Comp. Biochem. Physiol. A,* 81(4), 741–754, 1985.

Hill, A. V., The dimensions of animals and their muscular dynamics, *Proc. R. Inst. Great Britain,* 34, 450–471, 1950.

Hoel, D. G., Low-dose and species to species extrapolation for chemically induced carcinogenesis, in *Banbury Report No. 1: Assessing Chemical Mutagens: The Risks to Humans,* McElheny, V., Ed., Cold Spring Harbor Laboratory, Cold Spring Harbor, NY, 1974, 135–145.

Hogan, M. and Hoel, D. G., *Extrapolation to man.* in *Principles of Toxicology,* Hayes, A. W. Ed., Raven Press, New York, 1982, 711–731.

Holt, J. P., Rhode, E. A., and Kines, H., Ventricular volumes and body weight in mammals, *Am. J. Physiol.,* 215(3), 704–715, 1968.

Huxley, J. S., On the relation between egg-weight and body-weight in birds, *J. Linn. Soc. London Zool.,* 36, 457–466, 1927.

Jansky, L., Total cytochrome oxidase activity and its relation to basal and maximal metabolism, *Nature,* 189, 921–922, 1961.

Jansky, L., Body organ cytochrome oxidase activity in cold-and-warm acclimated rats, *Can. J. Biochem. Physiol.,* 41, 1847–1854, 1963.

Kissel, J. C. and Robarge, G. M., Assessing the elimination of 2,3,7,8-TCDD from humans with a physiologically based pharmacokinetic model, *Chemosphere,* 17(10), 2017–2027, 1988.

Kleiber, M., Body size and metabolism, *Hilgardia,* 6, 315–353, 1932.

Kleiber, M., Body size and metabolic rate, *Physiol. Rev.,* 27, 511–541, 1947.

Kleiber, M., *The Fire of Life: An Introduction to Animal Energetics,* John Wiley & Sons, New York, 1961.

Kunkel, H. O., Spalding, J. F., de Franciscis, G., and Futrell, M. F., Cytochrome oxidase activity and body weight in rats and in three species of large animals, *Am. J. Physiol.,* 186, 203–206, 1956.

Lindstedt, S. L. and Calder, W. A., III., Body size and physiological time, and longevity of homeothermic animals, *Q. Rev. Biol.,* 56, 1–16, 1981.

Lindstedt, S. L., Allometry: body size constraints in animal design, in *Drinking Water and Health. Pharmacokinetics in Risk Assessment,* Vol. 8, National Academy Press, Washington, D.C., 1987.

Lutz, R. J., Dedrick, R. L., Matthews, H. B., Eling, T. E., and Anderson, M. W., A preliminary pharmacokinetic model for several chlorinated biphenyls in the rat, *Drug Metab. Dispos.,* 5(4), 386–396, 1977.

Mathieu, O., Krauer, R., Hoppeler, H., Gehr, P., Lindstedt, S. L., Alexander, R., Taylor, C. R., and Weibel, E. R., Design of the mammalian respiratory system. VII. Scaling mitochondrial volume in skeletal muscle to body mass, *Respir. Physiol.,* 44, 113–128, 1981.

McMahon, T., Size and shape in biology, *Science,* 179, 1201–1204, 1973.

Midwest Research Institute (MRI), Risk Assessment Methodology for Hazardous Waste Management, prepared for the U.S. Environmental Protection Agency and Council on Environmental Quality, Draft Final Rep., July 31, 1986.

Mordenti, J., Man versus beast: pharmacokinetic scaling in mammals, *J. Pharm. Sci.,* 75(11), 1028–1040, 1986.

National Research Council (NRC), Drinking Water and Health, *Vol. 6, National Academy Press, Washington, D.C.,* 1986.

National Research Council (NRC), Drinking Water and Health: Pharmacokinetics in Risk Assessment, *Vol. 8, National Academy Press, Washington, D.C.,* 1987.

Paustenbach, D. J., Andersen, M. E., Clewell, H. J., III, and Gargas, M. L., A physiologically-based pharmacokinetic model for inhaled carbon tetrachloride in the rat, *Toxicol. Appl. Pharmacol.,* 96, 191–211, 1988.

Pinkel, D., The use of body surface area as a criterion of drug dosage in cancer chemotherapy, *Cancer Res.,* 18(1), 853–856, 1958.

Prothero, J. W., Scaling of blood parameters in mammals, *Comp. Biochem. Physiol. A,* 67, 649–657, 1980.

Prothero, O., Heart weight as a function of body weight in mammals, *Growth,* 43, 139–150, 1979.

Ramsey, J. C. and Andersen, M. E., A physiologically based description of the inhalation pharmacokinetics of styrene in rats and humans, *Toxicol. Appl. Pharmacol.,* 73, 159–175, 1984.

Ramsey, J. C., Young, J. D., Karbowski, R., Chenoweth, M. B., McCarty, L. P., and Braun, W. H., Pharmacokinetics of inhaled styrene in human volunteers, *Toxicol. Appl. Pharmacol.,* 53, 54–63, 1986.

Reitz, R. H., Fox, T. R., Ramsey, J. C., Quast, J. F., Langvardt, P. W., and Watanabe, P. G., Pharmacokinetics and macromolecular interactions of ethylene dichloride in rats after inhalation or gavage, *Toxicol. Appl. Pharmacol.,* 62, 190–204, 1982.

Reitz, R. H., Nolan, R. J., and Schumann, A. M., Development of multispecies, multiroute pharmacokinetic models for methylene chloride and 1,1,1-trichloroethane (methyl chloroform), in *Pharmacokinetics in Risk Assessment,* Vol. 8, National Academy Press, Washington, D.C., 1987, 391–409.

Rubner, M., Ueber den einfluss der körpergrösse auf stoff- une kraftwechesel, *Z. Biol.,* 19, 535–562, 1883.

Schein, P. S., Davis, R. D., Carter, S., Newman, J., Schein, D. R., and Rall, D. P., The evaluation of anticancer drugs in dogs and monkeys for the prediction of qualitative toxicities in man, *Clin. Pharmacol. Ther.,* 11, 3–40, 1979.

Schmidt-Nielsen, K., Energy metabolism, body size and problems of scaling, *Fed. Proc.,* 29, 1524–1532, 1970.

Schmidt-Nielsen, K., *Scaling: Why is Animal Size So Important?,* Cambridge University Press, Cambridge, 1984.

Smith, R. E., Quantitative relations between liver mitochondria metabolism and total body weight in mammals, *Ann. N.Y. Acad. Sci.,* 62, 403–422, 1956.

Stahl, W. R., Organ weights in primate and other mammals, *Science,* 150, 1039–1042, 1965.

Stahl, W. R., Scaling of respiratory variables in mammals, *J. Appl. Physiol.,* 48, 1052–1059, 1967.

Travis, C. C., Interspecies and dose-route extrapolations in *Drinking Water and Health: Pharmacokinetics in Risk Assessment,* Vol. 8, National Academy Press, Washington, D.C., 1987.

Travis, C. C., Tissue dosimetry for reactive metabolites, *Risk Anal.,* 10, 317–321, 1990.

Travis, C. C. and Bowers, J. C., Interspecies scaling of anesthetic potency, *Tox. Indust. Health,* 7, 4, 249–260, 1991.

Travis, C. C. and Hattemer-Frey, H., Pharmacokinetics and its application to risk assessment, in *Hazard Assessment of Chemicals,* Saxena, J., Ed., New York: Hemisphere, New York, 199, 39–82.

Travis, C. C. and White, R. K., Interspecific scaling of toxicity data, *Risk Anal.,* 8, 119–125, 1988.

Travis, C. C., White, R. K., and Arms, A. D., A physiologically based pharmacokinetic approach for assessing the cancer risk of tetrachloroethylene, in *The Risk Assessment of Environmental and Human Health Hazards,* Ed., Paustenbach, D. J., John Wiley & Sons, New York, 1989, 769–796.

Travis, C. C., White, R. K., and Ward, R. C., Interspecies extrapolation of pharmacokinetics, *J. Theor. Bol.,* 142, 285–304, 1990a.

Travis, C. C., Quillen, J. L., and Arms, A. D., Pharmacokinetics of benzene, *Toxicol. Appl. Pharmacol.,* 102, 400–420, 1990b.

U.S. Environmental Protection Agency (EPA), Health Assessment Document for Dichloromethane (Methylene Chloride), EPA/600/8–82/004F, 1985.

U.S. Environmental Protection Agency (EPA), Addendum to the Health Assessment Document for Tetrachloroethylene (Perchloroethylene): Updated Carcinogenicity Assessment for Tetrachloroethylene (Perchloroethylene, PERC, PCE), EPA/600/8–82/005FA, 1986.

U.S. Environmental Protection Agency (EPA), Update to the Health Assessment Document and Addendum for Dichloromethane (Methylene Chloride): Pharmacokinetics, Mechanism of Action and Epidemiology, EPA/600/8–87/030A, 1987.

Ward, R. C., Travis, C. C., Hetrick, D. M., Andersen, M. E., and Gargas, M. L., Pharmacokinetics of tetrachloroethylene, *Toxicol. Appl. Pharmacol.,* 93, 108–117, 1988.

White, L., Haines, H., and Adams, T., Cardiac output related to body weight in small mammals, *Comp. Biochem. Physiol.,* 27, 559–565, 1968.

Yates, F. E. and Kugler, P. N., Similarity principles and intrinsic geometries: contrasting approaches to interspecies scaling, *J. Pharm. Sci.,* 75(11), 1019–1027, 1986.

Young, J. D., Ramsey, J. C., Blau, G. E., Karbowski, R. J., Nitschke, K. D., Slaughter, R. W., and Braun, W. H., Pharmacokinetics of inhaled or intraperitoneally administered styrene in rats, in *Proc. 10th Inter-American Conf. on Toxicology and Occupational Medicine,* Deichmann, W. B., Ed., Elsevier, New York, 1978, 297–310.

New Models in Skin Research

Chapter 25

HUMAN SKIN XENOGRAFTS TO ATHYMIC RODENTS AS A SYSTEM TO STUDY TOXINS DELIVERED TO OR THROUGH SKIN

Gerald G. Krueger and Lynn K. Pershing

TABLE OF CONTENTS

0-8493-7357-3/93/$0.00 + $.50
© 1993 by CRC Press, Inc.

I. INTRODUCTION

The concept that human skin can be transplanted to athymic rodents to study various aspects of skin physiology, pathology, pharmacology and toxicology has been tested in some detail and was the subject of a recent review.[94] That review emphasized pharmacologic aspects; however, because the concepts and methodologies are essentially the same when these systems are used to study the toxic effects of topical application of agents on skin it is of necessity that this manuscript borrows very heavily from this recent review.

Athymic mutants have been described for both mice and rats. In each the athymic mutant has an associated skin defect. This is manifested as a much diminished number of anagen hair in the case of the athymic rat and few to no hair in the anagen phase in the congenitally athymic mouse. In each, the hair that is present is fine, short, and brittle. This abnormal pelage caused these mutations to be called nude rats and nude mice respectively.[1-3] A hairless athymic guinea pig has also been described; however, that mutant appears to have been lost to the experimental community.[4]

The nude mouse mutant was discovered by N.R. Grist of Glasgow in 1962[5] and was more completely described and termed nude (nu) by Flannigan in 1966.[6] He noted that this mutation was inherited as autosomal recessive, had a shortened life span, and abnormal hair. Two years later this mutant was recognized as being athymic by Pantelouris.[6,7] A remutation at the nude locus was identified by Schultz et al. of the Jackson Laboratory, Bar Harbor, ME while back-crossing the nude mutation into several, well-characterized inbred strains. Genetic mapping of this new mutant, termed streaker (nu^{str}), localized the mutation to 20.6 ± 4.9 recombination units from rex on chromosome 11. Further crossing of this mutant with established outbred nude mice demonstrated that the gene causing the athymic abnormality and the abnormal hair formation in nude was allelic.[8] This has caused investigators to conclude that regulation of thymus development and epithelial development which leads to hair formation is genetically controlled by either the same gene or a group of very closely related genes.[2]

The nude rat was first noted in 1953, rediscovered and reported in 1977.[1,9] The description of the first nude rat mutation was in the Rowett strain. Another mutation gave rise to another strain of nude rats that was discovered and reported by Barrage in 1979.[1] That mutant has fewer hair than the Rowett mutant; however, the genes for the phenotypic changes noted with "nude" were, like the nude mouse, allelic.

Nude mice have a shortened life span unless they are derived and maintained in a germ-free status and have not acquired mouse hepatitis virus via vertical transmission.[10-12] Nude mice derived in this manner can have a life span that exceeds 12 months. Nude mice first described by Flannigan did not survive beyond 6 months,[6,10] typically undergoing "wasting" as a terminal event. This phenomenon appears to be inducible, as mice that have undergone manipulation, e.g., surgery, etc., have a hastened onset of "wasting". Years of experience in our laboratory, as well as that of others, suggest that this hastened "wasting" is largely secondary to activation of vertically transmitted viral hepatitis in mice.[10]

The nude rat is more hearty and our 10 years of experience show that they can be expected to survive for more than 1 year when maintained in a pathogen-free environment. Both sexes of homozygous nude rats are fertile. This contrasts with nude mice where only the female can be depended upon to be fertile.[1,12]

II. THE IMMUNE SYSTEM OF NUDE MICE AND NUDE RATS AND XENOGENEIC SKIN GRAFTS

The first transplantation of xenografts of human skin was accomplished with full-thickness foreskin engrafted onto the corpus carnosis of the lateral thoracic cage of nude mice by Reed and Manning in 1973.[13] This accomplishment was repeated by Rygaard in 1974.[14] Engraftment of split-thickness skin xenografts (mice and guinea pigs) to nude rats was initially described by Festing.[1] In this description he noted rejection of grafts after prolonged periods of engraftment.

The immune system of nude rats and mice has a limited capacity to respond to antigen.[1,6,8] Antibodies to xenograft antigens following engraftment of skin xenografts to nude mice have been reported; however, most typically this does not result in graft rejection. It does appear that the administration of fresh exogenous complement to these animals can lead to rejection of skin xenografts.[15,16] Despite this observation, the successful engraftment of human skin, as well as skin from other species to athymic mice, is well recognized.[5,17]

This is not the case for nude rats. We concur with Festing's observations, noting that nearly 90% of split-thickness skin grafts of human skin on nude rats will be rejected late in the course of engraftment, up to 42 days.[18] This rejection appears to be humorally mediated. However, grafted nude rats will maintain a human skin graft for the length of the observation period, 120 days, if treated with cyclosporine, 25 mg/kg/day for 21 days as a subcutaneous injection.[18] This treatment protocol has since been modified because of detrimental side effects, we now treat with a single 20 mg/kg dose of cyclosporine on the day of transplantation, with 5 to 10 mg/kg being administered subcutaneously every other day thereafter. This dose of cyclosporine maintains the grafts but does, reversibly, modify barrier function of skin, see below.

III. TRANSPLANTATION OF HUMAN SKIN TO ATHYMIC RODENTS

A. NORMAL AND MODIFIED HUMAN SKIN

The greatest experience in transplanting human skin to congenitally athymic rodents has been with remnants of skin from elective surgery, abdominoplasties, face lifts, breast reduction, etc. Remnants are trimmed to a thickness similar to the skin of the mouse, approximately 300 to 500 μm, and transplanted to the lateral thoracic cage. At this thickness, adnexal structures are lost as engraftment ensues; however, an occasional hair follicle will persist and hair growth will be noted.[5,17]

Other types of skin that have been grafted onto athymic rodents are listed in Table 1. Skin grafts, thicker than 500 μm, can be grafted if derived from selected donor areas and are not more than 1 cm in diameter. Experience in our laboratory has demonstrated that greater than 50% of grafts of this size from skin of the face and scalp will survive, even when they are as thick as 1 mm. Adnexal structures can be expected to persist in such grafts post-engraftment. Grafts bearing sebaceous glands from remnants of facial skin not only persist, but are functional, i.e., they enlarge secondary to exogenously administered testosterone (see below).[19]

TABLE 1
Types of Human Skin Grafts on Athymic Rodents

Type of graft	Observation	Ref.
Split-thickness skin grafts	Source for grafts is readily available, success rate of engraftment is high, adenexal structures are lost	5 17
Grafts bearing sebaceous glands	Glands persist and are functional, i.e., respond to testosterone	19
Scalp grafts (3 mm plugs or 1 mm thick split-thickness)	Hair growth is observed even if dermal papilla are not included	20 (see text)
Fetal skin (8–19 weeks)	Maturation occurs, with full complement of appendages developing; Merkel cells are of epidermal origin	20 22 23
Rat human skin sandwich flap (RHSSF) on the athymic rat	Provides human skin that is viable, served by mammalian mediators and nutrients and an accessible isolated vasculature	50 51 52 53 67
Skin equivalent, i.e., an artificial dermis with an epidermis generated *in vitro*	System to analyze feasibility of methods for generating a *in vitro*, and methods to generate a diseased which can be transplanted to the *in vivo* environment of athymic rodents	24 25 26 27 32
Genetically modified skin	Human keratinocytes transfected with human growth hormone have persistence of gene after transplant to nude mice	28 29

Scalp skin can be transplanted as 3 mm plugs, or as split-thickness grafts (unpublished personal experience). When grafted at a thickness of approximately 0.5 mm or more, hair growth can be expected. Curiously, when 0.5 mm split thickness scalp skin is transplanted to the athymic rat being treated with cyclosporine, an immediate and continuous hair growth is noted despite the fact that most, if not all, of the dermal papillae have been left behind. This observation of immediate growth is not witnessed in the absence of cyclosporine (unpublished observations, G. Krueger and A. Gilhar, 1987).

Scalp skin grafts, 2 mm in diameter taken with a punch biopsy from patients with alopecia areata, have been successfully grafted onto nude mice.[20] This success prompted an unreported series of experiments which have been directed at the feasibility of transplanting human scalp grafts to nude mice as a system to explore the pathology, pharmacology, toxicology, and physiology of hair growth. The following are our conclusions.

First, scalp skin grafts, 3 mm in diameter, can be successfully grafted to nude mice by placing a small slit, ~5mm in length, in the skin overlying the lateral thoracic cage of the mouse and inserting the 3 mm graft into a subcutaneous pocket that is prepared by blunt dissection. As many as five 3 mm grafts can be placed in five slits and their subcutaneous pockets in an area that is ~1.5 cm in diameter.

Second, scalp skin from face lift remnants, as well as from scalp reductions, can be utilized for grafting. Prior to harvesting the grafts, the hair needs to be closely trimmed or shaved. Orientation is critical, as the grafts need to be taken parallel to the shaft of the hair. After the graft has been taken, and prior to transplantation, excess fat needs to be trimmed from the grafts. To assist in orienting the grafts pre-grafting as well as for optimal histological evaluation, i.e., transverse sectioning post-grafting, we have found it necessary to permanently mark the surface. This is accomplished with a tattooing instrument using carbon ink and tattooing the scalp skin before the grafts are taken. The tattoo provides a readily identifiable mark to assess engraftment and is useful to monitor for slippage of the graft beneath the host mouse skin with subsequent overgrowth by the adjacent mouse skin.

Third, success can be enhanced by attaching the 3 mm scalp grafts to an 8 mm disc of vapor-permeable adhesive prior to transplantation. This will facilitate affixing the graft to the surrounding skin and it helps prevents slippage of the graft to below the surface with loss of the graft secondary to overgrowth by mouse skin.

Fourth, as seen with hair transplants in human subjects, there are fewer hair follicles per graft after engraftment, relative to the number present at the time of grafting. However, this stabilizes by week 4. There is progressive growth of hair.

Fifth, the sex of the recipient mouse does not alter either the total number of hair follicles per graft or the number of terminal/anagen follicles per graft when grafts are from patients suffering from androgenetic alopecia.

Sixth, the grafts can be readily removed and transverse histologic sections from these grafts can be used to determine whether hair is in telogen or anagen growth phase.[21]

Fetal scalp skin, without histologic evidence of adnexal structures, (8 to 19 weeks of gestation) has been transplanted to nude mice. Following transplantation the development of a full complement of the adnexal structures inherent to the site of the donor tissue has been reported.[22] Using histologic assessment of this type of skin transplant on nude mice, investigators have recently shown that Merkle cells are of epidermal origin.[23] Access to embryonic skin is limited, thus this approach will have limited application.

Grafting just the epidermis appears possible. Recently a multi-step procedure has been described for engrafting donor epidermis directly onto mouse skin.[24,25]

It is intuitive that a very significant advance in clinical medicine would come with the ability to generate an artificial skin *in vitro*, also referred to as a skin-equivalent. This skin could be transplanted onto human subjects to replace burned skin, cover skin ulcers, or possibly even to correct a disorder inherent to skin. Such skin would, ideally, be non-immunogenic, functional, and persistent or slowly undergo resorption in the transplanted state. Current technology dictates that skin equivalents have an artificial dermis that can be impregnated with fibroblasts and covered with keratinocytes or an artificial barrier. The cellular constituents in such a skin-equivalent can be either allogeneic or, if they can be generated from the patient needing the artificial skin, autologous. In each scenario, there are questions as to the function, survival time, etc. of the artificial skin in the grafted state. Being able to observe such grafts in the *in vivo* state of the nude rat or mouse has and will answer at least some of the preclinical questions inherent to this technology.[24-27]

The athymic mouse has also been used as a system to explore the feasibility of using genetically modified keratinocytes to correct heritable disorders. Recent experiments have demonstrated that human keratinocytes transfected with human growth hormone can be transplanted to nude mice and that the transfected gene persists after transplantation. It is anticipated that the model system described herein can be used to determine the potential for successfully correcting selected heritable diseases, specifically, those that are characterized by a genetically defined protein deficiency and those that are characterized by a genetic aberration which leads to genetic abnormalities that are confined to the skin.[28-31]

B. DISEASED SKIN

Various skin diseases have been studied after transplantation to athymic rodents (see Table 2). To review each is beyond the scope of this chapter. The table is presented for general interest as there is little to no data relative to the effect of toxic substances on already diseased skin.

TABLE 2
Skin Diseases Studied Via Transplantation to Athymic Rodents

Disease studied	Principle finding	Ref.
Psoriasis	Uninvolved and involved skin are equally diseased; inflammatory system necessary for expression of lesion of psoriasis	68 69 70
Lamellar ichthyosis	Increased epidermal cell turnover and histology persist; defect is inherent to the epidermis	71
Epidermolytic hyperkeratosis	Disease features persist; defect is inherent to skin	72
Essential fatty acid deficiency (EFA)	EFA skin corrects on transplantation to nude mice; metabolic defects of skin can be corrected	72
Alopecia areata	Grafts from afflicted sites grow hair, thus the pathogenesis is humoral	20
Vitilgo	Involved split-thickness grafts repigment, epidermal grafts do not; thus the pathogenesis is humoral, the dermis contains melanocytes that spontaneously migrate into the afflicted epidermis	73 74
Melanoma	Metastatic implants grow more readily than primary tumors; the former have been used as a model for chemotherapy for melanoma	77 78
Warts	Warts do not develop in transplanted human skin escharified with HPV but do develop in foreskins impregnated under the renal capsule	79 80 81
Experimental dermatophytosis	Dermatophytosis was induced on human skin grafts, nude mouse skin is resistant to infection, some dermatophytes will more readily infect and persist than others	82
Neonatal lupus erythematosus; subacute cutaneous lupus	Injection of serum with anti-Ro (SSA) into mice with human skin grafts shows specific binding; this antibody is likely to be involved in the pathogenesis of these diseases	83 84

The ability to study human skin disease on a rodent has caused both the academic and the industrial investigator to want to exploit this system for toxicologic, therapeutic, and pathogenic studies. Unfortunately, split-thickness skin graft donation by diseased subjects is associated with significant morbidity, scarring, infection, etc. For this reason, it is difficult to predictably obtain sufficient quantities of diseased skin to do more than pivotal experiments. With the advent of systems described above it may be possible to generate diseased skin. This presumes that diseased fibroblasts and/or diseased keratinocytes in combination with and/or on an artificial dermal matrix would evolve into a skin that would have sufficient markers of disease so as to allow further insights into toxic, therapeutic, as well as pathogeneic, aspects of disease.[32]

IV. TECHNICAL ASPECTS OF TRANSPLANTING HUMAN SKIN TO ATHYMIC RODENTS

A. TRANSPLANTING SPLIT-THICKNESS SKIN GRAFTS

The methodology for transplanting split-thickness human skin has been described.[5,17] While the rate of successful engraftment with this type of skin is high, failures do occur.

More than 15 years experience has shown that most failures are technical and are secondary to one or more of the following.

First, grafts are placed too near or cross the dorsal spine. In this scenario many grafts do not vascularize sufficiently and subsequently undergo partial to complete necrosis.

Second, the graft is not held firmly against the recipient graft bed. This can be obviated by affixing the dressing more tightly to the animal, directly suturing the central part of the graft to the underlying musculature or increasing the pressure on the graft with a stint dressing.[27]

The third reason for failure is due to the thickness of the graft. Grafts less than 100 μm in thickness are difficult to obtain and generally do not survive after engraftment. Grafts thicker than 400 μm also do not survive well unless they are from the face or scalp.

Fourth, the donor skin is held under conditions which result in death of some or all components of skin. It is our opinion that the epidermis is more vulnerable, in this regard, than the dermis. It has been shown that split-thickness skin can be held in complete media under refrigeration for up to 3 weeks and still be successfully engrafted.[33] If the epidermis is not viable, it will be overgrown by mouse epidermis. In this scenario, because the existing human dermis is the substrate, the skin graft takes on the appearance of human skin. However, the skin is not pigmented. A non-pigmented graft suggests an overgrowth of murine epidermis (unpublished observations). This can be proven by preparing a cryostat section of a biopsy of such skin and analyzing same for species specific class I major histocompatibility antigens with routine immunohistologic techniques.[34]

Following transplantation, it is customary to witness a sloughing of part of the epidermis, identified by others as a "ghost".[5,13,17] Grafts can be expected to become hyperpigmented, will display some acanthosis of the epidermis, and will shrink in size such that after 60 to 90 days the graft is typically ∼ 60 to 70% of the size it was immediately after grafting.[5,17,35]

Another method for transplanting human skin to nude rodents is to place the graft subcutaneously and unroof it at a later date. In this technique, a subcutaneous space is generated by blunt dissection through an incision in the host skin. The split-thickness skin graft is slipped through the incision and placed in the subcutaneous pouch such that the dermis is in contact with the musculature of the body wall. The incision is closed and the graft is fixed to the animal's skin with a suture or a surgical clip through both the host and the donor skin. Seven to ten days later, the animal's skin overlying the split-thickness skin graft is removed, the macerated keratin cleaned from the surface of the donor skin, and the edges of the animal's skin secured to the graft. This approach has the advantage of ease, but does require two surgical procedures. It is an effective way of transplanting a large graft and insuring good nutrition.[5]

V. COMPARATIVE RESPONSES OF TRANSPLANTED SKIN

Inherent to transplantation are questions relating to whether the transplanted skin retains features inherent to the donor or acquires the phenotypic responsiveness of the host. It is clear, if the engraftment occurs without signs of tissue compromise, from gross and microscopic analyses that split-thickness skin grafts retain the morphologic phenotype of the donor.[5,17,34,36]

Responsivenss of skin grafts to stimuli and persistence of cell types unique to skin are highlighted in Table 3. The relative resistance of nude mouse skin, compared to human or pig skin, to undergo a wave of epidermal proliferation in response to the topical application of tetradecanoyl-phorbal myristate (TPA) is curious. The fact that donor responsiveness persists after grafting serves to reassure that the host response, at least in response to agents which induce and inhibit epidermal proliferation, is not transferred to the graft.[37-39]

TABLE 3
Responsive Studies of Transplanted Human Skin

Study	Outcome	Ref.
Comparative induction of epidermal pro- liferation with TPA on graft vs. host skin	The inherently less responsive mouse skin does not dic- tate a lesser response by the graft, i.e., donor response persists	37 38 39
Response to thermal injury	Blister forms after 60°C for 20 sec	43
Biology of Langerhans cells of trans- planted skin	Mouse Langerhans cells do not move into the human graft; human Langerhans cells persist for life of graft and respond to UVR and steroids as if still on human subject	85 86 87 88 89 90
Biology of endothelial cells of trans- planted skin	The donor endothelium is gradually replaced by endothe- lial cells of the host	40
Surgical wound of graft	Granulation tissue is rodent, after healing, epidermis and dermis are human	41
Response to carcinogens	Cytochrome P–450 in liver is induced when applied to human skin grafts, tumors of mouse origin develop at graft-host interface, application of modulators of carcin- ogen-DNA adducts to grafts causes cells to enter S phase	91 92 93
Export of molecular species of graft to ro- dent host	UVR induction of IL-6 in human graft causes a transient increase in the host blood; apoliprotein-E made by graft epidermis is found in host blood	31 32

The observation that the granulation tissue that forms in a wound of a human skin graft on a nude mouse is murine, but that the wound in the healed state has both an intact epidermis and dermis of human origin, speaks both to the response capability and the persistence of cells and tissue remaining donor in origin. The observation that the endothelium of the grafts is revascularized with cells from the murine host is not understood.[40,41] It can be interpreted variously, as evidence that endothelial cells break off, migrate and reimplant, or that there is a gradual but relentless take over by the murine cells, which, for unknown reasons, have a selective growth advantage.

The observation that a molecular species unique to human tissue (e.g., human IL-6 and apolipoprotein-E) appear in the circulation of the grafted rodent serves to further highlight the utility of these systems to study biologic, physiologic, and pathologic processes unique to skin. It also demonstrates that products from the avascular epidermis can move across the basement membrane, through the surrounding dermis and into the vasculature.[31,32]

VI. PHARMACOLOGIC STUDIES: NUDE MOUSE-HUMAN SKIN GRAFT MODEL

A. NORMAL SKIN: RESPONSE TO PHARMACOLOGIC AGENTS AND AGENTS WHICH INDUCE PATHOLOGY

1. Application of Toxic Compounds to Grafted Skin

There is a small but significant literature which describes events that occur in xenogeneic skin grafts on nude mice following the topical application of toxic substances. Human skin grafts on nude mice are reported to undergo blister formation in responses to heat, 60°C for 20 sec. Microvesiculation is also noted following the topical application sulfur mustard (mustard gas).[42,43] The topical application of sulfur mustard to human skin grafts results in the formation of vesicles (microblisters) after a latent period of several hours. The sequence of pathologic events, in skin grafts treated with sulfur mustard, the separation of the epidermis from the dermis, as well as epidermal death, has been used to hypothesize a molecular basis for sulfur mustard induced vesication.[43] This hypothesis suggests that exposure of skin to sulfur mustard results in a dose- and time-related decrease in NAD+ levels. This hypothesis is supported by studies of events in skin after sulfur mustard was applied to human skin grafts on nude mice where a time-related decrease in NAD+ was noted and where the NAD+ levels preceded and correlated with the severity of pathology. Evidence to support this hypothesis were the experiments which demonstrated that the sulfur mustard induced NAD+ loss was prevented by appropriate inhibitors.[44]

The grafting procedure does not appear to alter the absorption process across skin grafts. In a comparative study of the epicutaneous, subcutaneous, and intravenous administration of soman (a cholinesterase inhibitor), on mouse skin grafted vs. non-grafted skin, it was noted that the permeation characteristics were the same in each type of skin. When soman was applied to human skin grafts, less inhibition of blood cholinesterase activity was noted, demonstrating that human skin was less permeable to this agent than mouse skin.[43]

Because ethanolamine, a nucleophilic compound, accelerates the hydrolysis of some chemical warfare agents it is under consideration as an agent to decontaminate skin. Following the topical application of ethanolamine to mouse skin or human skin on the mouse very similar amounts of ^{14}C ethanolamine appear to be absorbed when expired CO_2 is analyzed. Only a small amount was lost from the skin by evaporation. The bulk of the dose remained in the epidermis for the 24-h period following application. Radiometric and enzymatic assays showed that ethanolamine enhances the hydrolysis of topically applied diisopropylfluorophosphate.[45]

Trivalent organic arsenicals are potent vesicants. Under light microscopy, it can be demonstrated that a series of events occur after the topical application of phenyl dichloroarsine, a trivalent arsenical. These changes can be visualized as early as 2 h after application and continue to increase in severity for 48 h. The nude mouse skin reacts more promptly to topical application of phenyl dichloroarsine than human skin grafted to the nude mouse; yet, the histologic changes are very similar.[46]

These results demonstrate that human skin grafted to the nude mouse can be used to monitor the pathology that toxic substances inflict on human skin. It is probable that grafted skin could be used to monitor both the efficacy of decontaminating the skin following exposure to toxic materials, and the capacity for neutralizing toxic materials after application.

2. Comparison of Permeation of Test Agents through Grafted Skin

The question of whether topically applied compounds permeate differently through human skin *in situ* than human skin transplanted to the nude mouse has been addressed by Reifenrath et al. Experiments demonstrate that permeation of a series of compounds through pig skin *in situ* was very similar to that of pig skin transplanted to the nude mouse r = 0.97. The same series of compounds applied to human skin grafted to nude mice correlated well with permeation data from the existing literature r = 0.78.[47] It is concluded that transplanted human skin has barrier properties that are quite similar to that of the normal *in situ* state.

B. METHODOLOGY AND CONSIDERATIONS OF CUTANEOUS HALF-LIFE OF DRUG/TOXINS IN SKIN

Topical treatment or accidental exposure operates under the assumption that drug or toxin applied topically partitions into the stratum corneum, remains within the skin at the

FRESH TISSUE SECTIONING TECHNIQUE
FOR DRUG LOCALIZATION IN SKIN

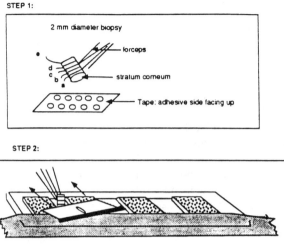

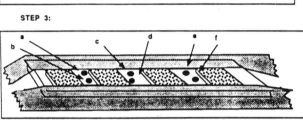

FIGURE 1.   Fresh skin sectioning technique. Step 1: Remove the stratum corneum from the core skin biopsy by applying the stratum corneum surface to a strip of adhesive, repeat × 10. Step 2: Place the epidermal side of the biopsy into cyanoacrylic cement and gently press this to the surface of a double adhesive tape applied to a glass microscope slide that is positioned between the two 110 μm thick glass coverslips shown in the figure. Shave the biopsy by gliding a razor blade over the coverslips and through the biopsy. Step 3: Re-dip the remainder of the core biopsy into cyanoacrylic cement and repeat step 3 until the core biopsy is completely sectioned.

appropriate location at sufficient concentration and for a duration of time sufficient to elicit a desired pharmacological action or undesired toxic response. Understanding the kinetics of disposition of a compound in the specific skin layers over time as a function of concentration represent the elements needed to make an experimental assessment of drug/toxin half-life in skin. As noted in the human skin on the nude mouse section (see above), human skin grafts on nude rodents can be used to make these kinds of assessments prior to clinical trials or accidental exposure.

Experiments with topical minoxidil and ketoconazole on human skin grafted to nude mice and with piroxicam on human skin of the rat-human skin sandwich flap (RHSSF), the latter to be described in the RHSSF section (see below), demonstrates the utility of these *in vivo* model systems for determining drug/toxin disposition in target skin layers and for determining drug half-life in skin. To quantitate drug/toxin in specific skin layers requires that a biopsy be taken from the treated area and that it be sectioned into wafers parallel to the skin surface (see Figure 1). This manual sectioning technique of fresh skin involves several steps.

First, after collecting a skin punch biopsy (2 to 8 mm diameter) from the treated skin site, it is tape-stripped ten times with Transpore™ tape (3M, St. Paul, MN) to collect the stratum corneum. Histology has confirmed that this tape-stripping method removes 90% of the stratum corneum.[48]

Second, the biopsy is dipped, epidermis-side down into cyanoacrylate cement and pressed against cellophane tape attached to a glass microscope slide positioned between two 110 μm thick glass microscope coverslips which act as a thickness guide. A single- edged stainless steel razor blade (Personna™) is used to shave the biopsy at a thickness dictated by the thickness of the glass coverslip (110 μm) to generate a skin wafer. The first skin wafer represents the remaining stratum corneum and all of the epidermis. The remaining skin wafers represent subsequent layers of the dermis.

Third, in the typical experiment, the ten strips are combined (the stratum corneum), these and the wafers are independently extracted and submitted to analytical methods to quantitate drug content. Quantitation of drug content in the various skin samples can be performed with radiolabeled drug or nonradiolabeled drug if adequate analytical methods are available. Data can be presented in a number of ways. We have found it useful to express the data as quantity per volume of tissue. This because the dermis represents a large volume of the tissue analyzed and volume as the denominator more accurately represents the actual concentration gradient of the solute *in vivo*.

C. GRAFTED HUMAN SKIN TO STUDY PERMEATION, DISTRIBUTION, AND BIOAVAILABILITY

In unpublished experiments, the permeation of six different minoxidil-containing preparations (extemporaneously prepared products, n = 5, vs. a non-extemporaneously prepared product (Rogaine®) through human skin grafts on nude mice was compared. A set volume of each product, spiked with ³H minoxidil, was applied to four different graft sources (remnant abdominoplasty skin). Four hours later, a 3-mm core of treated skin was taken (see Figure 1). The surface of the core was lightly swabbed with a dry Dacron™ applicator, the stratum corneum removed via ten tape-strippings and the remainder of the skin divided into 110 μm wafers, epidermis through dermis. Each wafer was solubilized and the amount of minoxidil in the various layers of the skin determined by liquid scintillation counting. This analysis was repeated 24 h later.

Chemical analysis of the six preparations revealed that one of the extemporaneously prepared compounds contained no minoxidil and two contained less than 0.3%, one contained 1.9%, one contained 2.36% minoxidil and Rogaine® contained 2% minoxidil. The distribution in skin, corrected for the volume of each compartment of minoxidil, in the various compartments of the skin 4 h after application is illustrated in Figure 2a. Contrast the product labeled MD HR CLIN PROD containing 2.36% minoxidil with the product labeled EXT PHAR PREP which contained 1.9% minoxidil. The increased concentration of minoxidil in MD HR CLIN PROD formulation does not provide an increased quantity in the stratum corneum. At 24 h, there is essentially no difference in the amount in the stratum corneum, epidermis, or dermis when the three products containing ∼ 2% minoxidil are compared (Figure 2b). If, however, one calculates drug half-life in the various skin components, appreciating the limitations of determining half-life in tissue with two time points, it is recognized that there are dramatic differences between the different products (Figure 3). The dermis is the probable target site for stimulating hair growth, as it contains the dermal papilla and the hair bulb, thus the quantity at this level of the skin is likely most critical.

A similar analysis has been conducted where the concentration of skin levels of ketoconazole were compared, over time following oral vs. topical administration of this drug to mice bearing human skin grafts. In these experiments, grafted mice were gavaged with ketoconazole (24 mg/kg) three doses/day for 3 days, or had 10 mg of a 2% ketoconazole/cm² applied as one dose/day for 3 days. At the end of the dosing regimens, residual drug

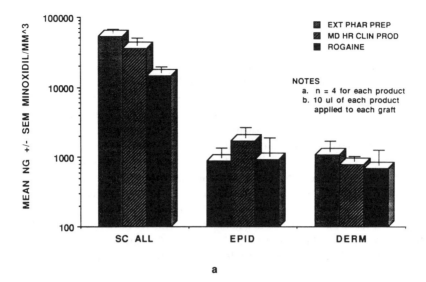

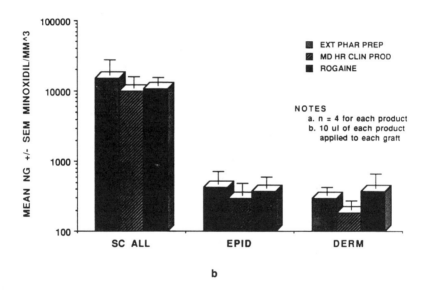

FIGURE 2. a. Comparison of the amount of minoxidil (ng) ± SEM in various compartments of skin, corrected for the volume (mm³) that compartment represents *in situ*, 4 h after three different minoxidil containing preparations were applied to human skin grafts on nude mice. b. Same as Figure 2a, except values are those present at 24 h after application of the minoxidil containing preparations.

was removed from treated sites and 3-mm biopsies obtained. These biopsies were tapestripped × 10 to remove the stratum corneum and sectioned into 110 μm wafers (see Figure 1). The stratum corneum and the wafers were extracted with organic solvents and the amount of ketoconazole determined by HPLC. Following topical administration, the concentration of ketoconazole in the stratum corneum was 20- 40-fold greater than in the epidermis-dermis confirming a concentration gradient through grafted skin. Drug was detected primarily in the stratum corneum after oral administration. Oral administration, corrected for dose, delivers 25 times more parent drug to the target site, the stratum corneum at 2, 4, and 24 h than a clinically relevant dose of topically applied ketoconazole. Bioavailability of the ketoconazole within the various layers was further quantitated by subjecting the extracted

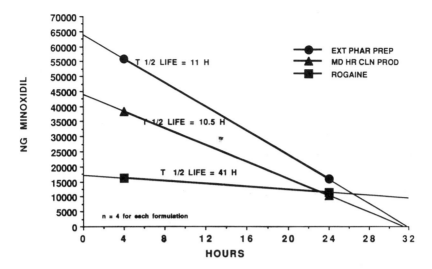

FIGURE 3. Comparison of the amount (ng) of minoxidil remaining in all layers of the treated skin at 4 and 24 h after topical application; half-life, presented as hours (H), was calculated from the values at the two time points.

tissue samples to an assay where ketoconazole inhibits the growth of yeast. Correlation between the amount of drug in the various skin layers and the inhibition as determined by this bioassay is good and suggests that this drug is not extensively metabolized to inactive intermediates in the skin.[49]

Experiments such as these demonstrate that human skin grafted to nude mice can be utilized to determine the pharmacokinetics of topical and oral drug delivery to skin, elimination drug as influenced by vehicle formulation, route of administration, and the potential for first pass metabolism. These same principles apply to the study of cutaneous toxins.

D. APPLICATION OF AGENTS TO SKIN GRAFTS BEARING ADENEXAL STRUCTURES

As noted above, skin bearing adenexal structures has been transplanted to nude mice and subjected to pharmacologic manipulation. Human skin containing sebaceous glands grafted to nude mice has been subjected to exogenous testosterone. In the treated group, sebaceous gland volume increased nearly fourfold which contrasted with a twofold decrease in the control group. Topical application of a known comedogen, Hallowax HN–34, to grafts that had been stimulated by testosterone, revealed twice as many microcomedoes in the Hallowax treated grafts as were seen in the vehicle treated graft at the end of the 6-week observation period.[19]

In another study the effects of drug on hair growth have been made. Because cyclosporine will induce hair growth in patients taking this drug, mice were grafted with alopecia areata skin and were divided into two groups to determine the effects of cyclosporine on hair growth.[20] One group received oral cyclosporine, 10 mg/kg/day, the other group did not receive cyclosporine. Results demonstrate that at the end of the 3-month observational period, cyclosporine treated animals had more hair per graft[(4.8)] than the nontreated group[(1.3)], $p < 0.01$. Further, the average length of hair was greater in the cyclosporine treated group than in the non-cyclosporine treated group, $p < 0.001$.[20] This and the foregoing experiment show that grafted adenexal bearing skin can be utilized to gain insights into the physiology and pathology of these specialized skin structures as well as the pharmacological management of disorders in these structures.

We conclude that these transplantation models provide systems to explore the effect of toxins on adenexal structures and functions such as hair growth and sebum production.

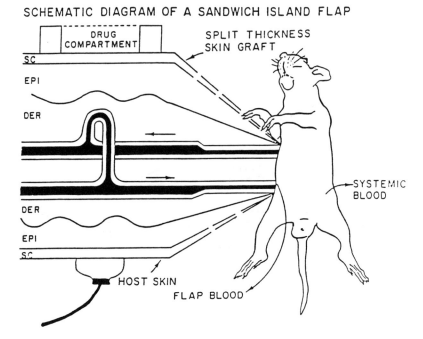

FIGURE 4.   Schematic of the rat/human skin sandwich flap (RHSSF).

VII. THE RAT-HUMAN SKIN SANDWICH FLAP (RHSSF)

A. INTRODUCTION

Compounds that are absorbed, following topical application, partition into and traverse the stratum corneum and diffuse across the epidermis and the papillary dermis until the capillary plexus is reached. At that point the compound, its metabolites, or both, either remain in the dermis or are transported to the rest of the body through the circulatory system. This scenario of absorption purposefully ignores compounds that might, because of the unique physicochemical properties partition only into the stratum corneum, epidermis, or dermis. An accurate assessment of the dynamic process of agents moving through the skin in the *in vivo* state, as a function of time, has been unavailable because of the lack of an appropriate model system. The need to understand percutaneous absorption *in vivo* for topical transdermal drug delivery, the rapid absorption of potentially toxic compounds, as well as the need for a more accurate assessment of absorption, metabolism, and compartmentalization of topically applied pharmaceutical agents fueled the development of the RHSSF by our laboratory. This flap, generated through a series of microsurgical procedures, is transplanted to the back of a congenitally athymic (nude) rat that has functional human skin on one side and rat host skin on the other. The flap is uniquely nourished by a single artery and drained by a single vein and has minimal collateral blood supply (Figure 4).[50–53] The vasculature is experimentally accessible and the skin is biologically functional and is served by mammalian nutrients and mediators. This model avoids many of the pitfalls inherent to *in vitro* models where nonviable tissue might result in erroneous conclusions. Similarly it avoids the problems of using fresh tissue in vitro where declining viability occurs as a function of time secondary to the lack of a physiologic blood supply. In typical *in vitro* testing the test compounds and/ or their metabolites must necessarily traverse the entire dermis, up to 1 cm, *in vitro* before entering the equivalent of the capillary bed *in vivo* which is immediately adjacent to the epidermis; the RHSSF avoids the potential problems associated with this type of *in vitro* analysis.[53]

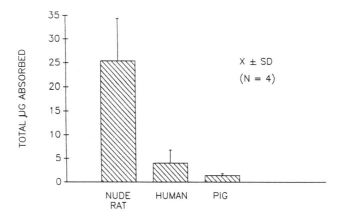

FIGURE 5. Total ^{14}C caffeine absorption across the grafted side of the skin sandwich flap, engrafted with nude rat, human, or domestic pig skin.

Because of its unique features the RHSSF has been proposed for:[51]

1. Quantitative analysis of percutaneous absorption process at the site of drug application
2. Disposition and kinetics of agents within the skin over time
3. The influence of cutaneous microcirculation on the disposition and clearance from the skin
4. Cutaneous metabolism
5. Comparison of *in vivo* and *in vitro* percutaneous absorption model systems
6. Mathematical modeling of the *in vivo* percutaneous absorption process
7. Correlative studies between this model system and human skin *in situ*
8. Bioavailability

B. THE RHSSF TO STUDY THE INFLUENCE OF SKIN SOURCE ON PERCUTANEOUS ABSORPTION

1. Type of Skin

Many investigators have noted dramatic differences in the extent of percutaneous absorption of topically applied compounds when various skin sources are compared. *In vitro* absorption of compounds across rabbit and rat skin is greater than that observed across human skin,[54–57] whereas pig skin and monkey skin are viewed as being more like human skin relative to permeation.[47,58,59] The skin sandwich flap model is conducive to grafting skin from various sources to a skin sandwich configuration, thus permitting a ready comparison of percutaneous absorption of a variety of compounds across different skin types *in vivo*. Absorption of ^{14}C caffeine in an ethanol vehicle across nude rat and human and domestic pig skin following the topical deposition of 36 μg/cm^2 over 4 h released that total micrograms μg absorbed across nude rat was greater (25.4 $\pm$ 9 μg [mean $\pm$ SD, n = 4]) than human (4.0 $\pm$ 2.5 μg) or domestic pig skin (1.4 $\pm$ 0.4) (Figure 5). These differences *in vivo* corroborate those observed *in vitro*.[51]

2. Variability in Percutaneous Absorption

A major question that confronts all who do clinical trials is "why is there such variability between subjects"? Is it inherent to the population or is it secondary to poor design by the investigator? To analyze the variation in percutaneous absorption of topically applied compounds across human skin, the human skin of the RHSSF was utilized in a series of experiments. Absorption was measured across the same flap on different days, the same

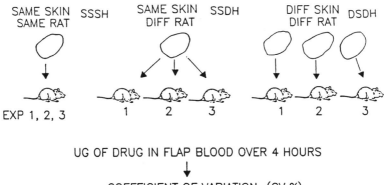

EXPERIMENTAL DESIGN

UG OF DRUG IN FLAP BLOOD OVER 4 HOURS
↓
COEFFICIENT OF VARIATION (CV %)
(SD/MEAN)*100

FIGURE 6.   Experimental design for elucidation of variability in percutaneous absorption.

skin grafted onto different rats and different skin sources grafted on to different rats (Figure 6).[60] With this experimental design, the influence of the skin source and animal host on the variability of percutaneous absorption of a test compound (caffeine) was elucidated. Variability [coefficient of variation (%)] in percutaneous absorption of caffeine was greatest (59%) with different skin sources grafted onto different rats and least (41%) when the same skin was used to generated different skin sandwich flaps (Figure 7). The variability decreased even more (30%) in rats grafted with the same skin and tested day to day. Variability in these *in vivo* experiments does not differ significantly from other *in vivo* studies using human subjects or *in vitro* studies with different skin sources.[47,61,63]

This variability did not demonstrate increasing or decreasing trends as a function of flap age. The inherent variability in skin penetration was substantiated by evaluating transepidermal water loss and cutaneous blood flow in these grafts. Both parameters demonstrated a variability of ∼30% in the same skin day to day. These data suggest that the host does influence barrier function and supports an inherent day to day variation in skin barrier function that cannot be eradicated. We speculate that this day to day variation simply reflects the dynamic nature of this continually replenished organ.

C. ALTERED PERCUTANEOUS ABSORPTION SECONDARY TO DRUGS
1. Oral Administration of Cyclosporine

Cyclosporine is used to prevent rejection of the human skin component of the sandwich flap on the athymic nude rat.[18] The influence of cyclosporine, (Sandimmune® intravenous formulation) administered orally to nude rats in their drinking water, on the percutaneous absorption of topically applied compounds across nude rat skin engrafted onto the nude rat, was investigated in a series of validation experiments.[64] The passive percutaneous absorption of topical ^{14}C benzoic acid across rat/rat skin sandwich flaps was increased fourfold following 2 weeks of this oral cyclosporine formulation (10 mg/kg/day) in the drinking water compared with no cyclosporine treatment. More recent experiments suggest that the effects are completely reversible after removal of cyclosporine for 2 days. The increase in the total amount of drug adsorbed was reflected in the flux measurement, with no significant alteration in cutaneous blood flow as measured with the laser Doppler velocimeter, or in barrier function as measured by transepidermal water loss. These experiments demonstrate that ingestion of one drug or its vehicle may influence the topical absorption of another drug.

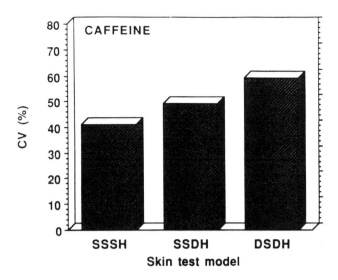

FIGURE 7. Coefficient of variation (CV%) in caffeine absorption across the human skin sandwich flap. SSSH = same human skin, same host; SSDH = same human skin grafted on different rats; DSDH = different human skin on different rats.

2. Topical Isotretinoin

Alteration in the permeability barrier can also be achieved with topical drug treatment. A single topical treatment of a human skin graft on the RHSSF with 1% isotretinoin in an alcohol vehicle resulted in a fourfold increase in the permeability coefficient (163 ng/cm^2) for topical progesterone in 5% ethyl acetate in PBS pH 7.2 vehicle 2 weeks later. When the permeability was measured in the same flap 4 weeks later, enhanced permeation was still noted (Figure 8). Performing the same progesterone percutaneous absorption experiment with this human skin flap 44 days after the isotretinoin pre-treatment demonstrated that 6 weeks were required for the progesterone permeability coefficient to return to control values. These data illustrate that a single dose of a topical drug can induce alterations in the skin barrier and that this effect can be long lasting.

D. INFLUENCE OF VEHICLE ON TOPICAL DRUG ABSORPTION

The influence of vehicle on transdermal drug delivery across human skin *in vivo* has also been investigated using the RHSSF and estradiol in a series of ethanol concentrations. Maintaining constant estradiol concentration in ethanol over the experimental time period was requisite to elucidating the influence of vehicle on transdermal estradiol flux across human skin *in vivo*. Therefore a special donor cell was created from these experiments. It has the following characteristics: lightweight, so as not to diminish cutaneous blood flow in the sandwich flap; minimal diffusional boundary layer secondary to a magnetic stirrer above the skin surface (S); capacity to maintain the temperature of the drug solution in the cell (Te); and a mechanism to deliver a constant concentration of solute to the exposed skin surface (CC). It is referred to as the STeCC cell[65] (Figure 9). Constant concentration of various ^{3}H estradiol solutions were infused at a rate of 1.0 ml/h through the STeCC cell via an infusion pump (Harvard Apparatus Biologicals, South Natick, MA). This resulted in steady state flux of ^{3}H estradiol in these vehicles across the human skin sandwich flap into the local flap blood within 30 to 60 min (Figure 10a, b).[66] Flux of tracer concentrations (0.31 μg/ml) of ^{3}H estradiol in 95 and 75% ethanol vehicles across the human skin sandwich flap was 10- and 30-fold less, respectively, than the flux of similar concentrations of ^{3}H estradiol in phosphate buffered saline pH 7.2 (PBS). The decreased ^{3}H estradiol flux observed

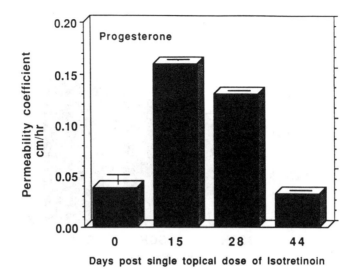

FIGURE 8.   Influence of isotretinoin pretreatment on the permeability of progesterone.

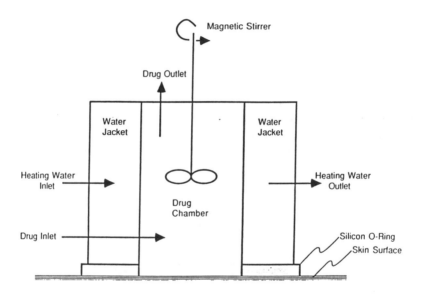

FIGURE 9.   Schematic of the STeCC donor cell.

with the ethanol vehicles directly reflected the tenfold decrease in the apparent skin-to-vehicle partition coefficients in those ethanol vehicles. Vehicle effects on the stratum corneum thickness and calculated diffusivitiy of the human skin were not significant.

Solubility of estradiol in these vehicles also influences the flux of the solute across human skin. Saturated solution of ^{3}H estradiol in 95% ethanol produce ^{3}H estradiol fluxes across human skin *in vivo* that are 30,000-fold greater than tracer solutions in the same vehicle. Further, saturated solutions of ^{3}H estradiol in 95% ethanol produce 3,000-fold greater flux across human skin sandwich flap than saturated solutions of ^{3}H estradiol in PBS. Thus, both the concentration of estradiol in the vehicle and the skin-to-vehicle ^{3}H estradiol partition coefficient are critical to the resulting flux of solute across human skin *in vivo*. Figure 11 demonstrates how knowledge of two physicochemical parameters, solute

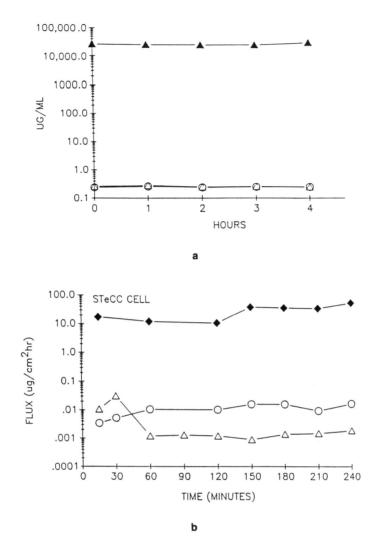

FIGURE 10. Delivery and flux of ^{3}H estradiol from three vehicles across the human skin sandwich flap using the STeCC donor cell. (a) Delivery of constant concentrations of ^{3}H estradiol in three vehicles to the human skin sandwich flap surface with the STeCC cell. (b) Flux of ^{3}H estradiol across the human skin concentrations in 95% ethanol, closed diamond = saturated concentrations of estradiol in 95%. ▲ = saturated solution of estradiol in 95% ethanol, △ = tracer estradiol concentration in 95% ethanol, ○ = tracer estradiol concentration in PBS.

concentration, and the apparent partition coefficient can be used to predict the actual *in vivo* flux of ^{3}H estradiol across the *in vivo* human skin sandwich flap.[66]

E. USE OF THE RHSSF TO DETERMINE HALF-LIFE OF DRUGS IN HUMAN SKIN

Use of the skin sectioning technique (see Figure 1), to quantitate disposition and cutaneous half-life in the RHSSF of a topically applied drug has been demonstrated with piroxicam, a nonsteroidal anti-inflammatory drug. Topical application of ^{3}H piroxicam (Pfizer Pharmaceutical Inc., Groton, CT) (64 μg/cm^2) in a gel vehicle to the surface of three different human skin sandwich flaps (surface area of 7.5 ± 0.1 cm^2 mean ± SD) for 8 h produces the drug profiles through the treated skin site shown in Figure 12a. In this experiment, the stratum corneum was not tape-stripped before skin sectioning and thus the drug content in

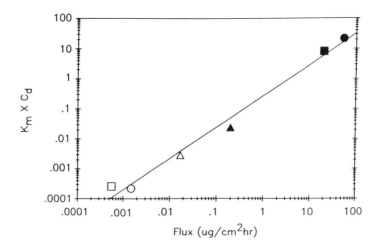

FIGURE 11.   Prediction of ^{3}H estradiol flux across the human skin sandwich flap using *in vitro* physicochemical parameters. K_m = partition coefficient, C_d = estradiol concentration solution.

the first section represent drug in both the stratum corneum and the epidermis. The majority of drug in the skin is located in the first three skin sections, with the greatest amount of ^{3}H piroxicam in the first skin section. Thereafter, the amount of drug decreases throughout the remaining skin layers, consistent with concentration gradient and diffusion principles.

As noted earlier the skin sectioning technique can be used to calculate drug half-life in skin (Figure 12b). At the end of the dosing interval (8 h), in the above experiments, residual ^{3}H piroxicam was removed from the skin surface with three saline-soaked cotton swabs and 2-mm diameter punch biopsies were taken from the treated human skin site at 8, 24, and 48 h. These biopsies were analyzed for total drug content. Total amount of ^{3}H piroxicam in the skin biopsy at 8 h was 130 ng, 2.5-fold greater than seen 16 h after the drug was removed and 9-fold greater than 40 h after drug removal. Plotting the log of drug content over time reveals that elimination over these time periods is linear (r = 0.9999) and estimates a cutaneous half-life of ^{3}H piroxicam of 10.8 h. Flux was sufficiently low so as to not allow detection of ^{3}H piroxicam in the venous drainage of the flap during the 4 h observational period following topical application.

VIII. SUMMARY

This review of the literature and our laboratories experience illustrates the use of the RHSSF and the orthotopically grafted human skin on the nude mouse (OHNM) to study toxicology in human skin, quantitate bioavailability of topical drug products, and to study the effect of vehicles on percutaneous absorption. This approach seems particularly useful when:

1.   The amount, distribution, and cutaneous half-life of a compound in human skin are important.
2.   There is a need for risk assessment following exposure to and through human skin.
3.   There is a need to develop a strategy as to the frequency and concentration of application.
4.   There is a need to evaluate detoxification procedures.
5.   There is a need to determine if the topically applied compound is metabolized or alters the metabolic capacity of skin.

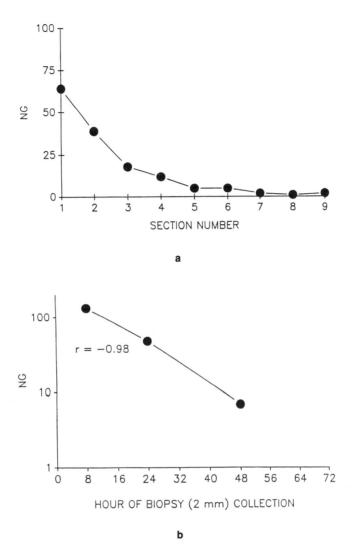

a

b

FIGURE 12. Skin concentrations of topical ^{3}H piroxicam in the human skin sandwich flap. (a) Concentration profile of ^{3}H piroxicam through human skin from a treated skin site. (b) Estimated cutaneous half-like of ^{3}H piroxicam in human skin.

REFERENCES

1. **Festing, M. W. F.,** Characteristics of nude rats, in *Proc. 3rd Int. Workshop on Nude Mice,* Reed, N. D., Ed., Gustav Fischer, New York, 1982, 41–49.
2. **Rigdon, R. and Packchanian, A.,** Histologic study of the skin of congenitally athymic "nude" mice, *Tex. Rep. Biol. Med.,* 32, 711–723, 1974.
3. **Eaton, G. J.,** Hair growth cycles and wave patterns in "nude" mice, *Transplantation,* 22, 217–222, 1976.
4. **Reed, C. and O'Donoghue, J. L.,** The hairless athymic guinea pig, in *Proc. 3rd Int. Workshop on Nude Mice,* Reed, N. D., Ed., Gustav Fischer, New York, 1982, 51–57.
5. **Black, K. and Jederberg, W.,** Athymic nude mice and human skin grafting, in *Models in Dermatology,* Lowe, N. and Maibach, H. I., S. Karger, Basel, 1985, 228–239.

6. **Hansen, C. T.,** The nude gene and its effects, in *The Nude Mouse in Experimental and Clinical Research,* Fogh, J. and Giovanaella, B. C., Eds., Academic Press, New York, 1978, 1–13.

7. **Pantelouris, E.,** Absence of thymus in a mouse mutant, *Nature,* 217, 370–371, 1968.

8. **Shultz, L. D., Bedigian, H. G., Heiniger, H.-J., and Eicher, E. M.,** The congenitally athymic streaker mouse, in *Proc. 3rd Int. Workshop on Nude Mice,* Reed, N. D., Ed., Gustav Fischer, New York, 1982, 33–39.

9. **Festing, M., May, D., Connors, T., Lovell, D., and Sparrow, S.,** An athymic nude mutation in the rat, *Nature,* 274, 365–366, 1978.

10. **Fujiwara, K.,** Spontaneous virus infections in nude mice, in *The Nude Mouse in Experimental and Clinical Research,* Fogh, J. and Giovanaella, B. C., Eds., Academic Press, New York, 1982, 1–18.

11. **Nomura, T. and Kagiyama, N.,** Importance of microbiological control in using nude mice, in *Proc. 3rd Int. Workshop on Nude Mice,* Reed, N. D., Ed., New York, 1982, 11–21.

12. **Ediger, R. and Giovanella, B. C.,** Current knowledge of breeding and mass production of the nude mouse, in *The Nude Mouse in Experimental and Clinical Research,* Fogh, J. and Giovanaella, B. C., Eds., Academic Press, New York, 1982, 16–28.

13. **Reed, N. and Manning, D.,** Long-term maintenance of normal human skin on congenitally athymic (nude) mice, *Proc. Soc. Exp. Biol. Med.,* 143, 350–353, 1973.

14. **Rygaard, J.,** Skin grafts in nude mice. III. Fate of grafts from man and donors of other taxonomic classes, *Acta Pathol. Microbiol. Scand.,* 82, 105–112, 1974.

15. **Gerlag, P., Capel, P., Berden, J. and Koene, R.,** Antibody response and skin graft rejection in the nude mouse, *Transplant. Proc.,* 9, 1179–1182, 1977.

16. **Koene, R., Gerlag, P., Jansen, J., Hagemann, J., and Wijdeveld, P.,** Rejection of skin grafts in the nude mouse, *Nature,* 251, 69–70, 1974.

17. **Krueger, G. and Briggaman, R.,** The nude mouse in the biology and pathology of skin, in *The Nude Mouse in Experimental and Clinical Research,* Fogh, J. and Giovanella, B. C., Eds., Academic Press, New York, 1982, 301–322.

18. **Gilhar, A. et al.,** Description of and treatment to inhibit the rejection of human split-thickness skin grafts by congenitally athymic (nude) rats, *Exp. Cell Biol.,* 54, 263–274, 1986.

19. **Petersen, M. A., Zone, J. J., and Krueger, G. G.,** Development of a nude mouse model to study human sebaceous gland physiology and pathophysiology, *J. Clin. Invest.,* 74, 1358–1365, 1984.

20. **Gilhar, A. and Krueger, G. G.,** Hair growth in scalp grafts from patients with alopecia areata and alopecia universalis grafted onto nude mice, *Arch. Dermatol.,* 123, 44–50, 1987.

21. **Headington, J. T.,** Transverse microscopic anatomy of the human scalp: a basis for a morphometric approach to disorders of the hair follicles, *Arch. Dermatol.,* 120, 449–456, 1984.

22. **Lane, A. T., Scott, G. A., and Day, K. H.,** Development of human fetal skin transplanted to the nude mouse, *J. Invest. Dermatol.,* 93, 787–791, 1989.

23. **Moll, I., Lane, A. T., Franke, W. W., and Moll, R.,** Intraepidermal formation of Merkel cells in xenografts of human fetal skin, *J. Invest. Dermatol.,* 94, 359–364, 1990.

24. **Green, H.,** Regeneration of the skin after grafting of epidermal cultures, *Lab. Invest.,* 60, 583–584, 1989.

25. **Barrandon, Y., Li, V., and Green, H.,** New techniques for the grafting of cultured human epidermal cells onto athymic animals, *J. Invest. Dermatol.,* 91, 315–8, 1988.

26. **Bosca, A. R., et al.,** Epithelial differentiation of human skin after grafting onto nude mice, *J. Invest. Dermatol.,* 91, 136–141, 1988.

27. **Boyce, S., Foreman, T., and Hansbrough, J.,** Functional wound closure with dermal-epidermal skin substitutes prepared in vitro, *Tissue Eng.,* 104, 81–86, 1988.

28. **Morgan, J. R., Barrandon, Y., Green, H., and Mulligan, R. C.,** Expression of an exogenous growth hormone gene by transplantable human epidermal cells, *Science,* 237, 1476–1479, 1987.

29. **Morgan, J. R. and Eden, C. A.,** Retroviral-mediated gene transfer into transplantable human epidermal cells, in *Clinical and Experimental Approaches to Dermal and Epidermal Repair: Normal and Chronic Wounds,* Barbal, A., Pines, E., Caldwell, M., and Hart, T. R., Eds., Wiley-Liss, New York, 1991, 417–428.

30. **St. Louis, D. and Verma, I. M.,** An alternative approach to somatic cell gene therapy, *Proc. Natl. Acad. Sci. U.S.A.,* 85, 3150–3154, 1988.

31. **Fenjves, E., Gordon, D., Pershing, L., Williams, D., and Taichman, L.,** Systemic distribution of apolipoprotein E secreted by grafts of epidermal keratinocytes: implications for epidermal function and gene therapy, *Proc. Natl. Acad. Sci. U.S.A.,* 86, 8803–8807, 1989.

32. **Krueger, G. G. and Jorgensen, C. M.,** Experimental models for psoriasis, *J. Invest. Dermatol.,* 95, 56S–58S, 1990.

33. **Cram, A. E., Domayer, M. A., and Scupham, R.,** Preservation of human skin: a study of two media using the athymic (nude) mouse model, *J. Trauma,* 25, 128–130, 1985.

34. **Merrell, S. W., Shelby, J., Saffle, J. R., Krueger, G. G., and Navar, P. D.,** An in vivo test of viability for cryopreserved human skin, *Curr. Surg.,* 43, 296–300, 1986.

35. **Bruengger, A., Heilbronner, R., Anderegg, M., Hubler, M., and Rohr, H. P.,** Human skin grafts on nude athymic mice: a light microscopic stereological study, *Arch. Dermatol. Res.,* 276, 78–81, 1984.

36. **Manning, D., Reed, N., and Shaffer, C.,** Maintenance of skin xenografts of widely divergent phylogenetic origin on congenitally athymic (nude) mice, *J. Exp. Med.,* 138, 488–494, 1973.

37. **Chambers, D. A., Cohen, R. L., Sando, J. J., and Krueger, G. G.,** Resistance to TPA-induced mitogenesis in the nude mouse, *Exp. Cell Biol.,* 52, 97–102, 1984.

38. **Krueger, G. G., Chambers, D., and Shelby, J.,** Epidermal proliferation of nude mouse skin, pig skin, and pig skin grafts failure of nude mouse skin to respond to the tumor promoter 12-O-tetradecanoyl phorbol 13-acetate, *J. Exp. Med.,* 152, 1329–1339, 1980.

39. **Krueger, G. G. and Shelby, J.,** Biology of human skin transplanted to the nude mouse, *J. Invest. Dermatol.,* 76, 506–510, 1981.

40. **Demarchez, M., Hartmann, D. J., and Prunieras, M.,** An immunohistological study of the revascularization process in human skin transplanted onto the nude mouse, *Transplantation,* 43, 896–903, 1987.

41. **Demarchez, M., Hartmann, D. J., Herbage, D., Ville, G., and Prunieras, M.,** Wound healing of human skin transplanted onto the nude mouse, *Dev. Biol.,* 121, 119–129, 1987.

42. **Papirmeister, B., Gross, C. L., Meier, H. L., Petrali, J. P., and Johnson, J. B.,** Molecular basis for mustard-induced vesication, *Fundam. Appl. Toxicol.,* 5, S134–S149, 1985.

43. **van Genderen, J., Mol, M. A. E., and Wolthus, O. L.,** On the development of skin models for toxicity testing, *Fundam. Appl. Toxicol.,* 5, S98–S111, 1985.

44. **Gross, C. L., Meier, H. L., Papirmeister, B., Brinkly, F. B., and Johnson, J. B.,** Sulfur mustard lowers nicotinamide adenine dinucleotide concentrations in human skin grafted to athymic nude mice, *Toxicol. Appl. Pharmacol.,* 81, 85–90, 1985.

45. **Klain, G. J., Reifenrath, W. G., and Black, K. E.,** Distribution and metabolism of topically applied ethanolamine, *Fundam. Appl. Toxicol.,* 5, S127–S133, 1985.

46. **McGown, E. L., Ravensway, T. V., and Dumlao, C. R.,** Histologic changes in nude mouse skin and human skin xenografts following exposure to sulfhydryl reagents: arsenicals, *Toxicol. Pathol.,* 15, 149–156, 1987.

47. **Reifenrath, W., Chellquist, E., Shipwash, E., Jederberg, W., and Krueger, G.,** Percutaneous penetration in the hairless dog, weanling pig, and grafted athymic nude mouse: Evaluation of models for predicting skin penetration in man, *Br. J. Dermatol.,* 111, 123–135, 1984.

48. **Pershing, L. K., Silver, B. S., Krueger, G. G., Shah, V. P., and Skelley, J. P.,** Feasibility of measuring the bioavailability of topical betamethasone diproprionate in commercial formulations using drug content in skin and a skin blanching bioassay, *Pharm. Res.,* 9, 45–51, 1992.

49. **Pershing, L. K. and Corlett, J.,** In vivo cutaneous kinetics of ketoconazole in human skin following oral and topical administration, *J. Invest. Dermatol.,* 96, 580 (abstr.), 1991.

50. **Krueger, G. et al.,** The development of a rat/human skin flap served by a defined and accessible vasculature on a congenitally athymic (nude) rat, *Fundam. Appl. Toxicol.,* 5, S112–S121, 1985.

51. **Pershing, L. and Krueger, G.,** Human skin sandwich flap model for percutaneous absorption, in *Percutaneous Absorption,* Bronaugh, R. L. and Maibach, H. I., Eds., Marcel Dekker, New York, 1989, 397–414.

52. **Pershing, L. K., Huether, S., Conklin, R. L., and Krueger, G. G.,** Cutaneous blood flow and percutaneous absorption: a quantitative analysis using a laser doppler velocimeter and a blood flow meter, *J. Invest. Dermatol.,* 92, 355–359, 1989.

53. **Wojciechowski, Z. et al.,** An experimental skin sandwich flap on an independent vascular supply for the study of percutaneous absorption, *J. Invest. Dermatol.,* 88, 439–446, 1987.

54. **Bartek, M. J., LaBudde, J. A., and Maibach, H. I.,** Skin permeability in vivo: comparison in rat, rabbit, pig and man, *J. Invest. Dermatol.,* 58, 114–123, 1972.

55. **Chowhan, Z. T. and Pritchard, R.,** Release of corticoids from oleaginous ointment bases containing drug in suspension, *J. Pharm. Sci.,* 64, 754–759, 1975.

56. **Tregar, R. T.,** The permeability of mammalian skin to ions, *J. Invest. Dermatol.,* 46, 16–23, 1966.

57. **Tregar, R. T.,** The permeability of skin to albumin, dextrans, and polyvinyl pyrrolodone, *J. Invest. Dermatol.,* 46, 24–27, 1966.

58. **Bronaugh, R. L.,** Determination of percutaneous absorption by in vitro techniques, in *Percutaneous Absorption,* Bronaugh, R. L. and Maibach, H. I., Eds., Marcel Dekker, New York, 1985, 239–258.

59. **Bronaugh, R. L. and Maibach, H. I.,** Percutaneous absorption of nitro-aromatic compounds, *J. Invest. Dermatol.,* 84, 180–183, 1985.

60. **Pershing, L. K., Conklin, R. L., and Krueger, G. G.,** Assessment of the variation in percutaneous absorption across the skin sandwich flap, *Clin. Res.,* 88, 511, 1987.

61. **Feldman, R. J. and Maibach, H. M.,** Absorption of some organic compounds through the skin in man, *J. Invest. Dermatol.,* 54, 399–404, 1970.

62. **Bronaugh, R. L. and Stewart, R. F.,** Methods for percutaneous absorption studies. IV. The flow-through diffusion cell, *J. Pharm. Sci.,* 74, 64–67, 1985.

63. **Franz, T. J.,** The finite dose technique as a valid in vitro model for the study of percutaneous absorption, *Curr. Probl. Dermatol.,* 7, 58–68, 1978.

64. **Pershing, L. K., Conklin, R. L., Manning, C. A., and Krueger, G. G.,** Effects of oral cyclosporine therapy on percutaneous absorption, in vitro versus in vivo, *Pharm. Res.,* 3, 53S, 1986.

65. **Silcox, G., Parry, G., Bunge, A., Pershing, L., and Pershing, D.,** Percutaneous absorption of benzoic acid across human skin. II. Prediction of an in vivo, skin-flap system using in Vitro parameters, *Pharm. Res.,* 7, 352–358, 1990.

66. **Pershing, L., Lambert, L., and Knutson, K.,** Mechanism of ethanol enhanced estradiol permeation across human skin in vivo, *Pharm. Res.,* 7, 172–177, 1990.

67. **Pershing, L. and Krueger, G.,** New animal models for bioavailability studies, in *Pharmacology and the Skin,* Shroot, B. and Schaefer, H, Eds., S. Karger, Basel, 1987, 57–69.

68. **Krueger, G. G., Chambers, D., and Shelby, J.,** Involved and uninvolved skin from psoriatic subjects: are they equally diseased?, *J. Clin. Invest.,* 68, 1548–1557, 1981.

69. **Krueger, G.,** A perspective of psoriasis as an aberration in skin modified to expression by the inflammatory/ repair system, in *Immune Mechanisms in Cutaneous Disease,* Norris, D. A., Ed., Marcel Dekker, New York, 1989, 425–445.

70. **Fraki, J., Briggaman, R., and Lazarus, G.,** Uninvolved skin from psoriatic patients develops signs of involved psoriatic skin after being grated onto nude mice, *Science,* 215, 685–687, 1982.

71. **Briggaman, R. A. and Wheeler, C. E.,** Lamellar ichthyosis: long term graft studies on congenitally athymic nude mice, *J. Invest. Dermatol.,* 76, 567–572, 1976.

72. **Briggaman, R.,** Localization of the defect in skin diseases analyzed in the human skin graft-nude mouse model, *Curr. Probl. Dermatol.,* 10, 115–126, 1980.

73. **Gilhar, A., Pillar, T., Eidelman, S., and Etzioni, A.,** Vitiligo and idiopathic guttate hypomelanosis, *Arch. Dermatol.,* 125, 1363–1366, 1989.

74. **Orme, R. L., Piepkorn, M. W., and Krueger, G. G.,** Pigmentary changes in normal and vitiliginous human skin grafted to the nude mouse, *Clin. Res.,* 34, 20A, 1986.

75. **Grimwood, R. E. et al.,** Transplantation of human basal cell carcinomas to athymic mice, *Cancer,* 56, 519–523, 1985.

76. **Grimwood, R. E., Glanz, S. M., and Siegle, R. J.,** Transplantation of human basal cell carcinoma to C57/BALB/C bg/bg-nu/nu (beige-nude) mouse, *J. Dermatol. Surg. Oncol.,* 14, 59–62, 1988.

77. **Barr, L. H., Gartside, R. L., and Goldman, L. I.,** The biology of human malignant melanoma, *Arch. Surg.,* 114, 221–224, 1979.

78. **Bellet, R. E. and Mastrangelo, M. J.,** Malignant melanoma: investigations in the nude mouse, in *The Nude Mouse in Experimental and Clinical Research,* Fogh, J. and Giovanaella, B. C., Eds., Academic Press, New York, 1982, 511–519.

79. **Cubie, H. A.,** Failure to produce warts on human skin grafts on nude mice, *Br. J. Dermatol.,* 94, 659–665, 1976.

80. **Brown, D. R., Chin, M. T., and Strike, D. G.,** Identification of human papillomavirus type 11 E4 gene products in human tissue implants from athymic mice, *Virology,* 165, 262–267, 1988.

81. **Kreider, J. et al.,** In vivo transformation of human skin with human papillomavirus type 11 from condylomata acuminata, *J. Virol.,* 59, 369–376, 1986.

82. **Kakutani, H. and Takahashi, S.,** Experimental chronic dermatophytosis in human skin grafted to nude mouse: inoculation of Trichopyhtons and histopathological evaluation, *J. Dermatol.,* 15, 230–240, 1988.

83. **Lee, L. A. et al.,** An animal model of antibody binding in cutaneous lupus, *Arth. Rheum.,* 29, 782–783, 1986.

84. **Lee, L. A., Gaither, K. K., Coulter, S. N., Norris, D. A., and Harley, J. B.,** Pattern of cutaneous immunoglobulin G deposition in subacute cutaneous lupus erythematosus is reproduced by infusing purified anti-ro (SSA) autoantibodies into human skin-grafted mice, *J. Clin. Invest.,* 83, 1556–1562, 1989.

85. **Czernielewski, J., Demarchez, M., and Prunieras, M.,** Human langerhans cells in epidermal cell culture, in vitro skin explants and skin grafts onto ''nude'' mice, *Arch. Dermatol. Res.,* 276, 188–292, 1984.

86. **Czernielewski, J., Vaigot, P., and Prunieras, M.,** Epidermal Langerhans cells—A cycling cell population, *J. Invest. Dermatol.,* 84, 424–426, 1985.

87. **Czernielewski, J. and Demarchez, M.,** Further evidence of the self-reproducing capacity of langerhans cells in human cells, *J. Invest. Dermatol.,* 88, 17–20, 1987.

88. **Krueger, G. G., Daynes, R., and Emam, M.,** Biology of langerhans cells: selective migration of langerhans cells into allogeneic and xenogeneic grafts on nude mice, *Proc. Natl. Acad. Sci. U.S.A.,* 80, 1650–1654, 1983.

89. **Krueger, G. G. and Emam, M.,** Biology of Langerhans cells: analysis by experiments to deplete Langerhans cells from human skin, *J. Invest. Dermatol.,* 82, 613–617, 1984.

90. **Krueger, G., Daynes, R., Roberts, L., and Emam, M.,** Migration of Langerhans cells and the expression of la on epidermal cells following the grafting of normal skin to nude mice, *Exp. Cell Biol.,* 52, 97–102, 1984.

91. **Das, M. et al.,** Carcinogen metabolism in human skin grafted onto athymic nude mice: a model system for the study of human skin carcinogenesis, *Biochem. Biophys. Res. Commun.,* 138, 33–39, 1986.

92. **Graem, N.,** Epidermal changes following application of 7,12-dimethylbenz[a]anthracene and 12-o-tetradecanoylphorbol-13-acetate to human skin transplanted to nude mice studied with histological species markers, *Cancer Res.,* 46, 278–284, 1986.

93. **Yohn, M. D., Lehman, T. A., Kurain, P., Ribovich, M., and Milo, G. E.,** Benzo[a]pyrene diol epoxide I modification of DNA i human skin xenografts, *J. Invest. Dermatol.,* 91, 363, 1988.

94. **Krueger, G. G. and Pershing, L. K.,** Skin xenografts and their utility in pharmacologic research, in *Pharmacology of the Skin,* Mukhtar, H., Ed., CRC Press, Boca Raton, FL, 1991, chap. 4.

Chapter 26

THE ISOLATED PERFUSED PORCINE SKIN FLAP

Jim E. Riviere

TABLE OF CONTENTS

0-8493-7357-3/93/$0.00 + $.50
© 1993 by CRC Press, Inc.

I. INTRODUCTION

Penetration with subsequent absorption of topically exposed chemicals through the skin is a major route for systemic effect. This has been appreciated for many years with pesticides[1,2] and numerous *in vitro* and *in vivo* animal models have been developed to predict *in vivo* human exposure. Recently, quantitative strategies using pharmacokinetic models have been applied to this problem. This area has been well reviewed in earlier sections of this text. Most of these experimental approaches used to date have assumed that the rate limiting barrier for the passive absorption of topically applied chemicals is the stratum corneum and physiochemical models for describing compound absorption by diffusion across a lipid domain (intercellular lipids of the stratum corneum) suffice. However, recently, workers have begun to appreciate that for certain chemicals, percutaneous absorption may be more complex, and total transdermal flux may be dependent upon both diffusion and partitioning phenomenon in the stratum corneum, cutaneous metabolism, the vasoactivity of the penetrating compound, and chemical affinity for specific tissue constituents. There are also inherent limitations to the *in vitro* systems used which preclude studying many of these effects. These topics, as well as general principles of percutaneous absorption, have been extensively reviewed in various recent books and monographs.[3-7] These reviews and other chapters of this text should be consulted for in-depth explanations of many of the points outlined in the present chapter.

II. ADVANTAGES OF AN INTACT VASCULARIZED *IN VITRO* SKIN FLAP MODEL

Skin is a complex tissue whose roles in mammals include:

1. Providing a barrier to the outside environment
2. Temperature regulation (sweating and altered blood flow)
3. Thermal insulation
4. Neurosensory functions
5. Immunological recognition
6. Endocrine factor release
7. Metabolism
8. External chemical communication
9. Shielding the body from harmful ultraviolet radiation

This functional heterogeneity is manifested in anatomical complexity. However, when skin models are developed for studying percutaneous absorption, often only the barrier function is considered. For many compounds, this has fortunately been acceptable since the stratum corneum is the rate limiting barrier. However, for other compounds, this assumption may not be valid if more complex partitioning, cellular binding, vascular or metabolic factors intervene.

Classic *in vitro* approaches utilize sections of skin (or model membranes) taken parallel to the surface. These may be composed solely of stratum corneum, or also include epidermis, dermis, and even subcutaneous tissue (full thickness). The thinner membranes are collected in hairless animal species in order to prevent holes in the preparation. These membranes are mounted in a diffusion cell and compound is placed on the surface of the skin. The underlying dermis is then bathed with a perfusate which ranges in complexity from saline to plasma. Static systems are those in which the reservoir is not changed while flow-through systems perfuse the reservoir continuously with fresh perfusate.

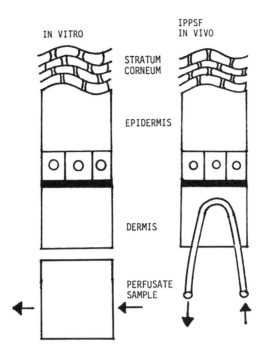

FIGURE 1. Simplified structure of skin illustrating lack of capillary bed in *in vitro* models, requiring xenobiotic to penetrate a non-physiologic dermal barrier before reaching perfusate.

The major difference between these *in vitro* models and skin *in vivo* is the pathway that an absorbing compound must follow to be absorbed into the systemic circulation. As can be appreciated from the schematic depicting this difference in Figure 1, penetrating compound must traverse the stratum corneum, epidermis and only a small amount of dermis before being taken up into the cutaneous capillary network. In the *in vitro* system, the capillaries are not present and additional dermis must often be traversed before reaching perfusate. This approach assumes that the microvasculature functions as a perfect sink, an assumption that may not be valid since a small fraction of some absorbed compounds may reach underlying subcutaneous fat and connective tissue.[8] Therefore, the transit times of compounds through the dermis will be significantly different. In thick preparations with abundant dermis, the dermis may in fact function as the receptor phase for certain chemicals.[9]

Additional variables which may be encountered *in vivo* include metabolism by the viable epidermis, dermis, or adnexial structures; altered capillary perfusion due to vasoactivity of the penetrating compound; or binding to specific cell types seen in viable dermis. Even recently developed human organ culture membranes suffer from this lack of anatomical complexity.

As has been previously mentioned, for many compounds, transport through the stratum corneum is the rate limiting barrier to percutaneous absorption. The postulated pathway for penetration is through the intercellular lipids and rate of penetration within a series of similar compounds generally correlates to the compound's lipid partition coefficient.[10–14] This is true for most polar and non-polar compounds. For very hydrophilic compounds, absorption across the skin may occur through "shunts" (e.g., hair follicles, sweat ducts), but the small fraction of skin surface area covered by these pathways severely limits the magnitude of the transdermal flux possible. There is also considerable debate as to whether "shunt" pathways even exist. For very lipid soluble compounds, the rate of penetration through skin may decrease from that calculated for diffusion alone because of the tendency for the penetrating chemical to partition at the stratum corneum-epidermis interface.[11] In fact, rather than acting as a barrier, the stratum corneum may function as a "sponge" for very lipid soluble penetrants and subsequent penetration through the viable epidermis may be the true rate limiting barrier.[9]

The cutaneous metabolism of xenobiotics has also recently received considerable attention.[9,15-20] It is now clear that the skin is capable of metabolizing a significant number of compounds after topical exposure. In order to fully understand and predict percutaneous absorption, both diffusional and metabolic processes must be integrated into any predictive pharmacokinetic model since the parent chemical's transit time through the various skin compartments may be important in determining the fraction metabolized and the total transdermal flux. Example of such modeling approaches were proposed by Kao and Carver[19] using both compartmental and physiologically based perfusion pharmacokinetic models.

As is one of the theses of the present text, there has been a general consensus that future risk assessment strategies will be heavily dependent upon recently developed quantitative pharmacokinetic strategies for predicting percutaneous absorption. These have been extensively reviewed in the recent literature.[21-26] A typical approach used is to simultaneously model, using linear compartmental pharmacokinetics, penetration through skin and systemic disposition. Refinements utilize physiological pharmacokinetic models to estimate systemic disposition. A major problem confronting these approaches is that sufficient data is often not available to uniquely identify any of the model parameters since the relevant biological samples cannot be collected to uniquely identify these parameters. However, if sufficient samples could be collected, then these approaches would be ideally suited for making interspecies extrapolations since species differences in skin penetration could be separated from species differences in systemic distribution and elimination. The major success obtained with these approaches have been in studies of zero-order transdermal delivery systems where drug release from patches is the rate limiting step. Such approaches would lend themselves to incorporation of cutaneous metabolism as previously discussed.

If one were to design an *in vitro* model which would be optimal for deriving pharmacokinetic parameters of percutaneous absorption, it should be able to respond similarly to the *in vivo* situation. This would include having sufficient "anatomical" and "physiological" complexity that pathways and processes which determine rate of penetration *in vivo* are similar in the *in vitro* model. As has been outlined, many *in vitro* models lack this complexity. For many compounds, when experiments are carefully conducted, classic *in vitro* models can make reasonable approximations to human *in vivo* exposure.[17-29] They can provide estimates of the penetration rate constants across the stratum corneum. However, if a quantitative understanding of the absorption and cutaneous distribution of a toxicant is desired; or if the purpose of the investigation is to probe the relative contributions of competing kinetic events in overall percutaneous absorption (e.g., diffusion, partitioning, binding, metabolism), then more sophisticated models are required that can address these many facets. The development of isolated perfused skin models meet many of these needs.

Isolated organ perfusions have been a useful tool for pharmacological and toxicological studies of the kidney, liver, lung, intestine, and heart. These organs have lent themselves to perfusion techniques because of both an easily isolatable vascular supply and a "closed" anatomical structure (encapsulated or surrounded by serosa) amenable to perfusion. The skin does not possess either of these attributes. Early attempts at perfusing skin did not create closed vascular systems and were never optimized for detailed absorption or disposition studies.[30-32] These limitations may be overcome if a two-stage reconstructive surgical procedure is utilized to create a "closed" skin preparation with an easily isolatable vasculature. When such a tubed, pedicle flap is employed, skin may be perfused under ambient environmental conditions without concern for dermal dehydration. Such preparations may then be harvested by cannulating the artery which eliminates the confouding influence of systemic

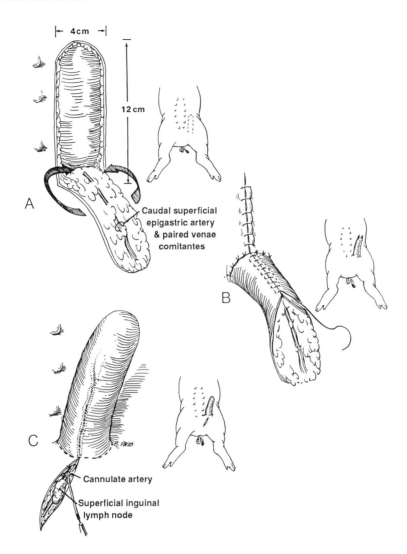

FIGURE 2. Surgical procedure for creating IPPSFs. (A) A single pedicle axial pattern skin flap is raised and (B) tubed completing the Stage One procedure. Two days later, (C) the superficial epigastric artery is cannulated (Stage Two) and transferred to the perfusion chamber in Figure 3.

metabolic processes. The superficial epigastric artery perfusing the abdominal skin of swine is ideally suited for this purpose.[33] Use of swine is also advantageous since the pig is an acceptable animal model for studying percutaneous absorption of many compounds.[4,18] Such a model optimized for percutaneous absorption studies, the isolated perfused porcine skin flap (IPPSF), was developed in our laboratory.

III. THE ISOLATED PERFUSED PORCINE SKIN FLAP

A. EXPERIMENTAL TECHNIQUE

The IPPSF is a single-pedicle, axial pattern tubed skin flap obtained from the ventral abdomen of female weanling swine.[33-38] Two flaps per animal, each lateral to the ventral midline, can be created in a single surgical procedure. As depicted in Figure 2, the procedure involves two steps; creation of the flap in Stage I and harvest in Stage II. Briefly, pigs

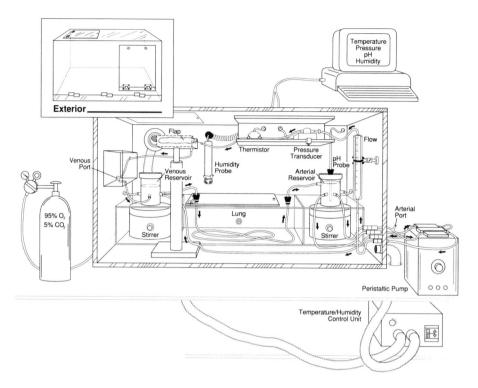

FIGURE 3. Schematic of the perfusion chamber used to maintain a viable IPPSF at controlled temperature, humidity, pressure and perfusate flow.

weighing approximately 20 to 30 kg are premedicated with atropine sulfate and xylazine hydrochloride, induced with ketamine hydrochloride and inhalational anesthesia is maintained with halothane. Each pig is prepared for routine aseptic surgery in the caudal abdominal and inguinal regions and a 4 × 12 cm area of skin, known from previous dissection and *in vivo* angiography studies to be perfused primarily by the caudal superficial epigastric artery and its associated paired venae comitantes, is demarcated. Following incision and scalpel dissection of the subcutaneous tissue, the caudal incision is apposed and sutured and the tubed skin flap edges trimmed of fat and closed. Two days later, a second surgical procedure is used to cannulate the artery and harvest each of these skin flaps. The 2-day period between flap creation and harvest was determined to be optimal from the standpoint of lack of overall flap leakiness, normal histologic appearance and vascularization and animal housing economics.[33,37] The IPPSF is then transferred to the perfusion chamber illustrated in Figure 3. The remaining wound is flushed and allowed to heal and pigs are generally returned to the housing facility. The references cited above should be consulted for complete details.

The isolated perfusion apparatus is a custom designed temperature and humidity regulated chamber made specifically for this purpose. A computer monitors perfusion pressure, flow, pH, and temperature. Flexibility is afforded in the experimental design by allowing both temperature and relative humidity to be maintained at specific set points (normally 37°C and 60 to 80% RH). The perfusion media is a Krebs-Ringer bicarbonate buffer (pH 7.4, 350 mOsm/kg), containing albumin (45 g/l), and supplied with glucose (80 to 120 mg/dl) as the primary energy source. Albumin is included in the perfusate to maintain oncotic pressure in the IPPSF, to mimic *in vivo* plasma so that protein binding kinetics are similar, and to facilitate absorption of lipid soluble penetrants. Since the IPPSF cannot be a sterile

MEAN GLUCOSE UTILIZATION (N=12)

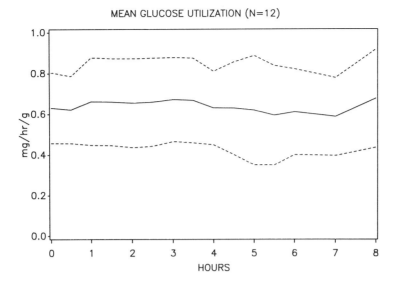

FIGURE 4. Mean (+ -SD) glucose utilization over an 8-h experimental period in 12 control IPPSFs.

organ preparation, antimicrobials (penicillin G and amikacin) are included to prevent bacterial overgrowth from the microflora normally present on the skin surface. Heparin is included to prevent coagulation in the skin flap's vasculature from residual formed blood elements.

Normal perfusate flow through the skin flap is maintained at 1 ml/min/flap (3 to 7 ml/min/100 g) with a mean arterial pressure ranging from 30 to 70 mmHg. These values are consistent with *in vivo* values reported in the literature. Both recirculating and non-recirculating (single pass) configurations are possible.

B. ASSESSMENT OF VIABILITY

To date, our laboratory has perfused over 1700 IPPSFs. Biochemical viability has been assessed by monitoring glucose utilization (arterial-venous glucose extraction), lactate production, enzyme leakage, perfusate flow, pressure, vascular resistance, and pH.[35,37,39,40] Glucose utilization from a series of control flaps is depicted in Figure 4 and is the parameter primarily used during an experiment to monitor flap viability. Note the relative stability of this parameter over an experiment. In a series of 255 flaps, mean ($\pm$ SD) glucose utilization was 0.8 ± 0.2 mg/h/g flap. Viability for 8 to 12 h has been confirmed through extensive light and electron microscopic histological studies.[38] A final sensitive real-time indicator of IPPSF viability and function is the vascular resistance profile (pressure/flow) over the course of an experiment. This remains constant unless toxicity occurs or vasoactive drugs are administered. When vasodilators such as tolazoline are administered, vascular resistance decreases while vasoconstrictors such as norepinephrine increase vascular resistance.

Finally, as a further indicator of biological relevance, cutaneous toxicants perturb these biochemical and physiological parameters and result in specific histopathological changes.[41,42] For example, when a toxin such as sodium fluoride is infused or the vesicant 2-chloroethyl methyl sulfide is applied topically, glucose utilization decreases. In fact, the IPPSF has been shown to be one of the few *in vitro* model systems which produce gross blisters similar to those seen *in vivo* after application of cutaneous vesicants. This is particularly pertinent in percutaneous absorption studies where the penetrant is a direct cutaneous toxicant since the absorption profile may be affected by the toxicity induced.

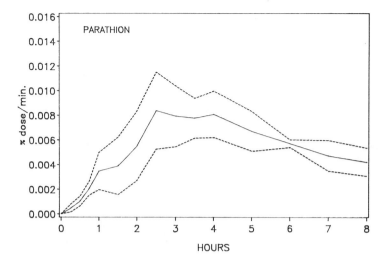

FIGURE 5. Percutaneous absorption profile of parathion in four IPPSFs (mean ± SD) after passive topical application of 40 μg/cm².

C. PERCUTANEOUS ABSORPTION

Percutaneous absorption studies are conducted by placing the study chemical on the surface of the IPPSF and assaying venous effluent over time for absorbed compound. Compound may be applied neat or diluted in vehicle under ambient or occluded conditions. Various types of patches or transdermal delivery systems may also be employed. A relatively large dosing area of up to 10 cm² is available for compound application and is a major advantage of this system since prototype human transdermal delivery systems may be studied. IPPSFs are usually allowed to equilibrate for 1 to 2 h prior to compound application to insure biochemical and physiological stability. Both passive and active (iontophoresis) transdermal delivery systems have been extensively studied. When experiments are designed to model uptake of compound from perfusate into skin (systemic distribution, outward transdermal migration, inverse penetration, etc.), drug is added to the arterial reservoir and infused into the IPPSF. Analysis of the venous efflux profile allows one to then study the kinetics of cutaneous uptake. The original studies should be consulted for further details.[34,43–47] Figures 5 and 6 depict typical percutaneous absorption profiles of topically applied compounds in the IPPSFs in ethanol over a 5 cm² demarcated area of skin. In Figure 6, lidocaine hydrochloride was administered by transdermal iontophoresis and venous skin flap effluxes monitored by HPLC. These figures illustrate the adaptability of IPPSF to study different transdermal delivery strategies.

The application of the IPPSF which clearly demonstrates its usefulness in studying percutaneous absorption is the work done on iontophoretic drug delivery.[42,44,45] These studies clearly demonstrated that co-administration of vasoactive compounds could significantly alter the transdermal flux of lidocaine, supporting the assertion that the microvasculature may play an important role in modulating drug absorption. Classic *in vitro* diffusion studies using excised human skin did not demonstrate this effect, which was also seen *in vivo*.

Additional studies have been conducted evaluating the cutaneous biotransformation of parathion seen during percutaneous absorption.[43] This cytochrome P-450 mediated metabolism could be completely blocked by pre-dosing infusion of a non-toxic dose of a cytochrome P-450 inhibitor (1-aminobenzotriazole). Occlusion resulted in a different metabolic profile. Metabolism studies have also been conducted looking at the cutaneous activation of Vitamin D.[48] These studies lay the experimental and theoretical groundwork for integrating metabolic processes into a comprehensive pharmacokinetic model.

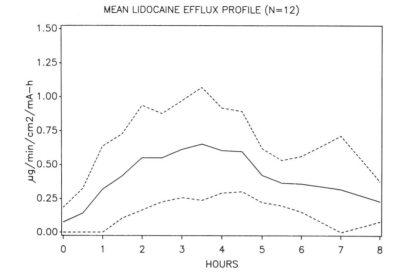

FIGURE 6. Percutaneous absorption profile of lidocaine in 12 IPPSFs (mean ± SD) after active iontophoretic delivery.

IV. AN INTEGRATED PHARMACOKINETIC APPROACH

The strength of the IPPSF is that venous efflux profiles may be continuously monitored over time, making the system well suited for pharmacokinetic analysis. Dosing can be monitored and skin samples are available after dosing to quantitate both nonabsorbed and penetrated dose. Since these data are collected in an individual IPPSF, it can directly be incorporated into a pharmacokinetic model.

The approach we have taken is to use the measured IPPSF venous efflux profile of an absorbed drug to probe the kinetics of absorption and cutaneous disposition. Once when a pharmacokinetic model is fit to these data, the absorption profile of a compound may then be extrapolated as a function of time. Compartmental pharmacokinetic models were initially utilized to model these profiles.[34] Significant curvefits to a single model could be obtained for a test series of seven compounds (DFP, malathion, parathion, testosterone, progesterone, caffeine, benzoic acid), and the correlation between 6-day IPPSF extrapolated and *in vivo* absorption[49] was excellent. However, because of the nature of the IPPSF, more information is available then could be easily utilized in a compartmental pharmacokinetic model. The approach presently utilized[46,47] is illustrated in Figure 7.

The strength of this model is that physiologically relevant mathematically identifiable pharmacokinetic models may be fit to these venous efflux profiles to study the kinetics of compound penetration and distribution within skin and to predict *in vivo* disposition. This model takes advantage of the unique sampling sites available in an isolated perfused organ preparation. A mathematically identifiable model was initially developed to model cutaneous uptake experiments where drug (cisplatin, carboplatin, tetracycline, doxycycline) was infused into the cutaneous artery supplying the flap and distribution to skin modeled by assaying arterial and venous drug concentrations.[48,50] This model was then expanded to study percutaneous absorption by including the skin surface as a compartment.[47]

The model is formulated with four basic "compartments".

1. The cutaneous capillary bed which receives drug from the arteries with a flux of J_{01} and loses drug into the veins with an overall flux of J_{15},

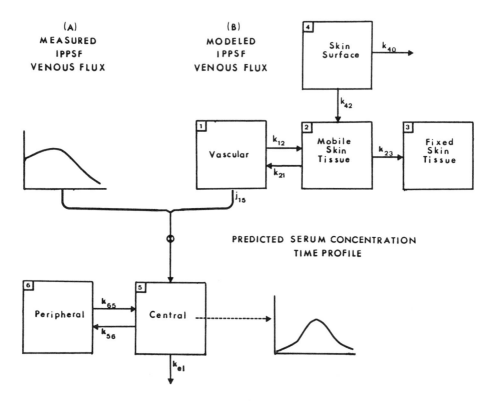

FIGURE 7. Schematic illustrating the pharmacokinetic strategy for integrating experimental IPPSF profiles (A) or modeled IPPSF venous flux profiles (B) with a systemic pharmacokinetic model to predict *in vivo* serum concentration-time profiles.

2. The nonvascular tissue of skin which receives topically penetrated drug at a rate of K_{42}
3. A skin depot which tends to sequester chemical (e.g., covalently bound cisplatin in perfusion experiments)
4. The skin surface

Except for flow through the capillary bed, all other movement of compounds are described by linear first-order fractional rate constants (K_{12}, K_{21}, K_{42}, K_{23}, K_{42}, K_{40}). Note that in this diagram, the rate constant returning drug from compartment 3 (K_{32}) is zero, allowing compartment 3 to function as a sink. The rate of surface loss (K_{40}) may also be estimated in an experiment and has been shown to correlate to the vapor pressures of these compounds reported in the literature. The volumes of the vascular and extracellular space of the IPPSF have been determined in independent radiolabel marker infusion studies[40] which provide an estimate of actual physiological volumes in this model.

Using this model and extrapolating to 6 days, the correlation to *in vivo* absorption using the seven compounds described above was excellent (R = 0.973) and improved over that provided by the compartmental model. More importantly, the correlation of IPPSF predicted to literature reported values of absorption of these compounds (and carbaryl) in humans[51–54] was excellent as seen in Figure 8. These data support the previous evidence that for certain compounds, the pig (and by extension the IPPSF) is an excellent model for predicting human absorption.

It must be stressed that the approach we have adopted to model percutaneous absorption profiles is based on optimizing experimental sampling points which are uniquely available

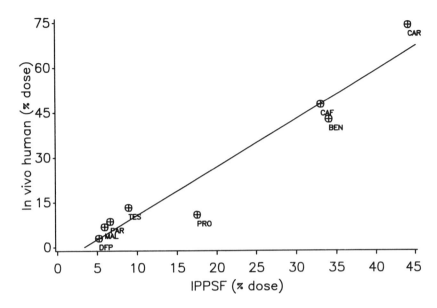

FIGURE 8. IPPSF predicted (——) vs. literature reported + *In vivo* human forearm absorption data for a series of eight compounds.

in the IPPSF. Our approach is to estimate the parameters to these models using the observed venous effluxes and then integrate physiologic data where available (i.e., estimated vascular volumes, vapor pressures, etc). The resulting rate constants and estimated distribution volumes may then be correlated to physiochemical and biological factors. This is in contrast to pure physiological pharmacokinetic approaches which directly incorporate physiochemical and biological constants in a model. However, this approach may also be easily adopted in the IPPSF.

The goal of our modeling studies to date has been to generate an approach whereby the venous efflux profiles of a number of diverse compounds could be described by a single pharmacokinetic model scaled to skin flap perfusate flow. This provide a mechanism by which interspecies extrapolations can be made partially on the basis of an existing database of interspecies and interregional estimates of cutaneous blood flow.[55] As illustrated in Figure 7, this skin **efflux** profile would then serve as an **input** profiled into a systemic pharmacokinetic model allowing serum concentration-time profiles to be predicted. In the absence of any pharmacokinetic model, the measured IPPSF profiles could also serve as a direct input into a systemic model. Although a compartmental model is illustrated in Figure 7, a physiological model could also be employed to describe systemic disposition. This general approach using observed IPPSF venous efflux profiles has been successfully utilized to predict the serum concentrating profile of arbutamine.[56]

Future efforts will be aimed at improving the biological and physiological relevance of our models, primarily by studying more compounds and forming structure-activity correlates. Cutaneous biotransformation compartments will directly be incorporated into these models as will perfusate protein binding. This latter phenomenon has been neglected in most percutaneous absorption studies. Similarly, data obtained from *in vitro* diffusion cell studies may also be incorporated into the models which increases the power of the IPPSF experiments to estimate parameters not accessible to other experimental approaches. Experimental conditions may be easily manipulated (e.g., temperature, humidity, perfusate composition) to study their effects on absorption or distribution parameters. In fact, the main purpose of utilizing pharmacokinetic models in these studies is to provide experimental endpoints (estimated kinetic parameters) which could be correlated to physiological events in an effort to increase our understanding of the biology of percutaneous absorption. This approach also results in ease of extrapolation to the *in vivo* situation. This approach allows *in vivo* variability to be divided into dermal (IPPSF) and systemic components.

V. CONCLUSIONS

Isolated perfused skin preparations provide a novel experimental approach to investigate compound absorption and distribution in the skin. They are particularly useful when compound absorption is a function of competing kinetic events; diffusion, partitioning, binding or metabolism. If a compound is vasoactive or induces direct toxic effects, their utility is obvious. They, and other similar models such as the rat/human skin sandwich flap[57] (see preceding chapter) are also ideally suited for developing pharmacokinetic models which can both increase our understanding of compound penetration and distribution in skin as well as provide a firm basis for predicting *in vivo* disposition in man and other species. Data from other *in vitro* approaches can easily be incorporated into pharmacokinetic models used in these systems. Additionally, the IPPSF is an alternative *in vitro* human animal model since all experiments are conducted after harvest from the pig. The pig can then be returned to its prior existance. For drugs or toxicants with direct cutaneous activity, this model is also amenable to formulating linked pharmacokinetic - pharmacodynamic models. The effects of altered skin physiology and dermatologic disease may also be experimentally studied and the *in vivo* consequences simulated using pharmacokinetic extrapolations. These are especially relevant when previous normal experiments have already been conducted validating the correspondence between IPPSF and *in vivo* results for a specific compound. Finally, there are no inherent limitations to creating isolated perfused skin flaps in other species in cases where porcine skin is not a suitable model for man. In fact, our laboratory routinely creates isolated perfused equine skin flaps[58] for pharmacologic studies of drug penetration in horses. Circumstances are easily foreseen when primates models may be appropriate.

In conclusion, the IPPSF is a novel *in vitro* model for studying many problems in dermatopharmacology. It is not designed to replace either existing *in vitro* or *in vivo* studies, but rather should serve as a bridge between these two approaches. In cases where *in vivo* human studies are impossible (e.g., carcinogens, lethal toxins, chemicals of unknown toxicity), this model may increase our ability to conduct realistic risk assessments. Likewise, when the behavior of a chemical is known to be similar in the IPPSF and man, then the number of human studies required may be reduced by using the IPPSF as a surrogate model. The utilization of pharmacokinetic techniques makes this approach feasible and should provide a firmer foundation for integrating data collected in other models systems (*in vitro* human and animal, *in vivo* animal studies) into human risk assessment strategies.

REFERENCES

1. **Honeycutt, R. C., Zweig, G., and Ragsdale, N. N.,** *Dermal Exposure Related to Pesticide Use,* American Chemical Society, Washington, D.C., 1985.
2. **Wang, R. G. M., Franklin, C. A., Honeycutt, R. C.,and Reinert, J. C.,** *Biological Monitoring for Pesticide Exposure,* American Chemical Society, Washington, D.C., 1989.
3. **Barry, B. W.,** *Dermatological Formulation: Percutaneous Absorption,* Marcel Dekker, New York, 1983.
4. **Bronaugh, R. L. and Maibach, H. I.,** *Percutaneous Absorption,* 2nd ed., Marcel Dekker, New York, 1989.

5. **Scott, R. C., Guy, R. H., and Hadgraft, J.,** *Prediction of Percutaneous Penetration,* IBC Technical Services, London, 1990.
6. **Kemppainen, B. W. and Reifenrath, W. G.** *Methods for Skin Absorption,* CRC Press, Boca Raton, FL, 1990.
7. **Schaefer, H., Zesch, A., and Stuttgen, G.,** *Skin Permeability,* Springer-Verlag, New York, 1982.
8. **Marty, J. P., Guy, R. H., and Maibach, H. I.,** Percutaneous penetration as a method of delivery to muscle and other tissues, in *Percutaneous Absorption,* Bronaugh, R. L. and Maibach, H. I., Eds., Marcel Dekker, New York, 1985, 469–487.
9. **Kao, J.,** Validity of skin absorption and metabolism studies, in *Methods for Skin Absorption,* Kemppainen, B. W. and Reifenrath, W. G., Eds., CRC Press, Boca Raton, FL, 1990, 191–212.
10. **Elias, P. M.,** Epidermal lipids, barrier functions and desquamation, *J. Invest. Dermatol.,* 80, 44, 1983.
11. **Guy, R. H. and Hadgraft, J.,** Physiochemical aspects of percutaneous penetration and its enhancement, *Pharm. Res.,* 5, 753, 1988.
12. **Flynn, G. L.,** Mechanisms of percutaneous absorption from physiochemical evidence, in *Percutaneous Absorption,* 2nd ed., Bronaugh, R. L. and Maibach, H. I., Marcel Dekker, New York, 1989, 27–52.
13. **Scheuplein, R. J.,** Mechanism of percutaneous absorption. I. Routes of penetration and the influence solubility, *J. Invest. Dermatol.,* 45, 334, 1965.
14. **Scheuplein, R. J.,** Mechanism of percutaneous absorption. II. Transient diffusion and the relative importance of various routes of skin penetration, *J. Invest. Dermatol.,* 48, 79, 1967.
15. **Bickers, D. R.,** Drug, carcinogen, and steroid hormone metabolism in skin, in *Biochemistry and Physiology of the Skin,* 2nd ed., Goldsmith, L. A., Ed., Oxford University Press, New York, 1983, 1169–1186.
16. **Bronaugh, R. L., Stewart, R. F., and Storm, J. E.,** Extent of cutaneous metabolism during percutaneous absorption of xenobiotics, *Toxicol. Appl. Pharmacol.,* 99, 534, 1989.
17. **Guy, R. H., Hadgraft, J., and Bucks, D. A. W.,** Transdermal drug delivery and cutaneous metabolism, *Xenobiotica,* 17, 325, 1987.
18. **Kao, J.,** The influence of metabolism on percutaneous absorption, in *Percutaneous Absorption,* 2nd ed., Bronaugh, R. L. and Maibach, H. I., Eds., Marcel Dekker, New York, 1989, 259–282.
19. **Kao, J. and Carver, M. P.,** Cutaneous metabolism of xenobiotics, *Drug Metab. Rev.,* 22, 4, 1990.
20. **Potts, R. O., McNeill, S. C., Desbournet, C. R., and Wakskull, E.,** Transdermal drug transport and metabolism. II. The role of competing kinetic events, *Pharm. Res.,* 6, 119, 1989.
21. **Guy, R. H. and Hadgraft, J.,** The prediction of plasma levels of drugs following transdermal application, *J. Control Release,* 1, 177, 1985.
22. **Guy, R. H. and Hadgraft, J.,** Mathematical models of percutaneous absorption, in *Percutaneous Absorption,* 2nd ed., Bronaugh, R. L. and Maibach, H. I., Eds., Marcel Dekker, New York, 1989, 13–26.
23. **Guy, R. H., Hadgraft, J., and Maibach, H. I.,** Percutaneous absorption in man: a kinetic approach, *Toxicol. Appl. Pharmacol.,* 79, 123, 1985.
24. **Fisher, H. L., Most, B., and Hall, L. L.,** Dermal absorption of pesticides calculated by deconvolution, *J. Appl. Toxicol.,* 5, 163, 1985.
25. **McDougal, J. N., Jepson, G. W., Clewell, H. J., MacNaughton, M. G., and Andersen, M. E.,** A physiological pharmacokinetic model for dermal absorption of vapors in the rat, *Toxicol. Appl. Pharmacol.,* 85, 286, 1986.
26. **McDougal, J. N., Jepson, G. W., Clewell, H. J., Gargas, M. L., and Andersen, M. E.,** Dermal absorption of organic chemical vapors in rats and humans, *Fundam. Appl. Toxicol.,* 14, 299, 1990.
27. **Bronaugh, R. L., Stewart, R. F., and Congdon, E. R.,** Methods for *in vitro* percutaneous absorption studies. II. Animal models for human, *Toxicol. Appl. Pharmacol.,* 62, 481, 1982.
28. **Reifenrath, W. G., Chellquist, E. M., Shipwash, E. A., Jederberg, W. W., and Kreuger, G. G.,** Percutaneous penetration in the hairless dog, weanling pig and grafted athymic nude mouse: evaluation of models for predicting skin penetration in man, *Br. J. Dermatol.,* 11, 123, 1984.
29. **Wester, R. C. and Maibach, H. I.,** *In vitro* percutaneous absorption and decontamination of pesticides in humans, *J. Toxicol. Environ. Health,* 16, 25, 1985.
30. **Halprin, K. M. and Chow, D. C.,** Metabolic pathways in perfused dog skin, *J. Invest. Dermatol.,* 36, 431, 1961.
31. **Hiernickel, H.,** An improved method for *in vitro* perfusion of human skin, *Br. J. Dermatol.,* 112, 299, 1985.
32. **Kjaersgaard, A. R.,** Perfusion of isolated dog skin, *J. Invest. Dermatol.,* 22, 135, 1954.
33. **Bowman, K. F., Monteiro-Riviere, N. A., and Riviere, J. E.,** Development of surgical techniques for preparation of *in vitro* isolated perfused porcine skin flaps for percutaneous absorption studies, *Am. J. Vet. Res.,* 52, 75, 1991.
34. **Carver, M. P., Williams, P. L., and Riviere, J. E.,** The isolated perfused porcine skin flap (IPPSF). III. Percutaneous absorption pharmacokinetics of organophosphates, steroids, benzoic acid and caffeine, *Toxicol. Appl. Pharmacol.,* 97, 324, 1989.

35. **Riviere, J. E., Bowman, K. F., Monteiro-Riviere, N. A., Carver, M. P., and Dix, L. P.,** The isolated perfused porcine skin flap (IPPSF). I. A novel *in vitro* model for percutaneous absorption and cutaneous toxicology studies, *Fundam. Appl. Toxicol.,* 7, 444, 1986.

36. **Riviere, J. E., Bowman, K. F., and Monteiro-Riviere, N. A.,** The isolated perfused porcine skin flap: a novel animal model for cutaneous toxicologic research, in *Swine in Biomedical Research,* Tumbleson, M. E., Ed., Plenum, New York, 1986, 657–666.

37. **Monteiro-Riviere, N. A.,** Specialized technique: isolated perfused porcine skin flap, in *Methods for Skin Absorption,* Kemppainen, B. W. and Reifenrath, W. G., Eds., CRC Press, Boca Raton, FL, 175.

38. **Monteiro-Riviere, N. A., Bowman, K. F., Scheidt, V. J., and Riviere, J. E.,** The isolated perfused porcine skin flap (IPPSF). II. Ultrastructural and histological characterization of epidermal viability, *In Vitro Toxicol.,* 1, 241, 1987.

39. **Riviere, J. E., Bowman, K. F., and Monteiro-Riviere, N. A.,** On the definition of viability in isolated perfused skin preparation, *Br. J. Dermatol.,* 116, 739, 1987.

40. **Williams, P. L. and Riviere, J. E.,** Estimation of physiological volumes in the isolated perfused porcine skin flap, *Res. Commun. Chem. Pathol. Pharmacol.,* 66, 145, 1989.

41. **King, J. R. and Monteiro-Riviere, N. A.,** Cutaneous toxicity of 2-chloroethyl methyl sulfide in isolated perfused porcine skin, *Toxicol. Appl. Pharmacol.,* 104, 167, 1990.

42. **Monteiro-Riviere, N. A.,** Altered epidermal morphology secondary to lidocaine iontophoresis: *in vitro* and *in vivo* studies in porcine skin, *Fundam. Appl. Toxicol.,* 15, 174, 1990.

43. **Carver, M. P., Levi, P. E., and Riviere, J. E.,** Parathion metabolism during percutaneous absorption in perfused porcine skin, *Pest. Biochem. Physiol.,* 38, 245, 1990.

44. **Riviere, J. E., Sage, B., and Monteiro-Riviere, N. A.,** Transdermal lidocaine iontophoresis in isolated perfused porcine skin, *Cutan. Ocular Toxicol.,* 8, 493, 1990a.

45. **Riviere, J. E., Sage, B. S., and Williams, P. L.,** The effects of vasoactive drugs on transdermal lidocaine iontophoresis, *J. Pharm. Sci.,* 80, 615, 1991.

46. **Williams, P. L. and Riviere, J. E.,** Definition of a physiologic pharmacokinetic model of cutaneous drug distribution using the isolated perfused porcine skin flap (IPPSF), *J. Pharm. Sci.,* 78, 550, 1989.

47. **Williams, P. L., Carver, M. P., and Riviere, J. E.,** A physiologically relevant pharmacokinetic model of xenobiotic percutaneous absorption utilizing the isolated perfused porcine skin flap (IPPSF), *J. Pharm. Sci.,* 79, 305, 1990.

48. **Bikle, D. D., Gee, E., Halloran, B. P., and Riviere, J. E.,** Production of 1,25 dihydroxyvitamin D by pig skin *in situ, J. Invest. Dermatol.,* 82, 404, 1989.

49. **Carver, M. P. and Riviere, J. E.,** Percutaneous absorption and excretion of xenobiotics after topical and intravenous administration to pigs, *Fundam. Appl. Toxicol.,* 13, 714, 1989.

50. **Williams, P. L. and Riviere, J. E.,** Effect of hyperthermia on cisplatin (CDDP) disposition to isolated perfused skin, *Int. J. Hyper.,* 6, 923, 1990.

51. **Feldman, R. J. and Maibach, H. I.,** Percutaneous absorption of steroids in man, *J. Invest. Dermatol.,* 52, 89, 1969.

52. **Feldman, R. J. and Maibach, H. I.,** Absorption of some organic compounds through the skin of man, *J. Invest. Dermatol.,* 54, 399, 1970.

53. **Wester, R. C. and Maibach, H. I.,** Relationship of topical dose and percutaneous absorption in rhesus monkey and man, *J. Invest. Dermatol.,* 67, 518, 1976.

54. **Wester, R. C. and Maibach, H. I.,** Animal models for percutaneous absorption, in *Models in Dermatology,* Vol. 2, Maibach H. I. and Lowe, N. J., Eds., S. Karger, Basel, 1985, 159–169.

55. **Monteiro-Riviere, N. A., Bristol, D. G., Manning, T. O., Rogers, R. A., and Riviere, J. E.,** Interspecies and interregional analysis of the comparative histological thickness and laser Doppler blood flow measurements at five cutaneous sites in nine species, *J. Invest. Dermatol.,* 95, 582, 1990.

56. **Riviere, J. E., Williams, P. L., Hillman, R., and Mishky, L.,** Quantitative prediction of transdermal iontopheric delivery of abutamine in humans using the *in vitro* isolated perfused porcine skin flap (IPPSF), *J. Pharm. Sci.,* 81, 504, 1992.

57. **Wojciechowski, Z., Pershing, L. K., Huether, S., Leonard, L., Burton, S. A., Higuchi, W. I., and Krueger, C. G.,** An experimental skin sandwich flap on an independent vascular supply for the study of percutaneous absorption, *J. Invest. Dermatol.,* 88, 439–446, 1987.

58. **Bristol, D. G., Riviere, J. E., Monteiro-Riviere, N. A., and Bowman, K. F.,** The isolated perfused equine skin flap: Metabolic parameters, *J. Vet. Surg.,* in press.

Epidemiologic Studies of Route of Exposure and Health Effects

Chapter 27

PERSPECTIVES ON ASSESSMENT OF RISKS FROM DERMAL EXPOSURE TO POLYCYCLIC AROMATIC HYDROCARBONS

Christopher J. Borgert, Stephen M. Roberts, Robert C. James, and Raymond D. Harbison

TABLE OF CONTENTS

0-8493-7357-3/93/$0.00 + $.50
© 1993 by CRC Press, Inc.

I. SOURCES AND EXPOSURES

Polycyclic aromatic hydrocarbons (PAHs) are a diverse class of chemicals characterized by two or more fused aromatic rings that contain only carbon and hydrogen atoms. PAHs find their way into the environment primarily by the incomplete combustion of organic materials. Because combustion of organic materials occurs in such diverse settings throughout society, PAHs are ubiquitous. Polycyclic aromatic hydrocarbons are released into the environment in a number of common materials, including soot, coal tar, tobacco smoke, petroleum, air pollutants, and cutting oils (Klaassen et al., 1986; page 172). They are found in wood smoke, gasoline and diesel exhaust, and are synthesized by some plants, including human food crops (International Agency for Research on Cancer (IARC), 1983). From such sources, PAHs easily reach air, water, soil, sediments, and aquatic organisms. Exposure to PAHs is extensive in modern, industrialized countries.

In addition to environmental sources, PAHs are encountered in a number of common domestic items. Cooking foods at high temperatures, especially grilling and smoking meats, increases the PAH content of the food. Combustion of fuels in home heating and cooking appliances contributes to domestic PAH levels. Humans expose themselves directly to PAHs in cigarette smoke, mineral oils, skin care products and shampoos, medicines, dyes, and plastics. PAHs are found in nearly all soils, where their concentrations vary depending upon whether the local surroundings are rural, urban, or agricultural. Contact with any of these media — fuel residues, household products or contaminated soils — may result in considerable dermal exposure to PAHs. The National Academy of Science has estimated that a member of the general population has an average daily intake of 0.21 μg PAHs from air, 0.03 μg from water, and 1.6 to 16 μg from food (National Academy of Science, 1986).

II. TOXICITY AND CARCINOGENICITY

More than 200 PAHs have been identified. A number of specific toxicities are associated with exposure to PAHs. PAHs have been shown to suppress immune responses, induce liver microsomal enzymes, and many are carcinogenic in animal models. Benzo[a]pyrene (BaP) is one of the most potent animal carcinogens of the class. Human exposure to this compound is widespread. It is estimated that annual airborne emission of BaP exceeds 900 tons in the U.S. alone. Of the PAHs tested only benz[a,h]anthracene has been shown to have greater carcinogenic potency than BaP (Krewski et al., 1989), although others, including perylene, cyclopenta(cd)pyrene, and fluoranthene have been shown to have greater mutagenic potency than BaP (Slaga et al., 1981). With the exception of benz[a,h]anthracene, all other PAHs are much less potent carcinogens than BaP, and some lack carcinogenic potency altogether. For purposes of assessing risk to human health, the carcinogenicity of PAHs is of primary toxicologic interest.

Exposures to PAHs are nearly always to complex mixtures and almost never to a single compound. Unfortunately, carcinogenicity testing has been completed for relatively few PAHs. Studies used in determining potency levels of various PAHs include long-term animal feeding studies, animal skin-painting tests, mutagenicity tests, and structure activity relationship (SAR) analyses (Klaassen et al., 1986). Skin painting studies in animals may have particular relevance to risks of adverse health effects in humans exposed to PAHs by dermal contact in the home, workplace, or in the general environment.

Immunological effects of PAHs were reported as early as 1952 when Malmgren and co-workers showed that mice treated with 3-methylcholanthrene, 1,2-benzanthracene or 1,2,5,6-dibenzanthracene produced an attenuated serum antibody response to sheep erythrocytes (Malmgren et al., 1952). Later, 3-methylcholanthrene injected into newborn mice was shown to be associated with severe damage to the thymus gland, stunted growth and development of thymomas (Yasuhira, 1964).

Subsequent studies confirmed that a long-term reduction in antibody responsiveness can follow administration of 7,12-dimethylbenz[a]anthracene and BaP to mice (Ward et al., 1986; Stjernsward, 1966). Mice exposed to BaP showed depressed responses to B- and T-cell mitogens (Dean et al., 1983). Exposure of pregnant female mice to BaP resulted in severe depression of antibody responses in pups immediately after birth (Urso and Gengozian, 1980). This suppression persisted for well over one year and was accompanied by an increase in incidence of tumors in adults. 3-Methylcholanthrene (3-MC), another carcinogenic PAH, may suppress formation of cytotoxic T-cells (Wodjani and Alfred, 1984). 7,12-Dimethylbenz[a]anthracene produced long-term immunosuppression of antibody production, natural killing and cytotoxic T-lymphocyte cytotoxicity (Ward et al., 1986). Thus, 7,12-dimethylbenz[a]anthracene appeared to affect all phases of immunologic protection in mice, including mechanisms of tumor resistance. Immunologic suppression of tumor resistance mechanisms by PAHs tends to correlate with the carcinogenic potency of the chemical. Immunosuppressive PAHs tend to be carcinogenic, but non-immunosuppressive congeners are not (Ward et al., 1985).

III. METABOLIC ACTIVATION

PAHs are not themselves carcinogenic, but are metabolized in several tissues, including skin, liver, and lung to molecular species capable of forming adducts with cellular DNA. DNA adduct formation has been identified as the lesion responsible for tumorigenic transformation of cells. The enzymes responsible for metabolism of PAHs are the cytochrome P-450 dependent mixed-function oxidases, the most important enzyme system involved in the metabolism of chemical carcinogens. For PAHs, it is the cytochrome P-450 I family (also called P-448) that is most closely associated with metabolism to carcinogenic molecular species (Ioannides and Parke, 1987), a process often referred to as metabolic activation. Cytochrome P-450 I enzymes catalyze specifically the oxidation of rigid, planar molecules (Phillipson et al., 1982). For several groups of chemical carcinogens, metabolism by cytochrome P-450 I isoforms results in production of reactive intermediates rather than detoxified metabolites (Ioannides and Parke, 1987; Lewis et al., 1986; Lewis et al., 1987). In addition to activating PAHs, the cytochrome P-450 I family is uniquely capable of activating aromatic amides and amines such as 2-acetylaminofluorene (Astrom and DePierre, 1985) and beta-naphthylamine (Hammons et al., 1985).

Epoxidation is the first activating oxidation reaction. Arene-epoxides are electrophilic and can spontaneously form covalent bonds with nucleophilic centers in biological macromolecules such as DNA, RNA, and protein. The position and stereochemistry of the epoxidation varies among PAHs. For BaP, the intermediate formed is a 7,8-epoxide. This reaction is catalyzed by P-450 IA1 isoform. Following hydrolysis by epoxide hydrolase, a second oxidation results in the 7, 8-diol 9,10 epoxide, often termed a "bay-region" dihydrodiol epoxide due to the physical positions of carbons 9 and 10. This second oxidation is catalyzed by a different microsomal enzyme, epoxide hydrolase. For most PAHs characterized, formation of specific enantiomers of diol epoxides, principally the (S,R,R,S) form, produces the ultimate carcinogenic metabolite (Platt et al., 1990). For a number of PAHs, the bay region diol epoxide is presumably the molecular species that forms mutagenic adducts with cellular DNA (Slaga et al., 1981). Although not all polycyclic aromatic structures contain a bay region, epoxidation by cytochrome P-450 I can nonetheless produce reactive molecular species. For example, metabolites of indeno[1,2,3-cd]pyrene from rat liver microsomes are mutagenic in *S. typhimurium* assays. In PAHs that lack a bay region, such as indeno[1,2,3-

cd]pyrene, formation of K-region epoxides appear to be responsible for adduct formation and mutagenicity (Bucker et al., 1979; Rice et al., 1985).

Metabolism is thus the critical event in conversion of a PAH to an ultimate carcinogen (Pelkonen and Nebert, 1982). The scheme of metabolic activation by way of diol epoxide formation appears to hold for many PAHs, including benzofluoranthrenes and methyl-substituted congeners. However, epoxides and diol epoxides are not the final products of metabolic transformation. Epoxide hydrolases further metabolize these reactive species to nonreactive molecules, thus completing the detoxification process. Conjugation of metabolites by glutathione then facilitates elimination from the organism. Non-carcinogenic PAHs are acted upon by these same enzymes, but do not form metabolites that react as readily with cellular DNA. Anthracene and benzo(e)pyrene, for instance, display virtually no carcinogenic potency, yet are planar molecules and substrates for cytochrome P-450 I enzymes as are benzo(a)pyrene and some methyl derivatives of benzo(a)anthracene, which are potent carcinogens (IARC, 1983; Rosenkranz and Poirier, 1979). For all PAHs studied, the greater share of biotransformations lead to detoxified metabolites that are readily conjugated and excreted. Hence, few adducts are formed due to the efficient inactivation and removal of reactive metabolites such as PAH epoxides. This explanation is plausible because the cytochrome P-450 IA1 family constitutes less than 5% of the total cytochrome P-450 content of uninduced animals (Luster et al., 1982; Pickett et al., 1981). Because of low basal cytochrome P-450 I A1 levels, the activation pathway is a minor contributor to overall biotransformation. The small amounts of reactive metabolites formed in uninduced animals are readily conjugated and excreted.

Post-mitochondrial metabolites from various species of uninduced animals have often failed to produce reactive mutagenic metabolites in the Ames test. However, strong positive responses are observed when the post-mitochondrial fractions are obtained from animals that have first been treated with cytochrome P-450 I inducing agents such as Aroclor 1254. Good correlations are seen between cytochrome P-450 I activity and mutagenicity of BaP when epoxide hydrolase activity is unchanged. A positive result in the Ames test is only an indication that a chemical can be converted to genotoxic products. It is not an assay of carcinogenicity in mammalian tissues. Several relatively non-carcinogenic PAHs, such as benzo[e]pyrene (BeP), are mutagenic in the Ames test. It must be emphasized that the Ames assay employs microsomal preparations from rats treated with strong cytochrome P-450 I inducers such as Aroclor 1254. If the compound itself is not a strong inducer of P-450 I, a positive reaction in the Ames assay will most likely be countered with a negative reaction in mouse skin tumorigenesis assays because formation of reactive metabolites remains a minor pathway in the absence of cytochrome P-450 I induction (Ayrton et al., 1990). Because of this caveat, the Ames assay may overestimate the carcinogenic potency of non-inducers of cytochrome P-450 I. Unfortunately, discrepancies are not absent from results obtained using the mouse skin assay. This model suffers from two drawbacks (Ayrton et al., 1990). First, test compounds are applied as a single dose, or the interval between doses is sufficiently long that any induced activity has returned to baseline before the following dose is given. Hence, activation is catalyzed by basal levels of cytochrome P-450 I A1, which, as has been mentioned, are very low. A second drawback is that concentrations of the test compound applied are too high to be achieved in tissues following exposures by other routes. This system may be adapted to improve its predictive capability by using repeated administrations of lower doses of test compounds (Ayrton et al., 1990). For good predictive value, an assay must be able to measure not only the ability of the chemical to form reactive products, but also the ability of the chemical to induce metabolic enzymes responsible for its conversion to reactive products (Ayrton et al., 1990).

IV. INDUCTION OF XENOBIOTIC TRANSFORMING ENZYMES

PAHs are known to be specific inducers of the very enzymes that carry out their metabolic activation. For instance, 3-methylcholanthrene (3-MC) treatment induces hydroxylation of both 3-MC and benzo(a)pyrene in rat liver microsomes. In contrast to other inducers, such as phenobarbital, cytochrome P-450 induction by PAHs is not accompanied by large increases in liver weight, protein and phospholipid synthesis, and NADPH-cytochrome P-450 reductase induction. The pattern of enzyme induction is a specific biological effect characteristic of a class of inducers. Phenobarbital induction is representative of one principal induction pattern whereas induction by PAHs or 2,3,7,8-tetrachlorodibenzo-*p*-dioxin (TCDD) is representative of another pattern. Both phenobarbital and PAH-inducible enzymes are typically undetectable in livers of uninduced cytochrome animals. The mechanism of cytochrome P-450 induction by PAHs has been well characterized and involves binding of the hydrocarbon to a specific receptor protein. Recent studies suggest that two distinct proteins may function as separate receptors, one for TCDD congeners and another for PAHs (Raha et al., 1991.). Translocation of the receptor-ligand to the nucleus leads to transcriptional activation of target genes for cytochrome P-450 I apoproteins. No such receptor has yet been identified for phenobarbital induction. Other biotransformation enzymes induced by 3-MC-type inducers include cytosolic glutathione-S-transferases and microsomal UDP-glucuronosyltransferases and epoxide hydrolase. These enzymes perform important detoxification reactions and compete with cellular DNA for epoxide substrates.

V. CONSIDERATIONS FOR ASSESSING RISK FROM DERMAL CONTACT

Exposure is clearly one of the most important aspects of assessing risk to human health from any chemical. Exposure to contaminated soil can occur by a number of different routes, and soil is a medium with which the skin has frequent contact. In addition to contaminated soil, dermal contact with chemicals in industrial settings is another potentially common route of exposure. Topical application of medicines has become a widely used route of administration for some drugs. Skin absorption following such exposures is now recognized as an important route of entry into the human body for a number of drugs and chemicals. Because of the concern over health hazards posed by contaminated industrial and waste disposal sites, risk assessment methodologies have been formulated for potential exposure to chemicals in soils (Kimbrough et al., 1984; Paustenbach et al., 1986; Hawley, 1985). For assessment of risks from dermal exposure to PAHs, two of the most influential factors are the type of medium in which PAHs are found and the duration of exposure to that medium. The amount of exposure to chemicals in soil is dependent on the amount of soil adhering to the skin and the extent to which the bare skin of an individual comes into direct contact with soil.

Crude oil, a common source of PAHs, is a potential soil contaminant. In order to study the dermal bioavailability of PAHs from contaminated soil, Yang and co-workers applied crude oil containing 100 ppm ^{3}H-BaP to soil and measured percutaneous absorption of BaP from that soil compared to percutaneous absorption from crude oil alone (Yang et al., 1989). Results of *in vitro* experiments using freshly isolated rat skin were compared with those from *in vivo* application of soil or crude oil to the shaved backs of rats. The soil used was first sieved to obtain a fraction with particle size less than 150 μm for preparation of a 1% crude oil-fortified soil. Preliminary experiments revealed that most of the BaP was contained in the silt and clay fraction (particle size less than 50 μm), consistent with earlier studies that found the bulk of soil organic carbon in fine-particle fractions (Brady, 1984). Soil organic carbon has been shown to be the primary sorbent for lipophilic compounds, and silt and clay fractions are much more effective sorbents than the sand fraction (Karickoff et al., 1979). Following application of soil to skin sections and removal of excess soil, it was the silt and clay fractions that remained adhered to the skin surface (Yang et al., 1989). For both *in vivo* and *in vitro* experiments, 9 mg was found to be the minimum amount of soil that covered one cm^2 of skin surface with a monolayer of soil. This is a larger adherence value than the < 1 mg/cm^2 reported for dust on human skin (Lepow et al., 1975; Roels et al., 1980; Que Hee et al., 1985). These differences in soil adherence may be due to differences in skin texture between rats and humans, to differences in types of soil used or soil moisture, or to differences in methods of measuring soil adherence (Yang et al., 1989). The amount of BaP absorbed from soil through skin slices was 1.3 ng in 96 h regardless of whether 9 mg or 56 mg of crude-fortified soil was applied. These values corresponded to 8.4 or 1.3% of the applied dose of BaP, respectively. When the same dose of BaP was applied in crude oil alone, in the absence of the soil matrix, approximately 38% of the applied dose was absorbed in 96 h (Yang et al., 1989). No statistically significant difference in absorption was seen between *in vitro* and *in vivo* experiments using these concentrations of BaP in 1% crude-fortified soil. Based on both *in vivo* and *in vitro* findings, the rate of dermal uptake for BaP from 1% crude-fortified soil matrix was estimated to be 0.2 ng/cm^2 skin/day (Yang et al., 1989). BaP absorption through skin was so low that no change in the rate occurred over the 96 h absorption period. The constant slope of the absorption curve seen with both 9 mg soil/cm^2 of skin and 56 mg soil/cm^2 of skin indicated that the small amount of BaP absorbed came entirely from the monolayer of soil in direct contact with the skin. The amount absorbed represented only 10% of the BaP in the monolayer. In contrast, the slope of the absorption curve for BaP applied in crude oil alone was much steeper in the first 72 h of the exposure period compared to the curve for crude-fortified soil. After 72 h the rate of absorption from crude oil began to slow (Yang et al., 1989). No experiments were conducted to determine the rate of absorption of BaP from heavier applications of crude oil or from crude oil containing higher concentrations of BaP.

Percutaneous absorption of BaP has been studied *in vitro* using human cadaver skin and *in vivo* using Rhesus monkey skin (Wester et al., 1990). Human cadaver skin was dermatomed to 500 μm and placed in flow-through cells with human plasma as the receptor fluid on one side of the cell. ^{14}C-BaP in acetone was applied to sieved soil (80 mesh) to produce a soil mixture containing 10 ppm BaP. BaP was then applied to skin in equal amounts either in acetone vehicle or in 40 mg of soil matrix for 24 h. BaP readily penetrated human skin from acetone vehicle (approximately 18% of the applied dose) (Wester et al., 1990), but penetrated skin much less from the soil matrix (approximately 1% of the applied dose) (Wester et al., 1990). BaP partitioned into plasma receptor fluid very poorly from either acetone vehicle or from the soil matrix (0.04 and 0.08% of applied dose, respectively). Approximately 96% of the radioactivity was recovered from soap and water surface washes of the skin following the application of BaP in soil, but an average of 22% of the applied dose remained bound to the skin when BaP was applied in acetone (Wester et al., 1990). *In vivo* percutaneous absorption of BaP was determined by the ratio of urinary excretion following topical application to that following intravenous injection. Urinary excretion of BaP following i.v. injection in rhesus monkeys was approximately 7% of the administered dose. This value is similar to that seen in humans and is consistent with values determined from monkeys (Wester et al., 1990). Percutaneous absorption of BaP *in vivo* averaged 51% of the applied dose from acetone vehicle but only 13% from soil.

A number of factors can influence percutaneous absorption studies and thereby affect the absolute numbers obtained in such experiments (Hawkins and Reifenrath, 1984). When comparing percutaneous absorption data relative to man, the species used as well as technical manipulations can increase skin absorption so that values obtained are not applicable to humans. An analysis of the percutaneous absorptions from soil matrix versus those from vehicle alone may allow one to consider together the two studies described above (Wester et al., 1990). Yang et al. (1989) reported a maximal *in vitro* percutaneous absorption for BaP from a 1% crude oil-fortified soil matrix of 8.4% of the applied dose. This represented a reduction of about 26% as compared with absorption from crude oil alone. A reduction ratio may be calculated by dividing the percutaneous absorption from soil (8.4%) by the absorption from crude oil alone (38.1%). The reduction ratios for rat skin were found to be 0.22 *in vitro* and 9.2%/35% = 0.26 *in vivo* (Yang et al., 1989; Wester et al., 1990). Applying this analysis to human and monkey skin, the ratio was found to be 1.4%/23.7% = 0.06 *in vitro* and 13.2%/51% = 0.26 *in vivo* (Wester et al., 1990). Therefore, both studies found *in vivo* percutaneous absorption of BaP from a soil matrix to be about 26% that from an organic vehicle (Wester et al., 1990).

Results of these studies point to factors that should be considered when assessing risks to human health from dermal contact with PAHs. Consideration should be given to the nature of the soil contamination. Soil containing pockets of oil or liquid organic residues could influence the absorption of PAHs profoundly compared with absorption from soil matrix. In such cases, the use of typical soil matrix factors would greatly underestimate, by roughly a factor of 4, the potential for dermal absorption of PAHs. In addition, short exposure durations may limit the potential for absorption of PAHs from soil more than from organic liquids. Wester et al. (1990) showed that whereas a soap and water wash removed more than 95% of the BaP from skin exposed to the chemical in soil, less than 64% could be washed from skin exposed to BaP in acetone. Results of these studies should be considered when determining exposure assumptions for health risk assessments. Recently, the USEPA has recommended that a soil matrix factor of 0.01 be used to estimate dermal absorption of organic compounds from contaminated soils (USEPA Region IV, personal communication).

VI. METABOLIC ACTIVATION AND ADDUCT FORMATION IN SKIN

Percutaneous absorption of BaP was found to be greater *in vivo* than *in vitro* in both studies discussed above. For a number of chemicals, metabolic viability of the skin has been shown to be an important determinant of percutaneous absorption. For BaP, this relationship has been found to be particularly strong (Kao et al., 1984). Drug-metabolizing enzymes have been found to be the critical factors responsible for differences in percutaneous absorption between viable and nonviable skin (Kao et al., 1985). In fresh skin slices from six species, permeation by [14C]BaP was accompanied by extensive "first pass" metabolism. [14C]BaP and a full spectrum of its metabolites were detected in receptor fluid from viable skin (Kao et al., 1985). Permeation through nonviable skin slices was negligible. Inclusion of KCN, a metabolic inhibitor, in the culture medium resulted in decreased permeation of BaP through mouse skin. Of the species tested, penetration was greatest through mouse skin: about 10% of a topically applied dose permeated through mouse skin in 24 h. Penetration was from 1 to 3% through human, rabbit, rat and marmoset skin, and only 0.1% through guinea pig skin (Kao et al., 1985). The remainder of the topically applied radioactivity was recovered in the cultured tissue, indicating that BaP had entered the skin slices but had not passed completely through in the 24 h incubation time. Studies with human skin exposed topically to [14C]BaP showed that of the radioactivity that permeated through the skin, 90% was extracted into ethyl acetate from nonviable skin slices, representing unchanged BaP. In contrast, less than 50% of radioactivity in the culture medium from viable skin slices was extractable into ethyl acetate. In the organic extract from the culture medium of viable human skin, 18% of total radioactivity was found in BaP, and all major classes of BaP metabolites were identified (BaP diols, other polar BaP metabolites, BaP phenols and BaP quinones) (Kao et al., 1985). Digestion of nonviable human skin slices following topical application of [^{14}C]BaP revealed that 87% of the radioactivity remained in parent compound, whereas in digests of viable human skin only 57% remained in BaP and the rest was attributed to BaP metabolites (Kao et al., 1985). Similar observations were made in experiments using skin from other species. From the media of rat, mouse, and marmoset skin preparations, approximately 70% of the radioactivity remained in aqueous fractions following extraction with ethyl acetate. Chromatographic analysis showed that all major classes of BaP metabolites were present. BaP diols and polar metabolites predominated. Metabolism of chemicals during cutaneous passage may represent a general phenomenon (Bronaugh et al., 1989).

Factors that affect the percutaneous absorption of chemicals have been reviewed (Wester and Maibach, 1983). Skin can be described most simply in terms of three basic layers, the epidermis, the dermis, and the subcutaneous fat (Wester and Maibach, 1983). The stratum corneum is the outermost layer of the epidermis, and is composed of dead, inactive cells. The stratum corneum is the principal barrier to absorption of most chemicals and materials with which the skin comes into contact. Below the stratum corneum lies the epidermis. Data indicate that the epidermal layer contains most of the drug-metabolizing enzymes and is the most metabolically active skin layer. The epidermis contains a complete complement of drug-metabolizing enzymes, including a cytochrome P-450 mixed function oxygenase system. These enzymes in skin are inducible as they are in liver. Enzyme activities have been compared in skin and liver (Wester and Maibach, 1983). Cutaneous enzyme activities are usually reported based on enzyme activities in whole skin homogenates, and have traditionally been considered to be lower than hepatic activities. However, most of the enzyme activity in skin is localized to the epidermis, and the epidermis comprises only 2.5 to 3% of the total skin volume. Using epidermal volume to calculate the metabolic activity of skin, the skin appears to be from 80 to 240% as active a drug-metabolizing organ as liver (Wester and Maibach, 1983). Chemicals that penetrate the stratum corneum into the epidermis may be subject to metabolic activities equal to or greater than those found in liver due to a concentration effect.

Most metabolism probably inactivates or detoxifies chemicals that penetrate the stratum corneum. However, this is assuredly not the case for PAHs. Epoxide hydratase activity has been compared in skin microsomal fractions from mouse, rat, and humans (Oesch and Golan, 1978). Epoxide hydratase from each species was able to hydrate all epoxides tested. Hydratase activity toward the K-region epoxides of several different PAHs varied similarly among skin microsomal fractions from the three species. The activity of epoxide hydratase from human skin showed little pH dependence and showed inhibition characteristics similar to its hepatic counterpart. No variations in activity were seen with regard to age or sex of donor, but a regional correlation was suggested. Differences in cutaneous drug metabolism with body region may have interesting implications in light of possible regional variations in absorption of chemicals through the stratum corneum (Ayrton et al., 1990).

In addition to epoxide hydratases, cytochrome P-450 dependent monooxygenases have been studied in skin. Microsomes prepared from whole skin, dermis, and epidermis of mice all metabolized BaP to phenols, quinones, and diols (Das et al., 1986). Metabolite formation by microsomes from non-pretreated (control) whole skin and dermis was comparable. Metabolite formation by control epidermal microsomes was approximately 2.5-fold greater. Pretreatment of mouse skin with the cytochrome P-450 inducer 3-methylcholanthrene (3-MC) increased metabolite formation in microsomal fractions from all three tissues, but the increase was greatest in epidermal microsomes (Das et al., 1986). In addition, the *in vivo* covalent binding of [^{3}H]BaP,[^{3}H]BaP-7,8-diol, and 7,12-[^{3}H]dimethylbenz[a]anthracene to DNA was greater in epidermis than in whole skin or dermis, and this binding was enhanced more than twofold by pretreatment with 3-MC. Products of metabolic activation by mixtures of epidermal microsomes were more mutagenic in *Salmonella* tester strains than products from mixtures of microsomes from either whole skin or dermis (Das et al., 1986). Studies on enzyme induction in skin have been extended and induced enzymes have been identified and purified. Topical application of benz[a]anthracene to mouse skin was shown to induce a twofold increase in skin cytochrome P-450 content with concommitant increases in BaP hydroxylation, 7-ethoxycoumarin *O*-deethylation, and acetanilide 4-hydroxylation (Ichikawa et al., 1989). A major form of cytochrome P-450 dependent monooxygenase was purified from the microsomal fraction of benz[a]anthracene-induced mouse skin. One-dimensional polyacrylamide gel electrophoresis showed single-band purification of the enzyme. The enzyme was immunochemically cross-reactive with antibody raised against rat liver cytochrome P-450 $_{MC-I}$. This enzyme shared several characteristics with rat liver cytochrome P-450 $_{MC-I}$. When reconstituted with NADPH-cytochrome P-450 reductase, it catalyzed efficiently the hydroxylation of BaP and *O*-deethylation of 7-ethoxycoumarin. Like the activity of the rat liver cytochrome P-450, the activity of the enzyme purified from mouse skin was inhibited specifically by 7,8-benzoflavone and cytochrome P-450$_{MC-I}$ antibody (Ichikawa et al., 1989). It is suggested that this enzyme plays an important role in the activation of carcinogenic PAHs in skin. Other studies have suggested that mutational activation of the Ha-ras oncogene by PAH-DNA adduct formation in mouse skin may be a critical step in transformation of mouse cells and formation of skin tumors (Bizub et al., 1986). Because the skin contains the enzymatic machinery necessary to activate PAHs and is a site of tumor production in response to application of PAHs, a biologically sound assessment of carcinogenic risk from PAHs should consider the skin as a separate target organ of toxicity.

Formation of PAH-DNA adducts has been measured in skin following topical exposure to fuels and lubricating oils (Carmichael et al., 1990). Engine lubricating oils are known to accumulate PAHs during operation of both gasoline-powered and diesel-powered internal combustion engines. Levels of 19 PAHs were, on average, 22 times higher in used lubricant oil from gasoline engines than in used lubricant oil from diesel engines. Levels of PAH-DNA adducts in mice treated with used oil were up to 60 times greater than levels formed in the skin of mice treated with unused oil as measured by ^{32}P-postlabeling. Adduct levels in the lung tissue of these mice were similar to levels found in skin. In general, adduct formation in both lung and skin correlated with engine use and PAH content of the oil (Carmichael et al., 1990). These results have been corroborated with mutagenicity studies in *Salmonella* (McKee and Plutnick, 1989; Granella and Clonfereo, 1991) and dermal carcinogenicity studies of mouse skin (McKee and Plutnick, 1989). Diesel engine oils were not mutagenic in these studies, even after extended use as lubricants in engines. In contrast, oils that lubricated gasoline engines showed higher PAH content, mutagenicity, and dermal carcinogenicity with increased engine use. These phenomena may have important implications for some human occupational exposures to motor oils and high temperature lubricating oils. For some petroleum-derived liquids, such as kerosene, diesel fuel, and heating oil, complete carcinogenic action in mouse skin may depend upon epigenetic mechanisms related to skin irritation (McKee et al., 1989).

The carcinogenicity of PAHs may be influenced by the mixture of compounds to which skin is exposed. Benzo[e]pyrene (BeP) is a weakly carcinogenic congener of BaP and has been shown to modify the initiating activity of potent carcinogens such as BaP and 7,12-dimethylbenz[a]anthracene (DMBA) (Smolarek and Baird, 1986). BeP is a component of cigarette smoke and is found in fossil fuel combustion residues. BeP appears to have converse effects on the tumor initiating activity of BaP and DMBA, inhibiting strongly the effect of DMBA in mouse skin but increasing slightly the activity of BaP. Levels of DNA adduct formation correlate with tumor promoting potency in the mouse skin model and are affected similarly by BeP. To investigate the mechanism by which BeP-inhibits carcinogenic activation by DMBA, hydrocarbon-DNA adducts were measured in Syrian hamster embryo cell cultures following 24-, 48-, and 72-h exposures to 1 μg/ml DMBA in culture media along with various concentrations of BeP. Total binding of DMBA to DNA was inhibited approximately 4-fold by high doses of BeP, while binding was increased at low doses of BeP. BeP altered the stereoselectivity of adduct formation as well, a factor that may have significant effects on carcinogenicity (Smolarek and Baird, 1986). Other studies have shown that co-incubation of BaP and coal-derived complex organic mixtures decreases the metabolism and mutagenic activity of BaP. Four of five different mixtures were found to decrease the tumor-initiating activity of BaP (Springer et al., 1989). The PAH and nitrogen-containing fractions of these mixtures appeared to be more effective inhibitors of BaP tumor-initiating activity than aliphatic and hydroxy-PAH fractions. Stereoselectivity of BaP-DNA adduct formation was altered by these mixtures as was seen for the effect of BeP on DMBA-DNA adduct formation.

The animal data discussed above illustrate several important points regarding dermal exposure to PAHs. First, PAHs are absorbed into skin in amounts sufficient for measurement of effects not only at the site of application, but also in tissues distant from the exposed organ. Second, metabolic activation of PAHs occurs in skin at rates sufficient to produce levels of DNA adducts in skin similar to those seen in other well-characterized PAH-metabolizing tissues. Formation of diol epoxides in skin may be an important consideration in assessment of health risks to humans from dermal PAH exposure in both occupational and domestic settings. Third, mixtures of PAHs may present cancer potencies that cannot be ascertained simply by considering the effects of individual constituents in mouse skin tumor assays or in the Ames assay. On one hand, weak inducers that are not mutagenic in the Ames assay may be capable of reacting with tissue DNA if inducing components in mixtures provide for activation. On the other hand, the cancer potency of inducers, such as BaP or DMBA, have been shown to be attenuated by administration in mixtures (Springer et al., 1989; Smolarek and Baird, 1986).

VII. OCCUPATIONAL EXPOSURES

Occupational exposure to PAHs can involve a substantial dermal component. Skin swipes may provide a method to monitor occupational exposure to PAHs (Wolff et al., 1989). Skin-swipe samples and breathing zone air measurements were obtained from roofers during removal of an old coal-tar roof and application of a new asphalt roof. Levels of PAHs on skin swipes taken immediately after the workday correlated well with air concentrations of PAHs (Wolff et al., 1989). Levels of PAH-DNA adducts correlated well with PAH exposures in coke oven workers. Adducts were reduced 60% when subjects wore masks during work. The remaining 40% of adducts may represent exposure via dermal absorption from handled materials and airborne dusts. PAH-serum albumin adducts have been measured immunologically in foundry workers and roofers (Lee et al., 1991). Although occupational exposures often are associated with a large inhalation component of PAH intake, dermal absorption of PAHs in oils and grease may be equally important exposure routes.

VIII. ASSESSMENT OF RISK

All chemicals are toxic at some dose; thus, the key issue in risk assessment is not establishing toxicity itself, but rather in defining the levels of exposure which cause undesirable effects. For purposes of assessing risk to human health, USEPA currently categorizes chemicals as either carcinogenic or non-carcinogenic. The non-carcinogenic effects

TABLE 1
RfD Values for Some Non-Carcinogenic
PAHs

Chemical	Acceptable level of daily intake (mg/kg/day)
Acenapthene	6.0×10^{-2}
Anthracene	3.0×10^{-1}
Benzo[g,h,i]perylene	4.0×10^{-3}
Dibenzofuran	4.0×10^{-3}
Fluoranthrene	4.0×10^{-2}
Fluorene	4.0×10^{-2}
2-Methylnaphthalene	4.0×10^{-3}
Naphthalene	4.0×10^{-3}
Phenanthrene	4.0×10^{-3}
Pyrene	3.0×10^{-2}

of chemicals with known toxicity are assessed using the hazard index approach. Briefly, the hazard index approach compares the average daily intake for each chemical to a published acceptable level of daily intake (ALDI) for chronic or subchronic exposure (i.e., Chronic and Sub-Chronic Reference Dose or RfD). The USEPA defines the RfD as "an estimate (with uncertainty spanning perhaps an order of magnitude) of the daily exposure to the human population (including sensitive subgroups) that is likely to be without appreciable risk of deleterious effects during a portion of the lifetime, in the case of a subchronic RfD, or during the lifetime, in the case of a chronic RfD." These values represent the highest chronic exposure level not causing adverse effects (i.e., no observable adverse effect level [NOAEL]).

Essentially, the method assumes that a "threshold" level exists below which no adverse effects are produced (Paustenbach, 1989; Paustenbach, 1992). For most chemicals, each RfD value, or ALDI, contains a safety factor which accounts for the uncertainty associated with extrapolation of animal data to man. To assess the overall potential for a chemical to cause non-carcinogenic effects using the hazard index, a person's chronic daily intake (CDI) is divided by the RfD. The sum of the ratios for all chemicals to which a person may be exposed is called the hazard index. If the hazard index is less than one, the chemicals are considered unlikely to represent a risk to human health. If the hazard index is greater than one, the chemicals are segregated according to the target organ toxicities they cause. Separate hazard indices are then calculated for each toxicological effect. Finally, if the hazard index for any toxicological effect is still greater than one, exposure to the chemicals may represent a human health hazard.

The hazard index approach is used for assessment of risks that might be encountered from exposure to non-carcinogenic PAHs. USEPA lists many PAHs as non-carcinogenic, and has derived RfDs for non-carcinogenic health effects displayed in animals (U.S.EPA, 1991). RfD values for some non-carcinogenic PAHs are listed in Table 1.

Because exposure to PAHs is always to complex mixtures rather than to a single PAH, several methods have been used to assess cancer risks associated with exposure to mixtures of PAHs. These methods include assumptions that all PAHs are equivalent in cancer potency to BaP, that some PAHs are carcinogenic and equivalent in cancer potency to BaP, or that only BaP is carcinogenic. Since only benz[a,h,]anthracene has been shown to be as carcinogenic as BaP, the assumption that all PAHs are equivalent in potency to BaP would grossly overestimate cancer potencies of PAH mixtures. In fact, some data suggests that mixtures of PAHs may antagonize the carcinogenicity of BaP (Springer et al., 1989). On the other

TABLE 2
Relative Potencies of Carcinogenic PAHs

	Relative potency to benzo[a]pyrene				
PAH	ICF Clement (1988)	USEPA Region VI	Rugen et al. 1989	Deutsch-Wenzel et al. 1983	USEPA Region IV
Benzo[a]pyrene	1.0	1.0	1.0	1.0	1.0
Benzo[a]anthracene	0.145	0.0134	0.0045	—	0.1
Benzo[b]fluoranthene	0.140	0.08	0.036	0.11	0.1
Benzo[k]fluoranthene	0.066	0.0044	—	0.03	0.1
Benzo[g,h,i]perylene	0.022	—	—	0.01	—
Chrysene	0.0044	0.0012	—	—	0.01
Dibenzo[a,h]anthracene	1.11	0.69	0.6	—	1.0
Indeno[1,2,3-c,d]pyrene	0.232	0.0171	0.060	0.08	0.1

From Deutsch-Wenzel, R.P., Brune, F., Grimmer, G., Dettbarn, G., and Misfeld, J., *J. Natl. Cancer Inst.*, 71, 539, 1983; ICF-Clement, Interim Final Report, EPA Contract No. 68-902-4403, April 1, 1988; Rugen, P.J., Stern, C.D., and Lamm, S.H., *Reg. Toxicol. Pharmacol.*, 9(3), 273; Personal communication, U.S. EPA, Region IV, 1992; Personal communication, U.S. EPA, Region VI, 1991.

hand, assuming that only BaP is carcinogenic could underestimate the cancer potency of PAH mixtures. Some PAHs may be non-carcinogenic when administered alone only by virtue of their inability to induce bioactivating enzymes. These compounds may have increased cancer potency in the presence of PAHs that induce bioactivating enzymes. Indeed, some non-carcinogenic PAHs were mutagenic in the Ames assay when preincubated with microsomes from animals induced by Aroclor-1254 (Ayrton et al., 1990), indicating that these PAHs can be metabolized to genotoxic intermediates. To be biologically plausible, assessment of health risks to humans must employ a relative potency approach rather than assuming that all PAHs are equivalent in potency to BaP, or that only BaP is carcinogenic, and should adjust cancer potencies based upon the **effective carcinogenicity.**

Several methods have been proposed for adjusting cancer risks associated with exposure to mixtures of PAHs (Brown, 1989; Chu and Chen, 1992; Krewski et al., 1989; U.S. Environmental Protection Agency (USEPA), 1986). The USEPA has adopted a system to assign relative cancer potency values for several PAHs. These are based on a structure activity relationship approach to estimate the relative carcinogenic potency of individual PAHs. Critical structural features are associated with carcinogenic potential of PAHs. These have been summarized (Tong, 1989) and include:

1. Molecular size and shape
2. Presence of bulky substituents
3. Lack of substitutions in specific regions of the benzo ring
4. Presence of substitutions in specific regions of the benzo ring

The comparative potencies of potentially carcinogenic PAHs to BaP from four different sources are presented in Table 2. These values were derived by applying various methods of analysis to data from chronic animal feeding studies, mutagenicity tests, skin painting tests, and structure activity analyses (Klaassen et al., 1986).

For chemicals listed by USEPA as having carcinogenic potential, the current regulatory approach assumes that the dose response curve intercepts the ordinate at zero. In other words, there is no threshold for adverse biological effects and there is some risk associated with all doses greater than zero. However, effects seen in animal studies are invariably produced with high doses of the carcinogen. Several mathematical models have been developed to extrapolate an estimate of effects plausible at low doses in humans from these high-dose

TABLE 3
Weight of Evidence Classification and Slope Factors for
Potentially Carcinogenic Chemicals of Interest

Chemical	+ Weight of evidence	Slope Factor (mg/kg/day) $^{-1}$	
		Oral	Inhalation
Benzo[a]pyrene	B2	7.3 E + 00	6.10E + 00
Benzo[a]anthracene	B2	1.06E + 00	8.85E − 01
Benzo[b]fluoranthene	B2	1.02E + 00	8.54E − 01
Benzo[k]fluoranthene	B2	4.81E − 01	4.03E − 01
Benzo[g,h,i]perylene	D	1.61E − 01	1.34E − 01
Chrysene	B2	3.21E − 02	2.68E − 02
Dibenzo[a,h]anthracene	B2	8.10E + 00	6.77E + 00
Indeno[1,2,3-c,d]pyrene	B2	1.69E + 00	1.42E + 00
Benzene	A	2.90E − 02	2.90E − 02

data in animals. The USEPA uses a mathematical model known as the linearized multistage model (LMS) to extrapolate from high doses to very low doses. The LMS yields a 95% upper confidence limit on the predicted response at a specific dose rather a "maximum chance" of the response occurring.

To assess potential lifetime cancer risk posed by exposure to a suspected carcinogen by any route, average chronic daily intakes (CDIs) are calculated for an exposure scenario and are multiplied by the slope factor for that chemical. Multiplying the CDI by the slope factor gives a ratio indicative of the cancer risk posed by that exposure scenario. The resulting cancer risks are expressed in scientific notation (e.g., 1.0E-04 to 1.0E-07) and refer to additional lifetime cancer risks of one cancer in 10,000 persons, one cancer in 100,000 persons, one cancer in 1,000,000 persons or one cancer in 10,000,000 persons, respectively.

For purposes of assessing the risks posed by a mixture of potentially carcinogenic chemicals in soil, it is usually assumed that risks from all potentially carcinogenic chemicals in soil are additive and that there is no antagonism of the carcinogenic response. The concentrations of BaP and other potentially carcinogenic PAHs are summed to produce values for "total carcinogenic PAHs" in soil.

The oral and inhalation slope factors for BaP are then used to calculate cancer risks for all carcinogenic PAHs as a group (Table 3). Slope factors have yet to be derived for the dermal route of exposure. As discussed above, dermal contact with PAHs present the potential for unique and significant exposures to these compounds.

An approach more biologically relevant than those currently taken would be to consider the dermal route of exposure differently than other routes. In the absence of reference doses and slope factors specific for dermal exposures, one approach that can be taken is presented in Table 4. Oral reference doses and slope factors were used to estimate dermal values. This approach would probably produce overly conservative estimates of risk from dermal exposure to PAHs, but is more biologically sound than simply applying a single soil matrix absorption factor for dermal exposures to all chemicals. What is actually needed are dermal-specific values based upon the weight of evidence from dermal toxicity and absorption studies. Factors that should be considered include the absorption characteristics of the medium in which PAHs are encountered as well as the potential for specific dermal toxicity. Current approaches may overestimate the systemic carcinogenic potential of PAHs, but do not adequately address the potential for dermal carcinogenicity. Data available at this time provide limited information regarding effects of PAH mixtures on carcinogenicity. Hopefully, results of future research will provide additional information that can be used to rationally assess the dermal carcinogenicity of mixtures of PAHs found in the environment.

TABLE 4
Dermal Reference Doses and Slope Factors for PAHs

Chemical	Gastrointestinal absorption factor[a]	Dermal reference dose[b] Subchronic	Chronic	Dermal slope[c] factor
Anthracene	0.8	2.40E + 00	2.40E − 01	—
Benzo(a)anthracene	0.8	—	—	8.47E − 01
Benzo(a)pyrene	0.8	—	—	5.84E + 00
Benzo(b)fluoranthene	0.8	—	—	8.18E − 01
Benzo(k)fluoranthene	0.8	—	—	3.85E − 01
Benzo(g,h,i)perylene	0.8	—	—	1.28E − 01
Benzoic Acid	1	4.00E + 00	4.00E + 00	—
Chrysene	0.8	—	—	2.57E − 02
Dibenzo(a,h)anthracene	0.8	—	—	6.48E + 00
Fluoranthene	0.8	3.20E − 01	3.20E − 02	—
Fluorene	0.8	3.20E − 01	3.20E − 02	—
Indeno(1,2,3-c,d)pyrene	0.8	—	—	1.35E + 00
Naphthalene	0.8	3.20E − 02	3.20E − 03	—
Phenanthrene	0.8	—	—	—
Pyrene	0.8	2.40E − 01	2.40E − 02	—

[a] Gastrointestinal (GI) absorption factors are from the ATSDR Toxicological Profile for PAHs.
[b] Dermal reference doses were calculated by multiplying the oral reference dose by the gastrointestinal absorption factor.
[c] Dermal slope factors were calculated by dividing the oral slope factor by the gastrointestinal absorption factor.

REFERENCES

Astrom, A. and DePierre, J. W., Metabolism of 2-acetylaminofluorene by eight different forms of cytochrome P-450 and P-448 isolated from rat liver, *Carcinogenesis,* 6, 113, 1985.

Ayrton, A. D., McFarlane, M., Walker, R., Neville, S., Coombs, M. M., and Ionnides, C., Induction of the P-450 I family of proteins by polycyclic aromatic hydrocarbons: possible relationship to their carcinogenicity, *Toxicology,* 60, 173–186, 1990.

Bizub, D., Wood, A. W., and Skalka, A. M., Mutagenesis of the Ha-ras oncogene in mouse skin tumors induced by polycyclic aromatic hydrocarbons, *Proc. Natl. Acad. Sci.,* U.S.A., 83(16), 6048–6052, 1986.

Brady, N. C., *The Nature and Properties of Soils,* Macmillan, New York, 1984.

Bronaugh, R. L., Stewart, R. F., and Storm, J. E., Extent of cutaneous metabolism during percutaneous absorption of xenobiotics, *Toxicol. Appl. Pharmacol.,* 99, 534–543, 1989.

Brown, J. P., Objective ranking of airborne polynuclear aromatic hydrocarbons and related compounds based on genetic toxicity, presented at the Annual Meeting of the Air and Waste Management Association, unpublished.

Bucker, M., Glatt, H., Platt, K., Avnir, D., Ittah, Y., Blum, J., and Oesch, F., Mutagenicity of phenanthrene and phenanthrene K-region derivatives, *Mutat. Res.,* 66, 337–348, 1979.

Carmichael, P. L., Jacob, J., Grimmer, G., and Phillips, D. H., Analysis of the polycyclic aromatic hydrocarbon content of petrol and diesel engine lubricating oils and determination of DNA adducts in topically treated mice by 32P-postlabelling, *Carcinogenesis* 11(11), 2025–2032, 1990.

Chu, M. L. and Chen, C. W., Evaluation and estimation for potential carcinogenic risks of polycyclic aromatic hydrocarbons, unpublished.

Das, M., Bickers, D. R., and Mukhtar, H., Epidermis: the major site of cutaneous benzo[a]pyrene and benzo[a]pyrene 7,8-diol metabolism in neonatal BALB/c mice, *Drug Metab. Dispos.,* 14(6), 637–642, 1986.

Dean, J. H., Luster, M. I., Boorman, G. A., Lauer, L. D., Luebke, R. W., and Lawson, L. D., Immune suppression following exposure of mice to the carcinogen benzo(a)pyrene but not the non-carcinogenic benzo(e)pyrene, *Clin. Exp. Immunol.,* 52, 199–206, 1983.

Granella, M. and Clonfereo, E., The mutagenic activity of polycyclic aromatic hydrocarbon content of mineral oils, *Int. Arch. Occup. Environ. Health,* 63(2), 149–153, 1991.

Hammons, G. J., Guengerich, F. P., Weis, C. C., Beland, F. A., and Kadlubar, F. F., Metabolic oxidation of carcinogenic arylamines by rat, dog and human hepatic microsomes and by purified flavin-containing and cytochrome P-450 monooxygenases, *Cancer Res.,* 45, 3578, 1985.

Hawkins, G. S. and Reifenrath, W. G., Development of an in vitro model for determining the fate of chemicals applied to the skin, *Fundam. Appl. Toxicol.,* 4, S133–S144, 1984.

Hawley, J. K., Assessment of health risk from exposure to contaminated soil, *Risk Anal.,* 5, 289–302, 1985.

IARC Monographs on the evaluation of carcinogenic risk of chemicals to man. Polynuclear aromatic compounds, *IARC Monogr.,* 32, 1983.

Ichikawa, T., Hayashi, S., Noshiro, M., Takada, K., and Okuda, K., Purification and characterization of cytochrome P-450 induced by benz[a]anthracene in mouse skin microsomes, *Cancer Res.,* 49(4), 806–809, 1989.

International Agency for Research on Cancer (IARC), Polynuclear aromatic compounds, Part 1: Chemical, Environmental and Experimental Data, Lyon, France, 1983.

Ioannides, C. and Parke, D. V., The cytochromes P-448 — A unique family of enzymes involved in chemical toxicity and carcinogenesis, *Biochem. Pharmacol.,* 36, 4197, 1987.

Kao, J., Patterson, F., and Hall, J., Skin penetration and metabolism of topically applied chemicals in six mammalian species, including man: an in vitro study with benzo[a]pyrene and testosterone, *Toxicol. Appl. Pharmacol.,* 81, 502–516, 1985.

Kao, J. Y., Hall, J., Shugart, L. R., and Holland, J. M., An in vitro approach to studying cutaneous metabolism and disposition of topically applied xenobiotics, *Toxicol. Appl. Pharmacol.,* 75, 289–298, 1984.

Karickoff, S. W., Brown, D. S., and Scott, T. A., Sorption of hydrophobic pollutants on natural sediments, *Water Res.,* 13, 241–248.

Kimbrough, R. D., Falk, H., Stehr, P., and Fries, G., Health implications of 2,3,7,8-tetrachlorodibenzodioxin (TCDD) contamination of residential soil, *J. Toxicol. Environ. Health,* 14, 47–93, 1984.

Klaassen, C. D., Amdur, M. O., and Doull, J., *Toxicology. The Basic Science of Poisons,* Macmillan, New York, 1986.

Krewski, D., Thorslund, T., and Withey, J., Carcinogenic risk assessment of complex mixtures, *Toxicol. Ind. Health,* 5, 851–867, 1989.

Lee, B. M., Yin, B. Y., Herbert, R., Hemminki, K., Perera, F. P., and Santella, R. M., Immunologic measurement of polycyclic aromatic hydrocarbon-albumin adducts in foundry workers and roofers, *Scand. J. Work Environ. Health,* 17(3), 190–194, 1991.

Lepow, M. L., Bruckman, L., Gillette, M., Markowitz, S., Robino, R., and Kapish, J., Investigations into sources of lead in the environment of urban children, *Environ. Res.,* 10, 415–426, 1975.

Lewis, D. F. V., Ioannides, C., and Parke, D. V., Molecular dimensions of the substrate binding site of cytochrome P-448, *Biochem. Pharmacol.,* 35, 2179, 1986.

Lewis, D. F. V., Ioannides, C., and Parke, D. V., Structural requirements for substrates of cytochromes P-450 and P-448, *Chem. Biol. Interact.,* 64, 39, 1987.

Luster, M. I., Lawson, L. D., Linko, P., and Goldstein, J. A., Immunochemical evidence for two 3-methylcholanthrene-inducible forms of cytochrome P-448 in rat liver microsomes using a double antibody radioimmunoassay procedure, *Mol. Pharmacol.,* 23, 252, 1982.

Malmgren, R. A., Bennison, B. E., and McKinley, T. W., Jr., Reduced antibody titers in mice treated with carcinogenic and cancer chemotherapeutic agents, *Proc. Soc. Exp. Biol. Med.,* 70, 484–488, 1952.

McKee, R. H. and Plutnick, R. T., Carcinogenic potential of gasoline and diesel engine oils, *Fundam. Appl. Toxicol.,* 13 (3), 545–553, 1989.

McKee, R. H., Plutnick, R. T., and Przygoda, R. T., The carcinogenic initiating and promoting properties of a lightly refined paraffinic oil, *Fundam. Appl. Toxicol.,* 12(4), 748–756, 1989.

National Academy of Science, *Drinking Water and Health,* National Academy Press, Washington, D.C., 1986.

Oesch, F. and Golan, M., Specificity of human, rat and mouse skin epoxide hydroatase towards K-region epoxides of polycyclic hydrocarbons, *Biochem. Pharmacol.,* 27, 17–20, 1978.

Paustenbach, D. J., Important recent advances in the practice of health risk assessment: implications of the 1990s, *Regul. Toxicol. Pharmacol.,* 10, 204–243, 1989.

Paustenbach, D. J., A survey of health risk assessment, in *The Risk Assessment of Environmental and Human Health Hazards: A Textbook of Case Studies,* Paustenbach, D. J., Ed., John Wiley & Sons, New York, 1992, 27–124.

Paustenbach, D. J., Shu, H. P., and Murray, F. J., A critical examination of assumptions used in risk assessments of dioxin contaminated soil, *Regul. Toxicol. Pharmacol.,* 6, 284–307, 1986.

Pelkonen, O. and Nebert, D. W., Metabolism of polycyclic aromatic hydrocarbons: etiologic role in carcinogenesis, *Pharmacol. Rev.,* 34, 189, 1982.

Phillipson, C. E., Ioannides, C., and Parke, D. V., Studies on the substrate binding site of liver microsomal cytochrome p-448, *Biochem. J.,* 207, 51, 1982.

Pickett, C. B., Jeter, R. L., Moris, J. and Lu, A. Y. H., Electrochemical quantitation of cytochrome P-450, cytochrome P-448 and epoxide hydrolase in rat liver microsomes, *J. Biol. Chem.,* 256, 8815, 1981.

Platt, K. L., Schollmeier, M., Frank, H. and Oesch, F., Stereoselective metabolism of dibenz(a,h)anthracene to trans-dihydrodiols and their activation to bacterial mutagens, *Environ. Health Perspect.,* 88, 37–41, 1990.

Que Hee, S. S., Peace, B., Clark, C. S., Boyle, J. R., Bornschein, R. L., and Hammond, P. B., Evolution of efficient methods to sample lead sources, such as house dust and hand dust, in the homes of children, *Environ. Res.,* 38, 77–95, 1985.

Raha, A., Reddy, V., Xu, L. C., Houser, W. H., and Bresnick, E., Presence of the 4S polycyclic hydrocarbon-binding protein in H4-II-E cells, *Toxicology,* 66, 175–186, 1991.

Rice, J. E., Coleman, D. T., Hosted, T. J., Jr., LaVoie, E. J., McCaustland, D. J., and Wiley, J. C., Jr., Identification of mutagenic metabolites of indenol[1,2,3-cd]pyrene formed in vitro with rat liver enzymes, *Cancer Res.,* 45(11), 5421–5425, 1985.

Roels, H. A., Buchet, J. P. Lauwerys, R. R., Bruaux, P., Claeys-Thoreau, F., Lafontaine, A., and Verduyn, G., Exposure to lead by the oral and the pulmonary routes of children living in the vicinity of a primary lead smelter, *Environ. Res.,* 22, 81–94, 1980.

Rosenkranz, H. S. and Poirier, L. A., Evaluation of the mutagenicity and DNA-modifying activity of carcinogens and non-carcinogens in microbiol systems, *J. Natl. Cancer Inst.,* 62, 873, 1979.

Slaga, T., Gleason, G., Mills, G., Ewald, L., Fu, P., Lee, H., and Harvey, R., Comparison of the skin tumor-initiating activities of dihydrodiols and diolepoxides of various polycyclic aromatic hydrocarbons, *Cancer Res.,* 40, 1981–1984, 1981.

Smolarek, T. A. and Baird, W. M., Benzo(e)pyrene-induced alterations in the stereoselectivity of activation of 7,12-dimethylbenz(a)anthracene to DNA-binding metabolites in hamster embryo cell cultures, *Cancer Res.,* 46(3), 1170–1175, 1986.

Springer, D. L., Mann, D. B., Dankovic, D. A., Thomas, B. L., Wright, C. W., and Mahlum, D. D., Influences of complex organic mixtures on tumor-initiating activity, DNA binding and adducts of benzo[a]pyrene, *Carcinogenesis,* 10(1), 131–137, 1989.

Stjernsward, J., Effect of noncarcinogenic and carcinogenic hydrocarbons on antibody-forming cells measured at the cellular level *in vitro, J. Natl. Cancer Inst.,* 36, 1189–1195, 1966.

Tong, P., Development of Estimated Carcinogenic Relative Potencies for Polycyclic Aromatic Hydrocarbons (PAHs) U.S. Environmental Protection Agency, Washington, D.C., 1989.

U.S. EPA, Integrated Risk Information Services (IRIS), Cincinnati, OH, 1991.

U.S. Environmental Protection Agency (USEPA), Guidelines for the Health Risk Assessment of Chemical Mixtures, Federal Register, 1986.

Urso, R. and Gengozian, N., Depressed humoral immunity and increased tumor incidence in mice following *in utero* exposure to benzo(a)pyrene, *J. Toxicol. Environ. Health,* 6, 569–576.

Ward, E. C., Murray, M. J., and Dean, J. H., Immunotoxicity of nonhalogenated polycyclic aromatic hydro-carbons, in *Immunotoxicology and Immunopharmacology,* Dean, J. H., Luster, M. I., Munson, A. E., and Amos, H., Ed., Raven Press, 1985, 291–304.

Ward, E. C., Murray, M. J., Lauer, L. D., House, R. V., and Dean, J. H., Persistent suppression of humoral and cell-mediated immunity in mice following exposure to the polycyclic aromatic hydrocarbon, 7, 12-dimethyl-benz[a]anthracene, *Int. J. Immunopharmacol.,* 8, 13–22, 1986.

Wester, R. C. and Maibach, H. I., Cutaneous pharmacokinetics: 10 steps to percutaneous absorption, *Drug Metab. Rev.,* 14, 169–205, 1983.

Wester, R. C., Maibach, H. I., Bucks, D. A. W., Sedik, L., Melendres, J., Liao, C., and DiZio, S., Percutaneous absorption of [14C]DDT and [14C]benzo[a]pyrene from soil, *Fundam. Appl. Toxicol.,* 15, 510–516, 1990.

Wodjani, A. and Alfred, L., Alterations in cell-mediated immune functions induced in mouse splenic lymphocytes by polycyclic aromatic hydrocarbons, *Cancer Res.,* 44, 942–945, 1984.

Wolff, M. S., Herbert, R., Marcus, M., Rivera, M., Landrigan, P. J., and Andrews, L. R., Polycyclic aromatic hydrocarbon (PAH) residues on skin in relation to air levels among roofers, *Arch. Environ. Health,* 44(3), 157–163, 1989.

Yang, J. J., Roy, T. A., Neil, W., and Mackerer, C. R., In vitro and in vivo percutaneous absorption of benzo[a]pyrene from petroleum crude-fortified soil in the rat, *Bull. Environ. Contam. Toxicol.,* 43, 207–214, 1989.

Yasuhira, K., Damage to the thymus and other lymphoid tissues from 3-methylcholanthrene and subsequent thymoma production in mice, *Cancer Res.,* 24, 558–563, 1964.

Chapter 28

THE PARADOX OF HERBICIDE 2,4-D EPIDEMIOLOGY*

Gregory G. Bond and Ralph R. Cook

TABLE OF CONTENTS

*   This review was sponsored by a grant from the Industry Task Force II on 2,4-D Research Data. The views presented are those of the authors and should not be attributed to the Task Force collectively or to its member companies.

0-8493-7357-3/93/$0.00 + $.50
© 1993 by CRC Press, Inc.

I. INTRODUCTION

The herbicide 2,4-dichlorophenoxyacetic acid (2,4-D) has been used for the control of weeds and broadleaf plants in agriculture, forestry, and rights of way for more than 40 years. It is one member of a family of compounds referred to as the chlorophenoxys. Structurally similar chemicals include the butyric acid (2,4-DB) and the propionic acid (2,4-DP) derivatives; 4-chloro-2-methyl-phenoxyacetic acid (MCPA) and its propionic acid derivative (MCPP); and 2,4,5-trichlorophenoxyacetic acid (2,4,5-T) and 2-(2,4,5-trichlorophenoxy) propionic acid (2,4,5-TP or Silvex).

During the past decade, questions have arisen regarding the potential for chlorophenoxy herbicides to cause certain forms of cancer in humans. This began with a series of case reports[1,2] and case-control studies[3-6] from Sweden in the late 1970s and early 1980s which focused on increased risks of two diverse groups of neoplasms referred to, respectively, as soft-tissue sarcomas and malignant lymphomas. The studies were controversial because of suspected methodological problems.[7-9] Nevertheless, the hypotheses generated by them spawned additional research by investigators on three continents. The findings from the epidemiology studies have been strikingly inconsistent, presenting reviewers with a paradox: does one believe the studies which apparently show an association or those which do not?

This chapter reviews the evidence accumulated thus far from the analytic epidemiologic studies, particularly as it relates to the herbicide 2,4-D, and attempts to place in context with the known toxicology of this compound. The strengths and limitations of epidemiology for identifying potentially hazardous agricultural chemicals are also highlighted.

II. USES OF 2,4-D AND POTENTIAL FOR HUMAN EXPOSURE

A. 2,4-D PRODUCTION AND CONTAMINATION

2,4-D is produced by the condensation of 2,4-dichlorophenol and monochloroacetic acid. During the manufacture of the precursor chlorophenol, high temperatures and alkaline conditions can favor the production of chlorinated dibenzodioxins and dibenzofurans as unwanted contaminants. As a consequence, the derivative herbicide itself may sometimes contain these impurities in measurable quantities.

Because the toxicology profile of 2,4-D differs substantially from that of its contaminants, the types and levels of impurities present in the formulations are important considerations. The di- and trichlorodibenzo-*p*-dioxins have been detected at extremely low levels in some samples of 2,4-D produced in North America; however, the most toxic dioxin isomer, 2,3,7,8-tetrachlordibenzo-*p*-dioxin (2,3,7,8-TCDD), which gained notoriety as a contaminant in some samples of 2,4,5-T, has never been detected in 2,4-D products with analytical methods sensitive to 1 ppb.[10] Moreover, beginning in 1989, the EPA has required U.S. manufacturers and suppliers to analyze 2,4-D at a limit of detection of 0.1 ppb,[11] and to our knowledge still none reportedly has been found.[12] By comparison with 2,3,7,8-TCDD, the di- and trichlorodibenzo-*p*-dioxins are several orders of magnitude less toxic on both an acute and chronic basis.[13]

Although the acid is the parent compound, nearly all 2,4-D formulations in use contain either the water soluble amine or alkali salts or the ester derivative, which is readily dissolvable in organic solvents. It should be noted that trace amounts of nitrosamines have been found in the amine salts of 2,4-D; however, it has been concluded that the amounts found in 2,4-D formulations pose, at most, a negligible risk to human health.[14]

B. USES FOR 2,4-D HERBICIDES

2,4-D is by far the predominant phenoxy herbicide used in North America. During 1989, approximately 55 million pounds were applied to control unwanted weeds and vegetation.

TABLE 1
North American Phenoxy Market Segments
2,4-D Volume

Segment	Active ingredient LBS (M)	% of total
Spring wheat and barley	20,110	36.2
Range and pasture	6,474	11.7
Winter wheat	4,440	11.7
Turf	6,160	11.1
Industrial vegetation	4,870	9.0
Corn	4,850	8.7
Fallow	3,520	6.4
Aquatics	1,100	2.0
Sorghum	900	1.6
Other	576	1.4
Total	55,000	100

It is preferred to other herbicides because of its low cost and general efficacy. Furthermore, it has a low acute toxicity in humans and other animals, does not bioconcentrate, and has an environmental half-life of from days to weeks.[14]

Table 1 presents some estimates of the amount of 2,4-D applied in various market segments. Its major uses are in the control of broadleaf weeds in cereal crops such as wheat and barley. Substantial amounts are also used on range and pasture, powerline and roadway rights of way, forestry, and turf. Consumers are likely to be familiar with 2,4-D because for many years it has been the active ingredient in most lawn treatments for dandelions and other weed species. It is also widely used in public health programs to suppress poison ivy and allergenic plants such as ragweed.

A map showing the estimated amount of 2,4-D used by county in the U.S.[15] is presented as Figure 1. Not surprisingly, it is most heavily used in the Midwest farm-belt, particularly in the major wheat-growing states of Kansas, Oklahoma, North Dakota, and Montana. A substantial amount is also used in Florida for aquatic purposes and for weed control in sugarcane.

2,4-D is frequently mixed with other phenoxy and non-phenoxy herbicides to gain a greater spectrum of weed and vegetation control. Historically, it was often mixed with 2,4,5-T for use in forestry and rights of way treatments, and it has been well publicized that a variation of this particular formulation was used as a defoliant, code name Agent Orange, during the Vietnam conflict.[16] The mixed exposures of applicators have important implications for interpreting epidemiology studies and will be discussed in greater depth further on.

C. POTENTIAL FOR HUMAN EXPOSURE

Understanding potential sources of exposure is critical to any assessment of possible health risks from 2,4-D. Monitoring studies have shown the individuals with the opportunity for the highest exposures to 2,4-D are those who handle herbicides on the job, specifically workers involved in manufacture, formulation, or application. Although homeowner use of 2,4-D formulations for lawn and garden weed control has not been nearly as well studied, the daily doses and total lifetime exposures associated with this infrequent activity are estimated to be much lower than those received occupationally. In 1984, the W.H.O. concluded that the general population outside of use areas has no detectable 2,4-D exposure based on drinking water and retail food monitoring programs conducted since 1970, which have not detected any residues.[17]

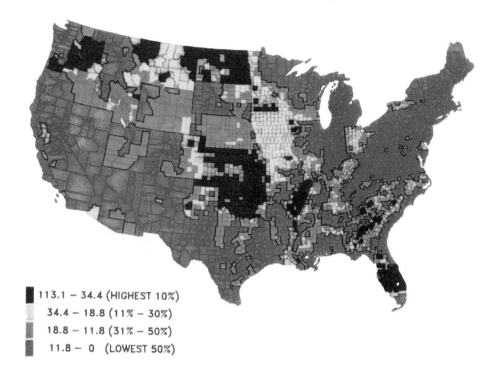

113.1 – 34.4 (HIGHEST 10%)

34.4 – 18.8 (11% – 30%)

18.8 – 11.8 (31% – 50%)

11.8 – 0 (LOWEST 50%)

FIGURE 1.   Pounds of 2,4-D used per square mile by U.S. county in 1982.

Field studies conducted on occupational groups suggest that the level of exposure is a direct function of the rate at which 2,4-D formulations are applied.[14] Thus, individuals who spray concentrated material from backpacks on rights of way have the highest estimated daily personal exposures (3.4 to 4.9 mg/day), followed by helicopter and airplane application personnel (0.005 to 1.04 mg/day), farmers driving spray rigs (0.48 mg/day), and hand and tank commercial lawn sprayers (0.29 mg/day). Variations in estimated exposures within a particular category are presumably attributable to differences in workplace conditions and work practices (i.e., type and amount of protective equipment worn, and personal hygiene). Estimates of the total lifetime doses are related directly to the number of days per year and the number of years of spraying activities.

The metabolic and excretory pathways of 2,4-D have been studied in both human volunteers and occupational groups.[14,17] 2,4-D is rapidly excreted in the urine and has a half-life of 18 to 20 h. Within 4 days, 75 to 90% of the absorbed dose is eliminated. The majority is excreted as the original material without undergoing any metabolic alteration. Some 2,4-D conjugates have been identified, but they are considered less toxic than the parent compound.

Studies of percutaneous absorption and urinary excretion in adult male volunteers have indicated that, while only about 4 to 8% of the 2,4-D dose applied to the skin is absorbed, dermal absorption is the principle route of exposure accounting for approximately 90% of the total dose.[17–18] Greater than 90% of the total skin exposure is estimated to occur via the hands.[19] By contrast, inhalation accounts for only about 2% of the total dose. Experiments under field conditions have shown that exposures can be greatly reduced by observing some standard hygienic practices of wearing gloves and showering and changing clothes following spraying activities. The industry is currently testing alternative delivery mechanisms which, if successful, would eliminate the handling and mixing of the concentrated products and thus greatly reduce the opportunity for applicator exposure.

III. SUMMARY OF TOXICOLOGY STUDIES ON 2,4-D

Epidemiologic findings must be interpreted in context with what is known about the biology of the agent under investigation, particularly the evidence from experimental studies. There have been numerous toxicology studies conducted on 2,4-D and it is not the authors' intent to review them all. Excellent review articles have been published elsewhere and the reader is referred to them for a more in depth discussion.[14,17,20–22] Instead, the more recent studies which impact the question of carcinogenicity are presented and discussed.

A. MUTAGENICITY STUDIES

2,4-D has been rigorously evaluated for mutagenicity in a variety of *in vivo* and *in vitro* test systems. The available evidence has been interpreted as equivocal to negative.[14,22] Clearly, 2,4-D does not exhibit the gene-damaging potential of a classic mutagen. For example, it is not mutagenic in the Ames *Salmonella* test nor in *E. coli,* two commonly used *in vitro* test systems. Tests in other, less commonly used *in vitro* systems have produced mixed results. The *in vivo* mutagenicity tests have been predominantly negative. In particular, 2,4-D has not been mutagenic in *in vivo* studies of lymphocytes or bone marrow.

B. IMMUNOLOGICAL EFFECTS

A few investigators have suggested that the chlorophenoxy herbicides may produce a carcinogenic response secondary to a suppression of the immune system mediated by their dioxin contaminants.[23] Although TCDD has been shown to cause immunosuppression in some animal studies,[24] there is no evidence of it causing similar effects in highly exposed humans.[25–26] The dioxin isomers which have been found at low levels in some samples of 2,4-D are orders of magnitude less biologically active and toxic than TCDD.[13]

There are few reports of any effects of 2,4-D itself on the human immune system. Recent dermal and oral studies evaluated the potential immunotoxicity of 2,4-D *n*-butylester in female mice.[27–28] Overall, the authors concluded that 2,4-D esters are unlikely to have any major immunotoxicological significance. Whether these findings extend to other 2,4-D formulations is unknown. Regardless, there is no evidence to suggest that any carcinogenic response of the phenoxys could be due to immune suppression induced either by the parent compounds or by any contaminants.

C. ONCOGENICITY BIOASSAYS

Several long-term studies with rats and mice were conducted a number of years ago to assess the oncogenic effects of 2,4-D, but those studies are now considered inadequate because they do not meet current experimental bioassay guidelines. In response to a request by the EPA for additional studies to support the reregistration of 2,4-D in the U.S., the Industry Task Force on 2,4-D Research Data recently sponsored two long-term cancer bioassays. In one study, groups of male and female B6C3F1 mice were fed 0, 1, 15, or 45 mg/kg/day of 2,4-D for 2 years.[29] (For perspective, the top dose tested exceeds the highest estimated occupational exposure by a factor of about 600, and was chosen because at dose levels above 50 mg/kg/day the excretion of 2,4-D via the kidney becomes saturated.) No excess incidence of tumors was observed in any sex or dose group.

In the second study, groups of male and female Fischer 344 rats were fed 0, 1, 15, or 45 mg/kg/day of 2,4-D for 2 years.[30] The study showed an increased incidence of astrocytomas in male rats in the high dose group. No increase in brain tumors was noted in the female rats or at any other site in either males or females. These data have been the subject of review by a number of experts. According to Dr. Adalbert Koestner, a recognized authority in neuropathology, the astrocytomas in the study did not conform to published biological characteristics of chemically induced brain tumors.[31] His evaluation concluded that the astrocytomas in the rat study were not related to treatment.

Another group of scientists, convened at the request of the Ontario Minister of the Environment, concluded that "While it is not possible to discount this evidence for carcinogenicity, the characteristics generally attributed to a brain carcinogen were not present in this experiment.[14]" The EPA's Scientific Advisory Panel concluded that both the rat and mouse studies were adequate tests of carcinogenicity, but that the astrocytoma data were equivocal and recommended that a repeat study of the high dose male rats be conducted as a test of the astrocytoma hypothesis.[32] The EPA rejected the recommendation of its Scientific Advisory Panel and concluded that both the rat and mouse studies were inadequate and must be repeated because, in their opinion, the studies did not achieve a maximum tolerated dose. The Industry Task Force is currently repeating both the mouse and rat studies at higher doses.

Still another expert group, convened by the Harvard School of Public Health, has speculated that doses of 45 mg/kg/day or more of 2,4-D given to male rats may saturate the capacity of both the kidney and the choroid plexus of the brain to excrete 2,4-D, leading to tissue accumulations, especially in the brain.[22] They suggested that, if the astrocytoma incidence is increased in the repeat studies, then a pharmacokinetic study should be done to test this hypothesis. If confirmed, then the findings at higher doses may not be relevant for the lower exposures that humans are likely to receive.

IV. THE ROLE OF EPIDEMIOLOGY IN HAZARD IDENTIFICATION

Epidemiology, because it is the study of humans and the types and levels of exposures that they actually receive, can serve as a critical check on the appropriateness of animal models for predicting human health effects associated with chemicals like herbicides. If the animal models are correct, and human exposures have been controlled, then epidemiology studies of manufacturers, formulators, and applicators should not find evidence of herbicide-related health effects. On the other hand, if the animal models are not appropriate or exposures have not been well controlled, then epidemiology can provide important warnings that changes are needed to reduce or eliminate exposures. Once those changes have been implemented, only epidemiology can evaluate their effectiveness.

Many people consider epidemiology and toxicology to be analogous disciplines, the only difference being the species of interest. This is true in only a limited sense and the differences are important. Epidemiology is an observational science. It is complicated because the conditions of the study are not under the direct control of the investigator. As a consequence, epidemiologists must deal with issues that are often alien to experimentalists. This includes the non-random selection of subjects into exposure groups, which can lead to a potential selection bias; the imprecision with which both human exposures and diseases are measured and recorded, possibly resulting in misclassification bias; the fact that humans are exposed to a variety of potentially hazardous agents, many of which can play a plausible role in disease causation (i.e., confounding bias); and the role of random variation in producing findings. Only after the epidemiologist has successfully dealt with these issues, can a causal relationship between an exposure and disease be seriously entertained.

Criteria have been developed to assist in evaluating whether an association observed in an epidemiology study might represent a causal relationship (Table 2).[33] None is sufficient on its own and only the criterion of temporality (i.e., the exposure must have preceded the disease in time in order to be a cause) is necessary. The concept of consistency deserves special consideration. If the association is consistently observed across many studies of different designs, performed by different investigators, then it is unlikely that it is spurious and a causal interpretation is enhanced. On the other hand, a lack of consistency dictates

TABLE 2
Causation

Strength of association
Consistency
Specificity
Temporality
Biological gradient
Biological plausibility
Coherence of the evidence

TABLE 3
Spectrum of Epidemiologic Study Designs

Design	Causal implication
Descriptive (hypothesis-generating)	
Anecdotal evidence; case histories; ecologic	Speculative
Analytical (Hypothesis-Testing)	
Cross-sectional; case-control	Suggestive
Retrospective cohort; prospective cohort	Highly suggestive
Experimental (clinical trials)	Firm

that the finding be interpreted cautiously and be further scrutinized. The application of these criteria is almost never straightforward and requires an in-depth evaluation of the strengths and weaknesses of the individual studies as well as a weighing of all the relevant evidence.

Among epidemiologic studies, there is a continuum of designs (Table 3).[34] At one extreme, closest to the experimental design, are clinical trials and prospective cohort studies. At the other extreme are case histories and anecdotal evidence. In between are the case-control and retrospective cohort studies — the two designs which constitute the majority of epidemiologic studies of suspected environmental hazards. The confidence that may be placed in any study varies from firm at the experimental end to speculative at the other. The continuum largely reflects the degree to which the various designs allow the investigator to control for selection, misclassification, and confounding bias. It is also important to note that some studies are designed primarily to generate hypotheses while others are designed to test hypotheses.

There is no point in dwelling on the extremes of the spectrum. It should be obvious that anecdotal evidence and case histories lack formal control groups or rigorous information collection and thus provide a basis only for suggesting hypotheses. At the other extreme, clinical trials are unlikely to be done to evaluate possible adverse human health effects caused by herbicides. Cross-sectional studies are frequently used to study diseases which have a high prevalence, and thus are unsuitable for examining cancer risks. More attention will be directed at the ecological, case-control, and cohort designs as these constitute the majority of epidemiology studies done thus far which impact agrichemicals.

Ecological studies, also sometimes referred to as geographical correlational studies, use exposure and disease data that have been classified in reference to an ecological unit such as a census tract, city, or county.[35] These units may, for example, be categorized into urban vs. rural, supplied by surface water vs. supplied by ground water, or high herbicide use vs. low herbicide use. Possible relationships between these characteristics and certain diseases are suggested by comparing mortality or morbidity rates among the ecologic units.

An advantage to the ecologic design is its low cost. Because they frequently make use of readily available data, ecologic studies can be done by virtually anyone who has access to a library and a computer. In some instances, this has led to abuse by persons whose sole intent is advocacy.

The major limitation of ecological studies is the tendency for associations that are observed at the level of the ecologic unit to fall apart when they are analyzed at the level of individual subjects — a phenomenon referred to as the "ecologic fallacy". As a consequence, the findings of such studies are properly viewed as hypothesis generating rather than hypothesis testing. This means that the results of ecologic studies alone are not an appropriate basis for policy making. Instead, the findings call for confirmation or refutation using more sophisticated epidemiology study designs.

Case-control studies are frequently employed to evaluate risk factors for diseases of low incidence like individual types of cancer.[36] They involve sampling the population according to who does and does not have a particular disease of interest. These cases and controls, or their next of kin, are then interviewed about their past occupational and personal exposures. The frequency of exposure among the cases is then compared with that among the controls to find differences which may give clues about the causes of the disease.

The main advantages of case-control studies are their efficiency for studying rare cancers and the opportunity they present for examining many potential causal factors simultaneously. Theoretically, when done correctly, they should produce the same results and conclusions as cohort studies. The disadvantages relate mainly to the increased opportunity for selection bias and for misclassification of exposure. The human memory is fallible, and because cases are interviewed after the disease has already been diagnosed, there is the possibility that they may be stimulated or motivated to recall past exposures to a different degree than healthy controls, thus leading to an artificial association between exposure and disease — a phenomenon referred to as "recall bias".[37] Because of the potential for such problems, some epidemiologists are very mistrustful of case-control studies and prefer to depend on the larger, more cumbersome cohort studies. Others will accept the results of case-control studies but will accord them less weight than the results of cohort studies.

Closest to the experimental studies is the cornerstone of observational epidemiologic research, the prospective cohort study. With this design, a population at risk is defined whose members have the potential of being exposed to the agent of interest and who could develop disease. This population is first examined to eliminate any persons who already have the disease. The remaining individuals are classified into those who are exposed and those not exposed to the agent. Both groups are followed prospectively to separate those who develop the disease from those who do not. The risk of developing the disease after exposure, referred to as the risk ratio, is calculated as a ratio of the disease rate among the exposed to the disease rate among the unexposed.

An advantage of cohort studies is the documentation of exposure prior to the occurrence of any adverse health effects in the population studied. This virtually eliminates the possibility of a differential misclassification of exposure among diseased and disease-free subjects. Furthermore, the availability of detailed work history records and industrial hygiene monitoring data permit a more precise and valid estimation of exposures than is possible with the case-control methodology.

The disadvantage of the prospective design is the amount of time that is required to wait for enough cases of the disease to develop in order to detect or confidently rule-out a difference in disease rates between the exposed and the unexposed. This is particularly true for a disease with a long latency period such as most types of cancer. Often the waiting can be avoided if historical records can be used to identify a group exposed in the past. This approach, referred to as the historical (or retrospective) cohort study, is often used to investigate the risk of disease in occupational groups. Its validity depends upon the extent to which past exposure data are available and reliable.

A lack of statistical power for detecting or ruling-out an increased risk of uncommon diseases can be a problem even in historical cohort studies. That is unless very large cohorts of workers can be identified or the findings from smaller studies can be pooled. This pooling of data, referred to as meta-analysis, is rapidly being embraced by epidemiologists as a systematic method for weighing evidence from multiple studies and offers hope for resolving some controversies in the face of apparently inadequate or conflicting studies.[38]

The evaluation of the relationship between pesticide exposures and latent diseases, like cancer, is complicated. Unlike acute health effects, latent ones are often determined by many factors which interact with each other in a manner that is complex, making the isolation of a single cause, or several causes, a formidable task. For instance, a cancer that is likely to have been caused by a pesticide can be clinically and pathologically indistinguishable from one caused by tobacco, alcohol, or other agents. Additionally, the longer the elapsed time from exposure to effect, the more difficult it is to demonstrate a causal association. Time has other effects. For the case-control study, the fallibility of human memory may influence the replies to questions posed to the cancer patient and the healthy control. Records can be destroyed, cohorts disperse, technology and exposure change.

A frequent and legitimate criticism of epidemiology studies on pesticides is the inadequacy of exposure estimation for individual subjects.[39] Information on the types of pesticides applied and the level and duration of exposure are critical to identifying associations. Yet this type of data is rarely available. Complicating matters are the mixed exposures that virtually all pesticide applicators receive. They may apply herbicides one day, and fungicides or insecticides the next. This can pose a problem in epidemiology — referred to as "collinearity" — which can effectively preclude the isolation of a single causal factor from among several competing candidates. Furthermore, field studies show wide variation in absorbed doses depending upon individual work and hygienic practices. Ideally, the investigator would like to have biological monitoring data to determine the absorbed dose for each individual subject over the entire period of their exposure so as to permit dose-response analyses. Although the availability of biomarkers of exposure is improving, they are virtually nonexistent to estimate exposures from many years ago.

Despite such problems, epidemiology has made important discoveries of latent effects from occupational exposures, including several involving pesticides. Historical cohort mortality studies of pesticide manufacturing workers, orchard sprayers, and copper smelter workers firmly established arsenic as a cause of human lung cancer, even though it could not be shown to produce cancer in laboratory animals.[40] The evidence on arsenic was convincing because of the high risk ratios that were found across many studies of workers exposed under a variety of very different occupational scenarios.

Overexposure to the nematocide 1,2-dibromo-3-chloropropane (DBCP) was convincingly linked to azospermia and oligospermia in pesticide manufacturing and formulation workers because of the strength of the association, enough of the workers were affected at one facility that they themselves detected the problems, its appearance in multiple workforces, and the fact that similar effects had been observed under experimental conditions in laboratory animals.[41] Recovery of spermatogenesis following the removal of exposure has further solidified the causal relationship.[42]

It must be recognized that we may not again encounter risk ratios as large as those that were seen with arsenic and DBCP. Consequently, in the future if may be even more difficult to firmly establish causal associations with epidemiology. Yet the studies will continue, fueled in part by the past successes like those with arsenic and DBCP and by the public's demand for greater assurances of pesticide safety.

TABLE 4
Methodologic Features and Summary of Results of Case-Control Studies Relating Soft-Tissue Sarcoma, Hodgkin's Disease or Non-Hodgkin's Lymphoma to Phenoxy Herbicide Use

Study	Cases	Controls	Interview	Exposure classification	Odds ratio (95% CL)
48	82 STS, population-based	92 with other types of cancer	Telephone	Duration and latency of phenoxy use	1.3 (0.6–2.5)[a]
52	690 NHL population-based	1245 general population	Personal	Ever/never used 2,4-D	1.2 (0.9–1.8)
47	133 STS 121 HD 170 NHL, population-based	948 general population	Telephone	Frequency, duration, and latency for herbicides Ever/never use specific phenoxys	0.9 (0.5–1.6) 0.9 (0.5–1.5) 2.2 (1.2–4.1)
51	83 NHL, population-based	168 with other types of cancer, 228 general population	Telephone	Duration and latency of phenoxy use	1.3 (0.7–2.6)
49	128 STS 576 NHL, population-based	694 general population	Personal	Duration and latency of phenoxy use Ever/never use specific phenoxys	0.8 (0.5–1.2) 1.1 (0.8–1.4)
50	68 STS (37 men, 31 women), population-based	158 general population	Personal and postal	Not exposed, uncertain exposure, certainly exposed to phenoxys	Males 0.91 (0.6–10.3)[a] Females 2.42 (0.6–12.4)[a]
53	201 NHL, population-based	725 general population	Telephone	Frequency, duration, and latency for specific herbicides, including 2,4-D	1.5 (0.9–2.4)

V. SUMMARY OF CASE-CONTROL STUDIES

Three case-control studies from Sweden initially suggested a link between phenoxy herbicide/chlorophenol exposure and soft-tissue sarcoma (STS), Hodgkin's disease (HD), and non-Hodgkin's lymphoma (NHL).[3–5] Those studies have been controversial because of suspected bias in the selection of cases and controls, potential recall and interviewer bias, and confounding.[7,8,43] Subsequently, the authors have attempted to address some of the criticisms via a series of replicate studies,[6,44–46] but several important questions remain. More importantly perhaps, they did not report any quantitative exposure pertaining to 2,4-D, and thus the studies are not directly relevant to the discussion at hand. Thus, it is more informative to focus on the studies done elsewhere which have been motivated by the original work from Sweden.

There have been several follow-up case-control studies.[47–53] A summary of their methodologic features and major findings is presented in Table 4. None of the four studies of STS has confirmed the association suggested by the Swedish investigators,[47–50] nor has there been any support for a link with HD.[47]

The newer studies have provided conflicting evidence regarding a possible association between phenoxy herbicides and an increased risk of NHL.[47,49,51–53] Hoar and colleagues reported a statistically significant sixfold risk of NHL related to the frequent use (i.e., >20 days/year) of herbicides by Kansas farmers.[47] Most farmers reported the use of chemicals in several of the herbicide subgroups, and statistically significant associations were reported for having ever used the phenoxyacetic acids, triazines, amides, trifluralin, and nonspecified herbicides. Data relating to the frequency and duration of use of specific herbicide types were not collected; therefore, dose-response relationships with 2,4-D or other phenoxy herbicides could not be rigorously explored. However, because 2,4-D was the herbicide reported most often, the authors focused their discussion on it.

More recently, Hoar Zahm and co-workers[53] reported the results from a similar study they conducted in eastern Nebraska. The questionnaire employed in this study included some improvement over the one that was used for the Kansas study which enabled the authors to evaluate duration and frequency in relation to specific herbicide types such as 2,4-D. Overall, a statistically non-significant odds ratio of 1.5 was reported for farmers reporting having mixed or applied 2,4-D. The risk of NHL was reported to increase with the frequency of 2,4-D use, rising to an odds ratio of 3.3 among those subjects who reported more than 20 days of use per year. However, none of the individual odds ratios, nor the trend, reached statistical significance.

Cantor and Blair found little evidence of an association between 2,4-D and NHL in a study they conducted in Iowa and Minnesota.[52] Frequency or duration of herbicide use was not collected from their subjects at the time of the original study, but based on the findings from the Kansas study they decided to reinterview a sample of subjects from Iowa to obtain it.

Woods et al. examined the risk of NHL associated with 2,4-D in western Washington and found an odds ratio of 0.73.[49] They further reported that the risk of NHL did not increase with increasing duration or estimated intensity of phenoxy herbicide exposure. No analyses of frequency of exposure were reported.

Finally, Pearce and colleagues studied NHL and exposure to sprays containing 2,4-D and 2,4,5-T in New Zealand.[51] No risk estimates were provided specifically for 2,4-D. The authors reported no association between exposure to phenoxy herbicides and NHL, even for subjects with probable or definite exposure of at least 5 days more than 10 years prior to diagnosis (OR = 1.4; 90% CI = 0.7 to 2.5). Prompted by the findings of Hoar et al., Pearce recently reanalyzed his data looking for an association between NHL and frequency or duration of phenoxy herbicide use and reported finding none.[54]

VI. SUMMARY OF COHORT STUDIES

Cohort studies have focused on workers who either manufactured the phenoxy herbicides[55–59] or applied these products.[56,60–69] Table 5 presents a summary of the major features of each of those studies. We have excluded from consideration any studies which focused on workers who only manufactured the precursor chlorophenols[70–79] as well as studies of groundtroops in Vietnam[80–87] whose exposure to phenoxy herbicides and their contaminants has not been substantiated.[16,38]

The majority of the cohort studies have been of mortality rather than cancer incidence. Thus, the assumption is made that the case-fatality rates do not differ between the exposed and comparison populations. Diagnostic information is restricted to statements appearing on the death certificate as coded to the rubrics of the International Classification of Diseases (ICD), because the same type of data forms the basis of the general population mortality statistics used for comparison. Classification problems associated with both the STSs and

TABLE 5
Methodologic Features of Historical Cohort Studies of Manufacturing Workers and Applicators Potentially Exposed to Phenoxy Herbicides

Study	Cohort	Comparison	Exposure classification
75	204 men engaged in 2,4,5-T[a] production (1950–1971)	Manufacturing workers U.S. general population	Duration and latency of 2,4,5-T exposure
59	884 men engaged in 2,4,5-T production (1955–1977)	U.S. general population (PMR[b] analysis)	Ever/never 2,4,5-T exposure or decedents only
57	4563 chemical workers (1947–1981), 940 of whom produced phenoxys	Denmark general population	Ever/never exposed to phenoxys, mainly MCPA[c]
56	5784 men engaged in production or spraying of phenoxys (1947–1975)	England and Wales general population	Graded exposure (high, low background) to mainly MCPA
55	878 men engaged in production of 2,4-D[d] (1945–1982)	U.S. general population, and internal referents	Cumulative dose of 2,4-D, process area, duration, and latency

Applicators

Study	Cohort	Comparison	Exposure classification
60	348 railway men who sprayed amitrole and phenoxys (1957–1972)	Sweden general population	Duration and latency
62	142 forestry workers who sprayed phenoxys (1954–1967)	Sweden general population	Foremen vs. others
63	1926 men who sprayed 2,4-D or 2,4,5-T (1955–1971)	Finland general population	Frequency and latency, type of employer
64	1278 airforce personnel from Operation Ranch Hand (1962–1971)	Multiple, including 6171 matched airforce personnel	Gallons of herbicides by occupational class
61	1222 forestry tradesmen who sprayed phenoxys (1950–1982)	Ontario, Canada general population	Duration
65	69,513 male farm operators in Saskatchewan (1971)	Saskatchewan general population	Acres sprayed for weeds
66	354,620 Swedish men in agriculture and forestry (1960)	Sweden general population	Six subcohorts defined on the basis of level of phenoxy herbicide use
67,68	20,245 Swedes licensed to use pesticides (1965, 1976)	Sweden general population	Years since license issued, year of license
69	25,945 Italian men licensed to use pesticides (1970–1974)	Italy general population	Birth cohort, village clusters defined by similar agricultural activity

[a]   2,4,5-Trichlorophenoxyacetic acid.
[b]   Proportionate mortality ratio.
[c]   4-Chloro-2-methylphenoxyacetic acid.
[d]   2,4-Dichlorophenoxyacetic acid.

NHL present a special handicap to their study. As a consequence, the power of cohort mortality studies to detect an increased risk of STS or NHL may be reduced.

Each of the studies was of the cohort design, except one which used a proportionate mortality method.[59] A few studies supplemented death certificates with cancer incidence data.[56,57] Mixed exposures to several products were common and the predominant phenoxy herbicide varied among the studies. The principal phenoxy exposure was often not specified;[60–64] in some studies it was 2,4,5-T,[58–59] in others it was 2,4-D,[55,65] and in still others it was MCPA.[56–57,66–69] In the majority of studies, the exposure and follow-up periods were allowed to overlap to some extent; however, supplemental analyses were done assuming latency periods of various lengths.

TABLE 6
Summary of Results from Cohort Studies of Workers Exposed to Phenoxy Acid Herbicides

Cause of Death

Study	All cancer sites		STS		NHL	
	Obs	Exp	Obs	Exp	Obs	Exp
Manufacturers						
75	1	3.6	0	*	0	*
59	9	10.9	1[a]	*	0	*
57	41	41.6	1	0.5	0	0.9[b]
56	297	314.0	1	0.9	2	5.5
55	26	22.6	0	*	2[c]	*
Applicators						
60	6	5.6	0	*	0	*
62	5	6.4	0	*	0	*
64	26	36.5	0	0.1	0	0.8
64	6	10.6	0	0.2	0	0.4
61	18	16.4	0	*	0	*
65	3198	3953.2	18	20.2	103	112.1
66	—	—	331	366.8	—	—
67,68	558	648.8	6	6.4	21	20.8
69	631	877.8	—	—	45[d]	31.8

Note: obs = observed number of events in study cohort. exp = expected number of events in study cohort.

[a] Reclassified a poorly differentiated carcinoma (Reference 73).
[b] Expected value reported for male workers only.
[c] Both of these cases were additionally exposed to TCDD and/orH/OCDDs and are included among the five cases reported by References 72 and 75.
[d] Includes totals for Hodgkin's lymphoma.
* Expected values not reported in the original reports.

A summary of the observed and expected number of events from all cancers combined, cancer of connective and other soft tissue and NHL is presented in Table 6. Nearly all of the studies showed phenoxy herbicide-exposed subjects to have experienced cancer mortality rates at or below those of the comparison groups. Several of the cohorts had elevated mortality from cancer at one or more sites. For instance, Lynge[57] reported statistically significant excesses of lung and rectal cancer among men employed in phenoxy herbicide operations, Coggon et al.[56] found excess nasal carcinoma, and Bond and co-workers[55] reported an excess of cancer of other and ill-defined sites. Because of the multiple comparison problem, it is not uncommon for cohort studies to display an elevated rate of cancer at one or more sites simply due to chance. The lack of a common tumor type among the studies argues against a casual relationship between phenoxy herbicides and the major types of cancer.

Evaluating the risk of the rarer types of cancer is more problematic owing to the small numbers of observed and expected events. For example, one case of STS originally reported among a cohort of 2,4,5-T production workers[59] was reclassified as a poorly differentiated carcinoma upon reexamination of pathologic materials.[73] Lynge[57] reported five cases of STS vs. 1.8 expected (RR = 2.72; 95% CI = 0.88 to 6.34) among a diverse group of male chemical workers, and no cases among female workers (0.75 expected). Yet, only one of those five cases had been assigned to chlorophenoxy operations, and his total chemical plant employment was limited to 3 months. Lynge also reported seven cases of malignant lymphoma vs. 5.4 expected among all male workers, and one case among female workers vs. 1.2 expected. If we restrict our consideration to those workers who were assigned to chlorophenoxy operations, there were no cases of lymphoma observed vs. 0.9 expected among the males and an unspecified number in the females.

Coggon et al.[56] reported one confirmed case of STS compared to 0.6 expected among workers whose jobs entailed potential exposure to phenoxy herbicides, principally MCPA. No further cases of STS were registered among the surviving members of the cohort. Two subjects died from NHL, and one from HD as compared to approximately nine expected lymphoma deaths in the total cohort, and six expected in the subset with likely exposure to the phenoxy herbicides.

Bond et al.[55] reported no cases of STS among 878 chemical workers who manufactured or formulated 2,4-D products, but did observe two cases of NHL among workers who additionally had been assigned to produce the precursor chlorophenols. Both of the cases of NHL occurred after very short latency periods which makes it unlikely that they were 2,4-D-related.

None of the remaining studies reported any cases of STS or NHL; however, the cohorts were small and so few cases of these cancers were expected.

Several large linkage-type studies[65–69] are also deserving of consideration. The authors were able to assemble cohorts of pesticide applicators from census data or from licensing bodies and link them to regional cancer incidence and/or mortality records to estimate cancer risks. These studies have greater statistical power than the cohort studies discussed above, but their exposure data is very limited and is a profound weakness.

Wigle and colleagues[65] linked records of almost 70,000 Saskatchewan males identified as farmers from the 1971 Canadian Census of Agriculture to mortality records complete through 1985. Among all farmers, small deficits in STS and NHL were found. Data on direct exposures of farmers were not available; however, farmers were categorized according to the number of acres they reported spraying for the control of insects or weeds, and dollars spent for pesticides or fuel and oil used for farm purposes. On farms of less than 1000 acres, the authors reported significant positive trends of NHL risk and negative trends for lung cancer risk with increasing numbers of acres sprayed for weed control and increasing fuel and oil expenditures. These findings did not hold on farms of 1000 acres or more. The authors estimated that 2,4-D constituted over 90 and 75%, by weight, of all herbicide active ingredients used agriculturally in Saskatchewan during the 1960s and 1970s, respectively.

Using a similar approach, Wiklund et al.[66] studied more than 350,000 Swedish men identified from a national census as having been employed in agriculture or forestry. Linkage was made to a central cancer registry. The study found no increased risk for STS among either the total group of workers or among any of six subsets defined by presumed level of phenoxy herbicide exposure. Moreover, no time-related increase in these cancers was observed despite the greatly increased use of phenoxy acid herbicides between 1947 and 1970.

These same authors linked records of more than 20,000 licensed pesticide applicators in Sweden to the central cancer registry.[67–68] An estimated 72% of the applicators had exposure to phenoxy herbicides, mainly MCPA, MCPP, and 2,4-DP but also 2,4-D and 2,4,5-T. No excess incidence of STS, NHL, or HD was observed in the total group of applicators nor in subsets defined by latency or year of licensing.

Corraro and colleagues[69] studied more than 25,000 farmers from the southern Piedmont of Italy who were licensed between 1970 and 1974 to buy and use pesticides classified as highly acutely toxic. Once again, data on pesticide exposures of individual workers were not available, nor was there mention made of the types or patterns of phenoxy herbicides used. The cohort was linked to a regional hospital discharge database to identify cases of cancer of the bone, connective tissue or skin, brain or other parts of the nervous system, or lymphatic or hematopoietic tissue tumors. The analysis ignored bone and connective tissue cancers after a quality control check found large discrepancies in these diagnoses. Overall, non-statistically significant increased risk ratios were seen for tumors of the skin and lymphatic tissue. The excess skin cancer appeared mainly among the oldest farmers, whereas the risk of cancer of lymphatic tissue was observed to have been independent of birth cohort. A cluster of villages with the highest percent of arable farming land appeared to have the highest risk of lymphatic tissue cancer. In view of potential biases introduced by the use of hospital discharge diagnoses, the authors urged a cautious interpretation of their findings.

VII. WEIGHT OF THE EVIDENCE EVALUATION

The foregoing discussion has highlighted the available toxicology and epidemiology data on the phenoxy herbicides, particularly as they relate to 2,4-D. Any synthesis of these data must acknowledge the many inconsistencies and contradictions, and provide some rationale for weighing all of the individual pieces of evidence, whether they provide support for or against the phenoxy herbicide/cancer hypothesis.

On balance, the evidence from experimental studies does not suggest that 2,4-D would cause cancer in humans. Indeed, 2,4-D does not behave as a classical mutagen, nor has it produced tumors in multiple species tested. Although it would not be unprecedented for a chemical which tested negative in standard animal bioassays to cause cancer in humans, nevertheless it would be unusual. The few exceptions (i.e., arsenic and benzene) have caused cancer in humans in the same organ systems in which acute toxic effects were observed in the rodent studies. By contrast, we are not aware of any acute or chronic effects observed in rodents exposed experimentally to 2,4-D which would predict that it might cause STS, HD, or NHL in humans.

Multiple case-control and cohort studies conducted throughout the world have failed to confirm the associations between phenoxy herbicide use and increased risks of STS or HD reported originally from Sweden. Thus, the epidemiologic evidence of an association between phenoxy herbicide exposure and either STS or HD is very weak and will not be discussed here further.

Several case-control epidemiology studies[47,53] suggest a relationship between the frequent use of herbicides by farmers and an increased risk of NHL. Other case-control studies[49,51-52] provide no support for such an association. The individual studies are not equivalent. Differences exist in the (1) opportunity for exposure to the underlying populations; (2) distribution of other risk factors for NHL among the populations; (3) methodology employed to gather and analyze the data which may have resulted in the underestimation of the strength of the association in some reports (e.g., non-differential misclassification of exposure) or overestimates in others (e.g., differential misclassification of exposure); and (4) precision with which the studies could estimate an association between phenoxy herbicides and NHL. The various possible explanations have been systematically evaluated and none is an obvious candidate.[89]

It is doubtful whether further case-control studies can resolve the herbicide/NHL question, at least until better estimates of exposure can be developed. Indeed, it has not been established that interviews with subjects or their next of kin can provide valid or reliable data on pesticide use. Intuitively, it would seem difficult for subjects, and particularly their next of kin, to recall the types of pesticides applied over a lifetime, the frequency and duration of spraying activities, and use of any protective equipment. Only very recently have the prerequisite methodological studies designed to answer these questions been initiated. We look forward to the results of studies currently underway at the University of

Minnesota and at the National Cancer Institute which are evaluating the reliability and validity of pesticide use data obtained from interviews.

It is also clear that the studies which have reported a herbicide/NHL link have not actually isolated an association specifically with 2,4-D. Indeed, use of 2,4-D as ascertained in those studies may be serving as a surrogate for some other factors such as the chemicals (e.g., solvents, or other herbicides) that were often mixed with the formulations or other exposures that are highly correlated with herbicide use (e.g., exposure to allergens from the noxious weeds themselves).

Generally speaking, the cohort studies which have been done of the workers who manufactured, formulated, or applied the phenoxy herbicides have been reassuring. Such workers have typically experienced cancer rates equal to or below those of similar groups with no exposure to the phenoxy herbicides. Thus, it is apparent that the continued use of these chemicals according to label directions should not pose a major public health problem. The cohort studies should be given special consideration, because the exposures received by the underlying populations have been more intense, frequent, and better documented than those of the majority of subjects considered exposed in the case-control studies.

Of course, a weakness of the cohort studies has been their limited statistical power for detecting or ruling-out modest increases in risk for some of the more uncommon types of cancer such as NHL. Exceptions to this are the large linkage-type studies conducted by Wigle,[65] Wiklund[66–68] and Corraro.[69] However, the gains in statistical power achieved by those authors were offset by their use of non-validated, surrogate measures of exposure. As a consequence, those studies have somewhat limited value for confirming or refuting a possible role played by specific agents such as phenoxy herbicides. A logical next step might be to conduct nested case-control studies of NHL within these large cohorts. Yet, the ultimate success of those studies will still depend upon an ability to classify exposures accurately.

When all of the evidence is evaluated against the criteria that epidemiologists use to judge causation, one must conclude that a causal association between exposure to phenoxy herbicides and NHL has not been established. Two important elements, consistency and biological plausibility, are missing. Misclassification, confounding, or chance remain viable alternative explanations for the associations that have been observed in a few studies.

Excluding arsenic, for no other class of pesticides is there as much epidemiologic data available as there exists for the phenoxy herbicides. Yet, we are not able to state unequivocally that exposure to these chemicals does or does not cause some small amount of human cancer. This apparent contradiction illustrates very well the current limitations of epidemiological data for identifying human carcinogens among pesticides. The prospects for the future will depend upon the availability of sufficient numbers of exposed workers for study, and the development of more rigorous measures of individual exposures.

ACKNOWLEDGMENT

The authors wish to acknowledge secretarial support provided by Diana Diamond.

REFERENCES

1. **Hardell, L.,** Soft-tissue sarcomas and exposure to phenoxy acids: a clinical observation, *Larkartidningen,* 74, 2753, 1977.
2. **Hardell, L.,** Malignant lymphoma of histiocytic type and exposure to phenoxyacetic acids or chlorophenols, *Lancet,* 1, 55, 1979.

3. **Hardell, L. and Sandstrom, A.,** Case-control study: Soft-tissue sarcomas and exposure to phenoxyacetic acids or chlorophenols, *Br. J. Cancer,* 39, 711, 1979.

4. **Eriksson, M., Hardell, L., Berg, N. O., Mollner, T., and Axelson, O.,** Soft-tissue sarcomas and exposure to chemical substances. A case-reference study, *Br. J. Ind. Med.,* 38, 27, 1981.

5. **Hardell, L., Eriksson, M., Lenner, P., and Lundgren, E.,** Malignant lymphoma and exposure to chemicals, especially organic solvents, chlorophenols and phenoxy acids: a case-control study, *Br. J. Cancer,* 43, 169, 1981.

6. **Hardell, L.,** Relation of soft-tissue sarcoma, malignant lymphoma and colon cancer to phenoxy acids, chlorophenols and other agents, *Scand. J. Work Environ. Health,* 7, 119, 1981.

7. **Coggon, D. and Acheson, E. D.,** Do phenoxy herbicides cause cancer in man?, *Lancet,* 1, 1057, 1982.

8. **Cole, P.,** Direct testimony before the Environmental Protection Agency, FIFRA Docket Nos. 415ff, Exhibit 860, Nov. 6, 1980.

9. **Hardell, L.,** Cross-examination before the Environmental Protection Agency, FIFRA Docket Nos. 415ff, Exhibit 773, Sept. 29, 1980.

10. **Cochrane, W. P., Singh, J., Miles, W., and Wakeford, B.,** Determination of chlorinated dibenzo-p-dioxin contaminants in 2,4-D products by gas chromatography: mass spectrometric techniques, *J. Chromatogr.,* 217, 289, 1981.

11. **Federal Register,** Polyhalogenated dibenzo-p-dioxins and dibenzofurans. Testing and reporting requirements. Final rule, 52, 21412-21452, 1987 (see also EPA Data Call-in Notice for Analytical Chemistry Data on Polyhalogenated Dibenzo-p-dioxins/Dibenzofurans in 2,4-Dichlorophenoxyacetic acid and its salts and esters, June 15, 1989).

12. **Berry, D. L.,** Final report of the determination of halogenated dibenzo-p-dioxins and dibenzofurans in 2,4-dichlorophenoxyacetic acid, Analytical Sciences Laboratories, The Dow Chemical Company, Midland, Michigan 48667, 1989, submitted to EPA August 1, 1989.

13. **Kociba, R. J. and Schwetz, B. A.,** A review of the toxicity of 2,3,7,8-tetrachlorodibenzo-p-dioxin (TCDD) with a comparison to the toxicity of other chlorinated dioxin isomers, *Assoc. Food Drug Off. Q. Bull.,* 16, 168, 1982.

14. **Canadian Centre for Toxicology,** Expert Panel Report on the Carcinogenicity of 2,4-D, Canadian Centre for Toxicology, Guelph, Ontario, Canada, March 23, 1987.

15. **Gianessi, L. P.,** Use of selected pesticides in agricultural crop production for counties, Resources for the Future, Washington, D.C., 1989.

16. **Gough, M.,** *Dioxin, Agent Orange: The Facts,* Plenum Press New York, 1986.

17. **World Health Organization,** 2,4-Dichlorophenoxyacetic acid (2,4-D), Environmental Health Criteria, 29, 1984.

18. **Sauerhoff, M. W., Braun, W. H., Blau, G. E., and Gehring, P. J.,** The fate of 2,4-Dichlorophenoxyacetic acid (2,4-D) following oral administration to man, *Toxicology,* 8, 3, 1977.

19. **Grover, R., Franklin, C. A., Muir, N. I., Cessna, A. J., and Riedel, D.,** Dermal exposure and urinary metabolite excretion in farmers repeatedly exposed to 2,4-D amine, *Toxicol. Lett.,* 33, 73, 1986.

20. **Mullison, W. R.,** An Interim Report Summarizing 2,4-D Toxicological Research Sponsored by the Industry Task Force on 2,4-D Research Data and a Brief Review of 2,4-D Environmental Effects, Industry Task Force on 2,4-D Research Data, Washington, D.C., 1986.

21. **Council for Agricultural Science and Technology,** Perspectives on the Safety of 2,4-D, COMMENTS from CAST, Ames, IA, December 1987.

22. **Harvard School of Public Health,** The Weight of the Evidence on the Human Carcinogenicity of 2,4-D, Final Rep.—January 1990. Harvard University, Boston, MA.

23. **Zahm, S. H. and Vineis, P.,** Immunosuppressive effects of dioxin in the development of Kaposi's sarcoma and non-Hodgkin's lymphoma, *Lancet,* January 2–9, 1988, 55.

24. **Vos, J. G., Faith, R. E., and Luster, M. I.,** Immune alterations, in *Halogenated Biphenyls, Terphenyls, Naphthalenes, Dibenzodioxins and Related Products,* R. Kimbrough R., Ed., Elsevier/North Holland Biomedical Press, 1980.

25. **Evans, R. G., Webb, K. B., Knutsen, A. P., Roodman, S. T., Roberts, D., Bagby, J., Garrett, W. A., and Andrews, J. S., Sr.,** A medical follow-up of the health effects of long-term exposure to 2,3,7,8-tetrachlorodibenzo-p-dioxin, presented at Dioxin '87: 7th Int. Symp. on Chlorinated Dioxins and Related Compounds, October 4–9, Las Vegas (abstr.), available from the University of Nevada, Las Vegas, 1987.

26. **Reggiani, G.,** Acute human exposure to TCDD in Seveso, Italy, *J. Toxicol. Environ. Health,* 6, 27, 1980.

27. **Blakely, B. R.,** The effect of oral exposure to the n-butylester of 2,4-Dichlorophenoxyacetic acid on the immune response in mice, *Int. J. Immunopharmacol.,* 8, 93, 1986.

28. **Blakely, B. R. and Schiefer, B. H.,** The effect of topically applied n-butylester of 2,4-Dichlorophenoxyacetic acid on the immune response in mice, *J. Appl. Toxicol.,* 6(4), 291, 1986.

29. **Hazelton Labs,** Oncogenicity study in mice with 2,4-Dichlorophenoxyacetic acid (2,4-D), final rep., Vol. 1, Hazelton Laboratories America, Inc. Vienna, VA, prepared for the Industry Task Force on 2,4-D Research Data, 1987.

30. **Hazelton Labs,** Combined toxicity and oncogenicity study in rats, 2,4-Dichlorophenoxyacetic acid (2,4-D), final rep., Vol. 1, Hazelton Laboratories America, Inc. Vienna, VA, prepared for the Industry Task Force on 2,4-D Research Data, 1986.

31. **Koestner, A.,** Histological evaluation of brain sections obtained from F344 rats exposed to various doses of 2,4-D in a 2-year chronic oral toxicity study, final rep. to the Industry Task Force on 2,4-D Research Data, 1986.

32. **U.S. EPA,** 2,4-D, 2,4-DB, and 2,4-DP; Proposed Decision Not to Initiate a Special Review, Federal Register 53, 9590, March 23, 1988.

33. **Kleinbaum, D. G., Kupper, L. L., and Morgenstern, H.,** *Epidemiologic Research: Principles and Quantitative Methods,* Lifetime Learning Publications, Belmont, CA, 1982.

34. **Ibrahim, M. A.,** Direct testimony before the Environmental Protection Agency of the United States of America, Docket Nos. 415 et al., November 4, 1980.

35. **Morgenstern, H.,** Uses of ecologic analysis in epidemiologic research, *Am. J. Public Health,* 72, 1336, 1982.

36. **Schlesselman, J. J.,** *Case-Control Studies: Design, Conduct, Analysis,* Oxford University Press, New York, 1982.

37. **Sackett, D. L.,** Bias in analytical research, *J. Chronic Dis.,* 32, 51, 1979.

38. **Greenland, S.,** Quantitative methods in the review of epidemiologic literature, *Epidemiol. Rev.,* 9, 1, 1987.

39. **Blair, A., Zahm, S. H., Cantor, K. P., and Stewart, P. A.,** Estimating exposure to pesticides in epidemiologic studies of cancer, in *Biological Monitoring for Pesticide Exposure — Measurement, Estimation and Risk Reduction,* Eds. R.G.M. Wong, R. G. M., Franklin, C. A., Honeycutt, R. C. and Reinert, J. C., Eds., American Chemical Society, Washington, D.C., ACS Symp. Ser., 382, 38, 1989.

40. **Smith, A. H. and Bates, M. N.,** Epidemiological studies of cancer and pesticide exposure, in *Biological Monitoring for Pesticides Exposure — Measurement, Estimation and Risk Reduction,* Wong, R. G. M., Franklin, C. A., Honeycutt, R. C., and Reinert, J. C., American Chemical Society, Washington, D.C., ACS Symp. Ser., 382, 207, 1989.

41. **Whorton, M. D., Krauss, R. M., Marshall, S., et al.,** Infertility in male pesticide workers, *Lancet,* 2, 1259, 1977.

42. **Olsen, G. W., Lanham, J. M., Bodner, K. M., Hylton, D. B., and Bond, G. G.,** Determinants of spermatogenesis recovery among workers exposed to 1,2-dibromo-3-chloropropane, *J. Occup. Med.,* 32, 979, 1990.

43. **Colton, T.,** Herbicide exposure and cancer, *J. Am. Med. Assoc.,* 256(9), 1176, 1986.

44. **Hardell, L. and Eriksson, M.,** The association between soft tissue sarcomas and exposure to phenoxyacetic acids: a new case-referent study, *Cancer,* 62(3), 652, 1988.

45. **Persson, B., Dahlander, A. M., Fredriksson, M., Brage, H. N., Ohlson, C. G., and Axelson, O.,** Malignant lymphomas and occupational exposures, *Br. J. Ind. Med.,* 46, 516, 1989.

46. **Eriksson, M., Hardell, L., and Adami, H. O.,** Exposure to dioxins as a risk factor for soft tissue sarcoma: a population-based case-control study, *J. Natl. Cancer Inst.,* 82, 486, 1990.

47. **Hoar, S. K., Blair, A., Holmes, F. F., Coysen, C. D., Robel, R. J., Hoover, R., and Fraumeni, J.,** Agricultural herbicide use and risk of lymphoma and soft-tissue sarcoma, *J. Am. Med. Assoc.,* 256, 1141–1147, 1986 (see also correction, *J. Am. Med. Assoc.,* 256, 3351, 1986).

48. **Smith, A. H. and Pearce, N. E.,** Update on soft-tissue sarcoma and phenoxyherbicides in New Zealand, *Chemosphere,* 15, 1795, 1986.

49. **Woods, J. S., Polissar, L., Severson, R. K., Heuser, L. S., and Kulander, B. G.,** Soft tissue sarcoma and non-Hodgkin's lymphoma in relation to phenoxy herbicide and chlorinated phenol exposure in western Washington, *J. Natl. Cancer Inst.,* 78, 899, 1987.

50. **Vineis, P., Terracini, B., Ciccone, G., Cignetti, A., Colombo, E., Donna, A., Maffi, L., Pisa, R., and Ricci, P.,** Phenoxy herbicides and soft-tissue sarcomas in female rice weeders: a population-based case-referent study, *Scand. J. Work Environ. Health,* 13, 9, 1986.

51. **Pearce, N. E., Smith, A. H., Howard, J. K., Sheppard, R. A., Giles, H. J., and Teague, C. A.,** Non-Hodgkin's lymphoma and exposure to phenoxyherbicides, chlorophenols, fencing work, and meat works employment: a case-control study, *Br. J. Ind. Med.,* 43, 75, 1986.

52. **Cantor, K. P. and Blair, A.,** Agricultural Chemicals, Drinking Water, and Public Health: An Epidemiologic Overview, National Cancer Institute, Bethesda, MD, 1986.

53. **Zahm, S. H., Weisenburger, D. D., Babbitt, P. A., Saal, R. C., Vaught, J. B., Cantor, K. P., and Blair, A.,** A case-control study of non-Hodgkin's lymphoma and the herbicide 2,4-Dichlorophenoxyacetic acid (2,4-D) in eastern Nebraska, *Epidemiology,* 1, 349, 1990.

54. **Pearce, N.,** Phenoxy herbicides and non-Hodgkin's lymphoma in New Zealand: frequency and duration of herbicide use (letter), *Br. J. Industr. Med.,* 46, 143, 1989.

55. **Bond, G. G., Wetterstroem, N. H., Roush, G. J., McLaren, E. M., Lipps, T. E., and Cook, R. R.,** Cause-specific mortality among employees engaged in the manufacture, formulation or packaging of 2,4-dichlorophenoxyacetic acid and related salts, *Br. J. Ind. Med.,* 45, 98, 1988.

56. **Coggon, D., Pannett, B., Winter, P. D., Acheson, E. D., and Bonsall, J.,** Mortality of workers exposed to 2 methyl-4 chlorophenoxyacetic acid, *Scand. J. Work Environ. Health,* 12, 448, 1986.

57. **Lynge, E.,** A follow-up study of cancer incidence among workers in manufacture of phenoxy herbicides in Denmark, *Br. J. Cancer,* 52, 259, 1985.

58. **Ott, M. G., Holder, B. B., and Olson, R. D.,** A mortality analysis of employees engaged in the manufacture of 2,4,5-trichlorophenoxyacetic acid, *J. Occup. Med.,* 22, 47, 1980.

59. **Zack, J. A. and Gaffey, W. R.,** A mortality study of workers employed at the Monsanto Company plant in Nitro, West Virginia, *Environ. Sci. Res.,* 26, 575, 1983.

60. **Axelson, O., Sundell, L., Andersson, K., Edling, C., Hogstedt, C., and Kling, H.,** Herbicide exposure and tumor mortality: an updated epidemiological investigation on Swedish railroad workers, *Scand. J. Work Environ. Health,* 6, 73, 1980.

61. **Green, L. M.,** Mortality Analysis of Ontario Hydro Foresty Tradesmen Cohort 1950–1982, Health Services Department, Health and Safety Division, Ontario Hydro, Toronto, November 1986.

62. **Hogstedt, C. and Westerlund, B.,** Cohort studies of cause of death of forest workers with and without exposure to phenoxy acid preparations, *Lakartidningen,* 77, 1828, 1980.

63. **Riihimaki, V., Asp, S., and Hernberg, S.,** Mortality of 2,4-dichlorophenoxyacetic acid and 2,4,5-trichlorophenoxyacetic acid herbicide applicators in Finland, *Scand. J. Work Environ. Health,* 8, 37, 1982.

64. **Wolfe, W. H., Michalek, J. E., and Minet, J.,** An Epidemiologic Investigation of Health Effects in Air Force Personnel following Exposure to Herbicides. Mortality Update — 1989, USAF Technical Rep., Brooks Air Force Base, Texas, 1989.

65. **Wigle, D. T., Semenciw, R. M., Wilkins, K., Riedel, D., Ritter, L., Morrison, H. I., and Mao, Y.,** Mortality study of Canadian male farm operators: non-Hodgkin's lymphoma mortality and agricultural practices in Saskatchewan, *J. Natl. Cancer Inst.,* 82, 575, 1990.

66. **Wiklund, K. and Holm, L. E.,** Soft tissue sarcoma risk in Swedish agricultural and forestry workers, *J. Natl. Cancer Inst.,* 76, 229, 1986.

67. **Wiklund, K., Dich, J., and Holm, L. E.,** Risk of malignant lymphoma in Swedish pesticide appliers, *Br. J. Cancer,* 56, 505, 1987.

68. **Wiklund, K., Dich, J., Holm, L. E., and Eklund, G.,** Risk of cancer in pesticide applicators in Swedish agriculture, *Br. J. Ind. Med.,* 46, 809, 1989.

69. **Corrao, G., Calleri, M., Carle, F., Russo, R., Bosia, S., and Piccioni, P.,** Cancer risk in a cohort of licensed pesticide users, *Scand. J. Work Environ. Health,* 15, 203, 1989.

70. **Cook, R. R., Townsend, J. C., and Ott, M. G.,** Mortality experience of employees exposed to 2,3,7,8-tetrachloro-dibenzo-p-dioxin (TCDD), *J. Occup. Med.,* 22, 530, 1980.

71. **Cook, R. R., Bond, G. G., Olson, R. A., Ott, M. G., and Gondek, M. R.,** Evaluation of the mortality experience of workers exposed to the chlorinated dioxins, *Chemosphere,* 15, 1769, 1986.

72. **Cook, R. R., Bond, G. G., Olson, R. A., and Ott, M. G.,** Update of the mortality experience of workers exposed to chlorinated dioxins, *Chemosphere,* 16, 2111, 1987.

73. **Fingerhut, M. A., Halperin, W. E., Honchar, P. A., Smith, A. B., Groth, D. H., and Russell, W. O.,** Review of exposure and pathology data for seven cases reported as soft tissue sarcoma among persons occupationally exposed to dioxin-contaminated herbicides, in *Public Health Risks of the Dioxins, Lowrance, W. W.,* Ed., Rockefeller University, New York, 1984.

74. **Honchar, P. A. and Halperin, W. E.,** 2,4,5-Trichlorophenol and soft-tissue sarcoma, *Lancet,* 1, 268, 1981.

75. **Ott, M. G., Olson, R. A., Cook, R. R., Bond, G. G.,** Cohort mortality study of chemical workers with potential exposure to the higher chlorinated dioxins, *J. Occup. Med.,* 29, 422, 1987.

76. **Thiess, A. M., Frentzel-Beyme, R., and Link, R.,** Mortality study of persons exposed to dioxin in a trichlorophenol-process accident that occurred in the BASF AG on November 17, 1953, *Am. J. Ind. Med.,* 3, 179, 1982.

77. **Zack, J. A. and Suskind, R. S.,** The morality experience of workers exposed to tetrachlorodibenzodioxin in a trichlorophenol process, *J. Occup. Med.,* 22, 11, 1980.

78. **Bond, G. G., McLaren, E. A., Lipps, T. E., Cook, R. R.,** Update of mortality among chemical workers with potential exposure to the higher chlorinated dioxins, *J. Occup. Med.,* 31, 121, 1988.

79. **Zober, A., Messerer, P., and Huber, P.,** Thirty-four-year mortality follow-up of BASF employees exposed to 2,3,7,8-TCDD after the 1953 accident, *Int. Arch. Occup. Environ. Health,* 62, 139, 1990.

80. **Anderson, H. A., Hanrahan, L. P., and Jensen, M., et al.,** Wisconsin Vietnam Veteran Mortality Study, Wisconsin Division of Health, Madison, WI, 1985.

81. **Centers for Disease Control,** Post service mortality among Vietnam veterans, *J. Am. Med. Assoc.,* 257, 790, 1987.

82. **Fett, M. J., Adena, M. A., Cobbin, D. M., and Dunn, M.,** Mortality among Australian conscripts of the Vietnam conflict era. I. Death from all causes, *Am. J. Epidemiol.,* 125, 869, 1987a.

83. **Fett, M. J., Nairn, J. R., Cobbin, D. M., and Adena, M. A.,** Mortality among Australian conscripts of the Vietnam conflict era. II. Causes of death, *Am. J. Epidemiol.,* 125, 878, 1987b.

84. **Greenwald, P., Kovaszny, B., Collins, D. N., and Theriault, G.,** Sarcomas of soft-tissues after Vietnam service, *J. Natl. Cancer Inst.,* 73, 1107, 1981.

85. **Kang, H. K., Weatherbee, L., Breslin, P. P., Lee, Y., and Shepard, B. M.,** Soft-tissue sarcomas and military service in Vietnam: A case comparison group analysis of hospital patients, J. Occup. Med., 28, 1215, 1986.

86. **Kogan, M. D., and Clapp, R. W.,** Mortality among Vietnam Veterans in Massachusetts, 1972–1983, Massachusetts Office of Commissioner of Veterans Services, Agent Orange Program Massachusetts Department of Public Health, Division of Health Statistics and Research, Boston, MA, 1985.

87. **Lawrence, C. E., Reilly, A. A., Quickenton, P., Greenwald, P., Page, W. F., and Kuntz, A. J.,** Mortality patterns of New York State Vietnam veterans, *Am. J. Publ. Health,* 75, 277, 1985.

88. **Centers for Disease Control (CDC) Veterans Health Studies,** Serum 2,3,7,8-tetrachlorodibenzo-p-dioxin levels in U.S. Army Vietnam-era veterans, *J. Am. Med. Assoc.,* 260, 1249, 1988.

89. **Bond, G. G., Bodner, K. M., and Cook, R. R.,** Phenoxy herbicides and cancer: insufficient epidemiologic evidence for a causal relationship, *Fundam. Appl. Toxicol.,* 12(1), 172, 1989.

Chapter 29

A REVIEW OF EPIDEMIOLOGIC STUDIES WITH REGARD TO ROUTES OF EXPOSURE TO TOXICANTS

Anthony P. Polednak

TABLE OF CONTENTS

0-8493-7357-3/93/$0.00 + $.50
© 1993 by CRC Press, Inc.

I. INTRODUCTION

A. STUDIES OF HUMANS VS. ANIMAL EXPERIMENTS

In animal experiments with chemical and physical toxic agents, routes of exposure are controlled by the experimenter. In the evaluation of *in vitro* assays for genetic toxicity as correlates of carcinogenicity in animals, for example, routes of administration (i.e., dosed feed, gavage, inhalation, intraperitoneal injection, skin painting, and dosed water) for 73 chemicals were tabulated.[1] In animal experiments a maximum tolerated dose can be administered daily over an entire lifetime, for the purpose of testing for mutagenic, carcinogenic, or toxic effects. Yet the relevance of such experiments to the human situation has been questioned,[2] especially if one recalls Paracelsus' aphorism that nearly all substances are toxic in sufficient doses.

Animal experiments may provide clues regarding effective routes of exposure and their associations with health effects in humans. The problems involved in extrapolating the results of animal experiments to the human situation are complex and well known. Despite use of the same route of exposure, the results may vary greatly by species because of differences in metabolism and genetically modulated responses. One example is the toxicity and carcinogenicity of dioxin in animals vs. limited evidence for toxic effects (except for the skin condition known as chloracne) and carcinogenicity in humans.

Actual experiments of toxicity are rare in humans, for ethical considerations. Examples of exceptions include the injection of plutonium into 16 patients with "terminal" illnesses (or reduced life expectancy) to study the metabolism and short-term toxicity.[3] During World War II Nazi scientists deliberately exposed prisoners to the inhalation of phosgene gas ($COCl_2$), which had been used in chemical warfare during World War I, but it was decided for ethical reasons that such data should never be used.[4]

In epidemiologic studies of human exposure to toxicants, "natural" experiments may occur in which the routes of exposure are fairly well known. Examples from radiation epidemiology include the radium dial painters, who received internal exposure by ingestion of paint containing radium and external radiation exposure from working in a contaminated environment, and patients who received injections of thorium (thorotrast) which was used as a contrast medium for diagnostic radiology. In these instances the metabolism and distribution of the toxic agents are rather well known. For most epidemiologic studies, however, exposure is poorly defined in terms of the quality and quantity of toxic agents.

For humans one can distinguish between environmental exposure pathways or routes (such as air, water, and soil contamination) and routes of exposure (or dose) in individuals via ingestion, inhalation, dermal absorption, and other routes or mechanisms (see later). The history of epidemiology begins with Hippocrates, the father of both medicine and epidemiology, and his "Airs, Waters, Places". The title of this work implies the importance of different environmental pathways of exposure on human health. Lead colic or symptoms due to lead ingestion was known to Hippocrates. Sir P. Pott in 1775 reported the first occupationally related cancers involving scrotal cancer in chimney sweeps, related to exposure of the skin of the scrotum to soot; in this century soot was shown to contain carcinogenic polycyclic aromatic hydrocarbons. Modern epidemiologic studies of occupational and environmental exposures probably begins with ionizing radiation studies in the 1920s and 1930s and studies of asbestos workers, as well as smoking and lung cancer, in the 1950s.

Many epidemiologic studies involve crude surrogate indicators such as occupation or residence near a potentially hazardous point source, such as a toxic waste disposal site or an industrial plant. In relatively few studies, and more often with occupational than environmental epidemiologic studies, are actual exposures or doses examined in samples of the "exposed" populations, as discussed later. This is due in part to the high cost of human studies and the complexity of exposure situations.

Rarely, human studies lead to animal experiments, in search of consistency of toxic effects or of animal models for human diseases. Erionite and mesothelioma is an example, where the fibrous mineral was given to rats by inhalation after an association with mesothelioma was found in humans.[5] Studies of the metabolism and health effects of radium and plutonium in humans led to inhalation experiments with dogs and other animals.[6]

B. REVIEW OF EPIDEMIOLOGIC STUDY DESIGNS

Ecologic studies, or correlations between data on exposure and health outcomes in populations, are common in epidemiology as a first or exploratory stage of investigation, usually utilizing existing data bases. For example, data on potential exposure of populations to toxic agents found in water samples, or number and types of toxic waste sites, in different counties or other geopolitical units can be correlated with rates of specific diseases in those areas. Extrapolation of such correlation or regression analyses to individual persons is not possible. There is increasing awareness of the need to include in ecologic studies data on many variables, such as socioeconomic indicators and smoking survey results (if available)[7] along with exposure data for toxic agents other than the one(s) under study and on multiple pathways of exposure (e.g., occupational as well as non-occupational). These "extraneous" variables may be associated with both the particular exposure variable of interest and with the disease outcome (see later).

Two or more communities with evidence for different levels for ambient exposure to a given toxic agent are often compared in terms of both evidence for actual exposure to individuals and in relation to current health status or mortality rates. This design can lead to a cohort study or a case-control study.

The use of existing data bases for descriptive epidemiologic studies has been reviewed, for example, for toxicology studies related to birth defects, spontaneous abortions, and childhood cancer,[8] occupational sentinel events in several U.S. states,[9] and air pollution monitoring data or types of toxic waste disposal sites in California.[10]

Analytic studies in epidemiology are either case-control or cohort in design. In case-control studies a group of cases with the disease in question are compared with a group of controls without the disease from the same population, sometimes matched on demographic and other variables. Cases and control are compared in terms of exposure history, sometimes with attention to multiple routes of exposure, and sometimes involving estimated current levels of exposure assessed by biological monitoring. The measure of interest is an odds ratio which is an estimate of the relative risk of the disease or outcome in exposed relative to unexposed persons. Relative risks can be calculated only in a cohort or prospective study, in which all participants are free of any evidence for the disease of interest at the outset of the follow-up period. Two or more subgroups are defined with respect to exposure levels at the outset, and then followed-up for estimation of disease risks. The ratio of disease rates in the exposed to the unexposed group is the relative risk.

Briefly, the advantages of cohort studies relate to the ability to examine several disease outcomes, the occurrence of exposure to presumed etiologic factors prior to the development of overt disease (and hence the possibility of a causal relationship), and the better characterization of exposure information. Disadvantages relate to the cost involved, reflecting the large numbers of exposed and unexposed that must be followed to accrue sufficient numbers of rare events, and the long follow-up. Another problem concerns changes in exposure status during the follow-up period. Rarely, such changes can be monitored and adjusted for thorough analytic procedures, as in the case of uranium miners. Case-control studies can have greater statistical power with smaller sample sizes, and can be conducted in a shorter time period at lower costs than cohort studies.

In historical or retrospective cohort studies, investigators use historical records on both exposure and disease outcome for analyses beginning in the past and extending to the most recent date for which data on mortality or other health outcomes are available. "Nested" case-control studies are special substudies carried out within cohort studies, involving usually all cases of a specific disease and a sample of controls without the disease selected from the cohort. These nested studies are efficient especially in occupational setting where exposure data are available from records. With a limited number of cases and controls detailed exposure data can be obtained from records and carefully checked and edited prior to analysis.

Cohort studies have mainly involved occupational groups. Exceptions include follow-up studies of the Seveso, Italy population exposed to dioxin in 1976,[11] the Japanese atomic bomb survivor and the Mashallese Islanders exposed to fallout, and more recently the populations exposed to radioactive fallout from the Chernobyl accident and iatrogenic radiation exposure groups, including thymic irradiation and related exposures, and breast fluoroscopy.

C. PROBLEMS WITH EPIDEMIOLOGIC STUDIES

Epidemiologic studies are often less concerned with details of exposure status and routes of exposure than with basic problems of proper design and method, avoidance of biases, and interpretation of results. In the interpretation of causality of associations between exposures and disease outcomes, however, issues of the nature of the exposure (e.g., routes of exposure, metabolism and specific organs or tissues involved) and dose-response, as well as biological plausibility, must be addressed. In many studies the quality of exposure information precludes adequate assessment of causality.

Another problem in the interpretation of associations is the simultaneous exposure to several toxic agents, both in the workplace (e.g., polynuclear aromatic hydrocarbons, chromium, nickel and other metals, and asbestos in steel workers)[12] and in other situations (e.g., ambient air and drinking water), and also including direct or indirect exposure to cigarette smoke which contains a variety of toxic substances. Stratification and multivariate analysis are techniques that can assist in interpretation of effects independent of such variables as smoking or interacting with smoking, although statistical power may be limited due to reduced sample sizes in strata analyzed.

In natural experiments and many human exposure situations the size of the exposed population is determined by the nature of the exposure, in contrast to animal experiments. Statistical power may be limited, so that a negative result must be qualified by this limitation.

In recent years the issue of "weak" associations in epidemiologic studies has received attention. Relative risks of 1.1 to 1.5 are difficult to interpret. Especially in such situations of weak associations the problem of confounding must be recognized. Confounders are variables associated with both the exposure of interest (i.e., exposure to some specific toxic agent) and the disease outcome. For example, smoking rates may differ among different occupational groups or in different communities compared. The number of potential confounders that can be evaluated is usually limited and for many diseases the important risk factors are unknown and could be associated with the exposure being evaluated.

Misclassification of exposure status is a ubiquitous problem in epidemiologic studies due to the lack of adequate data on exposure. Such misclassification can obscure real associations between exposure and disease. In all epidemiologic case-control studies, and in certain cohort studies, there is a potential for bias in the reporting of exposure history. This also holds for studies involving self-reported symptoms or diseases as, for example, in surveys of populations near toxic waste disposal sites or other point sources of contamination.

Health outcomes in epidemiologic studies are varied and often depend upon the nature of the toxic substances and knowledge of their effects in animals. Studies dealing with cancer are affected by the long latency or time period between initial exposure and manifestation of clinical disease. Sometimes adjustments for estimated minimal latency periods are made in the analysis; that is, special analyses exclude the early years of follow-up during which cancers related to the exposure of interest are not expected to have appeared.

In epidemics or acute outbreaks of disease associated with food or water contamination with microorganisms, routes and levels of exposure can be more easily ascertained because of the nature of the exposure and the immediate appearance of disease.[5] In studies of long-term effects of toxic agents, obtaining information on past exposures is fraught with difficulties. Knowledge and recall of past exposures may be considerably inaccurate and also subject to bias associated with health status of the respondent or "surrogate" or proxy (usually, next-of-kin) interviewed.

As noted earlier, cohort studies are often complicated by changes in degree of exposure, or even routes of exposure, over time. In the case of radium dial painters or painters of watches with luminous paint containing radium, for example, the ingestion of radium through paint was greatly reduced by the prohibition of the pointing of brushes in the mouth in 1926, while external irradiation continued (see later). In many cohort studies the level of exposure changes over time because of changes in industrial processes or protection resulting from recognition of potential hazards. Other examples involve workers in the steel industry,[12] and radiologists, as well as workers exposed to such agents as asbestos and benzene, often due to changes in occupational standards. In some instances exposure information may be adequate to allow detailed analyses that take into account changes in level of exposure over time, as in studies of uranium miners. Often, however, such exposure data are crude and limited in extent or generalizability to the entire cohort (see later).

The reader might keep in mind these problems with epidemiologic studies while reviewing evidence from selected epidemiologic studies regarding routes of exposure, as discussed in the following sections.

II. ROUTES OF EXPOSURE IN EPIDEMIOLOGIC STUDIES

A. VARIETIES
1. Dermal, Inhalation, and Ingestion Routes

Among inorganic chemicals, skin absorption of lead is less important than ingestion and inhalation and in children ingestion of lead-containing paint is the main route (see later). Radium exposure in the past involved ingestion of luminous paint (see later) and "therapeutic" exposures, while recent exposures are due to ingestion of radium in water (from natural sources) and inhalation of radon gas, as well as from the burning of coal containing primordial radionuclides.

In a 1980 review of the epidemiology of environmental carcinogenesis, Maclure and MacMahon[13] observed that less information was available on organic vs. inorganic compounds (such as nickel, cadmium, chromium, asbestos, and arsenic). Since that time, considerably more interest is apparent in the toxicity and carcinogenicity of organic substances. Ingestion of organic contaminants in drinking water, for example, has received attention in epidemiologic studies of cancer since mid-1970s.[14]

Skin cancers (i.e., of the scrotum) related to exposure to organics (i.e., aromatic hydrocarbons) were mentioned earlier. Examples of organic chemicals considered in epidemiologic studies that are readily absorbed through the skin include phenols, organophosphate pesticides, carbon tetrachloride, and tetraethyl lead (i.e., lead used as an additive to gasoline).

TABLE 1
Examples of Occupational Epidemiologic Studies

Agent	Occupation(s)	Route(s) of exposure
Benzene	Rubber and other chemical workers; seamen	Inhalation, dermal (?)
Hydrocarbons, Soot	Chimney sweeps,	Dermal
	Coke oven worker	Inhalation
Phosgene	Various (also non-occupational)	Inhalation
Radium	Luminous dial painters	Ingestion; external irradiation
Uranium; radon gas	Uranium and other miners	Inhalation
Asbestos	Insulators, construction, shipbuilding, steelworkers	Inhalation
Wood dusts	Woodworkers, saw mill	Inhalation
Cotton dusts	Mill workers	Inhalation
Nickel oxides	Welders, refiners, etc.	Inhalation
Arsenic compounds	Workers at smelters	Inhalation; dermal
Pesticides and herbicides (organophosphates)	Pesticide applicators; aviation workers; farmers; industrial workers; Vietnam veterans	Inhalation; dermal

Phenoxy herbicides are chlorinated hydrocarbons that are readily absorbed by the respiratory and digestive systems in humans, but dioxins (e.g., 2,3,7,8-TCDD) are only incompletely absorbed. The pesticide DDT (dicholorophenyl tricholoroethane) is poorly absorbed through mammalian skin.

Exposure to organic solvents is assumed to occur mainly by inhalation, but in certain situations dermal exposure can be important. Animal experiments have shown that benzene is absorbed through the skin,[15,16] and in certain situations in humans dermal exposure could be important — e.g., in certain occupational situations and in direct contact with contaminated soils. On the other hand, ingestion has been the main route in methyl mercury poisoning, through contaminated food and water.

Physical agents such as ionizing radiation from external sources (X- and gamma rays, neutrons), and non-ionizing radiation (electromagnetic radiation), involve penetration of the whole body. With ionizing radiation, the skin may be the important target, especially for types of external radiation that do not penetrate deeply into the body (i.e., alpha and beta rays). Exposure to radon and the noble gases involves the inhalation route. Ultraviolet irradiation also involves mainly the skin, although systemic effects (such as on the immune system) may be involved.

In the majority of epidemiologic studies of humans, potential or actual routes of exposure are poorly defined. In some instances the route of human exposure can be inferred from the toxic agents involved, the nature of the exposure, and/or the associated health outcomes. The occurrence of nasal cancers in persons working in environments with wood dusts implies an inhalation route, while dermatitis in woodworkers is due to dermal exposure (Table 1).[17] The association between acute exposure to phosgene gas ($COCl_2$), mentioned earlier, and pulmonary edema is clearly due to inhalation (Table 1), as occurred among World War II workers exposed during the production of uranium chloride obtained by combining CCl_4 with uranium oxide.[16,19] Exposure to arsenicals, associated with both lung and skin cancers, also occurs by inhalation and dermal contact. Various agents associated with lung cancer, such as nickel oxides, radon gas, and arsenic compounds, clearly reflect inhalation exposures (Table 1).

Studies of drinking water and cancer assume that ingestion of water is the major route, although confounding due to occupational exposures and other pathways often cannot be ruled out. The ingestion of pesticides is a clearly established method of suicide, especially in Third World countries such as Sri Lanka.

Intentional painting of the skin, teeth, and other body parts with radium-containing paint occurred among workers in watch-dial painting factories in the early part of this century, and exposure of family members undoubtedly occurred after paint was taken home. Indirect and unintentional exposure by any route (i.e., dermal, ingestion, and inhalation) can occur among family members usually via contaminated clothing of a parent or spouse. Modern examples involve asbestos and lead exposure.[20] In late 1987, for example, a worker involved in the manufacture of lead buckles in Colorado had abdominal symptoms and a blood-lead concentration of 170 μg/dl (or almost seven times the level required to be reported), and his wife and three daughters all had elevated blood lead concentrations.[21] This case points out the importance of obtaining occupational histories in clinical examinations and of the surveillance of occupational diseases through "sentinel events" such as lead poisoning (see later).

Polychlorinated and polybrominated biphenyls (PCBs and PBBs) from the air, water, and food are stored in adipose tissue and released into other organs and tissues. Therefore, without detailed environmental sampling, which is rare in epidemiologic studies, routes of exposure are uncertain.

2. Other Routes: Transplacental and Lactational

Transplacental exposure to polychlorinated organic compounds (PCBs and PBBs) can occur during pregnancy, leading to "fetal PCB syndrome". Methylmercury also crosses the placenta. In mammalians exposure of infants can occur as a result of nursing or lactation if the mother's milk has been contaminated with such substances as dioxins, PCBs, PBBs, and DDT, leading to even greater exposures to PCBs and PBBs in offspring than via the transplacental route.[22]

Examples of transplacental exposure, such as to lead and PCBs, indicate that seemingly unusual routes of exposure are also important and that the potential for transplacental carcinogenesis and/or teratogenesis must be considered in epidemiologic studies.

3. Preconception Exposure

Maternal and paternal germ cells may receive damage from toxic-mutagenic agents, which can affect the health of offspring even though the offspring are not exposed to those agents. Maternal nondisjunction of chromosome during meiosis results in Down Syndrome in offspring and is related to maternal age, while increased paternal age (presumably reflecting mutations in germ cells) is associated with increased risk of certain dominantly inherited birth defects, but the role of parental exposures to toxic agents through occupational or other means is uncertain.

Interest has increased recently regarding the potential role of preconceptional paternal exposures to ionizing radiation and toxic chemicals in childhood cancers, especially leukemia and brain tumors. There is some limited support from case-control studies for the hypothesis that paternal germ cell damage, related to occupational exposures to certain organic compounds, might be involved in childhood brain cancers.[23] Provocative findings of the possible relationship between paternal preconception radiation exposure and childhood leukemia were based on small numbers,[24] and were not consistent with findings from the Japanese atomic bomb studies. The possibility of a role for paternal internal (vs. external) pathways of radiation exposure[25] is relevant to our discussion.

B. EXAMPLES

1. Occupational Studies

Some health outcomes are so clearly and predominantly linked to occupational exposures that the occurrence of these outcomes is a "sentinel" for such exposure. Lead poisoning was mentioned earlier with regard to exposure of family members of workers exposed to lead. Lead poisoning, silicosis, pneumoconiosis, and occupational burns are examples of "sentinel events (occupational)," as originally proposed by Rutstein and colleagues.[26] Mesothelioma due to asbestos exposure is another example of a disease (albeit rare) that is very strongly related to occupational exposure. In a recent report scrotal cancer was reaffirmed as a "sentinel health event (occupational)", although only a small number had occupations previously linked to scrotal cancer.[27]

Industrial exposures have long been recognized as important clues to toxic agents, beginning with the effects of lead and other metals, ionizing radiation in watch-dial painters (see later), and various chemicals such as arsenic and cadmium. This is due to the generally higher levels of exposure than in non-occupational settings (with rare exceptions) and the resulting high risks of disease, especially for cancers that are ordinarily rare. Inhalation and dermal exposures predominate (Table 1), but limited information is often available on dermal exposure.

Information of levels and routes of exposure, while generally better in occupational settings that in other environmental epidemiological studies, can still be limited. Evidence for poisoning with organophosphate pesticides, or example, was shown in crop-duster aviation mechanics in Nicaragua, but the lack of use of gloves, protective clothing, or respirators[28] precludes assessment of routes of exposure, because all routes could have been involved. Aerial pesticide applicators in the U.S. were identified for computerized Federal Aviation Administration medical examination records, but exposure information was not available and duration of follow-up was short; overall cancer mortality was not increased, although a slight excess for leukemia was found among pesticide applicators.[29] Positive evidence for an association between herbicide/pesticide exposure and cancer in humans is limited to two studies of non-Hodgkin's lymphoma in U.S. midwestern farmers reportedly exposed to phenoxy herbicides, based on small numbers and poorly documented histories of herbicide use.[30] While a higher risk of non-Hodgkin's lymphoma in U.S. veterans of Vietnam has been reported in a case-control study, there was no association with groups involved in the spraying of the herbicide Agent Orange and only 1 of 99 cases reported handling Agent Orange.[31,32]

Linkage systems have been developed between job titles/industry (such as aviation mechanics working in crop-dusting companies) and specific toxic agents such as asbestos.[33,34] These systems, however, do not consider the route of exposure but are intended as a first step in the development of epidemiologic studies aimed at identifying groups of occupations that involved potential exposure to the same chemical or physical agent.

Episodic releases of high levels of toxic substances can occur in occupational groups. One example is phosgene gas. Such exposures can be very difficult to quantify, especially if they occurred in the distant past. Reported symptoms may provide an index for high levels of exposure, by analogy with other reports involving acute high-level phosgene exposure, but lower-level exposure in the past may remain unknown.[18,19]

Occupational cohort studies often involve workers exposed in the distant past, and it is difficult to determine the nature, levels, and routes of exposure to chemical or physical agents. Often the chemical compounds involved can be identified through company records of chemical used and industrial hygiene surveys conducted. Blood and/or urine samples, often associated with air sampling data, are the traditional types of exposure data available from industrial hygiene surveys. These data are often spotty and selective, and interpretation of blood and urine sampling data requires information on the routes of exposure and solubilities of the toxic-carcinogenic agents.

Methods of monitoring, to differentiate inhalation from dermal exposure and to estimate such exposures, may be relatively limited and inaccurate. For example, limited data on urinary levels of uranium were available from a plant that operated during World War II, and air sampling data were used from this plant and other similar facilities to estimate possible amounts of uranium dust inhaled and the solubilities of the compounds involved.[35,36] Histories of dermal exposures, however, were unknown. Fat biopsies would be useful for assessing exposures to PCBs and dioxins, which concentrate in fat tissues, but such tests are prohibitively costly and present problems with compliance. Fortunately, serum or plasma levels may be highly correlated with fat levels in some situations.[37] These considerations apply in occupational and non-occupational situations (see later).

Because of the unavailability and/or limitations of industrial hygiene data, biomarkers of exposure are sought. Chromosome damage has been proposed as a ''dosimeter'' for ionizing radiation exposure.[38]

Since hemoglobin adducts have the same biological life span as hemoglobins in blood, they can provide a better index of recent exposure (including intermittent exposure) than measurements of metabolites in blood and urine. Methyl bromide has been shown to react with cysteine to form methylcysteine in hemoglobin. Studies of 14 methyl bromide workers suggested that methylcysteine (MeCys) in hemoglobin was a biological exposure index.[39] Oral and inhalation exposures to three monohalomethanes (methyl bromide, methyl chloride, and methyl iodide) in rats showed highest binding to hemoglobin for methyl bromide, and higher binding via the inhalation vs. the oral route.[40] These studies suggest that hemoglobin adducts should be explored further in human studies, including epidemiologic studies involving exposure to monohalomethanes. The genotoxic potential of certain monohalomethanes supports the need for such studies.

Hemoglobin adducts also have been used in assessing exposure to several carcinogens inhaled by passive smokers.[41] Passive smoking is being increasingly recognized as an important route of exposure in humans. Such exposures complicate studies of occupational groups and other populations.

Bronchopulmonary lavage is a relatively non-invasive method for detecting asbestos in lungs of workers, and the poor correlation with self-reported exposure in steel workers[42] indicated the potential value of such techniques in epidemiologic studies among workers who may be unaware of their exposure to specific agents inhaled and deposited in the lung.

2. Environmental Disasters

Exposure levels tend to be lower in non-occupational vs. occupational situations, but there are exceptions involving environmental disasters such as the Seveso (Italy) and the Bhopal (India) accidents (Table 2). Also, non-occupational exposures can be significant as in the case of lead exposure from the ingestion of paint and from other sources (see later).

Environmental disasters such as air pollution, at least in terms of their immediate health effects, are episodes, like acute outbreaks of infectious disease related to point sources of contamination. Yet long-term follow-up may be required for the consideration of late effects such as cancers. Environmental contamination may take decades to appear and be recognized, as in the case of methyl mercuric chloride poisoning in Minamata, Japan from a chemical plant that began operation in 1907 and dumped its wastes for decades until the causative agent for neurological problems in local residents was recognized in 1962. At Love Canal, chemicals were dumped for decades but unusual rains apparently contributed to the spread and detection of toxic chemicals in nearby houses; no evidence for effects on cancer rates was detected in the surrounding area.[43]

TABLE 2
Examples of Human Exposures from Environmental Disasters, Natural and Anthropogenic

Place, year	Toxic-physical agent	Route of exposure
Seveso, Italy 1976	Dioxin	Ingestion
Swiss town, 1984	Bromide gas	Inhalation
Bhopal, India 1984	Cyanide gas	Inhalation
Japan, 1968	PCBs (Yusho or rice-oil disease)	Ingestion
Taiwan, 1979	PCBs (Yucheng or rice-brain oil disease)	Ingestion
Marshall Island, 1954	Ionizing radiation	External irradiation; ingestion
Chernobyl, 1986	Ionizing radiation	External irradiation; ingestion
Japan, 1950s on	Mercury (Minamata disease)	Ingestion
Lake Nyos, Cameroon, Africa, 1986	Carbon dioxide (natural)	Inhalation
Petroleum (oil) spills		
Exxon Valdez, 1989	Polycyclic aromatic hydrocarbons	Inhalation and dermal[a]
Persian Gulf, 1991	Polycyclic aromatic hydrocarbons	Potential inhalation and dermal[a]

[a] Potential exposure to workers involved in clean-up operations.

The examination of mortality in a population exposed to 2,3,7,8-tetrachlorodibenzo-*p*-dioxin (TCDD) due to the accidental explosion in 1976 of a chemical plant producing tricholorophenol near Milan, Italy provides an example of a cohort study. The only established health effect at Seveso has involved about 200 cases of chloracne, or a disease "marker" of TCDD exposure, and only suggestive evidence for excessive mortality from certain cancers during the follow-up period.[11] Individual exposure was crudely estimated by a surrogate indicator, which was the area of residence at the time of the accident, and death rates were compared with a reference population in an area uncontaminated by the disaster. The greatest opportunity for exposure in the TCDD-contaminated areas was through the ingestion route, due to contaminated soils.

Bertazzi et al.[11] observed that the interpretation of the evidence at Seveso was "hampered by the short observation period, small numbers of deaths from certain causes, and poor exposure definition". These problems of latency periods, statistical power, and lack of information on exposure (including routes of exposure) in epidemiologic studies were mentioned earlier. A woman who died from cancer of the extrahepatic biliary system was the only person for whom measurements of fat concentration of TCDD were available; although high concentrations were found in this case, there are no data on the variability of fat concentrations in the population. More than 30,000 serum or plasma samples were taken from 1976 to 1985, and results may be highly correlated with adipose tissue levels,[37] but these data were not used in the mortality study reported.[11]

TCDD is a definite carcinogen in experiments with rats and mice involving oral administration and skin application, but the suggestive positive findings in the Seveso cohort did not include an excess of liver cancer (which has been associated with TCDD exposure in animals). Interpretation of the suggestive findings for soft-tissue sarcomas was complicated by the fact that the incidence of these cancers was relatively high in the area even before the accident.

Industrial plants and energy-producing power plants have been the sources of accidental high-level releases of toxic substances, as well as sources of continuous low-level releases into air, soil, and water. Examples include ionizing radiation released as gaseous and particulate matter from nuclear reactor accidents as in the Soviet Union in 1957 to 1958, in a minor accidental release at Three Mile Island in Pennsylvania in 1979, and at Chernobyl in 1989. Another example of an accident involving a chemical plant resulted in high levels of toxic bromide gas in a Swiss town of 25,000 inhabitants in 1984.[44] In massive spills of petroleum products (oil), evaporation of toxic-carcinogenic organics such as benzene is rapid, but workers involved in clean-up may be exposed by the inhalation and dermal routes.

Natural disasters involving "toxic" agents also occur, as in the cloud of CO_2 released from Lake Nyos in Cameroon (West Africa) which caused asphyxiation; there was no evidence of chemical burns or flash burns, although a few persons may have collapsed near a heat source and suffered skin burns indirectly.[45] This event recalls Paracelsus' aphorism that all substances can be "poisons" or toxicants.

In the case of the disasters in Japan and Taiwan involving cooking oils contaminated with PCBs during their production (Table 2), mothers presumably had ingested the contaminated oil during pregnancy. Hence, the route of maternal exposure was through ingestion, while offspring were exposed transplacentally in the maternal blood crossing the placental barrier.[46]

3. Point vs. Multiple Sources of Exposure

Point sources include industrial plants and toxic waste disposal sites. Industrial plants and power plants may release toxic substances at a rather continuous rate or through accidental, acute releases (discussed above).

Confounding due to occupational exposures in persons under study for ambient exposures in communities is a problem. In a pilot study of serum PCB levels in populations near three waste sites in Indiana, elevated serum levels could not be clearly associated with specific pathways of exposure and uptake; possible exceptions were persons with occupational exposures and those who reported salvaging metals from discarded electrical equipment (such as parts of capacitors).[47] In studies of populations near toxic waste sites, some persons may have been exposed to toxic agents due to occupations unrelated to the waste site, or to occupations involved in clean-up of these sites, as well as to recreational and other activities (such as salvaging) involving direct contact with the site, in addition to potential exposures via air, water, and soil pathways.

Epidemiologic studies of populations living "near" toxic waste disposal sites have involved limited data on potential exposures. Several ecologic studies have been reported, including one covering the entire U.S., concerning cancer mortality in counties with waste disposal sites.[48] Although statistical associations were reported, the lack of data on exposures from the sites and on confounders (such as smoking) indicates only that more in-depth epidemiologic studies are needed.

A case-control study of lung cancer in Niagara County involved identification of waste sites with potential for contamination due to migration of chemicals, or with radon gas detected at least at waste-site boundaries. Analyses were restricted to disposal sites where known or suspect lung carcinogens had been identified. The proxy exposure variable was residence in the same census tracts as those containing these selected dump sites, and no evidence was found for an association between such residence and lung cancer death.[49] No data could be collected on history of direct contact with specific waste sites through occupational or recreational exposures, because the sites were not identified until after the questionnaires had been completed.

In an extensive review of epidemiologic studies of toxic chemical disposal sites Upton et al.[50] concluded that lack of information on routes of contamination (soil, air, water, etc.) and exposure (ingestion, inhalation, dermal contact) has seriously limited the ability of such studies to assess health risks.

Routes or pathways of contamination or migration via ground water, surface water, and air have been included in hazard ranking systems for studies of toxic waste sites. Trichloroethylene is the most common toxic compound, along with other volatile organic chemicals such as benzene; metals and metallic compounds are also common. Direct contact exposures, either occupational or recreational, may be assessed through interviews in case-control or cohort studies. Volatile organics evaporate but may be concentrated in soils, and this varies with type of soil.

Smelters are an example of point sources for contamination of air, soil, and water in nearby communities and various toxic agents have been studied. Lead smelters are point sources for inhalation and ingestion of lead particulates, through environmental pathways of air, soil, and water contamination, but there are multiple pathways and routes of exposure to lead (see next section).

III. EXAMPLES OF EPIDEMIOLOGIC STUDIES WITH GOOD DATA ON ROUTES OF EXPOSURE

A. LEAD

In addition to occupational situations, lead exposure occurs to the general population from inhalation of automobile exhausts, from industrial point sources (via air and soil contamination), ingestion and inhalation of particles of lead-containing paint, and ingestion of contaminated drinking water, along with transplacental exposures. Ingestion of house dust or paint chips contaminated with lead, via the hand-to-mouth mode, has been known (since 1904) to be a major contributor to total body lead burden in children. As noted earlier, blood levels of certain toxicants such as lead are used to assess the extent of exposure.

The legacy of childhood exposure to lead paint through ingestion in contaminated households, however, continues to be a major public health problem, especially in poor urban areas with minority children.[51] Improvements in scientific knowledge of toxicity of lead, through comparisons of IQ levels in exposed and non-exposed children, have shown apparent effects on IQ at blood lead levels as low as 0.5 to 0.7 μmol/liter, vs. the standard of 2.9 of 25 years ago. The decline in blood lead levels of children in the U.S. and Sweden[52] may be largely due to the reduced use of lead additives in motor fuels and, hence, reduced inhalation.

The removal of lead from old buildings can result in the ingestion and inhalation of chips, fine dusts, and fumes (<1 μm diameter, and hence respirable), resulting in exposure of children. Parents and physicians must be made aware of the hazards of lead, its sources and routes of absorption.[53]

In adults ingestion from contaminated water is one route of exposure, and methods have been developed to estimate the amount of blood lead attributable to ingestion from this source.[54] Estimates of lead clearance from the body are based on studies in which volunteers were given known amounts of lead.

Lead from smelters is another pathway of exposure through contamination of air, soil, and water. Although exposures to lead are generally lower in non-occupational than in occupational situations, the geometric mean blood lead concentration was 0.77 μmol/L in the vicinity of a lead smelter in Yugoslavia vs. 0.25 in an unexposed town.[55] In Greece, children living near a lead smelter all their lives had higher blood levels of lead, a significantly higher prevalence of abnormal findings on neuromuscular tests, and poorer school performance than children from a less polluted area who were matched on socioeconomic and other variables.[56] These findings indicate that multiple environmental pathways and routes of exposure can result in significant lead exposure and behavioral and pathological effects in children, depending upon the location of the exposed populations and the environmental hazards involved. Both inhalation and ingestion are important routes of exposure in populations living near smelters and other industrial point sources. Engineering measures at these sources can result in reduction in exposure to surrounding populations.

Finally, occupational exposures to lead continue to be important. In a California study only about 1 to 3% of all lead-using industries had environmental or biological monitoring programs for lead, which seriously reduces the ability to eliminate occupational lead poisoning.[57] Thus, this occupational hazard recognized for 2000 years and demonstrated by epidemiologic studies to be of continuing relevance to health (mainly in terms of kidney disorders), still remains, along with the continuing hazard of lead exposure in children due to lead paint.

B. IONIZING RADIATION

In environmental and occupational epidemiology it is probably safe to conclude that more information is available on routes and levels of exposure, and on health effects, involved with exposures to ionizing radiation than for any other physical or chemical agents. For internal radiation exposure much information is available through decades of study on the retention, metabolism, and dose delivered to various organs and tissues by radium, uranium, and plutonium.[3]

In the first major report on the health effects of exposure to ionizing radiation, which appeared in the *Journal of the American Medical Association* in 1925,[58] Martland and colleagues noted cases of burns and dermatitis in radium chemists, aplastic anemia and cancers among radiologists, and osteomyelitis of the mandible among workers in luminous watch-dial factories. The skin of radium chemists was exposed to beta and gamma rays, while the bodies of radiologists were exposed to X-rays from sources used in the diagnosis of disease.[59] In radium dial painters ingestion of paint containing ^{226}Ra and ^{228}Ra resulted from the habit of pointing the paint brush in the mouth; this lead to osteomyelitis of the jaw or ''radium jaw''.

The radium dial painters, mostly young women, were also exposed to external radiation from the paint and from contaminated rooms and to inhalation of radon gas emanating from the radium as it decayed. In addition (as noted above), some women painted their teeth and various parts of their bodies with the luminous paint, and took the paint to their homes for painting objects. Using data from biological monitoring or estimation of radium retained in the body, and from estimates of external radiation in the workplace, estimates of radiation doses from internal and external irradiation have been made for such tissues as bone-marrow (relevant to leukemia risk)[60] and ovaries (relevant to effects of fertility and birth outcomes).[61] Health effects confirmed through cohort epidemiologic studies[62] include bone cancers and cancers of the tissues lining the paranasal sinuses and mastoid regions of the head; in the latter case, radon emanating from radium deposited in the bones of the head was responsible for carcinogenic effects.[3]

By the 1920s the administration of radium orally and parenterally, and of radon and its daughters orally, parenterally and by inhalation, for supposed therapeutic benefits flourished in the U.S.[63,64] Figure 1 is an example of an early medical advertisement, showing routes of exposure for ''therapeutic'' purposes involving injections, ingestion of radium water, and dermal application of a pad ''guaranteed'' to contain radium. Interestingly, inhalation of radon gas occurs not only through occupational exposure and ambient levels in homes from natural sources of radium-containing materials, but from the tradition of health ''spas'' continues; in 1987 people were paying to be exposed to radon gas in Montana's Merry Widow mine for purported therapeutic benefits.[65]

In view of evidence of carcinogenic effects in occupational studies of radium workers and patients treated with radium, attention turned to possible effects of lower levels of

FIGURE 1. Example of an early medical advertisement touting therapeutic benefits of radon and showing routes of exposure.

exposure from other sources. Exposure to radium and its decay or daughter products is currently related to phosphate mining and to emanation of radon in houses from soils and rocks containing radium. The much-debated issue of estimating numbers of lung cancers attributable to radon gas exposure, independent of smoking, attests to the continuing importance of ionizing radiation in epidemiology.[66] Both radium in water and radon gas have been suggested by ecologic studies to be related to leukemia, but ecologic studies are hypothesis-generating rather than confirmatory, and predicted risks from such studies (in the absence of confounding) would be much higher than expected from occupational epidemiologic studies.[67] A strongly positive finding in an ecologic study of the association between radium in well water, which was a drinking water source for only a small percentage of the population, and leukemia was reported in Florida[68] but not confirmed by another ecologic study in Iowa.[69]

The importance of obtaining actual exposure information on individuals, as opposed to the ecologic study approach, was highlighted by a study of cancer incidence in populations surrounding the Rocky Flats plutonium processing plant in Denver, CO. Negative findings regarding cancer incidence were consistent with the evidence that plutonium body burdens in tissues of persons living near the plant did not differ from levels in persons living elsewhere in Colorado or the U.S.[70] As noted earlier, occupational situations tend to involve higher levels of exposure than other situations such as ambient exposures in populations surrounding point sources of pollution. Occupational exposures to ionizing radiation at the same Rocky Flats plant were based on health physics records documenting internal exposure (presumably by inhalation and possibly ingestion of particles) and external exposure (gamma neutron, beta, and X-rays). Theoretically, dermal exposures could have occurred due to penetrating ionizing radiation and handling of radioactive materials, but there was no evidence for increased mortality from skin cancers.[71] Analyses by exposure level suggested some possible associations, as with lymphopoietic neoplasms, for the total cohort and for a subgroup with relatively higher body-burdens of plutonium, based on small numbers. Since insoluble plutonium particles are transported to tracheobronchial lymph nodes after inhalation, this association requires further study through additional follow-up. The issue of latency period (mentioned earlier) was addressed through separate analyses excluding an initial period of 10 years of follow-up, which resulted in higher estimates for certain cancers but not for lymphopoietic cancers.

Occupational exposures to radon gas, as in uranium and other miners, are much higher than those in the general population, but there is concern and controversy regarding such low-level exposures because of predicted health effects (mainly lung cancer) based on linear extrapolation of effects based on higher doses.[66]

Extrapolation of findings from the Japanese atomic bomb studies provides the major resource for predicting health effects in international epidemiologic follow-up studies of populations exposed to radioactive fallout for the Chernobyl disaster, and preliminary findings on chromosomal abnormalities suggest that pregnant women in some areas of Byelorussia may have received high external doses.[72] In West Germany ecologic studies suggest that radionuclides taken up by mothers through inhalation and ingestion may have resulted in fetal exposures and increases in infant death rates,[73] but the findings could be due to chance and better data on exposure are needed. International cooperation is needed in long-term epidemiologic follow-up of cancers (especially thyroid cancer and leukemia), as well as psychological effects which could influence other chronic diseases, in populations exposed to various levels of fallout, as well as to stress involved in evacuations.

IV. CONCLUSION

Ecologic studies will continue to provide suggestions for hypotheses to be tested. Improvement in design of ecologic studies involved more attention to potential confounders. Such careful studies are needed because of the generally low levels of exposure involved, the "weak" associations usually detected, and the potential for confounding due to exposures to other toxicants or other causal factors involved in the disease under study. Multiple computerized data sources can include information on potential pathways of exposure, as well as data from a wide variety of independent surveys such as those on smoking prevalence in geographic areas of the U.S. Such databases will provide little information on actual routes of exposure to toxicants. Ecologic studies, however, can point the way toward analytic epidemiologic studies that consider pathways and routes of exposure, or actual levels of exposure, in individuals.

There is widespread recognition that analytical epidemiologic studies need better data on routes and levels of exposure to toxicants, and interviews often can provide very limited data. More detailed data may be obtained on small samples of the groups under study, in order to establish potential routes and levels of exposure in the entire group. Biological monitoring can be prohibitively expensive for large-scale epidemiologic studies, such as those involving PCBs, but serum and fat levels of PCBs may be highly correlated. Biomarkers of exposure need to be used, such as chromosome damage as a predictor of ionizing radiation exposure and hemoglobin adducts for estimation of exposure to various organic substances.

Cooperation between epidemiologists and laboratory scientists is needed in the assessment of routes and levels of exposure as well as in interpreting the causality of associations in terms of consistency with findings from laboratory experiments, with consideration of possible differences in routes of exposure in the experiments (vs. the types of exposures involved in humans) as well as differences in metabolism of toxicants. International collaboration of scientists (including epidemiologists) and medical personnel, and access to databases on exposure and toxicity, are needed in responding to environmental disasters and environmental health problems.

REFERENCES

1. **Tennant, R. W., Margolin, B. H., Shelby, M. D., et al.,** Prediction of chemical carcinogenicity in rodents from in vitro genetic toxicity assays, *Science,* 236, 933, 1987.
2. **Abelson, P. H.,** Testing for carcinogens with rodents, *Science,* 249, 1357, 1990.
3. **Rundo, J., Failla, P., and Schlenker, R. A., Eds.,** Radiobiology of radium and the actinides in man, *Health Physics,* 44 (Suppl. 1), 1, 1983.
4. **Sun, M.,** EPA bars use of Nazi data, *Science,* 240, 21, 1988.
5. **Neutra, R. R.,** Counterpoint from a cluster buster, *Am. J. Epidemiol.,* 132, 1, 1990.
6. **Langham, W. H., Bassett, S. H., Harris, P. S., and Carter, R. E.,** Distribution and excretion of plutonium administered intravenously to man, *Health Physics,* 38, 1031, 1980.
7. **Buffler, P. A., Cooper, S. P., Stinnett, S., et al.,** Air pollution and lung cancer mortality in Harris County, Texas, 1979–1981, *Am. J. Epidemiol.,* 128, 683, 1988.
8. **Polednak, A. P. and Janerich, D. T.,** Use of available record systems in epidemiologic studies of reproductive toxicology, *Am. J. Ind. Med.,* 4, 329, 1983.
9. **Baker, E. L.,** Sentinel event notification system for occupational risks (SENSOR): the concept, *Am. J. Publ. Health,* 79 (Suppl.), 18, 1989.
10. **Frisch, J. D., Shaw, G. M., and Harris, J. A.,** Epidemiologic research using existing databases of environmental measures, *Arch. Environ. Health,* 45, 303, 1990.
11. **Bertazzi, P. A., Zocchetti, C., Pesatori, A. C., et al.,** Ten-year mortality study of the population involved in the Seveso incident in 1976, *Am. J. Epidemiol.,* 129, 1187, 1989.
12. **Blot, W. J., Brown, L. M., Pottern, L. M., et al.,** Lung cancer among long-term steel workers, *Am. J. Epidemiol.,* 117, 706, 1983.
13. **MacLure, K. M. and MacMahon, B.,** An epidemiologic perspective of environmental carcinogenesis, *Epidemiol. Rev.,* 2, 19, 1980.
14. **Fagliano, J., Berry, M., Bove, F., et al.,** Drinking water contamination and the incidence of leukemia: an ecologic study, *Am. J. Public Health,* 80, 1209, 1990.
15. **Tsuruta, H.,** Skin absorption of organic solvent vapors in nude mice in vivo, *Ind. Health,* 27, 37, 1989.
16. **Skowronski, G. A., Turkall, R. M., and Abel-Rahman, M. S.,** Soil absorption alters bioavailability of benzenes in dermally exposed male rats, *Am. Ind. Hygiene Assoc. J.,* 49, 506, 1988.
17. **Flechsig, R. and Nedo, G.,** Hazardous health effects of occupational exposure to wood dust, *Ind. Health,* 28, 107, 1990.
18. **Polednak, A. P.,** Mortality among men occupationally exposed to phosgene in 1943–45, *Environ. Res.,* 22, 357, 1980.
19. **Polednak, A. P. and Hudson, D.,** Mortality and causes of death among workers exposed to phosgene in 1943–45, *Toxicol. Ind. Health,* 1, 137, 1985.

20. **Knishkowy, B. and Baker, E. L.,** Transmission of occupational disease to family contacts, *Am. J. Ind. Med.,* 9, 543, 1986.

21. **Johnson, D., Houghton, K., Siegel, C., et al.,** Occupational and paraoccupational exposure to lead — Colorado, *Morbid. Mortal. Weekly Rep.,* 38, 338, 1989.

22. **Jacobson, J. L., Humphrey, H. E. B., Jacobson, S. W., et al.,** Determinants of polychlorinated biphenyls (PCBs), polybrominated biphenyls (PBBs), and dicholorodiphenyl trichloroethane (DDT) levels in the sera of young children, *Am. J. Public Health,* 79, 1401, 1989.

23. **Wilkins, J. R. and Sinks, T.,** Parental occupation and intracranial neoplasms of childhood: results of a case-control interview study, *Am. J. Epidemiology,* 132, 275, 1990.

24. **Gardner, M. J., Snee, M. P., Hall, A. J., et al.,** Results of case-control study of leukaemia and lymphoma among young people near Sellafield nuclear plant in West Cambria, *Br. Med. J.,* 300, 423, 1990.

25. **Anon.,** Childhood leukemia, radiation and the paternal germ cell, *Lancet,* 335, 447, 1990.

26. **Rutstein, D. D., Mullan, R. J., Frazier, T. M., et al.,** Sentinel health events (occupational): a basis for physician recognition and public health surveillance, *Am. J. Public Health,* 73, 1054, 1983.

27. **Weinstein, A. L., Howe, H. L., and Burnett, W. S.,** Sentinel health event surveillance: skin cancer of the scrotum in New York State, *Am. J. Public Health,* 79, 1513, 1989.

28. **McConnell, R., Anton, F. P., and Magnotti, R.,** Crop duster aviation mechanics: high risk for pesticide poisoning, *Am. J. Public Health,* 80, 1236, 1990.

29. **Cantor, K. P. and Booze, C. F., Jr.,** Mortality among aerial pesticide applicators and flight instructors, *Arch. Environ. Health,* 45, 295, 1990.

30. **Zahm, S. H., Weisenburger, D. D., Babbitt, P. A., et al.,** A case-control study of non-Hodgkin's lymphoma and the herbicide 2,4-dichlorophenoxyacetic acid (2,4-D) in eastern Nebraska, *Epidemiology,* 1, 349, 1990.

31. Selected Cancers Cooperative Study Group, The association of selected cancers with service in the U.S. military in Vietnam. I. Non-Hodgkin's lymphoma, *Arch. Intern. Med.,* 150, 2473, 1990.

32. **Suskind, R.,** The association of selected cancers with service in the U.S. military in Vietnam, *Arch. Intern. Med.,* 150, 2449, 1990.

33. **Hoar, S. K., Morrison, A. S., Cole, P., et al.,** An occupational and exposure linkage system for the study of occupational carcinogenesis, *J. Occup. Med.,* 22, 722, 1980.

34. **Hsieh, C. C., Walker, A. M., and Hoar, S. K.,** Grouping occupations according to carcinogenic potential: occupation clusters from an exposure linkage system, *Am. J. Epidemiol.,* 117, 575, 1983.

35. **Polednak, A. P. and Frome, E. L.,** Mortality among men employed between 1943 and 1947 at a uranium-processing plant, *J. Occup. Med.,* 23m, 169, 1981.

36. **Polednak, A. P., Keane, A. T., and Beck, W. H.,** Estimation of radiation doses to the lungs of workers at early uranium-processing plants, *Environ. Res.,* 28, 313, 1982.

37. **Mocarelli, P., Pocchiari, F., and Nelson, N.,** Preliminary report: 2,3,7,8-tetrachlorodibenzo-p-dioxin exposure to humans — Seveso, Italy, *Morbid. Mortal. Weekly Rep.,* 37, 48, 1988.

38. **Littlefield, L. G., Sayer, A. M., and Frome, E. L.,** Comparison of dose-response parameters for radiation-induced acentric fragments and micronuclei observed in cytokinesis-arrested lymphocytes, *Mutagenesis,* 4, 265, 1989.

39. **Iwasaki, K., Ito, I., and Kagawa, J.,** Biological exposure monitoring of methyl bromide workers by determination of hemoglobin adducts, *Ind. Health,* 27, 181, 1989.

40. **Xu, D., Peter, H., Hallier, E., et al.,** Hemoglobin adducts of monohalomethanes, *Ind. Health,* 28, 121, 1990.

41. **MacLure, M., Katz, R. R. A., Bryant, M. S., et al.,** Elevated blood levels of carcinogens in passive smokers, *Am. J. Public Health,* 79, 1381, 1989.

42. **Corhay, J. L., Delavignette, J. P., Bury, T., et al.,** Occult exposure to asbestos in steel workers revealed by bronchoalveolar lavage, *Arch. Environ. Health,* 45, 278, 1990.

43. **Janerich, D. T., Burnett, W. S., Feck, G., et al.,** Cancer incidence in the Love Canal area, *Science,* 212, 1401, 1981.

44. **Morabia, A., Selleger, C., Landry, J. C., et al.,** Accidental bromine exposure in an urban population: An acute epidemiological assessment, *Int. J. Epidemiol.,* 17, 148, 1988.

45. **Kling, G. W., Clark, M. A., Compron, H. R., et al.,** The 1986 Lake Nyos gas disaster in Cameroon, West Africa, *Science,* 236, 169, 1987.

46. **Jones, G. R. N.,** Polychlorinated biphenyls: where do we stand now?, *Lancet,* 2, 791, 1989.

47. **Stehr-Green, P., Ross, D., Liddle, J., et al.,** A pilot study of serum polychlorinated biphenyl levels in persons at high risk of exposure in residential and occupational environments, *Arch. Environ. Health,* 41, 240, 1986.

48. **Griffith, J., Duncan, R. C., Riggan, W. B., et al.,** Cancer mortality in U.S. counties with hazardous waste sites and ground water pollution, *Arch. Environ. Health,* 44, 69, 1989.

49. **Polednak, A. P. and Janerich, D. T.**, Lung cancer in relation to resident in census tracts with toxic-waste disposal sites; a case-control study in Niagara County, New York, *Environ. Res.*, 48, 29, 1989.

50. **Upton, A. C., Kneip, T., and Toniolo, P.**, Public health aspects of toxic chemical disposal sites, *Annu. Rev. Public Health*, 10, 1, 1989.

51. **Needleman, H. L.**, The persistent threat of lead: a singular opportunity, *Am. J. Public Health*, 79, 643, 1989.

52. **Schutz, A., Attewell, R., and Skerfving, S.**, Decreasing blood lead levels in Swedish children, 1978–1988, *Arch. Environ. Health*, 44, 391, 1989.

53. **Marino, P. E., Landrigan, P. J., Graef, J., et al.**, A case report of lead paint poisoning during renovation of a Victorian farmhouse, *Am. J. Public Health*, 80, 1183, 1990.

54. **Bois, F. Y., Tozer, T. N., Zeise, L., et al.**, Application of clearance concepts to the assessment of exposure to lead in drinking water, *Am. J. Public Health*, 79, 827, 1989.

55. **Murphy, M. J., Graziano, J. H., Popovac, D., et al.**, Past pregnancy outcomes among women living in the vicinity of a lead smelter in Kosovo, Yugoslavia, *Am. J. Public Health*, 80, 33, 1990.

56. **Benetou-Marantidou, A., Nakou, S., and Micheloyanis, J.**, Neurobehavioral estimation of children with life-long increased lead exposure, *Arch. Environ. Health*, 43, 392, 1988.

57. **Rudolph, L., Sharp, D. S., Samuels, S., et al.**, Environmental and biological monitoring for lead exposure in California workplaces, *Am. J. Public Health*, 80, 921, 1990.

58. **Martland, H. S., Conlon, P., and Knef, J. P.**, Some unrecognized dangers in the use and the handling of radioactive substances, *J. Am. Med. Assoc.*, 85, 1769, 1925.

59. **Merz, B.**, Multiple efforts directed at defining, eliminating excess radiation, *J. Am. Med. Assoc.*, 258, 577, 1987.

60. **Polednak, A. P.**, Long-term effects of radium exposure in female dial workers: differential white blood cell count, *Environ. Res.*, 15, 252, 1978.

61. **Polednak, A. P.**, Fertility of women after exposure to internal and external radiation, *J. Environ. Pathol. Toxicol.*, 4, 457, 1980.

62. **Polednak, A. P., Stehney, A. F., and Rowland, R. E.**, Mortality among women first employed before 1930 in the U.S. radium dial-painting industry, *Am. J. Epidemiol.*, 107, 179, 1978.

63. **Looney, W. B., Hasterlik, R. J., Brues, A. M., et al.**, A clinical investigation of the chronic effects of radium salts administered therapeutically, *Am. J. Roentgenol. Radium Ther.*, 73, 1006, 1955.

64. **Macklis, R. M.**, Radiothor and the era of mild radium therapy, *J. Am. Med. Assoc.*, 264, 614, 1990.

65. **Anon.**, How serious is the indoor radon health hazard?, *J. Am. Med. Assoc.*, 258, 578, 1987.

66. **Ennever, F. K.**, Predicted reduction in lung cancer risk following cessation of smoking and radon exposure, *Epidemiology*, 1, 134, 1990.

67. **Polednak, A. P.**, Leukemia and radium groundwater contamination (letter), *J. Am. Med. Assoc.*, 255, 901, 1985.

68. **Lyman, G. H., Lyman, C. G., and Johnson, W.**, Association of leukemia with radium groundwater contamination, *J. Am. Med. Assoc.*, 254, 621, 1985.

69. **Fuortes, L., McNutt, L. A., and Lynch, C.**, Leukemia incidence and radioactivity in drinking water in 59 Iowa towns, *Am. J. Public Health*, 80, 1261, 1990.

70. **Crump, K. S., Ng, T. H., and Cuddihy, R. G.**, Cancer incidence patterns in the Denver metropolitan area in relation to the Rocky Flats plant, *Am. J. Epidemiol.*, 126, 127, 1987.

71. **Wilkinson, G. S., Tietjen, G. L., Wiggs, L. D., et al.**, Mortality among plutonium and other radiation workers at a plutonium weapons facility, *Am. J. Epidemiol.*, 125, 231, 1987.

72. **Brennan, M.**, Medical effects of Chernobyl disaster, *Lancet*, 1, 1086, 1990.

73. **Luning, G., Schmidt, M., Scheer, J., et al.**, Early infant mortality in West Germany before and after Chernobyl, *Lancet*, 2, 1081, 1989.

Index

INDEX